UNITED STATES ARMY IN WORLD WAR II

Special Studies

THE WOMEN'S ARMY CORPS

AF713043

by

Mattie E. Treadwel

Book 1

CENTER OF MILITARY HISTORY

UNITED STATES ARMY

WASHINGTON, D.C., 1991

Library of Congress Catalog Card Number 53-61563

First Printed 1954--CMH Pub 11-8

For sale by the Superintendent of Documents, U.S. Government Printing Office Washington, D.C. 20402

COLONEL OVETA CULP HOBBY

UNITED STATES ARMY IN WORLD WAR
II Kent Roberts Greenfield, General Editor

Advisory Committee
(As of 1 May 1953)

James P. Baxter President,
Williams College

Brig. Gen. Verdi B. Barnes
Army War College

John D. Hicks
University of California

Brig. Gen. Leonard J. Greeley Industrial
College of the Armed Forces

William T. Hutchinson
University of Chicago

Brig. Gen. Elwyn D. Post
Army Field Forces

S. L. A. Marshall
Detroit News

Col. Thomas D. Stamps
United States Military Academy

Charles S. Sydnor
Duke University

Col. C. E. Beauchamp
Command and General Staff College

Charles H. Taylor
Harvard University

Office of the Chief of Military History

Maj. Gen. Albert C. Smith, Chief*

Chief Historian	Kent Roberts Greenfield
Chief, War Histories Division	Col. G. G. O'Connor
Chief, Editorial and Publication	Col. B. A. Day
Division Chief, Editorial Branch	Joseph R. Friedman
Chief, Cartographic Branch	Wsevolod Aglaimoff
Chief, Photographic Branch	Maj. Arthur T. Lawry

*Maj. Gen. Orlando Ward was succeeded by General Smith on 1 February 1953.

. . . to Those Who Served

Foreword

This history of the WAC is comprehensive and detailed. The author has written it not only from available records but also out of personal experience. She was a WAC staff officer, who, together with all the other Wacs, found herself in a man's army that was somewhat shocked by the advent of a women's corps in its midst.

It is usual for both newcomer and old resident to have suspicions of each other, but after the characteristic period of false starts prejudices disappear and confidence is established. So it was with the WAC and the Army.

This book stresses the misunderstanding, appropriately enough, since it affected many decisions reached at the policy-making level. The WAC did not always understand the Army—its customs and traditions, its organization and necessary chain of command. The Army did not always understand the WAC—its needs and temperament, and the many other things that man, being the son of woman, should have known but did not, much to his continued embarrassment.

Washington, D. C.
30 January 1953

ORLANDO WARD
Maj. Gen., U.S.A.
Chief of Military History

The Author

Mattie E. Treadwell, a native of Texas, holds a B.A. and an M.A. degree from the University of Texas. During World War II she was an officer, first in the WAAC and later in the WAC, holding such assignments as assistant to the Director WAC, assistant to the Air WAC Officer, and assistant to the Commandant, School of WAC Personnel Administration. She had the additional distinction of having been a member of the first class of women sent to the Command and General Staff School. While on active duty she attained the rank of lieutenant colonel.

From September 1947 to March 1952 Miss Treadwell was a historian in the Office of the Chief of Military History. Upon her departure she became Assistant Director, Dallas Regional Office, Federal Civil Defense Administration, in charge of women's activities and volunteer manpower, an office that she currently holds. Her present military status is that of a lieutenant colonel in the U.S. Air Force Reserve.

LEO J. MEYER
Colonel, Reserve Corps
Deputy Chief Historian

Washington, D. C.
30 January 1953

Preface

Soon after the end of hostilities, the decision was made to devote to the Women's Army Corps one volume of the Army's major historical series, U.S. ARMY IN WORLD WAR II. Although small by comparison with the size of the Army, the WAC at its peak strength of 100,000 constituted an enviably large group for study. Because of its 24-hour-a-day control of its personnel, the Army had access to information not easily obtainable by business or industry, concerning not only the women's job efficiency but their clothing and housing needs, and the effects of their employment upon their health, conduct, morale, and recreation.

For most of the war months, the potential importance of this material was not recognized, and little systematic effort was made to collect it. A number of Army commands had rulings against the collection of separate statistics for women, while others lacked either the time or the means to compile such material.

In postwar days, with renewed emphasis upon future planning, the present study was authorized in an attempt to pull together such evidence as remained. It was recognized that the experience of the relatively small group in World War II might provide a guide to any later and more extensive national mobilization of womanpower that might be necessary. Although no one possessed sufficient clairvoyance to predict the course of history, it was plainly evident that, in any future emergencies, the proper mobilization and employment of womanpower reserves might become a primary national issue.

The preservation of the wartime discoveries made in this field assumed added importance in view of the fact that no other American or British service has yet published a full official history of its women's corps. Significantly, comparison of the records of these groups reveals that the problems and achievements of each fall into a pattern so similar as to suggest a strong measure of predictability of the course of future groups. The Navy Department's draft narrative of the WAVES remains under classification, as do those of the Women Marines and the Army Nurse Corps. The story of the Air Forces women is included in the present volume, since the wartime Air Wacs were a part of the WAC.

The Army's discoveries in general appear valid and reliable, not only for militarized groups, but for most nonmilitary institutions or businesses which train or employ women. The observations on health, fatigue, accident rates, and psychological patterns should be a useful addition to current industrial studies. The discoveries in the fields of training, housing, clothing, feeding, and disciplining groups of women may present a fresh viewpoint to educational institutions. In particular, the conclusions on the leadership of women offer a clue to an explanation of the current misunderstandings and contradictory impressions on the subject.

It must be recognized at the outset that the problem of integrating women into an army was merely a part of the larger problem of their evolving status in civil life, accelerated by the industrial revolution and affecting every phase of modern society. Although the scope of this volume does not permit frequent comment upon the general place of women in society, few of the developments were without precedent. This was particularly true of the public skepticism and masculine hostility into which the WAC ran headlong in its first year. Admittedly, the Army had its share of a conservative element that had scarcely recovered from the shock of the mechanized horse when confronted with the militarized woman.

It should also be noted that the development and integration into the Army of a women's corps was at every turn a part of the larger development of the Army, and that few problems of the smaller group were unique. The Women's Army Corps, like this volume, must be viewed in perspective as one small facet of the larger entity.

While parallel, the problems of the employment of men and women were by no means identical in nature or solution. At the time of the organization of the Women's Army Auxiliary Corps early in World War II, the misapprehension was general that women could be treated exactly like men and that little research would therefore be required for the successful incorporation of womanpower into the Army. Some believed that the WAC, as a minor group within the Army, was in the same general category as other groups dubbed minorities by reason of race, creed, or color, for whom differences of treatment would be improper. In practice it was soon discovered, however, that while a soldier might wear the same design of clothing regardless of race or creed, the same could not be said regardless of sex. The same principle was shortly found true in the fields of medicine, conduct, recreation, recruiting, physical capacity, and others. While all authorities were agreed that equal treatment must be given to men and women in the Army, it was soon apparent that *equal* did not mean *identical* in every case. The Army was thus faced with the problem of what styles of garments, though not identical with those of men, gave equal comfort, fit, and military appearance; what medications and surgery, although not identical, promoted equally good health; what standards of conduct, well-being, recreation, and training would enable the military service to answer to the American public for the women in its keeping as conscientiously as it customarily did for the men. In most cases, by the end of the war, these problems were successfully solved or the key to the solution was known.

As the following pages will reveal, the final conclusions of the wartime heads of the women's services were far from optimistic concerning the dangers of employing women in the armed forces if their special needs were not constantly understood and dealt with by trained specialists and well-informed commanders.

Several major decisions concerning the scope and nature of this volume were dictated by the wide range of subjects it must cover, by the fact that its material was scattered through world-wide Army commands, and by the fact that only one writer-researcher could be assigned to the task. One such decision was that the approach must generally be on the level of policy and planning, rather than upon that of individual unit histories and statistics. Army commands employing Wacs activated and inactivated hundreds of companies, and sent thousands of women back and forth among them individually and in small groups, to an extent that would have required another volume to record. Even could statistics be included upon the locations and movements of such personnel, the significance for future planning would be small.

However, in the interests of proper emphasis and perspective, it should be noted that an account at this high level is not necessarily a complete picture of the Corps. The efficiency of a WAC unit in the field was often relatively untouched by the struggles concerning the nation's womanpower which raged over its head. A WAC unit could, and often did, exist happily for months without proper uniforms, training, or other advantages, no matter how distressing to the War Department such deficiencies might be.

A generally more unworried tone could be given this volume only if it were possible to place in a row, beside the headaches of headquarters, the approximately five hundred separate stories of field achievement, which by sheer weight would reduce the policy and planning problems to their proper proportion.

Another decision which affected the nature of the history was that it should include all possible material of assistance to future planners. A considerably shorter volume could have been produced by a rapid account of the Corps' formation, strength, employment, and achievements, with no indication of its problems, the private controversies they engendered, or the means by which they were surmounted. However, for those specialists whose assigned mission is the efficient employment of womanpower, or even for the general Army reader, such a surface analysis would have been of small value.

In preparing this volume, I have had the advice of almost all of the wartime leaders, men and women, of the Women's Army Corps. Col. Oveta Culp Hobby has not only commented upon the manuscript, but has answered specific questions and has given me generously of her time in discussing puzzling references. Lt. Col. Helen Hamilton Woods, WAAC preplanner and later Deputy Director, read and reread various drafts, and opened her Washington home to me for interviews with prominent participants including Congresswoman Edith Nourse Rogers. Dr. Betty Bandel has submitted a detailed commentary on each chapter in its draft form, and has answered innumerable questions concerning the 40,000 Air Forces women, whom she represented, and the Corps as a whole, of which she was Acting Deputy. Lt. Col. Katherine R. Goodwin has clarified many points regarding the Army Service Forces Wacs, for which she was advisor. Before her death, Dr. Jess Rice, the wartime Deputy Director, commented upon early parts of the manuscript and gave me many admonitions concerning historical technique, which I have endeavored to follow, as well as strictures against making her a heroine of the story, which I have endeavored to ignore in the interests of historical accuracy.

Gen. George C. Marshall has read and commented upon various passages, appealing on one occasion to the former Director to know if she had actually encountered all of the recruiting difficulties described.* Maj. Gen. Miller G. White, the Army man who, as G-1 of the War Department, worked most closely with WAC policy, has read all of the manuscript and added comments and excerpts from his diary. The Auxiliary Corps portion has been commented upon by, among others, Maj. Gen. John H. Hilldring, Brig. Gens. Don C. Faith and Thomas B. Catron, and Cols. Harold Tasker and Gilman Mudgett.

To the hundreds of other Army men and women—from general officer to private—who have given me their opinions—each chapter will make proper acknowledgment.

Within the Office of the Chief of Military History, I have received great assistance from the Chief Historian, Dr. Kent R. Greenfield, the Deputy Chief Historian, Lt. Col. Leo J. Meyer, and their entire staff, whose aid will be particularly acknowledged in the chapters concerned. Dr. Mae Link has contributed valuable research on the Army Service Forces, and Maj. Margaret Bacchus on the British services. For typing and preparing the manuscript I am indebted to Sgt. Amelia Madrak, Mrs. Lorraine Bonifant, Mrs. Lois Riley, and Mrs. Elizabeth Phillips and her staff.

My particular aid and counsel has been Miss Ruth Stout, the editor of this volume, who has provided much-needed help and encouragement as well as perspective, advice, and good judgment. We are both grateful to the Chief Editor, Mr. Joseph R. Friedman, for his sympathetic interest and advice, and to Mr. Allen R. Clark for care and precision in copy editing. I am also indebted to Mr. Clark for the comprehensive index. The work of selecting illustrations has been performed by Miss Margaret E. Tackley, who lent the project not only her technical skill but her experience as a wartime officer of the WAC, thereby avoiding the errors common to inexperienced judges of WAC photographs.

Credit for the successful planning and launching of the project belongs to the early staff of the Army's Historical Division, especially Col. Allen F. Clark, Col. John M. Kemper, Col. Allison R. Hartman, and the first Chief Historian of the Army, the late Dr. Walter L. Wright, Jr. And finally, the broad and constructive criticism, based on years of military experience, offered by Maj. Gen. Orlando Ward, Chief of Military History during the last months of the volume's preparation, was of great assistance in completing the work.

Washington, D. C. MATTIE E. TREADWELL
15 April 1953

*Her answer: "No, more."

Contents

PART ONE

Organization and Growth of a Women's Corps

PART TWO

World-Wide Employment

PART THREE

War Department Policy Concerning the Women's Army Corps

PART FOUR

Last Days of the Wartime WAC

Charts

Illustrations

Illustrations—*Continued*

Photographs are from the Department of Defense files.

PART ONE

ORGANIZATION AND GROWTH OF A WOMEN'S CORPS

CHAPTER I

From 1776 to World War II

On the hot and sticky morning of Monday, 20 July 1942, the green parade ground at Fort Des Moines, Iowa, was lush with grass, daisies, gnats, and members of the press. The photographers required the most supervision because of their tendencies toward photographing female underwear or latrine scenes. There were representatives of four press associations, nineteen newspapers, four foreign news organizations, six motion picture companies, and two photographic services, plus many well-known writers. The occasion was the opening of the United States Army's first training center for women.

Centuries of evolution in warfare and in society had been required to make possible this unprecedented event. From the days of the American Revolution to the early twentieth century, there could be found little serious consideration of a women's corps in the United States Army. For the pioneer woman, home defense was a readily acceptable activity, but it had no connection in the public mind with an organized military corps of women.

However, from the earliest days there were innumerable popular stories of females who had disguised themselves well enough to enlist in the Army as individuals. None of these stories could be verified from Army files, and they apparently occurred more often in legend than in fact, but there was evidence that not all were fictional. The American Congress went so far as to recognize the claims of one Revolutionary soldier, Deborah Sampson Gannett, by granting her husband a widow's pension.

Every succeeding war had its Molly Pitchers, most of whom made good copy for the young American press. Nineteenth century newspapers reveled in such headlines as SERVED BY HER LOVER'S SIDE, or THE DEAD SOLDIER WAS A WOMAN. According to these accounts, the Union Army inadvertently awarded some women the Kearny Cross and the Confederacy placed others on a Roll of Honor. One Confederate wife, declaring that she was "perfectly wild about war," was said to have donned a false mustache and successfully raised and commanded a regiment of recruits. Such stories were legion, but scarcely of help to later Army planners.[1]

Of more significance for the future were scattered cases in which, because of an actual need for women's skills, the Army employed groups of civilian women as nurses, laundresses, clerks, and emergency aides of many types, sometimes in uniforms of their own devising. It was not unusual, for example, that each regiment of the Braddock expedition was allowed some forty women employees, one ration

[1] (1) *Magazine of History*, Vol. XXV, pp. 33–34 (July 1917). (2) Ltr, TAG to Editor, *Rural New Yorker*, May 12, 1915. WD Rec AGO, Enl Br 7275–C, with 3132–C–1884, 2285035 National Archives. (See Bibliographical Note.) (3) Francis Henry Gribble, *Women in War* (London: Sampson Low, Marston & Co., Ltd., 1916).

per woman. In 1775 General Washington sponsored a bill that created a hospital department for the Army and allowed it to pay civilian nurses approximately twenty-five cents a day. In time of emergency, civilian women of prominence and reputation continued to nurse the sick, sew, operate canteens, and lend what assistance their skills permitted. However, for a reputable woman to accept such employment was often considered daring in view of the danger of confusion with the more numerous women camp followers, whose ill repute was apt to attach itself to any female employee. In later wars, the employment of women became more common; drill, mascot, and social groups were also organized, complete with uniforms and sometimes rifles.[2]

There was a very clear distinction in the Army's and the public's minds between such groups and a corps of women with soldiers' legal status, rights, and discipline. Just before the nation's entry into the first World War, the Army stated positively:

> No official record has been found in the War Department showing specifically that any woman was ever enlisted in the military service of the United States as a member of any organization of the Regular or Volunteer Army. It is possible, however, that there may have been a few instances of women having served as soldiers for a short time without their sex having been detected, but no record of such cases is known to exist in the official files. Women were often employed as laundresses and as nurses, but they were merely civilians while so engaged and were in no sense in the *military* service of the United States.[3]

Total War and the Industrial Revolution

Serious consideration of an official women's corps was scarcely possible before the twentieth century. Until then, war was not organized and mechanized to an extent that required more manpower than a nation could provide from among its men; the great supply systems and fixed headquarters of total war were yet to come. Also, women were skilled in few duties that would have been useful to an army even had it needed manpower, and few women felt it proper to practice even their traditional tasks of cooking and nursing outside the home.

Both reasons were swept away in the nineteenth and early twentieth centuries, when the industrial revolution that mechanized men's wars also taught women to work outside the home. Long before the Army began to consider the admittance of women, businesses and factories had employed them and had trained them as clerks, typists, telephone operators, and technicians. Some such fields were in fact taken over by women so completely that by the time of the first World War it was already difficult for the Army to find any number of competent male typists and telephone operators.

Nevertheless, there remained considerable room for doubt as to the value of woman as a military employee. Industry's experience had produced the general impression that women were suited only for work of limited responsibility; that "women won't work for women" and certainly that men would not; that a woman employee was handicapped by a weaker constitution and more frequent ailments. Business surveys as late as 1942 confirmed the fact that women supervisors were unpopular with employees because of alleged

[2] (1) Sir John William Fortescue, *History of the British Army* (London and New York: Macmillan and Co., 1890–1930), Vol. I. (2) Rapid City, S. D., *Daily Journal*, October 21, 1944.

[3] Ltr cited n. 1(2).

deficiencies in leadership. Army planners also noted delicately that a woman had a "physiological handicap which renders her abnormal, unstable, etc., at certain times." [4] Medical statistics indicated that women were in general smaller than men in stature, lighter in weight, less in average weight of skeletal muscles and heart, lower in basal metabolism, and with only about 60 percent the strength to lift, grip, or pull loads.[5]

It therefore was a matter for grave consideration whether an Army would be justified in accepting women into military positions where unreliability would be not merely uneconomical but disastrous, even though farm and factory were thereby stripped of their last man. In the relatively flexible civilian economy, a woman might soften in many ways her adjustment to difficult work, but in the rigid military framework, adaptation would be an all-or-nothing affair. There would be no easy absenteeism, home privacy and comforts, or choice of working conditions, nor yet the privilege of quitting a demanding job.

Even if these handicaps proved exaggerated, there was one obstacle that was not—the opposition of the American public. The Army did not operate in the vacuum of a dictatorship, but in a nation, society, and culture whose traditions must be considered. The course of public opinion regarding woman's place had by no means kept pace with her economic progress. The saying was still frequently heard that "woman's place is in the home," and it appeared certain that there would be great public opposition to placing women in soldier's jobs and in positions of rank and command. Army psychiatrists later noted that "in order for women to gain an active participation in military activities it was necessary for man to change his basic concept of the feminine role, to overcome his fear of 'women generals.' " [6] Army planners realized that such an obstacle existed, but it was not until after the establishment of a women's corps that its full extent was to be revealed.

The Army Nurse Corps

First to take the field against these obstacles was the Army Nurse Corps, whose development provided a close parallel to later WAC events. The admitted superiority of female nursing first caused acceptance of nurses as civilian employees, but in war after war it was found extremely difficult to maintain a civilian group within a military organization. Much inefficiency in Civil War medical care was believed due to "the lack of a single unified Nurse Corps with official status." In the Spanish-American War, it was noted that "unified direction and control within the Army framework itself was the only way to avoid administrative confusion and to assure maximum efficiency." [7]

Such full admission was delayed chiefly by popular opposition to military status for females, which led one early organizer to remark, "The nurse question is the women question; we shall have to run the

[4] Ltr, Maj L. W. McIntosh, Exec, Office Chief of Air Corps, to Maj Everett S. Hughes, G–1, 17 Apr 30. G–1/19835.

[5] Anna M. Baetjer, *Women in Industry; Their Health and Efficiency* (Philadelphia: W. B. Saunders Co., 1946). Prepared in the Army Industrial Hygiene Laboratory under auspices of the National Research Council.

[6] Maj Albert Preston, Jr., History of Psychiatry in the Women's Army Corps, 1946. SGO. Hereafter cited as Preston, Hist of Psychiatry in WAC.

[7] All references to the Army Nurse Corps, and to nursing, unless otherwise stated, are from a pamphlet published by the ANC, *The Army Nurse*. Copy in War College Library.

gantlet of those historic rotten eggs."[8] In 1901 Congress established the Army Nurse Corps, with somewhat the same status as the later Women's Army Auxiliary Corps (WAAC)—a military organization, but without Army rank, officer status, equal pay, or Army benefits such as retirement and veteran's rights. After the first World War its members were given relative rank and some retirement benefits, although pay and allowances still were not those of the men. Full military rank was not to be granted to nurses until 1944, a year after the WAAC had been legally admitted to full Army status and rank as the Women's Army Corps (WAC).[9]

Nevertheless, in placing nurses in a militarized and uniformed corps, the nation had taken one long step toward admission of women to full membership in the armed forces. Neither the public nor the Army was prepared to take further steps; the serious proposal for the establishment of the Women's Army Corps waited for the day when some great war that would almost drain the pool of American manpower should coincide with an availability of women workers trained in the modern skills an army required.

World War I

The first World War narrowly missed being that occasion. The first pinch of manpower shortage was felt chiefly by American industry and business, but by the end of 1918 the military services were also seriously concerned. The British, whose war effort was more nearly total, had already established women's auxiliaries in several of their services, and there was considerable evidence that had the war lasted a few months longer the United States might have done likewise. At the moment of the Armistice, the War Department General Staff was beset by serious proposals to this effect from both within and without the Army.[10]

One of the first of these proposals was initiated by the American Expeditionary Force in France. On 8 October 1917, General Pershing cabled a request for one hundred women telephone operators who could speak French and recommended that they be uniformed. This request was approved and the women were sent as civilian contract employees with privileges very similar to those of the Army Nurse Corps. Subsequently other groups were sent overseas, under varying contracts, by The Quartermaster General, the Ordnance Department, and the Medical Corps, but none had military status. In spite of extensive use of these groups, and of French women for unskilled work, the labor shortage of the AEF continued.[11]

At the same time, numerous civilian volunteer welfare groups, using over 5,000 American women, thrived and multiplied overseas in an un-co-ordinated manner, and appeared to the Army to be striving through competitive publicity to show what they had done for "the boys." While Army reports recorded almost no criticism of the conduct of these women, they did

[8] Statement attributed to Mrs. Bedford Fenwick, about 1887. Lucy Ridgely (Buckler) Seymer, *A General History of Nursing* (London: Faber and Faber, Ltd., 1933).

[9] The use of the abbreviations WAAC (later WAC) for the Corps and of Waacs (later Wacs) for its members was authorized, provided the full title of the Corps was used first, by Office Memorandum 21, 11 September 1942. SPWA 300, 1942 WAAC files. (See Bibliographical Note.)

[10] D. Colett Wadge, ed., *Women in Uniform* (London: Sampson Low, Marston & Co., Ltd., 1946).

[11] (1) Cablegram 276, par 15*b*, 8 Nov 17; (2) Memo, Dir of Women's Relations for G-1, 6 Aug 26, sub: Utilization of Women in Mil Serv. G-1/7000-2.

criticize the lack of orderly administration, the overlapping of duties, and the absence of any Army control. The confusion in the welfare groups was in striking contrast to the organized service given by members of the British Women's Auxiliary Army Corps, the largest of the British auxiliaries, whose services were lent to the AEF when the need for women's skills grew desperate. American Army officers in later years remembered favorably the discipline, efficiency, and *esprit de corps* of the British women's services.[12]

In still further attempts to solve the AEF's labor shortages, the commanding general of the AEF's Services of Supply, Maj. Gen. James G. Harbord, cabled repeatedly to the War Department in August of 1918 for authority to send a representative to the United States to recruit and organize a group of 5,000 women clerical workers to replace enlisted men. The representative was sent, followed by a proposal from the AEF that a women's service corps be organized as a part of the Army Service Corps. After considerable War Department debate as to whether the women should be enlisted, Civil Service, or uniformed contract employees, General Harbord was informed that, because of the change in draft age, 5,000 limited service men would be sent instead of the women. Concerning a women's corps, the War Department stated that it was not yet convinced of "the desirability or feasibility of making this most radical departure in the conduct of our military affairs." [13]

Meanwhile, strong supporting action for the AEF's effort came from posts and camps and from headquarters agencies in the United States. All Army and National Guard cantonments and camps had previously been forbidden to employ civilian women in any capacity except as nurses.[14] In the face of pleas from station commanders, this ruling was soon modified to authorize civilian employment of women "in essential work for which men employees cannot be obtained . . . taking only those of mature age and high moral character." [15] The War Department noted, "With careful supervision, women employees may be permitted in camps without moral injury either to themselves or to the soldiers." [16] Even with this concession, enough employees could not be procured, especially in isolated and uncomfortable stations, to perform all necessary duties. The Washington headquarters offices also found it difficult to obtain and hold an adequate number of female Civil Service employees.[17]

Therefore, the conclusion was reached almost simultaneously by several Army agencies, both in Washington and in the field, that a corps of women under military control would be the solution to these problems. As The Quartermaster General pointed out concerning laundresses:

> Every effort in the past of this office to provide a hired force of women at camps and cantonments has been unsatisfactory . . .

[12] (1) Memo cited n. 11(2), Table G, Rpts on Welfare Work, AEF; Tables A, B. (2) Senate Com on Mil Affairs, 77th Cong, 2d sess, *Hearings on S 2240*, 6 Feb 42, pp. 10–14.

[13] (1) Memo cited n. 11(2). Also: (2) Memo, CofS for TAG, Oct 18, sub: Enlmt of Women for Serv in France. WD Rec OCofS, 10730–19, National Archives. (3) Memo, TAG for CofS, 2 May 18, AG 011.2; approved by ACofS, 11 May 18, in Memo for TAG, 11 May 18. WD Rec AGO, WPD 10730–7, National Archives.

[14] Stf Memo, 8 Oct 17, in 1st Ind OQMG, 16 Oct 17. G–1/7000–2.

[15] Ltr, TAG to CGs of all National Guard and National Army Divs, 11 Dec 17. AG 230.2211 (Misc Div). See Memo cited n. 11(2).

[16] Memo, Lt Col Robert I. Rees, GSC, for CofS, 7 Dec 17. Copy in Memo cited n. 11(2).

[17] Memo 62, Col Ira L. Reeves, IGD, for ASW, 24 Aug 18. Copy in Memo cited n. 11(2).

only women of doubtful character show any inclination to remain as long as the voluntary system of employment is in vogue.[18]

The Quartermaster General thereupon recommended that legislation be secured to authorize the enlistment of women, ages 21–45, to be organized as the Women's Auxiliary Quartermaster Corps. Similar requests, giving similar reasons, were also made by the Inspector General's Department, the Chief of Engineers, the Operations Branch of the General Staff, and the Chief of Ordnance. The Chief of Ordnance estimated that the yearly turnover of civilian employees in his branch was approaching 84 percent. Ordnance and other branches went so far as to devise the uniform to be worn by enlisted women. At a meeting with representatives of The Quartermaster General, The Surgeon General, Signal Corps, Military Aeronautics, and the Corps of Engineers, it was decided that the uniform should be of "soft silver brown wool material," with a tan pongee blouse and brown Windsor tie, and that "no furs shall be worn with the uniform." [19]

These pleas did not receive favorable consideration by the War Department. Legislation to enlist "effective and able-bodied women" had in fact already been introduced in Congress in December of 1917, but had been returned to the House Military Affairs Committee by the Secretary of War with an expression of his disapproval. The memorandum upon which this opinion was based stated in unmistakable terms:

The enlistment of women in the military forces of the United States has never been seriously contemplated and such enlistment is considered unwise and highly undesirable . . . the action provided for in this bill is not only unwise, but exceedingly ill-advised.

War Plans Division noted in May of 1918:

Industrial conditions in the United States are not yet in such shape that it is necessary to undertake a line of action that would be fraught with so many difficulties.[20]

The War Department was equally unfavorable to an attempt by The Surgeon General to commission women doctors in the Medical Corps. The Judge Advocate General was unable to discover in the terminology of the National Defense Act any legal barrier to the appointment of women under that act. Nevertheless, the War Department's opinion was upheld on the grounds that only persons "physically, mentally, and morally qualified" could be appointed and that women doctors were obviously not physically qualified.[21]

The dismay of certain Army agencies at the lack of Army authority to enlist women was heightened by the Navy's action in placing the opposite interpretation upon its legislative authority and enlisting all the women it needed without further ceremony. As one Army planner later complained, "By enlisting Yeomen (F) the Navy Department ignored the Civil Service Commission and satisfied its needs regardless of the needs of others. They

[18] Memo, OQMG for TAG, 24 Apr 18, sub: Legislation Authorizing Enlmt of Women. WD Rec, OCofS, War College Div, 10730–7, 400.6 General, National Archives.

[19] (1) Memo cited n. 11(2), Tab F. (2) Series of five studies by Dr. Kristine Mann, Civilian Workers' Br, OCofOrd, 19 Sep to 28 Sep 18. (3) Memo, CofOrd for WPD, 26 Sep 18. WD Rec, OCofS, War College Div 10730–16 and 17, 342/8, National Archives.

[20] (1) HR 7112, 6 Dec 17. (2) Memo, Actg Chief, War College Div, for CofS, 22 Dec 17, sub: Bill Introduced in House of Representatives. Rejection was by Ltr, Newton D. Baker, SW, to Chmn, House Com on Mil Affairs, 26 Dec 17. WD Rec, OCofS, War College Div, 10730–1, National Archives. (3) Memo, Dir WPD for CofS, 8 May 18. WD Rec, OCofS, War College Div, 10730–7, AG 011.2, National Archives.

[21] Operations JAG 1917, I, 126. National Archives.

WOMEN DURING WORLD WAR I *served in France as telephone operators in a civilian capacity, and with the Navy and Marine Corps in the same status as men.*

wanted clerks and they got them."[22] Nearly 13,000 women enlisted in the Navy and Marine Corps on the same status as men and wore a uniform blouse with insignia. These women were the first in the United States to be admitted to full military rank and status.[23]

In addition to the pressure on the War Department from military advocates of the enlistment of women, there was undoubtedly some activity by organized women's groups. Prominent women in various sections of the country went so far as to organize semimilitary groups. In New York the Women's League for Self-Defense, five hundred strong and dressed in bloomers and puttees, drilled with rifles in the 66th Regimental Armory, and informed reporters that they had written Secretary of War Newton D. Baker and Maj. Gen. Leonard Wood "offering our service as the only strictly military women's organization in America."[24] The YWCA proposed that any more women sent to France be placed in a women's army corps like the British WAAC, with strict discipline and no individual billeting.[25] A similar proposal came from the American Council on Education, which concluded that military status would make for fewer resignations among trainees for war jobs.[26]

All such proposals came to an abrupt end with the cessation of hostilities on 11 November 1918. With a one-page sigh of relief the General Staff shelved the bulky documents that had set forth the arguments pro and con, and declared:

> In view of the present military situation it is believed no longer desirable that arrangements be made to form military organizations composed of women. . . . A continuance of the war would have required the United States in completing its program for the year 1919 to make a much more extended use of women . . . to replace men sent overseas or men shifted to heavy work which men alone can do.[27]

With this, the serious consideration of the establishment of a women's military group was relinquished to another generation of planners and to another war.

Twenty-three Years of Peace

The nation had hardly settled itself into postwar routine when it became evident that considerable political power was being wielded by the newly enfranchised female sex and its many well-organized national groups. It soon began to appear to the Army, as well as to many female leaders, that the mass of American women was dangerously susceptible to the charms of pacifism and other doctrines that advocated the abolition of the military as the best means of insuring peace.[28]

To stem this tide of opinion the Army sought to teach women voters more about

---

[22] Memo, Hughes for G-1, 21 Sep 28, sub: Participation of Women in War. G-1/8604-1.

[23] Memo cited n. 11(2). The legality of the Navy's action was held in "grave doubt" in a later decision of the Comptroller of the Treasury, but payment of the women was permitted in order not to interfere with naval efficiency. The Navy thereupon ceased enlisting women, but those already enlisted retained rights of service personnel and later of veterans. Memo, Maj Briant H. Wells, GS, for CofS, 22 Jun 17. WD Rec, OCofS, War College Div, 8196-7, National Archives.

[24] New York *American*, February 9, 1917.

[25] Memo, Dir WPD for CofS, 21 Oct 18, sub: Enlmt of Women for Serv in France. WD Rec, OCofS 10730-19, National Archives.

[26] Memo, Chmn, Com on War Serv Tng for College Women, American Council on Education, 18 Sep 18. WD Rec, OCofS 10730-16, National Archives.

[27] Memo, CofOpns Div GS for CofS,—Nov 18, sub: Orgn of WAAC and Equivalent Orgn in U.S. WD Rec, OCofS, War College Div, 10730-23, National Archives. Copy in Memo cited n. 11(2).

[28] Proceedings of Fifth Women's Patriotic Conference on National Defense, 29-31 Jan 30, Constitution Hall, Washington, D. C. Copy in G-1/9835.

its own nature and purpose. To this end, Secretary Baker in 1920 created a new position under the comprehensive title of Director of Women's Relations, United States Army, with an office in G–1 Division of the General Staff. The director was to maintain liaison between the War Department and the women of the country and to secure their co-operation by explaining to them that the Army was "a progressive, socially minded human institution" and that women voters should not "fanatically demand the dissolution of a ruthless military machine."[29] There was some consideration as to whether the director should have military status, but the Army decided to await developments.

The first incumbent of the position resigned after one year, for personal reasons and for what her successor believed to be "dissatisfaction with the lack of support given her."[30] The next appointee, Miss Anita Phipps, the daughter of an Army family, was also dissatisfied. There ensued a decade-long battle as to whether her position should degenerate into that of a supervisor of the Army's thirty-odd hostesses, or should be expanded into that of planner-in-chief for a women's Army corps, and first director.

Miss Phipps' difficulties, as set forth in periodic lengthy memoranda to the Secretary of War, centered around War Department failure to support her commitments and to give her military status. She thus, in her opinion, lost prestige in the eyes of powerful women's groups. She was embarrassed in the presence of Army and Navy nurses by her "undignified makeshift" of a uniform. Headquarters divisions and offices often failed to consult her when corresponding with women's organizations. General officers, when asked to speak before women's groups, ignored the careful tip sheets Miss Phipps had prepared for their guidance, and often alienated those whom she had cultivated with care. Her recommendations and staff studies were frequently filed without further action. One of her chiefs in G–1 Division went so far as to state baldly that her position should be abolished, since the male members of the General Staff were competent to plan the future utilization of women by the military service.[31]

In 1929, after almost ten years of work, Miss Phipps had nevertheless secured tentative approval of a plan that would unite the support of the most powerful women's groups behind the Secretary of War by means of a system of civilian aides to the Secretary—one chief aide, one in each corps area, and one in each state. Secretary of War Dwight F. Davis on 25 February 1929 publicly announced this plan to the press after a conference with women members of the League of Women Voters, American War Mothers, Daughters of the American Revolution, National Federation of Business and Professional Women's Clubs, and other such groups.

There at once arose a storm of letters from Senators and clergymen and from male civilian aides threatening to resign if women were also made civilian aides. Only fifteen days later a new Secretary of War, James W. Good, was in office and promptly sent telegrams to the women representatives, canceling the meeting of the nominating committee. This action, which in Miss Phipps' opinion greatly alienated the powerful women's groups,

[29] Memo, Dir Women's Relations for SW, 8 Jan 27. G–1/9562. Copy, with history of the office attached, in G–1/9835.

[30] *Ibid.*

[31] Memo, Brig Gen Campbell King, G–1, for CofS, 29 Sep 27, sub: Duty of Dir Women's Relations for U.S. Army. G–1/9835.

was followed in October by a final letter from Mr. Good to the women concerned, regretting that "the present is not a propitious time for appointing women civilian aides."[32]

Discouraged and in ill health, Miss Phipps in 1930 made a last appeal to the Secretary of War to define her duties and authority or to abolish her position. Nothing definite was done, and in 1931 her work was terminated by illness and by a new Chief of Staff, Gen. Douglas MacArthur, who overrode a favorable G–1 opinion and informed the Secretary of War that he considered her duties to be of no military value.[33]

Plans for a Women's Service Corps

The Director of Women's Relations left behind her the first complete and workable plan for a women's corps. Over a period of several years she compiled all evidence concerning the utilization of women during World War I, including the opinions of the British, the Congress, the Navy, and various Army commands. The heart of the plan was that the proposed women's service corps should be *in* the Army and not an auxiliary. She rejected the then-popular idea of enlisting women and giving them a uniform and a job but no military training, organization, discipline, housing, or required courtesies. This idea, she noted, had been tried and proved undesirable by both the Navy and the British. Instead, she proposed that women be fully trained and assigned only in units under the command of women officers, with not less than a squad at any station and no individual billeting allowed. All Army Regulations were to apply, plus special regulations for women as needed.[34]

By sending questionnaires to eight corps areas, three territorial departments, and eighteen chiefs of branches or similar services, Miss Phipps discovered that about 170,000 women would be wanted in wartime, although only six agencies favored giving them immediate military status.

In view of later debate about the proper jobs for women in the Army, it was significant that at this time the concept of a menial type of corps of low-grade personnel still loomed large in Army thinking. Not quite half of the requests were for clerks and stenographers; a few others were for small numbers of skilled workers such as draftsmen, dietitians, and telephone operators. The rest were chiefly for large numbers for relatively unskilled work: approximately 5,800 laundry workers, 7,000 cooks, 1,300 charwomen and janitors, 5,000 chauffeurs, 2,000 messengers, 11,000 laborers, 8,000 seamstresses, and so on.

The whole idea was rejected by the General Staff in August of 1926, although most divisions called it a "very splendid" study.[35] G–2 Division objected to the cost of housing, G–3 to difficulties in transportation and toilet facilities, and G–4 to the personnel policy involved. War Plans

[32] (1) Copies of ltrs and telgs are in G–1/9835 (1927). (2) Ltr, SW to Mrs. John Sippel, President, Gen Federation of Women's Clubs, 19 Oct 29. G–1/8604–3.

[33] (1) Memo, Miss Phipps for SW, 21 Oct 30; (2) Memos, G–1 for CofS, 13, 14 Nov 30; (3) Memos, OCofS, 5–6 Mar 31; (4) Ltr, CofS to SW, 10 Mar 31. All in G–1/9835 (1931).

[34] (1) Her voluminous preliminary notes and studies are still on file in several roughly bound volumes in G–1/7000. (2) Memo, Dir Women's Relations for G–1, 6 Aug 26. G–1/7000–2. For Navy and British experiences see: (3) Memo, OCofOrd for 3d ASW. WD Rec, OCofS, War College Div, 10730–16, National Archives. (4) Memo cited n. 25.

[35] Memo, G–1 for G–2, G–3, G–4, and WPD, 14 Aug 26, sub: Utilization of Women in Mil Serv, and atchd replies. G–1/7000–3.

Division suggested that the study be used only as a basis for further study and planning.

Throughout the comments of the entire General Staff ran the conviction that the mobilization of the United States would be a leisurely affair. G–4 Division noted: "To consider the development of a training organization for women workers in the beginning of a major emergency appears unthinkable.' [36]

A more powerful reason for private opposition among officers of the General Staff was noted by the Assistant Chief of Staff, G–1, Brig. Gen. Campbell King. General King wrote personally to the Chief of Staff that Miss Phipps seemed to wish to perfect within the Army an organization of women, headed by a woman, with a hierarchy of subheads plus influential civilian advisers, which would constitute "a powerful machine difficult to control and endowed with possibilities of hampering and embarrassing the War Department." [37]

Oddly enough, during a decade when neither a military corps nor a civilian advisory group could receive approval, a few women somewhat accidentally fell heir to the full blessings of Army status. When, in a more or less routine order, the supposedly all-male category of Army field clerks was blanketed into military service, it was found to include a handful of women employees. While this occurrence was scarcely noted on the planning level and was never considered a precedent for an organized women's corps, the few individuals concerned were later held by the Comptroller General to have been full-fledged members of the military service, with the same status as the Navy yeomanettes. Later, Congress amended the armed forces' legislation to place the word *male* before *persons*, thus effectually guaranteeing that neither the yeomanette nor the field clerk episode would be repeated without its sanction.[38]

The Hughes Plan

Male planners up to this time had not come to very close grips with the problem of planning for a women's corps. While students at the Army War College had studied a course entitled "Conservation by Utilization of Women in Industry, in Military Service, and in Welfare Work," it was ordinarily assumed that such "Conservation" would be most appropriately left to industry. The chief Army planner for a women's corps was appointed in 1928; he was Maj. Everett S. Hughes of G–1 Division, General Staff.[39]

Pointing out that nothing but fruitless conflict had resulted from previous arguments between extreme feminists on the one side and male die-hards on the other, all disagreeing endlessly over minor details, Major Hughes proposed acceptance of the fact that women would inevitably play a part in the next war—the more nearly total the war, the greater the part. No amount of wishful thinking could avert that necessity, powered as it was by social and economic trends beyond the nation's powers to reverse.

On the other hand, he stated, the Army should not attempt a detailed solution until the situation was known; it would be futile to waste time debating minor points,

[36] Memo, G–4 for G–1, 13 Sep 26. G–4/20390.

[37] (1) Memo, G–1 for CofS, 6 Jul 27. G–1/9835. (2) Memo, Lt Col George Grunert for G–1, 3 Feb 27. G–1/7000–2.

[38] The Comptroller General's decision is discussed in Ch. XII, below.

[39] Discussion of Hughes plan, unless otherwise indicated, is from Memo, Hughes for G–1, 21 Sep 28, sub: Participation of Women in War. G–1/8604–1.

such as whether women should enter the combat zone, until it was known what the combat zone of the next war might include. Before the time arrived to make these specific decisions, the women who were to help make them must be trained not merely in drill but in an understanding of Army thinking, a process that could not be achieved overnight. And the men who were to make the decisions must also be trained in understanding the problems of militarization of women. If the women who were to lead the new corps were ignorant, said Hughes, *"this ignorance, coupled with man's intolerance, may be fatal."* [40]

The Hughes plan contemplated that only women overseas or in danger zones would be militarized. The only advantage to the Army in militarizing other women would be that of being able to order them to perform certain unpopular duties such as laundry work, and the plan noted the fallacy of supposing that women of a higher type would enlist after they knew that only menial work was in store for them.

The need for military status in danger zones was strongly indicated by current arguments then before Congress concerning women who had worked in France and who were seeking compensation for the loss of their health. The War Department had already decided that such women were not entitled to veterans' gratuities, ruling in one case:

> Female telephone operators have no military status whatever . . . those serving with the American Expeditionary Forces [were] not even under civil service regulations. . . . It is not believed that Congress intended that a gratuity should be paid to any person not actually a member of the military establishment.[41]

Nevertheless, the sentiment in Congress favored those women whose disabilities were directly due to war service, and financial relief to prevent their destitution was voted in certain cases. At this time Congresswoman Edith Nourse Rogers of Massachusetts developed an interest in the problem which was to lead ultimately to her sponsorship, fifteen years later, of the WAAC bill.[42]

Major Hughes also stated that it was uneconomical and confusing to have separate organizations of men and women, and that qualified women could be integrated into the men's army, with a similar uniform and privileges. He argued against camouflaging rank by odd titles such as Deputy Controller, which spared the male ego but confused the employing agencies. "Why not take the whole step and do the thing right?" he wrote.

Major Hughes' prophetic efforts were embalmed with indorsements, laid out for observation for a period, and then buried so deep in the files that they were recovered only after the WAAC was six months old and War Department planners had already made most of the mistakes he predicted.[43] His study was sent to the Chief of

[40] Author's italics.

[41] Memo, Dir WPD for CofS, 18 Mar 19. WD Rec, OCofS, War College Div, 10730–26, National Archives.

[42] (1) House Com on Veterans Legislation, 69th Cong, 1st sess, *Hearings*, 6 Aug 26. (2) Interv with Mrs. Rogers, 14 Dec 45. (This form of citation is used throughout for interviews conducted by the author.)

[43] The study was recovered then only when Mrs. Oveta Culp Hobby, Director, WAAC, on temporary duty in England, was introduced to Maj. Gen. Everett S. Hughes, who demanded to know what G–1 had done with his masterpiece. Returning to Washington, she unearthed it but found no evidence that it had been known to the G–1 planners. Interv with Lt Col Helen Hamilton Woods, Deputy Dir WAC, 22 Oct 45.

Throughout the volume, the rank held by the officer at the time of the interview or of the discussion in the text is given. Since this is not a completely chronological account, an officer who appears in an early

Staff in 1928 and again, refurbished, in 1930.[44] At this time the new G–1, Brig. Gen. Albert J. Bowley, also distinguished himself in the field of prophecy by proposing an immediate campaign of education for Regular Army officers, since

> Successful co-operation between men and women during the next war will depend to a great extent on the attitude of the officers of the Regular Army toward the women of the country. To influence this attitude will require much time and discussion.[45]

A dejected-looking sheaf of handwritten scraps of paper indicated that the studies were carried back and forth from G–1 to the Chief of Staff to the Secretary of War to G–1, bearing notations of diminishing intensity, such as "Hold until Secretary of War decides"; "Hold until fall when women return to their homes after summer activities"; and, finally, merely "Hold." The last one in the series, dated 5 January 1931, stated: "General B. says may as well suspend; no one seems willing to do anything about it."

So ended the peacetime planning for a women's corps.

The Approach of World War II

On 1 September 1939, under the shadow of imminent hostilities, Gen. George C. Marshall was appointed Chief of Staff, and a month later planning for a women's corps was resumed. A staff study of the problem was shortly prepared by G–1 Division. The study did not reflect its predecessors in any way; planners were evidently not aware of the Phipps and Hughes studies.[46]

The most important idea in the new plan was that women must under no circumstances be given full military status. The Civilian Conservation Corps was hit upon as the model. The plan stated:

> The CCC has shown how persons may be grouped in units with a military form of organization, uniformed, given grades of rank, paid and cared for, employed under orders of Army Officers, administered by the Army's chain of command, and governed by War Department Regulations, without being members of the Army.

WAAC leaders were later to consider this analogy extremely unfortunate, since the CCC had been a peacetime organization employed in separate camps, whereas the WAAC was to be used in Army camps and overseas where a legal distinction in status would be far more difficult. The plan likewise proposed that some women would be set up in "quasi-military female organizations" similar to companies, while others would move about individually or in small groups. It was stated that women's probable jobs would include those of "hostesses, librarians, canteen clerks, cooks and waitresses, chauffeurs, messengers, and strolling minstrels." The plan was held in abeyance, and for another eighteen months the War Department did not commit itself.[47]

---

chapter as a lieutenant colonel subsequently may be referred to as a major or a captain. It should also be noted that, just as brigadier, major, and lieutenant generals are customarily referred to, after their initial mention, simply as General ———, lieutenant colonels are referred to similarly as Colonel ———. Colonel Hobby was the only full colonel in the wartime WAC.

[44] (1) Memo, Gen King, G–1, for CofS, 24 Aug 28. G–1/8604–1. (2) Memo, Brig Gen Albert J. Bowley, G–1, 28 Feb 30. G–1/8604–4.

[45] Memo, G–1 for CofS, 4 Jun 29, sub: Participation of Women in War. G–1/8604–2.

[46] Memo, Capt Williston B. Palmer for G–1, 2 Oct 39, sub: Women with the Army (Emergency). G–1/15839.

[47] (1) *Ibid.* (2) Personal Ltr, Col William Rose, AGO, to Col Lawrence Whiting, 11 Jun 40, forwarded to Hist Div WDSS, 11 Apr 45, sub: Compilation of WAC History. OCMH.

Meanwhile, numerous letters and telegrams began to be received from women's organizations and from individuals offering their services. G–1 Division now expressed mild anxiety lest any of these women arrive in a theater of operations before they could be organized in some civilian corps under military control. With the collapse of France in May and June of 1940 and the subsequent Battle of Britain, the offers of help redoubled. When in September of 1940 the first peacetime Selective Service Act became law, women's groups increased their demands that they be allowed to contribute to the nation's defense. By March of 1941 a change in tone could be noted in official statements. In a letter to an inquirer General Marshall stated:

> While the United States is not faced with an acute shortage of manpower such as has forced England to make such an extensive use of its women, it is realized that we must plan for every possible contingency, and certainly must provide some outlet for the patriotic desires of our women.[48]

The pressure of these patriotic desires was soon considerable. In Chicago, the Women's League of Defense, 17,000 strong, set itself up as an agency for the enrollment and classification of "women who can do anything helpful to replace a man in the event of war." In Los Angeles, the Women's Ambulance and Defense Corps trained women in military skills and persistently sought recognition as an Army-sponsored agency for training women officers. In Pittsburgh, the Memorial Gold Cross First Aid and Ambulance Corps enrolled 2,000 members. In Washington, D. C., the Green Guards stated that they had "contended, urged, and pleaded with the powers that be to include women in the national defense plan in some capacity." In Ohio, the Toledo Unit of the Willys-Overland Women's Motor Defense Corps proposed to train women for duties with the Army.

Many other such private organizations arose, most of them genuine, but a few spurious. G–2 Division was obliged to discontinue replies from the Chief of Staff to organizations that had previously tried to convince the public that they had Army sponsorship. During these months, private citizens without number also wrote to the Army and to Congress offering their services.[49]

More significant than this sporadic activity was evidence that Congress or national groups might act to set up women's units outside Army control, or with full military status. Congresswoman Edith Nourse Rogers reported herself ready to introduce a bill for full military status. The Air Corps requested a woman's volunteer defense corps for duty with its Aircraft Warning Service; as the War Department delayed, the Air Corps decided that a separate women's corps under its own control might be preferable. U.S. Army representatives in England also advised that a women's auxiliary would be necessary in the event that larger forces were sent overseas, since members of the British Women's Auxiliary Air Force and Auxiliary Territorial Services could not be borrowed by the U.S. forces in sufficient numbers because of the British manpower shortage.

From the White House came two infor-

[48] See entire file G–1/15839 (5–11–42), especially Memo, G–1 for CofS, 13 Aug 40; and Ltr, CofS to Dr. John W. Colbert, 19 Mar 41.

[49] Examples are from: (1) Collection of ltrs and documents in G–1/15839–1 (5–11–42); and (2) Rupert Hughes, "Shall We Have a Women's National Guard?" *Liberty*, March 8, 1941.

mal proposals. Mrs. Eleanor Roosevelt suggested that American women, like the British, be used in antiaircraft barrage work, a duty that was considered by some military planners as dangerously close to combat work. In a separate proposal she suggested a pool of women for service with the Army, Navy, and Marine Corps, as needed, but under the command of none of them; instead, it would possibly be under the Office of Civilian Defense, although members of Congress protested such action from the House floor because of the "miserable record of mismanagement" of that organization.[50]

Immediate impetus was supplied by Congresswoman Rogers, for many years an authority on legislation concerning women. Mrs. Rogers stated later, when asked her motives:

> My motives? In the first World War, I was there and *saw*. I saw the women in France, and how they had no suitable quarters and no Army discipline. Many dietitians and physiotherapists who served then are still sick in the hospital, and I was never able to get any veterans' compensation for them, although I secured passage of one bill aiding telephone operators. I was resolved that our women would not again serve with the Army without the protection men got.[51]

Mrs. Rogers later informed the House: "I have been nursing this measure along through the years."[52]

In the spring of 1941 Mrs. Rogers called upon the Chief of Staff and informed him that she intended to introduce a bill to establish a women's corps. General Marshall replied, "Give me a week to consider it," and afterwards asked that the week be extended to a month.[53]

During this month the War Department plunged into furious planning for a bill that the Army could safely sponsor. The planners' motives were made clear in a staff memorandum from the Assistant Chief of Staff, G–1, who wrote:

> Congresswoman Edith Nourse Rogers has been determined for some time to introduce a bill to provide a women's organization in the Army. We have succeeded in stopping her on the promise that we are studying the same thing, and will permit her to introduce a bill which will meet with War Department approval.
>
> Mrs. Roosevelt also seems to have a plan.
>
> The sole purpose of this study is to permit the organization of a women's force along lines which meet with War Department approval, so that when it is forced upon us, as it undoubtedly will be, we shall be able to run it our way.[54]

The resulting plan was the work of Col. James Taylor of G–1 Division and, later, of Maj. Robert W. Berry. It provided for a Women's Army Auxiliary Force (WAAF), definitely a civilian auxiliary and not part of the Army. The plan's authors believed that it would avoid the errors made during World War I. They stated:

> The War Department initially made no provision for the use of women in the last war with the exception of the Army Nurse Corps.

---

[50] (1) Ltr, AC to TAG, 19 Mar 41. AG 324.4 (3-19-41). (2) D/F, G-1 for AC, 8 May 41. G-1/15839-10, Pt 1. (3) Ltr, GHQ AF to CofAAF, 27 Dec 41, and subsequent documents. ACC 324.5 AWS (Women). AG 291.9 WAAC Sec 1, Pt 1. (4) [Henry G. Elliott] The Predecessor Commands, SPOBS and USAFBI, Pt I of The Administrative and Logistical History of the ETO, Hist Div USFET, 1946, MS, p. 255, OCMH. (5) Memos, Chief, Women's Interest Sec, WDBPR, for CofS, 14 Nov 41, and 9 Dec 41, sub: HR 4906. AG 291.9 WAAC Sec 1 Pt 1 (6-2-41). (6) *Congressional Record*, Vol. 88, No. 55, p. 2657, 17 Mar 42.

[51] Interv cited n. 42(2); also *Congressional Record*, Vol. 87, No. 100, p. 4639, 28 May 41.

[52] *Congressional Record* entry cited n. 50(6).

[53] Interv cited n. 42(2).

[54] Memo, Brig Gen Wade H. Haislip, G-1, for CofS, 29 Apr 41. G-1/15839-10. Also in WA 320 (5-29-41) DRB AGO.

> . . . However, the services of women were found to be so necessary overseas and in posts, camps, and bureaus in the United States that before the World War was over, a large group of women were serving with the Army in unorganized and uncoordinated groups, hastily and inefficiently recruited, under little if any discipline, and with no military status or recognition.[55]

Cited advantages of the plan included not only its greater controls but the fact that "it will tend to avert the pressure to admit women to actual membership in the Army." Cited disadvantages included not only the many special arrangements necessary for women's care but the fact that the corps would "inject many other unpredictable problems into military administration."

Most of the General and Special Staff Divisions, when the plan was submitted to them for comment, agreed with G-1's reasoning. The name was changed from WAAF to WAAC, at the suggestion of The Adjutant General, because a parallel with the earlier British name was believed desirable, although Mrs. Rogers and other women advisers objected because of the sound and the similarity to the word *wacky*. G-4 Division concurred "in time of war only." [56]

Although The Surgeon General concurred, the Director of the Army Nurse Corps, Maj. Julia Flikke, did not favor the bill, saying, "It is my opinion that the disadvantages outweigh the advantages . . . complications would arise between that organization and other existing organizations." She feared that it would inconvenience citizens by causing "a dearth of domestic help" during the war and that "this organization necessarily would be composed largely of married women who would find it difficult to comply with regulations because of home ties, and would always need special consideration and no doubt there would be many who would object to regimentation." [57] Only the Judge Advocate General proposed full military status, fearing—all too accurately, as it later proved—legal complications in an auxiliary.[58]

The Secretary of War rejected the plan at first sight, declaring it "premature," but upon learning the alternatives he acceded, being reassured by G-1 Division that "if the organization of this force is authorized by law, it is the intention of the War Department to develop it slowly and by trial and error. It is not the purpose of the War Department to rush into this matter on a large scale." [59]

H.R. 4906

On 28 May 1941 Mrs. Rogers introduced in the House of Representatives "A Bill to establish a Women's Army Auxiliary Corps for Service with the Army of the United States." [60] She accepted auxiliary status for the corps, saying:

> In the beginning, I wanted very much to have these women taken in as a part of the Army. . . . I wanted them to have the same rate of pension and disability allowance. I . . . realized that I could not secure that. The War Department was very unwilling to have these women as a part of the Army.[61]

[55] Memo, G-1 for CofS, 31 Mar 41. G-1/15839-10.

[56] Comments and concurrences are all in two bound files, G-1/15839-10, Pts 1 and 2.

[57] Memo, Supt ANC for Col Howard T. Wickert, 9 Apr 41, atchd to Memo, SGO for G-1, 12 Apr 41. SGO 322.5-1 WAAF.

[58] Bound files cited n. 56.

[59] Memo cited n. 54.

[60] HR 4906, 77th Cong, 1st sess, 28 May 41.

[61] *Congressional Record* entry cited n. 50(6).

She also correctly estimated Congressional temper as being at this time much opposed to any idea of full military status for women.

H.R. 4906, as prepared by Major Berry and other officers of G–1 Division, took fourteen printed pages to outline the ways in which the Corps did and did not differ from the Army. The WAAC was to be a corps of 25,000 women for noncombatant service; it was "not a part of the Army but it shall be the only women's organization authorized to serve *with* the Army, exclusive of the Army Nurse Corps." There followed a statement of its mission that was extremely significant, for it indicated a decision which was perhaps the most basic and vital of the Corps' history: the WAAC was established "for the purpose of making available to the national defense the knowledge, skill, and special training of the women of the nation." Deliberate emphasis was placed on the utilization of a small number of skilled high-grade workers, instead of large numbers of unskilled low-grade personnel, as contemplated in earlier plans. Planners noted:

> The problem in the United States is not primarily one of utilizing women in the military service for the purpose of releasing manpower, but is one of utilizing women to increase the efficiency of the Army.[62]
>
> Both educational and technical qualifications should be set exceptionally high to make of the projected organization an elite corps, in order that it may quickly attain the highest reputation for both character and professional excellence.[63]

Later developments were to make it appear that no one yet fully understood the tremendous implications which this decision would have in necessitating procedures different from those for men, who were of course enlisted without any such selectivity as to skills and character. In fact, the same planners in a list of possible duties included jobs manifestly unfit for recruits of skill and education, such as charwoman and laundry worker.

The bill neglected to spell out a point that had been clearly stated in earlier plans: the WAAC was to operate under Army command channels above the company level. Instead, command responsibility was publicly placed on a woman entitled Director, who was to "operate and administer the Corps in accordance with the normal military procedure of command." The Director was also to "advise the War Department" and "make recommendations as to plans and policies."

Endless legal distinctions were also attempted. Waacs were to receive medical service from the Army, but no medical or other benefits after discharge except those provided by the Employees Compensation Commission. They were considered military personnel for benefits of the Soldiers and Sailors Relief Act but not for benefits of the act granting re-employment rights, except for those who left Civil Service jobs. Most important, the extent to which Waacs could be disciplined was not clear. It was stated that Waacs were to be subject to the Articles of War "when applicable," but it was to develop that no one then or later ever knew when they were applicable except in overseas areas.

The bill was obliged to be specific on matters such as pay, allowances, grades and the numbers in each grade, travel pay, leave, and other points that were to be quickly outmoded by subsequent Army

[62] Memo cited n. 55.

[63] Memo, Morale Br for G–1, 7 Apr 41. MB 320.2 (3–31–41) WR. Copy with G–1/15839.

legislation, thereupon requiring additional WAAC legislation.

Still worse, the camouflaged grades were not strictly comparable to Army grades in either rank or pay and were to cause great confusion in finance and personnel offices when the attempt was made to replace men on Army Tables of Organization. Instead of lieutenants and captains, the WAAC had third, second, and first officers. A first officer was comparable to a captain in duties and insignia but drew the pay of a first lieutenant; a second officer drew pay somewhere between a first and second lieutenant's; and a third officer was comparable to a second lieutenant, with pay of $1,500.

Enlisted grades were still more confusing, with three grades of auxiliaries comparable to the Army's two grades of privates and three grades of leaders in place of the Army's five grades of noncommissioned officers. In addition there was to be a director, who was authorized the salary of $3,000, comparable to a major's, and a few assistant directors, who were to receive a captain's pay of $2,400.

This bill after its introduction immediately sank from sight; it was not to become law for a full year. It was referred routinely to the Bureau of the Budget, which did not reply for four months. Meanwhile, G-1 Division had reversed its previous stand and ventured the hope that the Chief of Staff would withdraw his support or at least not commit himself until the Bureau of the Budget was heard from.[64]

Nevertheless, the Chief of Staff, almost alone in the War Department, made it clear in the fall of 1941 that he had come to look on the bill with more than official approval. Mrs. Rogers noted that "General Marshall now became very enthusiastic about the bill." [65] Col. John H. Hilldring, soon to assume office as G-1, stated later:

> By the summer of 1941, General Marshall was intensely interested in the WAAC business. He foresaw a cycle of shortages; that of the moment was the supply shortage, in which the scarcity of supplies hindered mobilization, but he now became convinced that the bottleneck of the future would be that of manpower. He also considered the fact that war had become a complicated business which needed many civilian techniques, and that many of these were almost completely in the control of women. General Marshall asked me why we should try to train men in a specialty such as typing or telephone work which in civilian life has been taken over completely by women; this, he felt, was uneconomical and a waste of time which we didn't have. For example, the Army's telephone service had always had a reputation for being bad in spite of our superior equipment, and women operators could and did end all that.
>
> The Chief of Staff was also influenced by the fact that the ladies wanted in; he literally has a passionate regard for democratic ideals.[66]

General Marshall at this time told several of his staff members that, while he did not see any immediate need for large numbers of women in the Army, he would like to have authorization for a women's corps on his books so that, if the need for quick action should arise, the point of debate would be past.[67] This agreed with what he later wrote to Congress:

[64] Memo, Budget and Leg Plng Br for G-1, 18 Jun 41, sub: Rpt on HR 4906. AG 291.9 WAAC sec 1 pt 1 (6-2-41).

[65] Interv cited n. 42(2).

[66] Interv with Gen Hilldring, 17 Jan 46.

[67] *Ibid.;* also, statement was made to Mrs. Hobby and repeated by her later to a Congressional committee. House Com on Mil Affairs, 77th Cong, 2d sess, *Hearings on HR 6293*, 20-21 Jan 42, p. 33.

> I regard the passage of this bill at an early date as of considerable importance. In general, we have secured most of the legislation required for the complete mobilization of the Army so that we can go ahead with its development and definitely plan for the future. However, we lack Congressional authority for the establishment of a Women's Army Auxiliary Corps, and as a result can make no definite plans. Also, I am under continued pressure from many directions regarding this phase of our preparations.
>
> It is important that as quickly as possible we have a declared national policy in this matter. Women certainly must be employed in the overall effort of this nation. . . . We consider it essential that their status, their relationship to the military authority, should be clearly established.[68]

The Chief of Staff therefore took personal action to needle the agencies still brooding on the WAAC bill. Although the U.S. Civil Service Commission and the Employees Compensation Commission both had been persuaded by Mrs. Rogers to render favorable reports, the Bureau of the Budget informed the War Department on 7 October: "It is not believed that the enactment of the proposed legislation, at least at the present time, should be considered as being in accord with the program of the President." [69] The Bureau of the Budget stated, "There appears to be no need for a WAAC inasmuch as there will never be a shortage of manpower in the limited service field." [70]

In his attempts to get the measure past the Bureau of the Budget, the Chief of Staff employed a newcomer to the War Department, one of its civilian consultants, Mrs. Oveta Culp Hobby. Mrs. Hobby had, in the summer of 1941, been "virtually drafted" to come to Washington to set up the new Women's Interests Section of the War Department Bureau of Public Relations, a section designed to furnish soldiers' wives and mothers with information that would reassure them about the living conditions of their drafted relatives. She had agreed to stay for not more than four months, or until the agency was functioning properly; any longer absence from home she considered undesirable in view of the fact that she had a husband and two children and was coeditor and publisher of a newspaper.[71]

General Marshall reported himself impressed with Mrs. Hobby's work in organizing a meeting, at which he spoke, of the national presidents of the twenty-one largest women's organizations, designed to secure their good will and assistance to the Army.[72] He therefore asked her to give the G–1 planners any assistance on the public relations aspect of the WAAC bill that they might need. At the time of certain discussions with the President's wife, General Marshall himself sent a handwritten note to G–1: "Please utilize Mrs. Hobby as your agent to smooth the way in this matter through Mrs. Roosevelt." [73] As a result, Mrs. Hobby was thoroughly familiar with the legislation and was appointed

---

[68] Ltr, CofS to Hon. John W. McCormack, 6 Feb 42. CofS personal files.

[69] Ltr, Bur of Budget to SW, 7 Oct 41. AG 291.9 WAAC sec 1 pt 1 (6–2–41).

[70] Quotation from Memo, Maj Berry for Gen Haislip, 2 Dec 41. WA 042.2 (12–2–41), DRB AGO. See also Memo, Mrs. Hobby for CofS, 2 Dec 41, sub: HR 4906. CofS personal files.

[71] Senate Com on Mil Affairs, 77th Cong, 2d sess, *Hearings on S 2240*, 6 Feb 42, p. 23. The Senate suspected that the office had been set up to lobby for the WAAC, with which it actually had no connection. See also Ltr, CG, Pacific Ocean Area (Lt Gen Robert C. Richardson, Jr., formerly with WDBPR) to Col Hobby, 12 Oct 44. WDWAC 320.2. Also see undated MS draft, The Mission That Grew. WDWAC 314.7 (1945–46).

[72] Memo, Brig Gen Alexander D. Surles, Chief, WDBPR, for CofS, 13 Oct 41. CofS personal files.

[73] Memo, Chief, Women's Interests Sec for CofS, 14 Nov 41, sub: HR 4906. AG 291.9 WAAC sec 1 pt 1.

GEN. GEORGE C. MARSHALL *speaking to a meeting of the national presidents of the twenty-one largest women's organizations, 13 October 1941.*

the only female representative of the War Department in negotiations with the Bureau of the Budget and in later Congressional hearings.

In further efforts to get the WAAC bill past the Bureau of the Budget, General Marshall made a personal call and sent a highly favorable written report to the effect that "this legislation has great merit." Already overshadowed by the coming events at Pearl Harbor, the report on 25 November 1941 spoke of gaining "valuable time while time is available" in order to develop a small corps in an orderly and efficient fashion.[74]

On Thanksgiving eve General Marshall checked over with Colonel Hilldring the measures now completed to make the War Department ready for the gathering storm and, noting that WAAC legislation was still among the missing, directed him to omit no step to ensure its immediate passage. Colonel Hilldring recounted later: "General Marshall shook his finger at me and said, 'I want a women's corps right away and I don't want any excuses!' At that, I displayed considerable energy."[75]

Colonel Hilldring's energy was shortly reinforced by that of the Japanese at Pearl Harbor. On 11 December, four days after Pearl Harbor, the Bureau of the Budget withdrew its objections.[76] The way was now clear to ask Congress for the orderly

[74] Memo, DCofS for Dir Bur of Budget, 25 Nov 41, sub: Proposed Revised Rpt on HR 4906. G-1/15839-10.

[75] Interv cited n. 66.

[76] Ltr, Bur of Budget to SW, 11 Dec 41. G-1/15839-10. Also in AG 291.9 WAAC sec 1 pt 1.

and efficient development of a small women's corps capable of expansion in wartime.

When later criticism came to be leveled against the agencies and individuals who failed, in the five months now left before the passage of the bill, to anticipate and prevent every problem of the new organization's development, they were able to note in reply that they were attempting in one hundred and fifty days, and in the midst of events that dwarfed the WAAC in importance, to settle problems which twenty-three years of leisurely planning had not succeeded in solving. "Woman's ignorance" and "man's intolerance" had now come together in the fatal combination of Major Hughes' prophecy.

CHAPTER II

Establishment of the WAAC

From the date of Pearl Harbor onward, plans for a women's corps moved with a speed unequaled in the past decades. The Secretary of War sent his approval of the WAAC bill to Congress on Christmas Eve. By the last day of 1941 Mrs. Rogers had incorporated the War Department's proposed amendments into the bill and reintroduced it as H.R. 6293. The Navy Department occasioned some delay while it tried to persuade the War Department not to sponsor the corps. Finally the Assistant Secretary of War in person succeeded in persuading the Navy to withdraw "their objections to our endorsing it." Navy personnel chiefs flatly refused to make the project a joint one, informing Colonel Hilldring, "You are going to take a beating and we'll wait to see what happens." [1]

Congressional consideration was rapid but rough. "It was a battle," said General Hilldring later. "In my time I have got some one hundred bills through Congress, but this was more difficult than the rest of the hundred combined." [2] Opposition was felt more on the floor of the House, and in the cloakrooms, than in the Committees on Military Affairs. At the committee hearings Mrs. Rogers emphasized the protective and disciplinary aspects of the measure.

The only real issue at these hearings was that of militarization versus civil service. Army spokesmen offered reassurance that Waacs would be used only where civilians were unobtainable, or where security required military personnel, and would never under any conditions displace a Civil Service employee. One powerful argument was advanced by the Air Corps, which asserted that twenty-four-hour daily security in the Aircraft Warning Service could not be achieved with civilian volunteers and that, unless Waacs could be obtained, there would be actual danger to the east coast, especially the capital.[3]

Much of the work of steering the bill was personally handled by Col. Ira P. Swift of the War Department. The committees moved with what Mrs. Rogers felt was unprecedented speed. The House hearing on 20 January was followed by the committee's approval on 28 January, by

[1] (1) Ltr, SW to Chmn, House Com on Mil Affairs 24 Dec 41. AG 291.9 WAAC sec 1 pt 1. (2) Memo, ASW for Hilldring, 21 Dec 41. G-1/15839-10. (3) Interv with Maj Gen John H. Hilldring, 17 Jan 46. Confirmed by interv with Mrs. Rogers, 14 Dec 45, and by First Draft Narrative, U.S. Naval Administration in World War II: Bureau of Naval Personnel, Women's Reserve, 18 January 1946. Dir of Naval History. (Hereafter cited as WAVES Hist.)

[2] Interv with Gen Hilldring cited n. 1(3).

[3] All refs to com hearings are from: (1) House Com on Mil Affairs, 77th Cong, 2d sess, *Hearings on HR 6293*, 20-21 Jan 42; (2) House Rpt 1705, 77th Cong, 2d sess, 28 Jan 42; (3) Senate Com on Mil Affairs, 77th Cong. 2d sess, *Hearings on S 2240*, 6 Feb 42; (4) Senate Rpt 1051, Calendar 1086, 77th Cong, 2d sess, 9 Feb 42.

Senate hearings on 6 February, and committee approval on 9 February. There were no really angry remarks, although there were a few such as "You are going to have a few generals, aren't you? . . ." "No, the women 'generals' will remain at home"; or "Are you going to start a matrimonial agency?" Eventually the hearings ended in a blaze of gallantry on the part of the Southern senators.

Next, the bill stuck in the House Rules Committee for some days. General Hilldring said later: "In all my experience with legislation, I had never before had a bill stick in the Rules Committee, but they refused to report it out. I spent three hours talking to them; I never confronted a colder audience." Mrs. Rogers added, "The Rules Committee was rough." Its members eventually yielded, stating that they dared not oppose their opinions to the Chief of Staff's on the matter of measures required for the national defense.[4]

When the measure finally reached the House floor the real opposition developed. Members argued that a soldier would go forward in battle even if his buddy was shot down beside him, but if his buddy was a woman he would stop and render first aid. Others declaimed, variously:

> I think it is a reflection upon the courageous manhood of the country to pass a law inviting women to join the armed forces in order to win a battle.
>
> Take the women into the armed service, who then will do the cooking, the washing, the mending, the humble homey tasks to which every woman has devoted herself?
>
> Think of the humiliation! What has become of the manhood of America? [5]

It was General Hilldring's opinion that the bill was saved from immediate defeat at this time only by the personal support and prestige of the Chief of Staff. The Secretary of War likewise threw his support behind the measure. Nevertheless, it eventually became clear that, at best, passage would be delayed for some weeks while other more important war measures took precedence.[6]

The Appointment of WAAC Pre-Planners

Meanwhile, War Department planning was racing to keep ahead of the bill's estimated rate of progress. Forty-five days were lost in December and January before it was decided to secure an officer with wide experience and acquaintance in the War Department to establish and guide the new corps. He was to have the title of Pre-Planner, WAAC, and was to operate from a floating position in G–1 Division in order to secure necessary action from all other agencies. Such an officer was not immediately available in the War Department, and G–1 brought in from the field and temporarily assigned Lt. Col. Gilman C. Mudgett, ordering him to "build a fire under WAAC planning." [7] Colonel Mudgett was a Cavalry officer with twenty years' experience in the Regular Army, and an expert in Advanced Equitation. Most of his experience had been as a squadron officer with mounted units and as an instructor at the Cavalry School; he

[4] Intervs cited n. 1(3); also Memos, Chief, Budget and Leg Plng Br, for Secy, WDGS: (1) 11 Feb 42, sub: HR 6293; (2) 19 Feb 42, sub: WAAC; (3) 26 Feb 42, sub: Leg Proceedings. G–1/15839–10 (7–8–42).

[5] All from *Congressional Record*, Vol. 88, No. 55, 17 Mar 42.

[6] House Rpt cited n. 3(2).

[7] (1) Ltr, Col Mudgett to Miss Faustine Dennis, 11 Feb 42; (2) Ltr Orders, AG 210, G–1 Gen Stf (1–13–42). Both in Col Mudgett's Staybacks, 1942 WAAC files. (Hereafter cited as Mudgett Staybacks.) (3) Interv with Lt Col Harold P. Tasker, 20 Nov 42.

had never been assigned to the War Department until two weeks before he became WAAC Pre-Planner.[8]

Colonel Mudgett found that very little information could be furnished him by the War Department, most of that outdated.[9] Except for copies of the legislation, he was provided with little more than a statement that civilian employees were hard to get in such places as Pig Point, Va., and Matagorda Island, Tex.[10]

The problem at the time looked deceptively simple: The WAAC was to be a small organization, developed slowly. A director, and perhaps five assistant directors, and one hundred officer candidates would be appointed. Colonel Mudgett called the local YWCA to locate one hundred rooms for the officer candidates, who would attend a leisurely three months' course before enlisted women—called enrolled women or auxiliaries—were admitted. The War Department proposed to get its first enrolled women from among volunteers already working at Aircraft Warning Service stations, placing them in uniform and paying them for their rations and quarters so that they could continue to live at home—thus, incidentally, repeating the first mistake that British services had made. After the Aircraft Warning Service companies were filled, it was thought there would be ample time to consider the ten other companies which were to go to corps areas.[11]

Colonel Mudgett had one assistant, Mrs. Marjorie Fling, selected by Civilian Personnel Division because of her familiarity with War Department procedure and her desire to join the WAAC when it was established. Mrs. Fling set to work at once to compile WAAC Regulations, based on CCC Regulations and Canadian and British WAAC Regulations.[12]

On his first day of operation, the WAAC Pre-Planner, as directed by the War Department, distributed WAAC planning among the various staff agencies of the War Department, informing each of its responsibility:

To The Adjutant General's Office: WAAC recruiting. All administrative functions performed for the Army.

To The Quartermaster General: WAAC uniform, design, and procurement. WAAC insignia. WAAC equipment.

To the Judge Advocate General: Disciplinary regulations.

To The Surgeon General: Medical treatment for women.

To the Chief of Finance: Fiscal responsibility.

To G-3: WAAC training and training regulations. WAAC organization.

To G-4: Housing and supply policy. Tables of Basic Allowances. Burial.

To the Air Forces: Organization and training of Aircraft Warning Service units.

Each of these agencies was asked to designate a representative for WAAC planning. The instructions to these agencies did not specify their relationship to any future WAAC headquarters or who would have the final word in disputes.[13]

On the same day corps areas were asked

[8] Personnel Abstracts, WAC files, OCMH.

[9] He later noted, "As I look back, I would have given a great deal to have had access to the background material that you [Military History] have collected. . . . The widely scattered data were apparently on file, but unfortunately . . . not made available to me." Ltr, Mudgett to author, 19 Aug 49. OCMH.

[10] Pencil note, Swift to Mudgett. WA 600.9 (1-20-42), 1942 WAAC files.

[11] All statements unless otherwise specified are from Mudgett Staybacks. Original copies of most documents are in G-1/15839-10. See references to specific documents in following pages.

[12] (1) Memo, G-1 for Dir of Pers, 28 Jan 42. G-1/15839-10. (2) Interv with Mrs. Marjorie Fling Onthank, 16 Jan 46.

[13] Memos, G-1 for [each office], 4 Feb 42, sub: WAAC. G-1/15839. Replies are in G-1/15839-10 (7-8-42).

to plan for the use of one WAAC company each, and chiefs of branches were asked to comment on the possible later use of Waacs at their service schools. Letters to these commands included a statement on which the War Department was later to reverse itself: Waacs could be assigned to replace not only enlisted men but "civilian employees not in the Civil Service now used for purely administrative or housekeeping duties."[14]

It was not yet decided whether Waacs counted against a station's Troop Basis, or its civilian allotment, or both; most stations understandably assumed that Waacs constituted some sort of happy bonus. Also, the previous decision that the Army needed a small elite corps with high clerical skills was nullified by the authorization to use Waacs to replace non-Civil Service civilians, most of whom were unskilled. Consequently, corps area requisitions later that month asked for Waacs not only for skilled clerical and technical work but also for work that it was difficult to get civilians to accept: jobs as maids, charwomen, janitresses, cooks, mess attendants, messengers, hostesses, mail orderlies, housekeepers, and hospital attendants.

As a result of the double misunderstanding—that Waacs would be recruited for unskilled menial duties and that they would not count against personnel allotments—not too much significance could be attached to the corps areas' enthusiasm in at once requesting not ten but twenty-four WAAC companies. On the other hand, all of the chiefs of branches, including infantry, cavalry, and artillery, replied that they wanted none: "Strongly advise that they not be assigned." The Chief of Engineers asked to be excused from taking a unit. The Surgeon General refused to use Waacs in hospitals, saying, "Can see no use for a WAAC unit without displacing Civil Service employees."[15] No attempt was made to force a unit upon any reluctant command, since there would obviously not be enough Waacs to fill even the current requisitions.

In only a few days WAAC preplanning had the requisite fire under it. Uniform designs were being prepared by The Quartermaster General; The Adjutant General was converting Army blank forms and records for WAAC use; the Judge Advocate General was preparing tentative disciplinary regulations; the Signal Corps and the Air Corps were planning Aircraft Warning Service companies. Colonel Mudgett suggested that a female director and staff of three male officers be appointed at once, and action to secure these was begun by G–1. Within another ten days Colonel Mudgett had obtained rough drafts of WAAC Regulations, was working with G–4 on barracks plans, had consulted the Office of The Surgeon General on the enlistment physical examination, and had planned recruiting with The Adjutant General. In four more days he had made plans for press releases on the appointment of the Director and had secured G–3 approval of a training center Table of Organization.[16]

[14] D/F, G–1 to TAG, 5 Feb 42, sub: WAAC. G–1/15839–10. Branches included were Infantry, Cavalry, Field Artillery, Coast Artillery, Engineers, Air Corps, Armored Forces, Quartermaster Corps, and Medical Department.

[15] Tally of replies is preserved in Folder U, (Utilization), Col Tasker's files, in 1942 WAAC files. (Hereafter cited at Tasker files.)

[16] (1) Memo, Preplanner for Col Oscar B. Abbott, 6 Feb 42; (2) Memo, Preplanner for Maj Gen James A. Ulio, 11 Feb 42; (3) Memo, Preplanner for Lt Col Lester D. Flory, G–4, 11 Feb 42; (4) Memo, Preplanner for Col Harry D. Offutt, SGO, 13 Feb 42; (5) M/R, Conf with Col Harold N. Gilbert, AGO, 13 Feb 42; (6) Memo for Col Abbott, 19 Feb 42. All in Mudgett personal or stayback files, 1942 WAAC files.

Planning could not await the appointment of the Director, WAAC, and in fact planners deemed it wise "to be able to present [completed plans] to the Director, when appointed, in order to avoid the initiation of plans based entirely on the viewpoint of one individual." By 23 February 1942, when the future Director was added to the preplanning group, many plans were too far advanced for much change without controversy with the planning agency. By this date the WAAC Regulations were already written and approved by G-3 Division, as was the outline of instruction for the officer candidate school. Budget estimates had already been forwarded to the War Department, and the Finance Department had virtually completed all fiscal rules. A Table of Organization had been drawn up for WAAC Headquarters, showing the numbers and grades of persons required, as had similar tables for WAAC companies and platoons. The Army Recruiting Service believed itself prepared to conduct recruiting. G-4 Division was convinced that WAAC housing would present no great problem. Finally, a request had been submitted for the assignment of an experienced Regular Army officer to act as school commandant.[17]

Other matters awaited the Director's action. No location had been found for a training center; no uniforms or other clothing had been procured; no actual recruiting machinery was set up, although it was suggested that the Director make a transcontinental tour and personally select the first one hundred officer candidates. The future WAAC Headquarters had not been organized, although a major and two lieutenants had been ordered in for immediate assistance.[18]

All of the Pre-Planner's work, complete or incomplete, suffered from being based on the War Department's assumption as to the WAAC's future size, which was to be discarded later in the summer when manpower problems began to appear. It was supposed that the Corps would, during its first year, train only 10,600 auxiliaries and 340 officers—a miscalculation of some 600 percent. Most of the plans, therefore, shortly had to be revised or abandoned, but, as General Hilldring later noted, they were not without value, being in the same category as all military "anticipatory planning." [19]

Selection of a Director

G-1 Division demanded that all candidates for the position of WAAC Director be healthy, of an active temperament, between the ages of 30 and 50, with executive experience involving the successful management of both men and women assistants, and, most important, they must have had no previous affiliation with any "pressure group."

Mrs. Hobby, then of the Bureau of Public Relations, was asked to recommend women who met these requirements. She submitted a list of nine that included some of the nation's most successful career women—an advertising executive, a business manager, a bank president, several educators. Social and political leaders were not included. Congresswoman Rogers was also asked to recommend

[17] (1) Quotation from Memo for Col Flory, G-4, 11 Feb 42. Mudgett Staybacks. (2) Memo, Preplanner for Dir WAAC, 23 Feb 42. 1942 WAAC file, unnumbered. (3) Memo, CofS for G-1, 26 Feb 42. CofS/19481-60.

[18] Earlier draft of Memo cited n. 17(2).

[19] Ltr, Hilldring to Maj Gen Orlando Ward, Chief, Mil Hist, 5 Jan 51. OCMH.

candidates; she submitted only one name—Mrs. Hobby's. G-1 Division likewise submitted its own list of candidates to the Chief of Staff, and this included three names, headed by Mrs. Hobby's. Colonel Mudgett also approved of the recommendation, commenting later: "I have never known a finer executive, man or woman." [20]

Biographical data compiled by G-1 added that Mrs. Oveta Culp Hobby was at this time 37 years old, the wife of former Governor William P. Hobby of Texas, and the mother of two children. She had served for several years as parliamentarian of the Texas legislature, and, after her marriage, as newspaper and radio executive, publisher, lawyer, writer, president of the Texas League of Women Voters, and civic worker in numerous state and city organizations of both men and women. As of this date she had been for almost a year the chief of the Bureau of Public Relations' Women's Interests Section, which she had initiated and organized.

For a number of reasons the War Department's final choice fell upon Mrs. Hobby. She had the advantage over all other recommended candidates of already being familiar with WAAC plans and with War Department organization. Officers of G-1 Division wrote: "She has ability, vision, and is broad-minded enough to assemble a staff of capable assistants around her. [She is] already known to most of the key people in government and War Department circles." [21]

All acquaintances noted her personal energy, magnetism, sincerity, and idealism, and observed that a very considerable diplomatic ability on all matters was combined with a certain stubborn determination in pursuing major issues. Another asset, from the Army viewpoint, was that Mrs. Hobby was not a representative of any pressure group, as it was deemed essential that the Director owe allegiance only to the War Department. The Chief of Staff informed the Secretary of War that Mrs. Hobby was his choice for the position of Director solely because of her brilliant work in the Bureau of Public Relations and in negotiations for the WAAC bill, and that "I had never seen, or even heard of, Mrs. Hobby prior to this time, and she had no prior knowledge of the bill." He added:

> In all of these duties she displayed sound judgment and carried out her mission in a manner to be expected of a highly trained staff officer. She has won the complete confidence of the members of the War Department Staff with whom she has come into contact, and she made a most favorable impression before the Committee of Congress.
>
> . . . This Corps can be of great assistance to our military effort, and it can easily be a great embarrassment to the War Department. I therefore urge the appointment of Mrs. Hobby, with the request that the decision be made in advance of the completion of the legislation in order that the War Department can anticipate the burdens of organization so far as possible.[22]

Mrs. Hobby was not considered disqualified by the fact that she had children, since women with children were to be accepted for immobile units like the Aircraft Warning Service, where they could live at home.[23]

---

[20] (1) Memo, G-1 for CofS, 3 Feb 42, sub: Dir WAAC. G-1/15838. Three lists are Tables A, B, and C. (2) Ltr, Col Mudgett to author, 19 Aug 48. OCMH.

[21] Memo, Chief, Misc Div G-1, for Gen Hilldring, 29 Jan 42, sub: Suggested Candidates for Position of Dir WAAC. Mudgett Staybacks.

[22] Memo, CofS for SW, 18 Mar 42, sub: Chief of WAAC. WDCSA 291.9 (3-18-42).

[23] Interv with Mrs. Hobby, 29 Jun 48. Also see Ch. XXXV, below.

Meanwhile, a number of prominent political figures, including members of Congress, had acquired their own candidates,[24] and G-1 Division urged speed in the appointment of a director, since "efforts will be made by certain pressure groups to influence the Secretary of War in his appointment of the Director." [25]

At the end of February, when no announcement from the Secretary of War was forthcoming, the Chief of Staff took upon himself the responsibility of moving Mrs. Hobby from the Bureau of Public Relations to join the WAAC Pre-Planners, although in view of her lack of official status all planning done by her was subject to reversal later if another candidate was selected by the Secretary of War.[26]

Organization of Director's Headquarters

Mrs. Hobby now found herself in a strange status; she was the unannounced head of a nonexistent office which could not become WAAC Headquarters until passage of legislation at an uncertain future date, but which meanwhile must perfect very complete plans requiring the formal co-ordination of many War Department agencies. Her "staff" was one Cavalry lieutenant colonel and one civilian woman assistant. This little group began meeting in a corner of the Miscellaneous Branch of G-1 Division, but later in March acquired several rooms and a telephone in the wooden Temporary M Building.

On 27 February, a few days after Mrs. Hobby's arrival, the man who had been intended to be WAAC Pre-Planner, Lt. Col. Harold P. Tasker, arrived. Colonel Tasker was a Regular Army officer, with fifteen years' service in the Coast Artillery Corps and as instructor in mathematics at West Point. He had been retired for disability in 1939 but later recalled on limited service. Although he had never previously served in the War Department, all members of the new group stated that his outstanding ability quickly rendered him invaluable.[27]

With Colonel Tasker came two young Reserve lieutenants, just called to active duty, who had no Army experience.[28] For almost a month these six constituted the headquarters. About the middle of March the office acquired its senior member, Col. William Pearson, who was selected by The Adjutant General to be WAAC adjutant general. Colonel Pearson was the only member of the staff who had previously served in Washington. His service was not recent; he had been retired in 1936, after more than thirty-four years in the Regular Army, and had just been recalled to active duty. Shortly, four more officers arrived to complete the military staff. Two of these were lieutenant colonels who were intended to be school commandant and quartermaster, respectively, but both were replaced within a few months.[29] The third was Maj. J. Noel Macy of the Bureau of Public Relations, on temporary loan to handle publicity, and the fourth was another young Reserve lieutenant.[30]

Mrs. Hobby also secured the services of two women advisers. Mrs. Genevieve Forbes Herrick, a prominent writer and newspaperwoman, was designated adviser

[24] Ltrs, G-1/15839-10 (1942).

[25] Memo cited n. 20(1).

[26] Memo cited n. 22.

[27] Intervs with members of the group: Mrs. Hobby, Mrs. Woods, Mrs. Fling, Mrs. Genevieve Forbes Herrick.

[28] Lts. Charles L. Fleming and Henry Lee Munson.

[29] Interv with Col Tasker, 20 Nov 45.

[30] Lt. William W. Foulkes.

on public relations and the press. Mrs. Helen Hamilton Woods was named adviser on legislative matters, recruiting, and administration in general. Mrs. Woods, who was later to enroll in the WAAC, was the widow of New York Police Commissioner Arthur Woods, and the mother of three sons in the service. A civic leader and political worker, and a descendant of the first Secretary of the Treasury, she had had much experience with volunteer groups and had been employed by the Treasury in connection with the information program for defense bond sales.[31]

Assignment to Services of Supply

Planning was somewhat hampered when the little group of WAAC preplanners was immediately dropped two echelons in the War Department during the reorganization of 9 March 1942, one of the major upheavals of War Department history. At this time the entire General Staff was reduced in size and bereft of operating functions, and the many organizations reporting directly to it were swept away, leaving direct access to the General Staff a privilege of three new major commands—the Army Air Forces, the Army Ground Forces, and the Services of Supply (later renamed the Army Service Forces). Many of the chiefs of services who had formerly reported directly to the Chief of Staff—such as The Adjutant General, the Judge Advocate General, and The Quartermaster General—now were grouped under the command of the Services of Supply (SOS), which was to be the operating agency for the General Staff.

The WAAC was similarly transferred from G–1 Division to Personnel Division of the Services of Supply, and directives drafted by WAAC preplanners henceforth had to be signed by that office rather than by Brig. Gen. John H. Hilldring in G–1. Few of the transferred agencies were very happy about the change. WAAC planners in particular realized its disadvantages, since they no longer had the close guidance of the General Staff during the Corps' formative period, and could not reach it with any ideas that were disapproved by the Services of Supply. The move made it particularly difficult to work directly with the Army Air Forces or the Army Ground Forces.

The legality of the Director's assignment to this lower echelon was somewhat doubtful, since the wording of the WAAC bill made her adviser to the War Department. At the time of the move the Chief of Staff therefore directed Mrs. Hobby to come to him personally if she had any difficulty that she could not iron out by herself. This privilege Mrs. Hobby used infrequently, since she understood that the Services of Supply would scarcely be a comfortable location if she bypassed its channels whenever her proposals were disapproved.

From this time onward General Marshall was consulted only when crucial decisions were required. The new and smaller General Staff also obviously had less time to advise the planners. At one of its meetings, the WAAC was dismissed with the brief and overoptimistic note that its administration would be the same whether it was *in* or *with* the Army; the same meeting was devoted to dozens of more important matters involved in the world-wide combat situation. Nevertheless, WAAC planners united in giving great personal credit to the Assistant Chief

[31] Intervs with Mrs. Herrick, 9 Jan 46, and with Mrs. Woods, 14 Dec 45.

of Staff, G-1, General Hilldring, for the fact that in the following hectic weeks they actually did complete full plans, workable if not perfect. General Hilldring repeatedly stepped down from his echelon to give planners more of his personal time and advice than the small WAAC organization could have expected from its current importance to the war effort.[32]

The Services of Supply immediately delegated WAAC planning to the various SOS divisions, in much the same way that this once had been divided among War Department staff sections. For example, the SOS's Training Division received authority to establish policies for the training of all WAAC officers, auxiliaries, and units. In actual practice, the general nature of the subsequent WAAC plans was worked out in joint conferences, put in writing by the preplanners, and approved by various SOS Divisions.[33] It was not made clear what procedure would be followed if disagreements occurred.

The British Parallel

Mrs. Hobby's first step was to pause and assess the experience of the British and Canadian women's services, already well established.[34] During the first days of March she visited Canada, accompanied by Colonel Tasker and the commandant-elect, and was afforded the opportunity to talk to leaders of the Canadian WAC and of the Women's Division, Royal Canadian Air Force, as well as to British women officers then visiting Canada. She also later dispatched Major Macy to England to collect references and histories.

The visit did much to restore Mrs. Hobby's morale, which had been depressed by the opinion of many of her friends among older U.S. Army officers that a women's corps could not succeed. Every Canadian staff officer and post commander to whom she talked told her that he had experienced initial doubts but was now enthusiastic about employment of women in his command. British officers likewise reported the success of their women's services in spite of extensive early hardships.[35]

The histories of the several British and Canadian women's services presented patterns so nearly identical as to suggest a certain amount of inevitability, with the British, of course, one or two wars ahead. Before and during World War I, the British had set up a few tentative women's corps, on a civilian auxiliary basis. Manpower problems had proved so great, and the women such efficient workers, that by 1918 such groups were to be found in every branch of service. The only handicap had been the very considerable public gossip concerning the alleged immorality of women in France, which hampered recruiting and alarmed parents, although a Royal Commission of Enquiry reported after investigation that the charges were "mischievous and false."

Nevertheless, when the British women's services were revived in World War II, they repeated the whole unhappy pattern

[32] (1) Interv cited n. 29, and from study of whole series of documents and orders cited. (2) Min, Gen Council, 4 May 42.

[33] 1st Ind, Dir Tng SOS to WAAC Preplanners, 25 Apr 42, to Memo, Preplanners for Dir Tng SOS, 15 Apr 42, sub: Preliminary OCS. SPTRS 353 WAAC (4-15-42).

[34] Unless otherwise indicated this section is based upon, and quotations are from: (1) *Report of the Committee on Amenities and Welfare Conditions in the Three Women's Services, Presented to Parliament by Command of His Majesty, August, 1942*, (London: *HM Stationery Office, 1942*). Hereafter cited as *Conditions in the Three Women's Services*. Copy in 1942 WAAC files. (2) Wadge, *Women in Uniform*.

[35] Interv cited n. 29.

as far as public opposition was concerned. "Vague and discreditable allegations" were spread to such an extent that a Parliamentary committee investigation was ordered. Gossip alleged a high rate of illegitimate pregnancy, excessive drinking, and general promiscuity among servicewomen. The committee upon investigation found that the women's morality was actually better than that of the comparable civilian population, and deplored the "malicious and careless talk" that had again damaged recruiting and discouraged women members and their parents.

The British Auxiliary Territorial Services (ATS) and the Women's Auxiliary Air Force (WAAF) had also suffered in the first days of World War II from difficulties caused by rapid mobilization—shortage of uniforms, inadequacy of housing, improper medical attention, all intensified by the lack of trained and experienced women officers to care for the units. By 1942 many difficulties had been overcome, supply shortages had eased, gossip had been partially quelled, and the women's work had proved so valuable that their numbers had been tremendously increased and their jobs expanded from five or six to several hundred. Finally, when manpower conditions became more desperate and recruiting failed to fill needs, the British National Service Act was applied to women and female draftees were directed to fill the vacancies in the armed forces.

At the time of Mrs. Hobby's visit to Canada in March of 1942 the British services were thus well established and offered a valuable precedent to the WAAC in the United States. WAAC planners might have regarded their history with even more interest had they known that it could, without alteration, have been a summary of the WAAC's future, in all except the matter of eventual recourse to Selective Service.

One major effect of the Canadian visit upon the headquarters was that it became what its members called "public-relations conscious." All future plans came to be scrutinized in the light of their possible effect upon public sentiment, and extreme care was used to avoid any measure that might provoke an outburst of slander and gossip of the type that had hindered British recruiting and necessitated the drafting of women. This tendency was deplored by some Regular Army members of the staff, since it frequently raised the issue of whether certain established Army practices were adaptable to the WAAC.

Another conspicuous lesson in the history of the British services was that the civilian auxiliary organization was notably inefficient and friction provoking. Already two of the three major British women's services had been admitted to full military status. In Canada Mrs. Hobby noted that the Women's Division of the Royal Canadian Air Force, which was actually *in* the Air Force, had fewer administrative difficulties than the Canadian WAC, which was only an auxiliary *with* the Army. Most commanders heartily disliked separate command channels, separate rules, and separate status for any members of their station complements and, in fact, the Canadian WAC was to be admitted to full Army status within a few months, and the Women's Royal Canadian Naval Service, just being organized, never attempted to operate without full Navy membership.[36]

Mrs. Hobby therefore proposed that an amendment immediately be added to the

[36] *Ibid.*

WAAC bill before its passage, placing the WAAC *in* the Army and giving its members full military status and discipline. The War Department was not enthusiastic about this change and Mrs. Rogers objected also, on the assumption, which proved quite correct, that the attempt would merely delay the bill's passage and would not be accepted by Congress even after the delay.[37] Nevertheless, Mrs. Hobby insisted that the amendment at least be considered by the committee, and consideration was finally agreed upon, a step that was to delay passage of the bill until May.[38]

Recruiting Plans

The British precedent was also quickly reflected in other plans, primarily those for the selection of officers. Mrs. Hobby and Colonel Mudgett at once discarded previous plans for her to appoint assistant directors and to select one hundred officer candidates. The number of applications now on file from individuals with prominent social, political, or military figures for sponsors indicated that, were appointments direct, the WAAC might be forced to repeat the British ATS mistake of commissioning these women in such numbers that forced retirements would later be necessary. The ATS had come to the conclusion that selection by means of officer candidate boards and schools was the best way to produce women officers able and willing to care for their troops effectively. Accordingly, Mrs. Hobby formed a basic policy from which she was never thereafter to swerve[39]—that of Corps democracy: all officers, even assistant directors, were to be graduates of the officer candidate school and not direct appointees. After the first class, all officers would come from the ranks. Mrs. Hobby also decided at the same time that selection of the initial class should be by impartial Army recruiting machinery, and not by herself or any other individual. Negroes were to be included in the same proportion as in the Army.[40]

Preplanners hastily prepared and sent to the printers detailed instructions for corps areas to follow in selecting officer candidates. Mrs. Hobby secured the advice of a conference of eight prominent psychiatrists, as well as that of The Adjutant General's battery of test construction experts, to determine qualifications and to help devise a method of screening. All corps areas were alerted to expect instructions and application blanks by air mail, immediately upon passage of the bill, and were given a military schedule whereby date of passage was designated D Day. The first officer candidates were to be selected and at the school by D plus 47.[41]

The WAAC D Day now bore down upon the pre-planners. The WAAC bill passed the House of Representatives on 17 March by a vote of 249 to 86, authorizing an auxiliary corps only, and was sent to the Senate, which had promised to consider full Army status.[42]

Public interest mounted daily. Planners hid themselves away and answered the

---

[37] Interv with Mrs. Rogers cited n. 1(3).

[38] See text below for full discussion of War Department and other views in May committee hearings.

[39] See later chapters for discussion.

[40] (1) Interv cited n. 29. (2) *Conditions in the Three Women's Services*. (3) Memo, G-1 for CofS, 22 Mar 42. G-1/15839, in CofS 291.9.

[41] (1) Ltr, TAG to all Corps Areas, 16 Mar 42. AG 291.9 (3-16-42) E-R. (2) Memos, SOS for TAG, 17 Mar 42. G-1/15839-10. (3) Rough notes of conf of psychiatrists. 1942 WAAC files, unnumbered. (4) Interv cited n. 29. (5) See Ch. III, below.

[42] *Congressional Record*, Vol. 88, No. 55, p. 2657, 17 Mar 42.

telephone with a vague "Personnel Division," but were nevertheless plagued with telephone calls, visitors, letters, and unsolicited offers. Candidates sought guarantees of commissions because of their acquaintance with prominent people, or their dubious experience in bossing women; individuals sought to interest the WAAC in mobile laundry units or fur-lined overcoats.[43] Prospective recruits wrote, variously: "My husband is already drafted and my mother cannot afford to have me staying on . . ."; "My brother was killed on December 7"; "I have always wished that I were a man"; "I am a widow, with no dependents"; "At present I am working as a warder at the Women's Reformatory."[44]

This was only a foretaste of the merciless publicity that was to haunt the WAAC from its inception. In desperation the planners ordered thousands of acknowledgment cards thanking applicants for their offers and telling them that, if the bill became law, full details would be released through local recruiting stations.[45] Considerable embarrassment appeared in store if the WAAC D Day arrived before the War Department was ready with plans for it.

Search for a Training Center

For several frantic weeks it appeared impossible to find a training center. The idea of a CCC camp was considered and abandoned, since extensive additional construction and repair would have been necessary. Premature rejoicing was general early in March when the Tome School, a private institution in nearby Maryland, was found to be available. Unfortunately, G–3 Division ruled that no contract could legally be signed until the bill became law. The Army's complicated machinery of approval moved as far as it dared—budget estimates were made, an engineer survey for barracks space was authorized, acquisition of property was approved. After serious study of the relative fame of American heroines it was recommended that the school be named the Molly Pitcher School. Detailed grades were even allotted to the Third Service Command and personnel was being selected. All of these plans collapsed suddenly in the last week of March when the Navy acquired the Tome School.[46]

Immediate passage of the bill was now believed possible, with April occupancy of some school a necessity. Army engineers hastily surveyed Northwestern University and found it full of naval trainees; Lake Forest College had a capacity of only 400; the University of Chicago was filled by the Navy and the Signal Corps, and Ohio State by the Air Corps; Eastern Kentucky State Teachers College would not take Negroes.

As the search continued through other areas, it was found in general that all large universities were training so many Army and Navy technical specialists that even their gymnasiums and classrooms were converted to barracks; all small colleges

[43] Ltrs of application. 1942 WAAC files, unnumbered.

[44] *Congressional Record*, Vol. 88, No. 53, pp. 1078–81, Mar 42.

[45] Memo, Preplanners for TAG, 31 Mar 42, sub: Acknowledgment Cards for WAAC Applicants. Mudgett Staybacks.

[46] (1) Memo, G–3 for CofS, 19 Apr 42. WDGCT 291.9 (4–15–42). Copy in CofS 291.9. (2) Memo, G–1 for G–4, 7 Mar 42. G–1/15839–10. (3) D/F, Pers Div SOS to CofEngrs, 11 Mar 42. Same file. (4) Memo, Pers Div SOS for CofS, 13 Mar 42. Same file. (5) Memo, SOS for TAG, 28 Mar 42. Mudgett Staybacks. (6) Corps areas became service commands in the reorganization of 1942.

could accommodate only a few trainees; and most suitable resort hotels either had been taken over by the Air Forces or would have had to be condemned at a prohibitive price. Engineers as a last resort surveyed even state fair grounds, but found them either leased for government storage or not convertible. For a time the National Chautauqua Circuit at Jamestown seemed the best possibility, although isolated and in a cold climate.[47]

At last, in late April, the mechanization of the U.S. Cavalry made possible the use for Waacs of an old mounted Cavalry post, Fort Des Moines in Iowa. Fort Des Moines was near the geographical center of the United States, had no major defense projects in the area, would present no race and color difficulties, had suitable utilities to handle expansion to 5,000 population, and already had room for 1,000 and suitable administration, supply, and recreation buildings. Its solid red-brick barracks, needing only converted toilet facilities, surrounded an impressively large green parade ground. In addition, said the delighted planners, there were "nine large stables suitable for conversion to barracks."[48] Time was short and there was no inclination to quibble about the distance from WAAC Headquarters, the climate, or the previous equine occupants.

Immediate authority to start alterations at the post was sought, but could not be obtained until passage of the bill.[49] It was at least possible to bring into the planning group nine Army officers to be indoctrinated before they went to Fort Des Moines as a nucleus of the staff. To replace the officer previously selected as commandant, G-1 Division chose Col. Don C. Faith, a Regular Army officer of twenty-five years' service. He had been a National Guard instructor and a staff officer, and later a member of the War Department's G-4 and G-1 Divisions.[50]

The Uniform

The most troublesome problem remaining to the planners was that of the WAAC uniform, which soon assumed a difficulty out of all proportion to its importance. Procurement could not be undertaken until passage of the bill, but it was essential to have agreement on design and number of articles so that contracts could be signed with manufacturers at the moment the bill became law. At first, the problem of clothing only 12,000 Waacs appeared fairly simple, and The Quartermaster General, in one of the greater understatements of recorded military history, anticipated "no unusual difficulties." [51]

Three agencies sent representatives to planning sessions: The Quartermaster General's Office, the Philadelphia Quartermaster Depot, and WAAC Headquarters, although The Quartermaster General later decided that this last was a mistake, since "neither she [Mrs. Hobby] nor anyone on her immediate staff was qualified to make decisions." [52] As a matter of fact,

[47] (1) Personal ltr, Tasker to Lt C. M. Carr, Cooks and Bakers School, Ft Meade, Md., 21 Mar 42. Col Tasker's personal file, 1942 WAAC files. (2) M/R, 9 Apr 42, sub: Inspection of Available Sites. WA 600.9 (4-9-42). (3) M/R, 11 Apr 42, sub: Facilities. WA 600.9.

[48] Memo, G-1 for CofS, 22 Apr 42. G-1/15839. Colonel Mudgett successfully persuaded Des Moines newspapermen to ignore the conversion of stables to barracks for fear of the possible effect upon recruiting.

[49] Memo, G-1 for CofS, 13 May 42, sub: Estab of WAAC Tng Cen at Ft Des Moines, Iowa. WDGAP 352.01 WAAC. Approved by DCofS, 15 May 42.

[50] Personnel Abstracts, WAC file, OCMH.

[51] All references to WAAC clothing and equipment, unless otherwise noted, are from Erna Risch, *A Wardrobe for the Women of the Army* (QMC Historical Studies 12, October 1945).

[52] *Ibid.*, Chs. II and III.

none of the planning agencies included any specialists in women's clothing. The Quartermaster General had been responsible for the design and procurement of nurses' uniforms since World War I, but no great quantity had been involved, and the products, although they photographed well on a dress form (male), had admittedly always looked peculiar on the female figure.

Responsibility for the WAAC uniform program was delegated by The Quartermaster General, as a part-time duty, to Col. Letcher O. Grice of the Standardization Branch. Before Mrs. Hobby's arrival, Colonel Grice had secured some sketches by famous designers, and had suggested that the uniform be in two shades of blue. Because of the word *distinctive* in legislative authorization for a WAAC uniform, Colonel Grice was of the opinion that the uniform must be different in color and design from that of the Army or any other organization; even the two shades of blue must be different from that of the Army Nurse Corps. Blue was selected because New York designers informed him that blue dyestuffs would be most readily available; gray was rejected as too hard to match.

The representative of the Philadelphia Depot was not summoned by The Quartermaster General until some three months of planning were past and, upon arrival, expressed some annoyance in view of the depot's opinion that research and development was its rightful province. This officer also was not a specialist in women's clothing, and expressed a strong belief in "nothing fancier for Wacs than for combat soldiers," which threatened to leave the women equipped for office work in boots and coveralls.

The WAAC representatives at this time likewise included no clothing specialists and in any case had no authority to take action in the matter. Female members of Mrs. Hobby's staff faithfully wore sample undergarments while carrying on preplanning; male planners offered their best guesses in the matter, and the staff became accustomed, as one member noted, "to seeing Lt. F. stalk through the office with a cigar in one hand and a pair of pink panties in the other."

Mrs. Hobby at once announced her conviction that the WAAC uniform should be identical in color with that of the Army and as much like it in design as possible, especially in view of her pending attempt to place the WAAC in the Army. The battle of blue versus olive drab continued for some time and was finally resolved by the Philadelphia Depot, which wished to use olive drab and khaki material already procured, and deemed it madness to start procuring two more shades of blue. The materials, covert and barathea for winter and 8.2 khaki for summer, were also determined by depot stocks, although some, especially the khaki, were to prove too heavy for the proper fit of women's garments.

Sketches from well-known designers were considered—Mangone, Maria Krum, Russell Patterson, Helen Cookman, Mary Sampson. The jacket could be called a group product, since it incorporated desirable features from all designs—a lapel from one, a pocket from another. A belt for the jacket was on, off, and on again: it would help faulty female figures, said Mangone; it would rub holes in the jacket, said The Quartermaster General; it should be leather, said Maria Krum; cotton was cheaper, said The Quartermaster General. A rather narrow six-gore skirt was adopted after War Production Board restrictions on the use of material made pleats impossible, although all agreed that pleats would have looked better.

Slacks had originally been a part of the outfit, but they were eliminated as too troublesome to fit; culottes were considered and rejected by Mrs. Hobby as unsuitable for mechanics. Since at this time it was believed that the only outdoor work that Waacs could perform was in motor transport, no trousers except coveralls were believed necessary. Mrs. Hobby also desired that women wear skirts instead of slacks wherever possible, in order to avoid a rough or masculine appearance which would cause unfavorable public comment.

All agreed that a shirt with tie would be more military and dignified than one with an open collar, and the regulation khaki tie was chosen, although some designers favored an ascot.

The choice of WAAC headgear involved most controversy then and later. The Quartermaster General had suggested a stiff service cap like the men's cap for WAAC officers and an overseas cap for enlisted women, with a brimmed khaki summer hat, but Mrs. Hobby called for identical hats for officers and enlisted women, as being more democratic. The overseas cap was becoming, but was then being adopted by many women's volunteer groups and private service organizations. Mrs. Hobby believed that it was essential to have a WAAC cap that could not be mistaken for that of any civilian group or, more precisely, that it was essential that the Waacs not be blamed for any misconduct on the part of the thousands of civilian women soon to be wearing overseas caps. The firms of Knox and Stetson therefore submitted designs of "miner" hats, visor caps, and berets. A majority vote of the conference selected the visor cap as one that would shield the eyes, not blow off in parades, and be both distinctive and military.

A heavy topcoat was designed by Mangone, very similar in cut to the men's overcoat. In place of the men's field jacket, a light utility coat was designed by Maria Krum, resembling a hooded raincoat with button-in lining. A handbag with shoulder strap was authorized, since women's uniforms obviously had no pants pockets, and experiments with carrying necessities in breast pockets quickly produced a rule against even so much as a pack of cigarettes in that location.

Tan oxfords, tennis shoes, galoshes, and bedroom slippers were selected. Mrs. Hobby recommended plain pumps for dress shoes, but was outvoted on grounds of economy. She also desired lisle stockings for dress, but only rayons were available; and plain cotton stockings for work were chosen instead of the novelty-ribbed cotton she preferred.

There began to be apparent at this time a significant difference of opinion as to the number and type of garments to be issued. By accepted practice, the Army was obligated to furnish its members any articles it required them to wear. This clothing, in the Director's opinion, had to be judged according to accepted civilian custom for females, but in the opinion of most quartermaster representatives could most fairly be based on the amount and type of clothing received by men. Thus, concerning what were called "foundation garments," it was noted that some women required these in order to present a "neat and military appearance," yet could not be directed to wear them unless the garment was issued—an action without military precedent. Similarly, it appeared to be discrimination against men to issue free pajamas and bathrobes, yet it was also not desired to authorize nude female appearances in the various military installations that did not have connecting latrines. As for the required physical training, men received nothing similar to the seersucker exercise dress, yet it appeared undesirable for women to assemble on the drill field in nothing but panties. In this last case, it was luckily noted that Waacs, except for a few women drivers, did not receive the fatigue coveralls that men did; thus, the seersucker dress could be accepted as a nondiscrimatory substitute for men's coveralls. On most of the other items, authorization was not made until June and was withdrawn again in a few months.

The Quartermaster General reduced some of the requested allowances to the number authorized for men; the WAAC request for eight shirts was reduced to four. The request for six cotton dresses and six cotton aprons for cooks, and four coveralls for drivers, was reduced by The Quartermaster General to three, three, and two. In requesting two uniform jackets, preplanners pointed out that men received heavy wool shirts to be worn without a jacket, but that the WAAC's thin cotton shirts were designed for wear with the whole suit, and that "cleanliness, good health, and appearance" required two suit jackets. However, only one jacket was finally authorized, since men got only one.[53]

The Heraldic Section of The Quartermaster General's Office meanwhile had submitted designs for insignia. Designers were initially somewhat at a loss, since insignia usually portrayed the function of the corps concerned and no one knew exactly what the Waacs were to do, except that they would perform several Army jobs. A first attempt produced only a busy-bee-like insect, which Mrs. Hobby pronounced a bug, adding that she had no desire to be called the Queen Bee.

Designers then hit upon the idea of a head of Pallas Athene, a goddess associated with an impressive variety of womanly virtues and no vices either womanly or godlike. She was the goddess of handicrafts, wise in industries of peace and arts of war, also the goddess of storms and battle, who led through victory to peace and prosperity.[54] Accordingly, the head of Pallas Athene, together with the traditional *U.S.*, was selected for lapel insignia, cut out for officers and on discs for enlisted women.

An eagle for the cap was also designed, less intricate than the Army eagle and later to be familiarly known to Waacs, for reasons closely connected with its appearance, as "the buzzard." Since Army buttons could not be used for an auxiliary corps, the WAAC eagle was also to be imprinted on plastic buttons. Only the insignia of grade required no planning; it was to be the same as the Army's, with a tab lettered *WAAC* sewed under the chevrons.[55]

Housing Plans

Plans for housing were less urgent, since field companies would not be ready for assignment for several months. On the basis of British experience and of her conferences with psychiatrists, Mrs. Hobby

[53] All from: (1) Memo, OQMG for G-1, 3 Jun 42. SPQRR 421. Copy in SPWA 421. (2) Memo, Dir WAAC for OQMG, 5 Jun 42. SPWA 400.34 (3-21-42)(1) sec 1 (1943). (3) Memo, Lt. W. W. Foulkes for Dir WAAC, 20 Jun 42. Same file. (4) Memo, OQMG for USW, 16 Sep 42. Folder, WAC Recruiting, ASF Sp Coll DRB AGO.

[54] Personal ltr, Brig Gen Clarence Weems to *Field Artillery Journal*, January 1943. 1942 WAAC files WA 600.9.

[55] Memo, Dir Mil Pers SOS for G-1, 23 May 42. G-1/15839-10, in SPWA 421 (5-16-42) sec 1.

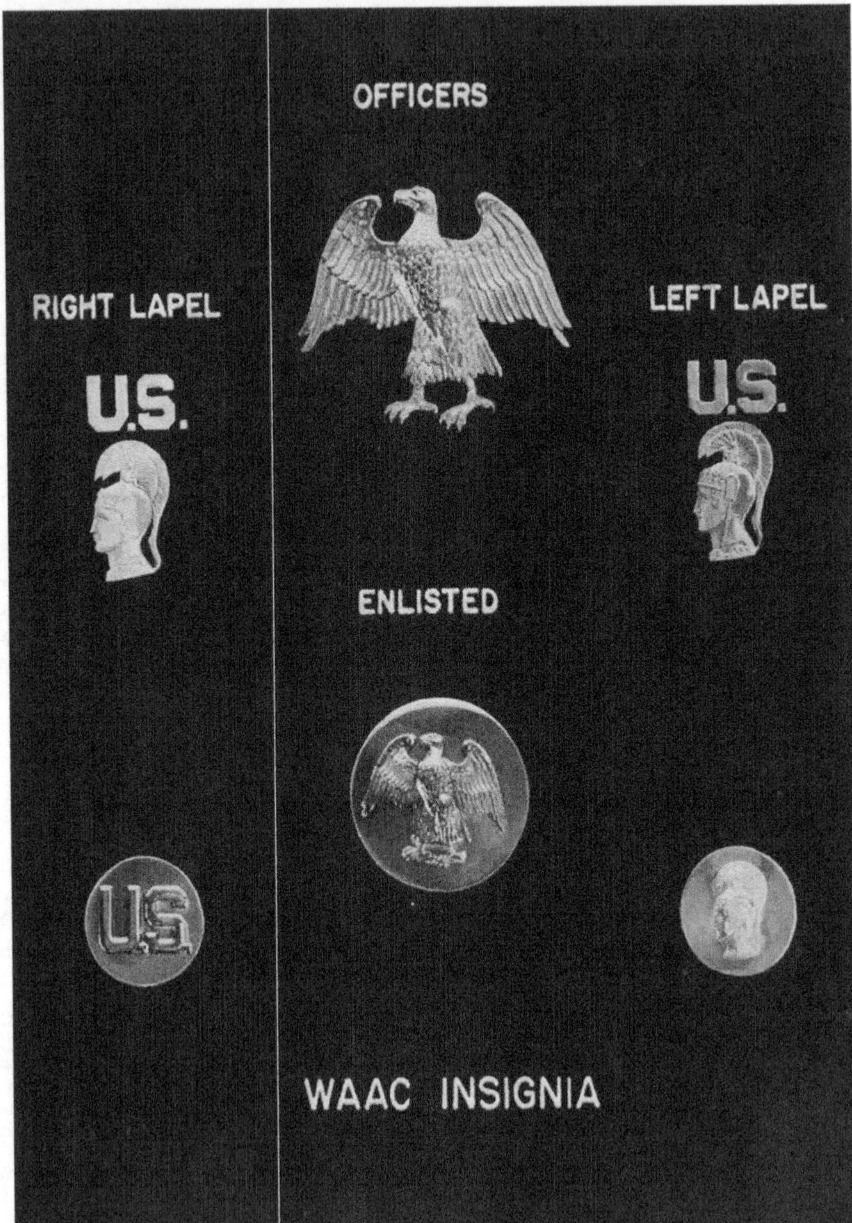
OFFICERS
RIGHT LAPEL
LEFT LAPEL
U.S.
U.S.
ENLISTED
U.S.
WAAC INSIGNIA

feared that women might not react to communal living exactly as did men. While men were accustomed from childhood to the tribal living of scout camps, gangs, teams, and clubs, no one could predict the results of the deprivation of privacy on women who were at all inclined to be nervous—particularly since most men's units remained only temporarily at any camp before overseas shipment, whereas a WAAC unit would settle down in its barracks for several years.

Mrs. Hobby therefore proposed that Waacs have dormitories of the type used to house civilian women workers, with two persons to a room. She was quickly overruled on this point on the grounds of economy and feasibility; housing plans had already been made that would use existing barracks, with minor modifications, or new construction like the old.

Allocation of Units

Only one major phase of planning remained: that of the composition of units and their allocation. It was decided that post headquarters companies would consist of 147 auxiliaries and 3 officers, and that Aircraft Warning Service companies would be somewhat smaller. Any unit under 50 members was deemed uneconomical. The problem was how to devise a fixed company with one Table of Organization that would suit all stations, since not all could use the same number of typists, drivers, or other workers.

Already the Army had realized that the fixed Table of Organization Company, while excellent for combat units, was not the proper means for allotting men to stations in the United States, where no two stations had exactly the same needs. Colonel Mudgett later noted, "The bulk allotment system would have saved us many a headache on this problem." Unfortunately for the WAAC, the bulk allotment system was not yet adopted by the Army, and preplanners were forced to try to set up an inflexible T/O unit that would meet the needs of all stations using Waacs. One table for filter companies and one for operation companies was set up. For the more difficult post headquarters unit, planners hit upon the idea of having five types of platoons: clerical, communications, service, machine records, and miscellaneous. It was believed that almost any station, large or small, could meet its needs by requesting the proper assortment of platoons in its WAAC company.[56]

Concerning the disadvantages of this system in eventual use, Colonel Faith said later:

> Post commanders didn't have the remotest idea as to what they wanted to use women for. . . . Their recommendations were studied, integrated into a type WAC company. . . . I am amazed that we did as well as we did. We set up a semi-rigid organization based upon ideas from the field, which in turn were based on poor guess-work, which the field accepted from us as being authoritative and scientific, which it was not. These

[56] (1) Ltr, Mudgett to author, 19 Aug 48. OCMH. (2) T/O 35-37, WAAC Filter Co, AWS, 4 Jul 42; T/O 35-67, WAAC Opns Co, AWS, Regional, 4 Jul 42; and T/O 35-12. WAAC Post Hq Co, 26 Aug 42. The Clerical Platoon had 10 stenographers, 30 typists, 15 postal clerks, etc.; the Communications Platoon included 18 switchboard operators; Service included 16 chauffers; Machine Records had various machine operators; Miscellaneous had housekeepers, motion picture projectionists, etc. For later modifications, see M/T 35-2018, 14 Dec 42, WAAC Post Hq Co; M/T 35-2026, 26 Nov 42, WAAC Bn; M/T 35-2137, WMC Photo Lab Co, Aerial; M/T 35-2022, 26 Nov 42, and change 1, 9 Jan 43, WAAC Hq Co; M/T 35-2122, 10 Dec 42, WAAC Post Hq Co, AAF. (3) D/F, G-3 to CG SOS, 11 Apr 42. WDGCT 291.9 (4-11-42). (4) D/F, G-1 to TAG, 6 May 42. WDGAP 320 WAAC.

tables provided for relatively few jobs for women—overhead, clerical, motor transport, service.[57]

Corps areas that had asked for Waacs were now requested to specify the assortment of platoons desired in a T/O unit, and planners optimistically proposed to recruit skills exactly to specification, since a woman with no skill or with some odd skill would not fit into any platoon and so could not be assigned. Only skilled women were sought; it was not proposed to waste time by taking women who needed technical training—except a brief course in army driving, army cooking, or army clerical work to help adapt civilian skills to the military.

Although supposedly soothed by the assurance that no WAAC post headquarters units would arrive for some time, or until further warning, many station commanders now became extremely nervous, and it was decided that Colonel Tasker or some other representative of the group must visit each designated station to prepare it for Waacs. In carrying out this indoctrination, Colonel Tasker later found many stations in a state of virtual siege, throwing up barbed-wire entanglements around WAAC areas and setting aside separate nights for Waacs to use post theaters and service clubs, so that men and women could be kept isolated from each other. Many post commanders were found to be much opposed to having Waacs on the post, and none of course had as yet any information as to the proper employment of such a unit.[58]

WAAC Regulations

The WAAC Regulations, which had been prepared in G-1 Division before Mrs. Hobby's arrival, were ready for publication as soon as the bill became law. As provided in the Rogers bill, the WAAC was set up as a separate command entity headed by the Director. The Corps was to be assigned to the Services of Supply, in which its headquarters was located, with its units only attached for duty to the Air Forces and other stations where they were employed.

The Director would at first command all of these units directly, requesting The Adjutant General to issue her orders for assignment, transfer, discharge, or other change in status. As soon as they could be trained, regional directors would be sent out to form an intermediate command echelon. It was decided that the regional directors would be nine in number and would be located in the headquarters of the nine corps areas in order that they might use the area facilities.

For a unit in the field, the station where it was located was responsible only for furnishing supplies, housing, and medical and dental care. As for the authority of Army section chiefs, the Regulations stated:

> Officers and noncommissioned officers of the Army under whom individuals or groups or units of the WAAC are assigned for work tasks have supervisory authority as they would with civilian employees generally, but have no disciplinary authority. Derelictions of duty will be reported to the WAAC officer commanding.[59]

The WAAC company commander and higher WAAC officers were specified as responsible for discipline, promotion, discharge, and other command matters.

---

[57] Speech to Meeting of Classification Offs. Min in SPWA 201.6 (9-18-43).

[58] Interv cited n. 35.

[59] WAAC Regulations (Tentative), 1942.

For a time there was consideration of removing these powers from WAAC Headquarters and delegating them to the regional directors or even to the company commanders, since for Army men powers such as discharge were already held at the station level. However, for men the correct procedures were set forth in detail in Army Regulations, thus insuring uniform action and individual justice Army-wide.

Because of the Waacs' civilian status, these Army Regulations could not always be applied, particularly in matters of discipline—in which court-martial was not possible except overseas—and of discharge—since certain types of discharge could be given only by courts-martial. Neither was it possible to make the first WAAC Regulations as full as Army Regulations, which had been built up over a period of years.

The first WAAC Regulations were therefore marked Tentative and were scheduled for amplification as soon as experience permitted. To permit such amplification, it was directed that all discharge cases and other personnel actions be sent to WAAC Headquarters for decision, until a sufficient body of experience existed to permit publication of detailed regulations. Until this time, Mrs. Hobby noted that it would be unsafe to allow action at the station level, since one woman officer would be made both accuser and judge, contrary to the American system.[60]

In the exercise of its command prerogatives, WAAC Headquarters proposed to be extremely cautious. It was directed that women be sent only to stations where housing and other arrangements had been checked upon by Colonel Tasker or another staff member; that no unit of less than fifty women be assigned, in order that supply and inspection would not be unduly difficult; and that no enlisted woman be sent to any station unless a WAAC officer was located there.[61]

By April several agencies had already begun to question this system. G-3 Division, which had previously approved the regulations, now declared in conference that Army commanders should have complete control of WAAC units on their stations, transferring them in the United States or overseas without informing the Director, and discharging, disciplining, or promoting, presumably under Army Regulations, since no other detailed ones existed. Also, G-3 proposed that the command of a company be divided among the different Army section chiefs for whom the women worked, rather than vested in the WAAC company commander.[62]

The Signal Corps concurred, fearing that the Director might use her command powers arbitrarily to remove women from vital communications work; it was asserted that the British Signal Corps had experienced many difficulties in employing servicewomen who were *with* and not *in* the Army. Air Forces representatives supported these views and also objected to the Corps' location in the Services of Supply.[63] If command authority was given to them, the Signal Corps proposed to place women at once in mixed tactical groups overseas, without a WAAC company com-

[60] These principles were incorporated in all subsequent WAAC Regulations. First ref to them seems to be Memo, Tasker for Dir WAAC, 21 May 42. Tasker Staybacks.

[61] (1) Senate Com on Mil Affairs, 77th Cong, 2d sess, *Hearings on HR 6293*, 1 and 4 May 42. (2) *Congressional Record*, Vol. 88, No. 81, p. 3821.

[62] Memo, G-3 for G-1, atchd to Memo, G-1 for Mil Pers Div SOS, G-3, G-4, WAAC, TAG, CofFin, JAG, SG, 29 Apr 42. WDGAP 320 WAAC (4-29-42).

[63] Memo, SigC for Dir Pers SOS, 4 Jun 42. G-1/15839-10 (7-8-42) WAAC. (2) Memo cited n. 60.

mander, and the Air Forces announced plans to assign an additional 10,000 women in mobile units of its Terrestrial Service.

It soon became clear that the heart of the difficulty was the auxiliary system itself. As predicted, it was plainly about to prove most objectionable to Army commanders if they could not at once exercise as full authority over women as over men, or if they were obliged to apply different regulations to men and women. On the other hand, to apply Army Regulations to a civilian auxiliary corps was clearly illegal. As for her own position as Director, Army advisers informed Mrs. Hobby that it would be unsafe to accept that office if G-3's plan prevailed, since she would be in the militarily impossible situation of having responsibility without authority. Thus, the WAAC's auxiliary status made her legally responsible for the women's command and well-being, yet she would actually be ignorant of where they had been transferred or what financial, supply, disciplinary, or other measures were being taken by station commanders.[64]

By 3 April the WAAC preplanners considered the situation so confused that they appealed directly to the Chief of Staff, stating: "Confusion exists as to the interpretation of the Bill which says: 'The Director shall operate and administer the Corps in accordance with normal military channels of command and administration.' "[65] The Chief of Staff upheld the WAAC preplanners by signing their draft of the WAAC Regulations and approving it for publication if the bill giving the WAAC auxiliary status should pass.

This verdict had the effect of uniting the Services of Supply with Mrs. Hobby in supporting the amendment to place the WAAC in the Army. The WAAC's immediate superiors in the Services of Supply in fact opposed the separate command system so strongly that they neglected to prepare the necessary War Department circulars that would make the WAAC Regulations binding upon the Army in the event that the amendment failed.[66]

Passage of WAAC Bill

In their drive to complete plans before passage of the legislation, all echelons were seriously handicapped by uncertainty as to the type of bill that would pass Congress. After the WAAC bill had passed all committees and the House, thus virtually assuring it of passage in some form, the Navy had introduced a bill to establish the WAVES *in* the Navy and not as an auxiliary.

True to prediction, the brunt of argument had been borne by the WAAC, and the WAVES bill passed the House without difficulty. It seemed only reasonable that Congress would approve the amendment to place women in the Army. The unhappy planners were therefore forced to make two sets of plans: one for an auxiliary and one for an Army corps. Two versions of regulations were prepared and approved by nine interested agencies. Fourteen different agencies were asked to reconsider their plans in the light of the possible adoption of the amendment, and some, like the Judge Advocate General,

[64] *Ibid.* (1) M/R, sub: Conf Concerning WAAC Admin, 24 Jun 42. Col Tasker's file of rough drafts, under "WAAC Hqs Org," 1942. (2) Memo, AG WAAC for CofAdm Serv SOS, 27 Jun 42. WA 210.31 (6-4-42).

[65] Memo, Preplanners for CofS, 3 Apr 42, sub: Responsibility for Opn of WAAC. G-1/15839-10.

[66] Memo, Dir Mil Pers SOS for G-1, 20 Apr 42, sub: Amendment to HR 6293. G-1/15839-10 WAAC (7-8-42).

replied despairingly that none of their previous plans would apply.[67]

In spite of support from the Services of Supply and Mrs. Hobby, the War Department was unable to push the amendment for equal status. Comment on the floor of the House of Representatives was unrestrained and caustic. Members objected strenuously to placing women in the Army because this would give them the disability benefits and pensions that men received. Some also feared that women generals would rush about the country dictating orders to male personnel and telling the commanding officers of posts how to run their business.

This opposition was not tempered in the least by the fact that the House had just passed and sent to the Senate the bill placing the WAVES in the Navy with all of these possible benefits. As members pointed out, Congress was now in the incongruous position of giving Waves the full protection of military status while forbidding them, in their bill, to go overseas, whereas the WAAC bill allowed women to go overseas without the protection of military status. Opinion in the Senate committee was divided, but members finally became convinced that great delay would result from any attempt to change the Rogers bill.

There was one last flurry when it seemed that the Director of Selective Service would secure Presidential approval to a plan to request the voluntary registration of all women. This step would have made little change in the WAAC plan, but might have considerably increased the prospect of recruits if women had believed the registration a prelude to forced service in farms and factories. However, on 4 May 1942 the President decided against the plan.[68]

It was shortly apparent that Congress would pass the Rogers bill immediately without any of the proposed amendments. The War Department and the WAAC planners waited, braced for the outburst of publicity and the need for furious action, which they knew must accompany the bill's passage. Their plans were complete, as far as human guesswork could foretell the future. Men who had never seen an enlisted woman and women who had never been one had together planned for the future welfare and efficiency of women who were to be enlisted to work for the Army.

On 14 May 1942, the Rogers bill was approved by the Senate, 38 to 27, and when signed the next day by the President became Public Law 554, An Act to Establish a Women's Army Auxiliary Corps for Service with the Army of the United States.

[67] (1) The nine: G-3, G-4, TAG, JAG, SGO, Fin, Mil Pers Div SOS, G-1, SW. See D/F, G-1 to [above], 20 Apr 42. Copy in Mudgett Staybacks. (2) The fourteen: G-4, Budget and Leg Plng Br, Fin, JAG, Req Div SOS, SGO, CofAdm Serv SOS, Tng Div SOS, TAG Enl Br, TAG, G-3, G-1, OQMG. See Memo for each, 24 Apr 42. G-1/15839-10. (3) Memo, JAG for Dir Pers SOS, 5 May 42. SPJGA 354.01.

[68] (1) D/F, G-1, 27 Apr 42, sub: Voluntary Registration of Women. G-1/15839-27; (2) Ltr, SW to Dir SS, 27 Apr 42, and reply, SS to SW, 8 May 42. G-1/15839-1 (5-11-42).

CHAPTER III

The WAAC's First Summer

On the rainy morning of 16 May 1942, Mrs. Oveta Culp Hobby took the oath of office as Director, WAAC. This event represented a neatly executed *coup d'etat* on the part of the Army, which, on the night of 14 May, had sent Col. Robert N. Young of the General Staff hurrying to the home of the Secretary of War with the letter of appointment in his hand. The Secretary demurred at the haste but was persuaded that this was the only way to avoid irresistible political pressure the next morning to appoint other candidates who would be disastrously unfamiliar with the Army's plans. Competition for the position was by now so open that newspaper columnists were speculating on various candidates' qualifications, or lack of them.[1]

Thus on 15 May there was simultaneous announcement of the President's signature and of the appointment of the Director. The appointment ceremony the next morning was witnessed by the Secretary of War, the Chief of Staff, Congresswoman Rogers, Governor Hobby, and a few others. Mrs. Hobby accepted, saying, "You have said that the Army needs the Corps. That is enough for me." The occasion was not particularly inspiring for participants: klieg lights blazed; the speeches had to be given over and over for cameramen; Mrs. Hobby was asked to raise her hand and repeat the oath several times. Furthermore, her wide-brimmed hat proved unreasonably difficult to photograph. Thus the first Waac was sworn in.[2]

From this moment forward, Director Hobby was faithfully credited by some millions of Americans with having personally initiated the Corps and dictated its every move. Although she had neither originated the idea, written the legislation, nor determined the Regulations, her office repeatedly had to answer inquirers such as an Army private, who wrote: "From whom or whence did she derive the idea of organizing such a unit?" [3]

The Press

There followed immediately a well-attended press conference, the first of many to come. The War Department realized the importance of the initial reac-

[1] Interv with Brig Gen Robert N. Young, CG MDW, 14 Dec 45. The syndicated column, "The National Whirligig," May 20, 1942. WDBPR Newspaper Clippings, WAC, DRB AGO.

[2] Letter of appointment was dated 15 May 1942. Memo, WAAC Hq for TAG, 18 May 42. WA 314.7 (6-2-41) sec 1, DRB AGO. Director Hobby's assigned serial number was OA-1. For others, the letter A was prefixed for auxiliaries, and L for officers. It was never specified what the L represented—presumably "Lady." The numbers from 100,000 to 199,999 were assigned to the First Service Command, 200,000 to 299,999 to the Second, and so on. Memos, TAG for Dir WAAC, 16 and 20 May 42. SPWA 341.

[3] Ltr, Pvt Frank Wetrich to OWI, 22 Feb 44, forwarded to Dir WAC. SPWA 095.

MRS. OVETA CULP HOBBY *is sworn in as the first Waac by Maj. Gen. Myron C. Cramer. General Marshall, second from left, and Secretary of War Henry L. Stimson were among witnesses of the ceremony.*

tion, and General Hilldring said, "This whole thing stands or falls on the next 60 days." [1]

The basic idea of the Corps which the Director and the Chief of Staff wished to impress upon the public was that of a sober, hard-working organization, composed of dignified and sensible women. Their work must be shown as real, just as the war was real and the nation's danger was real. Although there might be problems of feminine adjustment, the War Department did not intend to encourage frivolity.

This excellent but somewhat staid approach was not expected by even the most sanguine to offer any great appeal to the press. The Director's staff noted that "news" about women had long been patterned on three stereotypes of the American woman: (1) she was a giddy featherbrain frequently engaged in powder-puff wars and with no interest beyond clothes, cosmetics, and dates; (2) she was a henpecking old battle-ax who loved to boss

[1] Details, unless otherwise indicated, are from intervs with Mrs. Herrick, 9 Jan 46, Mrs. Eric Knight, 27 Feb 46, and Gen Hilldring, 17 Jan 46.

the male species; or (3) she was a sainted wife and mother until she left her kitchen, whereupon she became a potentially scarlet woman. Director Hobby had already taken issue with these views, saying, "Waacs will be neither Amazons rushing to battle, nor butterflies fluttering about." [5] One idea which she was particularly anxious to avoid was the familiar "morale purposes" accusation—that Waacs were not wanted to fill Army jobs so much as to provide companionship for soldiers. General Marshall had been deliberately careful, in urging Congress to pass the bill, to use solely the arguments concerning the usefulness of Waacs as workers, and never those that treated them chiefly as women.[6]

Here at this first conference was to begin a never-finished battle as to which idea of woman in the Army would prevail—whether the public would get stories mainly of her real work and useful jobs, or of her underwear, cosmetics, dates with soldiers, her rank-pulling, sex life, and misconduct. Certain sections of the press were expected to co-operate, in the interest of the war effort, by playing down sensational angles that might hinder recruiting. Since not all could be expected to do so, the Director and her public relations consultant, Mrs. Herrick, sat up the night before the conference listing all possible embarrassing questions that reporters might ask, and rehearsing and polishing the Director's replies so as to minimize frivolous matters and emphasize the serious purpose of the Corps. A comparison of their list with the questions asked the next day showed that they did not miss even one, and in fact thought of a few that the reporters overlooked.

At the press conference, the Director was supported by the Chief of Staff himself, General Marshall, and the head of the Bureau of Public Relations, Maj. Gen. Alexander D. Surles. A routine handout was devoted to recruiting and training plans, enlistment requirements, and other statistics. A description of the uniform, especially the underwear, for which reporters were eager, was tactfully withheld a few days because its procurement was not yet finally approved. Frustrated here, the reporters opened up the expected barrage of questions directed at Mrs. Hobby: [7]

Q. How about girdles?
A. If you mean, will they be issued, I can't tell you yet. If they are required, they will be supplied.

Q. Will Waacs be allowed to use make-up?
A. Yes, if it is inconspicuous.

Q. What do you consider that to be?
A. I hope their own good taste will decide.

Q. Nail polish?
A. If inconspicuous, yes.

Q. Will the women salute?
A. Yes, they will salute.

Q. Will they march and carry arms?
A. They will learn to march well enough to parade, but they will carry no guns.

Q. Will they be put in guardhouses?
A. No, no guardhouses.

Q. This is a burning question. Will officer Waacs permitted to have dates with privates?
A. [Here Mrs. Hobby turned to one of the generals [8] for a description of Army tradition. He explained that Army policy was that officers when not with troops, and enlisted men when off duty, might associate, but that he

[5] MS history of First WAAC Training Center. Folder, Supplementary Hist Material Sent to Washington Nov 1945. WD Library. (Hereafter cited as Hist of 1st WAAC Tng Cen.)

[6] Interv with Gen Hilldring, 17 Jan 46

[7] Compiled from Associated Press and United Press releases as well as syndicated columns for May 16, 1942. WDBPR Newspaper Clippings, WAC, DRB AGO.

[8] Associated Press said it was General Marshall; United Press said General Surles.

hadn't been to the field lately and didn't know how it worked for nurses.]

Some questions became even more pointed. "There was one male reporter in particular," reported a spectator, "who kept prodding the woman reporter next to him to ask more and more about illegitimate babies." To this query, Mrs. Hobby replied briefly that pregnant women, married or unmarried, would be discharged.

"The Director handled herself excellently," reported another observer. "She was very direct, forthright, composed, and as candid as could be. She made a good impression." The next day's newspaper stories confirmed this, although a few women columnists [9] pointed out that Mrs. Hobby's previous "unavailability" had not made her popular with newspaperwomen who wanted intimate details, and the Negro press attacked the appointment of a woman from Texas.[10] In general, the hundreds of clippings collected by the Bureau of Public Relations showed widespread praise for the Corps and the Director. A few spiced up the story with headlines such as "Petticoat Army" or "Doughgirl Generalissimo," [11] and some punsters were obviously unable to resist such opportunities as "Wackies," "Powder-magazines," "Fort Lipstick," and (concerning girdles), "It wouldn't do to let the fighting lassies get out of shape." [12]

Male columnists were the worst offenders, in spite of Damon Runyon's remark, "A lady in the uniform of her country will scarcely be an object of jesting on the part of a gentleman not in the same garb." [13] Some of the attacks contained savage comparisons to "the naked Amazons . . . and the queer damozels of the Isle of Lesbos." [14] Others, as expected, had a field day with the idea that woman's place is in the maternity ward. Said one: "Women's prime function with relation to war is to produce children so that the supply of men for fighting purposes can be kept up to par." Another: "Give the rejected 4-F men a chance to be in the Army and give the girls a chance to be mothers." Others became emotional over their assumption that all Waac recruits would be married women who had deserted infants.[15]

Thus began the fight to hold the line on public relations, which had wrecked recruiting for British women's services. Prematurely hopeful that the worst was over, the Director and her staff began their first day as a recognized and operating headquarters.

Establishment of WAAC Headquarters

"The next month was chaos," said a staff member later. "Every newspaper and magazine writer came down to see us. Important people of all sorts were also now willing to advise us." [16] The open bays of the office at Temporary M swarmed with new clerks, insistent visitors, and hurrying staff members; telephones buzzed and boxes of supplies were stacked in corners.

On the day after the Director's appoint-

[9] Columns mentioned n. 7.

[10] *Pittsburgh Courier*, May 23, 1942, and others. WDBPR Newspaper Clippings, WAC, DRB AGO.

[11] *Fargo* (N. D.) *Morning Forum*, May 17, 1942, and Niles (Mich.) *Star*, May 20, 1942, respectively.

[12] First three in Little Rock *Arkansas Democrat*, May 23, 1942. The last in the column "The Totem Pole," *Seattle Star*, May 20, 1942.

[13] The column, "The Brighter Side," May 22, 1942. WDBPR Newspaper Clippings, WAC, DRB AGO.

[14] The column, "Miami Story," Miami (Fla.) *News*, May 20, 1942.

[15] Editorials in New York *Daily News* and *Augusta* (Ga.) *Herald*, and a column in Oklahoma City *Daily Oklahoman*, May 1942, respectively. WDBPR Newspaper Clippings, WAC, DRB AGO.

[16] Interv with Mrs. Herrick, 9 Jan 46.

ment, the little headquarters pulled itself together and began to issue numbered memoranda. Now that there legally was a WAAC, the long-delayed funds could at last be allotted for a training center; the tentative Regulations could be published; the items of uniform could be approved for procurement and details released to the underwear-conscious press. More important, the still-nebulous group of preplanners could now be solidified into a real WAAC Headquarters, complete with charts and ready for rapid expansion. Thirty-seven civilians were also allotted, including another woman consultant, Mrs. Gruber, wife of Brig. Gen. William R. Gruber, to advise on military protocol and manage the Director's schedule. The office now totaled some fifty people.[17]

By any military standards this was hardly an impressive headquarters, scarcely the equal of that of a respectable Army regiment; and in spite of the exaggerated power that was attributed to it by the press and general public, its members by this time entertained few delusions of grandeur. As an insignificant civilian group, which included no general officer, they expected and got a low priority of attention as compared with the more vital phases of the combat effort, particularly in such matters as office space, cafeteria hours, entertainment allowances, and transportation.[18]

The office had at least a military-looking chief when, in June, there arrived the first WAAC uniform, especially made for Director Hobby. She promptly put it on, reported formally to the Chief of Staff, and came back wearing a colonel's silver eagles, which he had pinned on and directed her to wear. The relative rank of Director was thus established at that of colonel, although Congress had authorized pay equal only to that of a major. Both relative rank and pay were thus established as the same as that recently given to the Superintendent, Army Nurse Corps. The relative rank of full colonel was also necessitated by the grades of the Army officers on her staff, since Director Hobby would be superior in the WAAC chain of command to the commandant of the training center, Colonel Faith, as well as to her own staff, which now included Col. William Pearson as adjutant general and Col. Bickford Sawyer as finance officer.

No later Waacs ever suffered from uniform shortages as did Director Hobby at this time; there was in existence only one WAAC shirt, and in her travels her luggage contained little but an electric fan and an iron, which enabled her to wash, dry, and iron the shirt nightly.

Location in the Services of Supply

Meanwhile, the Services of Supply was considering a more important matter: the proper place in the headquarters chart to insert the square labeled WAAC. Competition for the honor of supervising the WAAC was not exactly brisk; it might be said that each agency outdid the other in striving to bestow this gift on its neighbor.[19]

[17] (1) Various Hq Memos in WA 300 (5-20-42) Office Memos, sec 1. (2) Memo, G-1 for CofS, 13 May 42. WDGAP 352.01 WAAC. (3) Memo, TAG for Dir WAAC, 26 May 42. AG 291.9 (5-15-42). (4) Memo, OQMG for G-1, 16 May 42, and Inds. SPQRE 420 WAAC. Copy in SPWA 421 (5-16-42) sec 1. (5) WD SO 129, 17 May 42; Memo, WAAC Ln Off for Pers Div SOS, 15 May 42. Tasker Staybacks.

[18] (1) Memo, W. Pearson for TAG, 19 Mar 42, and atchd answer. WA 029.21 (3-19-42). (2) Memo, Dir WAAC for SOS, 16 May 42, and atchd reply from TAG. WA 451 (5-16-42).

[19] Intervs with Lt Col Woods, 14 Dec 45, and Lt Col Betty Bandel, 15 Dec 45. (Both former Deputy Directors.)

At one time it seemed that Civilian Personnel Division would be forced to assume control, although this was obviously inappropriate, since WAAC administration would be conducted along military lines. Military Personnel Division, from whence the preplanners had operated, was also inappropriate, as it was strictly a policy-making agency and did not operate, while WAAC Headquarters would operate a training center and various units. The loser in this unique contest was eventually the Chief of Administrative Services, at that time Maj. Gen. John P. Smith. General Smith was shortly to be succeeded by Maj. Gen. George Grunert, who, as a lieutenant colonel, had in 1927 been one of the War Department staff officers to recommend postponement of plans for a women's service corps. The Administrative Services, a large subdivision of the Services of Supply, included many comparable operating agencies, such as the Army Specialist Corps, the Army Exchange Service, and the National Guard Bureau. WAAC Headquarters would now be removed from the Chief of Staff by three echelons: first, the Chief of Administrative Services; over him the Commanding General, Services of Supply; and over both on personnel matters the Assistant Chief of Staff, G-1. The new position was not advantageous in many ways, but it appeared to the Director's advisers that there was at the moment no alternative.[20]

The WAAC's new status as one of the Administrative Services was set forth to the field by the first important War Department Circular governing the WAAC—Circular 169 of 1 June 1942—which, after two weeks of the Corps' existence, at last made the WAAC legitimate. The circular was so brief that a second circular was clearly needed at once, although it did inform the Army that (1) the Director would command the Corps and (2) other Services of Supply offices would set WAAC policy. It gave no details either of command channels or of interoffice co-ordination; neither did it refer to the tentative WAAC Regulations. All WAAC plans and recommended policies were required to be sent to the Chief of Administrative Services "for reference to the appropriate staff division of the Services of Supply." The Chief of Administrative Services wrote: "While the WAAC is not a part of the Army, yet the entire administration of the Corps, including its organization and development, must be supervised and assisted by the Services of Supply." [21]

Director's Schedule

As the new Corps began the selection of officer candidates and the establishment of its training center, the work load upon the headquarters, never light, suddenly became so great as seriously to threaten its efficiency.[22] A fourteen-hour day and a seven-day week became standard in the Director's office, a schedule of course not uncommon throughout the War Department at this date. Even this was often exceeded; staff members reported that for days at a time Mrs. Hobby and her assistants worked every night until three, five, and sometimes seven in the morning, averaging only two or three hours of sleep each night, or merely going home for a shower and coffee before returning to work. Staff members worked under condi-

[20] Memo, Dir Mil Pers SOS for CofAdm Serv, 21 May 42. SPGAP/1619-49, in AG 291.9 WAAC (5-15-42).

[21] *Ibid.*

[22] This section from interviews, already cited, with staff members; or from Hq Memos cited n. 17(1).

tions of what they called "awful confusion" and pointed out that the developments of the following months could be seen in perspective only if it was understood that they occurred, not separately and in order, but simultaneously and in the midst of wildly growing national publicity that was in itself a full-time problem.

Director Hobby attempted to cope with the confusion by rigorously budgeting her time. To Mrs. Gruber was given the task of scheduling all persons wishing to see the Director, in the belief that "more speed can be made if she finishes one subject without interruption." Staff members prepared summaries of staff papers and kept a chart for the Director's daily briefing; Colonel Pearson also directed that the Director be left free for "formulation of basic policies." Unfortunately, at this stage everything was new and policy was inseparable from detail; the situation resolved itself into a struggle to meet the daily crises without crushing the Director with decisions. The War Department Bureau of Public Relations required the Director to hold a full two-hour press conference twice weekly. Also, interviews with the public were scheduled four times a week, thus leaving only a few days entirely free for military problems. Even this amount of time was highly unsatisfactory to members of the press and Congress and other citizens who now came to call, and some criticism resulted. One Congressman complained, "I can get to the President in five minutes, but I can't get to Mrs. Hobby at all."

The Director was aware of the problem and stated in an office memorandum:

> The successful launching of the WAAC is a public relations problem to probably a greater extent than any other War Department activity. We must establish a reputation for not giving the "brush-off" or "run-around" to those calling.

However, she felt little sympathy toward those who wanted merely a sight of her or some personal favor. She said later:

> Many of them were just curiosity-seekers; they could have got accurate information from any Army officer in our headquarters, but they wanted to sit and chat with that new curiosity, a woman in Army uniform.

More important than seeing such people, she felt, were her military responsibilities to the Army and to the women about to join the WAAC.

Operating Duties

These military duties, and not the public pressure, constituted the real problem of the heavy work load. The Director realized from the beginning that her duties were exactly the reverse of what they should have been; she said later: "We had foreseen this problem in Canada, and it was one of the reasons why we wanted Army status. Now, we were an Auxiliary and it was none of Army offices' legal business."

For example, the office set out to prepare Tables of Basic Allowances for the WAAC Training Center and for units, and was at once struggling with terms such as "T-Square, Maple, Plastic lined, one per publications section"; "Chart, Anatomical, Blood System, one per training center"; "Can, corrugated, with cover, 32 gal, one per 20 mbrs." Realizing that the WAAC's staff of eleven officers did not possess the necessary specialist training to prepare accurate tables of chemical, signal, medical, and other equipment and that the tables could be done quickly by the Army specialists who prepared tables for men, Colonel Pearson attempted to send the work to the Chief of Administrative Services for distribution to other Services of Supply offices. It was promptly returned to him with the statement that the only thing wanted for distribution to the SOS offices was the final plan or policy for approval.[23] Director Hobby noted further:

> We had to prepare our own budget for presentation to Congress—a complicated statistical matter—because the people who prepared Army budgets said it wasn't their responsibility. Then, three of us worked all one night until 9 the next morning writing the recruiting circulars which I had to present for approval to The Adjutant General; TAG wouldn't write them. We had to check all the blueprints for housing at every station—endless, bewildering blueprints.[24]

In every field of military responsibility, it was the same story. The technical work of the WAAC was done by an office that knew little of the method, and the policy approval was given to agencies that knew nothing of WAAC over-all needs. Staff members realized that, had Army status made it possible, just the opposite organization would have been most efficient. Under the circumstances, most of the summer's tasks were accomplished, if at all, only by a process of mutual irritation.

Colonel Pearson, the office's oldest and most experienced adviser, was likewise unable to obtain from the Services of Supply any clear-cut decision that would enable him to distinguish between policy matters which he must refer to other offices and routine decisions which he could make and put into effect without delay. Colonel Pearson finally forced the issue by sending direct to The Adjutant General for publication a minor decision which he considered routine—that of raising the age limit for the first officer candidate class from 45 to the legal limit of 50. Refusing to publish it, The Adjutant General sent it to Control Branch, SOS, for clearance; Control Branch sent it for comment to Military Personnel Division, which commented and sent it back to Control Branch, which a week later returned it to the WAAC, with directions to draft the circular instead of asking The Adjutant General to do it, and a brief verbal spanking: "In the future, please transmit all matters of this nature to the Chief of Administrative Services."

It was thus established that even simple WAAC decisions would probably be considered policy matters and thus require approval by several SOS branches before they could be put into operation. While such a procedure proved useful in avoiding errors, it became impossible for WAAC Headquarters to act with anything like speed and decisiveness in the emergencies bearing down upon it.[25]

Drastic expedients were suggested to the Director by her Army advisers, some involving the disbandment of WAAC Headquarters. The most plausible of these—although probably not legal—was to dismember the office and give the various pieces to the SOS divisions that handled the same matters for men, while the Director and a small office for control and inspection would soar upward for several SOS echelons and attach themselves to the office of the Commanding General, Services of Supply, Lt. Gen. Brehon B. Somervell. From this vantage point the Director could then heckle other offices when dif-

[23] Memo, Chief, Sup Div WAAC for CofAdm Serv, 1 Sep 42. SPWA 400.34 (3-21-43)(1) sec 1.

[24] Interv with Mrs. Hobby, 23 Aug 48.

[25] Correspondence cited is a series of papers under dates of 25 and 27 May, 11, 12, 13, 15, and 16 Jun 1942. AG 291.9 (6-2-41). Staff comments are from interv with Mrs. Onthank, 4 Sep 46.

ficulties occurred. "As it is now," the Army advisers wrote mournfully, "we have to ask help from all of these branches, and maybe get it and maybe don't, and get bounced all around unless we want to go to the top to help us out, and that is a bad system." [26] This plan, although eminently workable, was not be adopted until the WAAC became part of the Army.

In spite of the unresolved confusion and the demands upon them, staff members remained enthusiastic and optimistic; Director Hobby publicly thanked them for their loyalty and for putting so much into the struggle.[27]

Selection of the First Officer Candidates

From the public viewpoint, selection of the WAAC officer candidates was the only important matter that the WAAC now had on its hands. From the moment of passage of the bill, WAAC Headquarters had been besieged, to the detriment of its other duties, by long-distance telephone calls, telegrams, visitors, and letters of application, all asking commissions. Congressmen, Army heads, and public officials were swamped by demands from friends and constituents. Officers of all the armed forces sought to get commissions for their friends and relatives, while powerful pressure groups sought to name assistant directors. "Every important person had a candidate," said a woman staff member, "and they all wanted guarantees of commissions and important positions." [28]

Under its legislation, the WAAC could have commissioned many of its officers direct from civil life with appropriate rank, as the Army did and as the WAVES and other women's services were later to do. Mrs. Hobby's previous decision to grant no direct commissions was severely tested. Mrs. Rogers, the Corps' sponsor, urged her to commission several lieutenant colonels at once, as she feared that otherwise the WAAC leaders would have no prestige in their dealings with ranking Army officers.[29] However, it appeared that if even two or three women were commissioned directly, it would be impossible to refuse hundreds of others with prominent sponsors, and Director Hobby believed that Mrs. Rogers' objection concerning rank could be overcome by rapid promotion as soon as the women had demonstrated their ability.

It was General Hilldring's belief that the Director was strengthened in her original decision by the opinions of General Marshall.

> The Chief of Staff was a great democrat [said General Hilldring]. I told Mrs. Hobby how, at his direction, I had made a fight to get more officers from the ranks of enlisted men. I told her that he might not object to appointing a few civilian women for the top jobs, but the rest of the officers must come from the ranks. They must be the best women in the ranks even if this was not necessarily the best in the nation.[30]

Also, staff members in their discussions of this point gave some weight to what they called "the man-woman element," a factor which did not ordinarily operate in the selection of male officers for commissions, but which now made it difficult to judge with entire accuracy the motives behind demands for the commissioning of female friends and secretaries.[31]

The experiences of other nations indi-

[26] (1) Min, Stf Mtg, 24 Aug 42. (2) Memo, PRO for Dir WAAC, 9 Sep 42. SPWA 300.4.

[27] Unnumbered Memo, 1 Jun 42. WA 300 (5-20-42).

[28] List of some applicants in SPWA 029 (1942).

[29] Interv with Mrs. Rogers, 14 Dec 45.

[30] Interv with Gen Hilldring, 17 Jan 46.

[31] All interviews confirm this.

cated that, by its initial choice of officers, the entire future course of a women's corps might be determined. In England initial limitation of commissioned rank to socially or politically prominent women had left a corps without adequate troop officers. In some Continental countries commissions had been awarded chiefly to enable certain personnel to travel with ranking officers. It was recognized that, without firm leadership at this moment, the first American auxiliary might set out on a similar course. Finally, a General Staff decision was obtained that confirmed the Director's original policy of limiting the future WAAC leadership to those women who would enlist without guarantees, and making it possible to refuse all requests from high authorities without incurring charges of discrimination.[32] All applicants who stormed WAAC Headquarters were therefore told to apply at their local Army recruiting stations. Director Hobby's own civilian advisers, Mrs. Woods and Mrs. Fling, had to follow this procedure when they desired to enlist.

By 27 May, Army recruiting stations all over the United States were supplied with application blanks, and newspapers carried the story that the doors were now open for officer candidate applications.[33] The response was a rush that swamped recruiting stations and startled recruiters. The Adjutant General had optimistically sent each corps area 10,000 application blanks, a total of 90,000, although only 360 candidates were to be selected. Three days later, urgent calls had already come from the Second, Third, Fourth, and Sixth Corps Areas for 10,000 more each, and from the Fifth and Ninth for 5,000. Of the 140,000 blanks given out, it was assumed that many would not be returned, as the 4 June deadline gave little leeway on delayed applications. Some recruiting officers attempted to weed out the obviously unqualified and to give the scarce blanks only to eligibles.

In New York City, 1,400 women stormed the Whitehall Street office on the first day and stood in line from 8:30 to 5:00 o'clock. In five days over 5,200 had received application blanks in New York City alone, although only 30 women could be picked from the whole of New York State, New Jersey, and Delaware. In Washington 1,700 applied for the eight positions open to the District of Columbia. Numbers as reported by newspapers varied in different cities: Sheboygan—20; Lowell—65; Richmond—65; Baltimore—500; Portland—380; Sacramento—135; Minneapolis—600 given out, but only 50 returned; Butte—"a few cautious inquiries." Nothing could be told as yet of the type of woman who had applied. Newspapers featured an Indian woman in full tribal regalia, a sixteen-year-old who wanted to get away from home, a wild-eyed mother brandishing a pistol and demanding to get to the front. It was hoped that these were not typical applicants.[34]

Screening of Applicants

It had been decided that Army induction facilities could be used for women, with of course the necessary segregation

[32] Document not positively identified; possibly Memo, G-1 for CofS, 22 Mar 42, sub: WAAC. G-1/15839, in CofS 291.9.

[33] WDBPR Press Release, 26 May 42. All WDBPR Press Releases are in WDBPR file, DRB AGO.

[34] *New London Day* (Conn.), *New York World-Telegram*, *Washington Post*, *Sheboygan Press* (Wisc.), *Lowell Sun*, (Mass.), *Richmond Times-Dispatch*, Baltimore *Evening Sun*, *Portland Morning Oregonian*, *Sacramento Daily Union*, *Minneapolis Tribune*, all of June 2, 1942; Butte *Montana Standard*, May 24, 1942.

for physical examination.[35] This decision provided an extra and unscheduled screening device, for only women of some valor dared approach the Army offices. Members of the first class commented: "The recruiting station was the dirtiest place I ever saw." "It was in the post office basement next to the men's toilets." "I was whistled at by the selectees." "Everyone in the room turned to look as the Captain bawled out 'Are you one of them Wackies?' "[36] WAC recruiters were later to find the location and nature of Army recruiting facilities a major handicap in recruiting women.

Some 30,000 women braved this obstacle and filed applications. Their papers were reviewed by the stations, and those who were not obviously disqualified for age, citizenship, or other causes were summoned for an aptitude test. This test had been prepared by the Personnel Procedures Section of The Adjutant General's Office to eliminate 55 percent of all applicants.[37] It was attempted to make standard questions more appropriate for females, but some were not easily converted, and women were confronted by such problems as, "You are a first baseman, and after putting the batter out by catching the ball (not a fly) with your foot on the bag, you . . ."[38]

Those who passed the aptitude test were further screened by a preliminary local interviewing board of two women and an Army officer selected by the corps area. To avoid appointment of unqualified women to these boards, the preplanners had sent out instructions that members were to be local personnel directors, business executives, YWCA supervisors, and women of like standing. These local boards weeded out those applicants who were visibly unsuited by reason of character, bearing, or instability. Members were asked to consider the question: "Would I want my daughter to come under the influence of this woman?"[39]

The numbers passing both the test and the first board still hopelessly overcrowded Army medical facilities for the examination of women. It was therefore directed that only the top five hundred applicants in each corps area be given physical examinations. The examination was not entirely consistent throughout the corps area and it was evident that the Army still had far to go in working out a satisfactory physical examination for women, but it did eliminate those with the more obvious physical disqualifications.

There then arrived at each corps area headquarters an especially selected woman to be the Director's representative on the final screening board. Among the women selected were Dean Dorothy C. Stratton of Purdue, later head of the SPARS, and Dean Sarah Blanding of Cornell, later President of Vassar.[40] In spite of the pres-

---

[35] (1) M/R 1-7, Oct 40, sub: Reception of SS EM. (2) Folder, Selection Processing, Hobby files.

[36] Related to author by members of first officers' candidate class.

[37] (1) WDAGO Mental Alertness-1, X-Z, Mar 42, called MAT-1. (2) Speech, Col Tasker to Corps Area G-1 Cong, 12 Jun 42. WAAC Hist File, 1942 WAAC Files. (3) D/F, SOS to TAG, 11 Mar 42, sub: WAAC. G-1/15839-10.

[38] It proved very difficult to discover the correlation between the WAAC Mental Alertness Test (MAT) and the Army General Classification Test (AGCT). For details of testing, see Folder, Tests, in WAAC Planning files (unnumbered files now with Staff Rec Sec, DRB AGO), containing: (1) Memo, Dir WAC for TAG, 10 Aug 43, SPWA 201.6; (2) M/R, undated, 2d Lt Ruth W. Brainerd, sub: AGCT and MAT; (3) Memo, Opns and Tng Div AGO for Dir WAC, 16 Dec 43, AG 220.01 (12-16-43)OC-L.

[39] Memo, Hq SOS for TAG, 17 Mar 42, sub: WAAC. G-1/15839-10.

[40] All members listed in WDBPR Press Release, 30 Jun 42.

SCREENING OF APPLICANT *for first officer candidate class by local interviewing board at Fort McPherson, Georgia, 20 June 1942.*

sure of time, the final personal interviews were more extensive, and delved thoroughly into each candidate's record in school, college, and business to see if signs of leadership were evident.[41] The Director's representatives then returned immediately to Washington bearing the candidates' complete papers, pictures, and interview records. Their private comments varied with the corps areas. One said, "Decidedly disappointed in both number and general type of applicants"; another, "A complete cross-section from young things to professional women"; and a third, "I was impressed with the common denominator shared by the extremes in types—wanting to do organized effective war work." [42]

The final selection process was of unequaled intensity. An evaluating board of eleven prominent psychiatrists [43] searched the work histories for evidences of mental instability, choosing women who showed steady progress and acceptance of responsibility rather than frequent unexplained changes of jobs. Application forms also revealed data on parents and family life,

[41] Memo, Dir WAAC for Pers Div SOS, 2 Jun 42. G-1/15839-10 (7-8-42).

[42] Off-the-record comments, Folder, Selection Processing, Hobby files.

[43] Named in WDBPR Press Release, 30 Jun 42.

whether applicants had ever lived in clubs or dormitories, their travel, sports interests, and health. Answers to the essay question on "Why I Desire Service" were believed to be particularly revealing. There followed what Director Hobby called "four days of careful, almost prayerful, work of making the final selections."[44] The Director personally read each of the final papers. The 30,000 applicants had been reduced to the 4,500 who appeared before the final boards, and in the midst of selection procedure, word came from the Chief of Staff to pick not 360, but 1,300, the papers of the extra candidates to be kept for call to later classes as needed.[45]

The work was finally completed late on the night of 30 June. Within a week, corps areas had notified the 360 top-rated applicants, who thus had less than two weeks to close out their jobs, take the oath, and get to Fort Des Moines for the opening on 20 July.

The selected candidates had one characteristic in common: 99 percent had been successfully employed in civil life. Although there was no such educational requirement, 90 percent had college training and some had several degrees. Most fell in the age group of 25 to 39, although 16 percent were under 25 and 10 percent were 40 and over. About 20 percent were married, most of them to men in service, and there were some mothers, especially mothers of servicemen, though none with small children.[46] The candidates included many women who had held responsible jobs: a dean of women, a school owner and director, a personnel director, a Red Cross official, a former sales manager, and several editors, and there were many more who had been reporters, office executives, lawyers, social workers, Army employees, and teachers. In compliance with promises made to Congress, it had been necessary to choose a certain number from each corps area and from each state, regardless of density of population, with the result that many highly qualified women in certain crowded areas had to be turned down, while in certain sparsely settled states it was necessary to take almost every woman who applied.[47]

Special Groups of Candidates

Two other problems complicated the task of the selecting authorities: the selection of Negro officer candidates and the selection of Aircraft Warning Service candidates. The Negro problem was particularly delicate, for the Negro press had immediately attacked the choice of a Director, and the National Negro Council had written President Roosevelt protesting the appointment of a Southern woman.[48]

The opposition was unexpected, since the War Department had early informed Congress that it would train Negro WAAC officers and auxiliaries up to 10 percent of the Corps and Mrs. Hobby had declared her support of the plan; while the Navy Department's WAVES, although not headed by a Southerner, did not accept any Negroes at all for almost three years.

The real cause of the objections came out in later news stories, which revealed that Negro leaders had hoped to use the

[44] Speech at Howard University, 6 Jul 42. WDBPR Press Release, 6 Jul 42.

[45] Memo, Preplanners for G-1, 7 May 42. Tasker Staybacks.

[46] Some of the Aircraft Warning Service candidates later added to this group had young children, as it was supposed that they would live at home.

[47] (1) Figures given to Senate Com on Mil Affairs. (2) List in WDBPR Press Memo, 3 Sep 42.

[48] *Atlanta Daily World*, May 17, 1942.

new Corps to break the "traditional and undemocratic policy of the Army and Navy" of placing Negro personnel in separate units.[49] Their demand that the WAAC place Negro and white women in the same unit was rejected because the WAAC had been directed to follow the Army policy. Now the National Negro Council recommended to the Secretary of War that Dr. Mary McLeod Bethune and one other Negro leader be commissioned assistant directors, which was of course impossible under the no-direct-appointment policy. To make relations more precarious, it soon became apparent that not enough qualified Negro women had presented themselves as applicants for officer candidate school, and it was feared that WAAC failure to fill the quota of 10 percent would be interpreted by the Negro press as discrimination. Accordingly, Army officers made hurried trips to Negro colleges to attempt to recruit qualified women. Mrs. Bethune assisted in the selection of these candidates, not being within the age limit to attend the school herself.[50]

For the Aircraft Warning Service the Army Air Forces asked that it be allowed to put women in uniform without sending them to school, but after long consideration it was decided that AWS officers must graduate from the regular officer candidate school before donning the WAAC uniform. On the matter of selection the Director felt unable to offer opposition, since the women to be selected were already running the operations and filter boards. The Air Forces were therefore allotted eighty vacancies for the first officer candidate class, which brought that class to a total of 440. WAAC Headquarters suffered some misgivings about the absence of the usual tests. However, while choices were later found to include some women without responsible paid experience and with test scores in the lowest aptitude category, Group V, the majority made successful officers.[51]

There was one applicant for attendance at the school who gave the War Department more trouble than all others combined. This was Director Hobby herself, who had made up her mind to attend the school as an ordinary student in order to understand through her own experience the problems of a woman's adjustment to military training. She stood firm in this decision for some time while the expostulations of shocked advisers raged around her.[52] The Director carried this determination unaltered up through several strata of general officers and was stopped at last only by General Marshall himself, who told her sympathetically but with finality that the thing was impossible under the Army system of rank.

The First WAAC Training Center

Meanwhile, the situation at Fort Des Moines could scarcely have been termed tranquil. Immediately upon passage of the WAAC bill, Colonel Faith and his staff of ten Army officers shot out of Washington in the race to have the training center ready when the women arrived.[53] They reached Des Moines with less than two months to spare before opening date.

[49] *Chicago Defender*, May 23, 1942.

[50] (1) *Ibid.* (2) Copy of plan in Folder Recruiting, Tasker files, 1942 WAAC files. (3) Interv with Lt Col Woods, 21 Mar 46.

[51] Memo, AG WAAC for Dir WAAC, 3 Jun 42. Tasker Staybacks. Also from intervs.

[52] This incident related separately by former staff members present at the time: Mrs. Herrick, Mrs. Knight, and Mrs. Woods.

[53] Memo, G-1 for CofS, 13 May 42. WDGAP 352.01 WAAC. Approved 15 May 42.

Colonel Faith's basic plan in beginning the undertaking was to make the course so military that Army skeptics could find little to criticize. Reporters who interviewed him came away stating, "It will be no glamour girls' playhouse." They gathered that few concessions would be made to "feminine vanity and civilian frippery." [54] Colonel Faith deplored the press phrasing of the matter, and stated later:

> It was obvious to all of us that a considerable percentage of the Army was either opposed to the WAAC idea or very doubtful of its potential success. The best way to combat this attitude was to train the new soldiers in those qualities which the Army values most highly: neatness in dress, punctiliousness in military courtesy, smartness and precision in drill and ceremonies, and willingness and ability to do the job.[55]

The students were, however, to be taught the Army's reasons for the things they were required to do. "The American woman is the most co-operative human being on earth if she fully understands an order," Faith declared hopefully.[56]

Remodeling and new construction at Fort Des Moines began as soon as the induction station and reception center could be moved. Classrooms were erected, stables and plumbing converted, and separate housing provided for the male cadre. The rest of the staff, forty-one officers and 192 enlisted men, arrived about 30 May.[57] The director of training was especially selected by Colonel Tasker; he was Lt. Gordon C. Jones, a former instructor at The Citadel.

Direct communication had been authorized between Colonel Faith and the Director, and this proved indeed fortunate for the training center, which began to send by telephone, wire, and mail a steady stream of requests.[58] Dayroom furniture was needed, also a Special Services officer, two women doctors, a post utilities officer, pictures of uniforms for the local press, instruments for the band, and money for everything. Boxes of training literature had not arrived, nor had test forms.

A hostess and a librarian were needed for the service club, also a Red Cross field director. Highly skilled classification teams had to be found to interview and classify the women, who might be expected to have job experience not ordinarily encountered by the Army. The Chief of Finance "strenuously objected" to furnishing a fiscal officer. The Corps of Engineers had no capable post utilities officer to offer. The $20,000 already allotted to the Seventh Corps Area was so restricted that Colonel Faith could not use it to purchase many of the miscellaneous items needed. There was no money for entertaining the scores of prominent people who desired to view the school.

These assorted requests, coming as they did during the period when WAAC Headquarters was struggling to establish its own organization and policies and to select officer candidates, quickly began to reveal still more of the difficulties which the Director's office must face as an operating headquarters. Most of the fairly minor matters discussed with Colonel Faith could easily have been decentralized to the Seventh Corps Area had the WAAC been in the Army, but since it was a separate organization, all of its funds and

[54] (1) AP Release, June 1, 1942, quoted in *New York Times*, June 2, 1942. (2) Lee Carson, INS Release, May 19, 1942, in Los Angeles *Examiner*, May 20, 1942.

[55] Ltr, Faith to Ward, 18 May 50. OCMH.

[56] Press Releases cited n. 54.

[57] Memo cited n. 53.

[58] Personal ltrs, memos, and transcripts of tp convs in: (1) Folder T, Tasker files, 1942 WAAC files, dated 3, 8, 22, 25, 26, 27, 29 June, 3, 6, 20, 21, 28 July. (2) Tasker Staybacks, dated 3, 10 June, 6 July.

personnel allotments had to be handled separately and justified by the Director's office.

The Adjutant General had meanwhile directed various Army commands to select and send to Des Moines the men who would serve as instructors and cadre until Waacs could take over.[59] This personnel had no sooner reported than it was discovered that few if any of the instructors were qualified for the exacting and unprecedented task ahead. The director of training wrote:

> Out of 41 officers which we requested originally, we got one man who is fitted for the job at hand. The others are all doing their best to fill jobs far beyond their former experience. The new officers are all fine fellows (we are fortunate to have gotten no duds) but professionally. . . .[60]

As the time grew shorter, it became apparent that some deficiencies were inevitable. Requisitions for field and technical manuals, circulars, forms, and Army Regulations were only partially filled or back-ordered, with no assurance from harassed supply depots that they could be filled within the next six months. The post's two mimeograph machines ran continuously to duplicate missing material.[61] Funds for certain equipment did not arrive on time, and neither did some of the faculty. Colonel Tasker wrote, "The lieutenant who was to join your force cannot be located. Please let me know if others are missing." [62] Even ahead of these difficulties came the greatest problem of all, and an unexpected one—expansion.

Expansion Plans, Summer of 1942

In the midst of the WAAC's hurried preparations there occurred an event which was to bring far-reaching changes to all plans. On 13 June 1942 the Chief of Staff called into his office three Army officers representing the Chief of Administrative Services, the Office of the Director, WAAC, and Aircraft Warning Service.[63] General Marshall informed these men that in view of the approaching Army personnel shortage it was his desire that every effort be made to organize and train personnel of the Women's Army Auxiliary Corps at the earliest possible date. He felt it would be possible to have considerable numbers in active service a great deal sooner than existing plans provided. The staff officers were directed to formulate the necessary speed-up plans.

All British precedent indicated that such a move meant great difficulty for the WAAC. British observers had commented, "It is difficult for the general public to realize the strain created by the great expansion of the women's services, built up as they have been . . . from almost non-existent foundations." [64] Director Hobby and the preplanners had fervently hoped for at least "time for trial and error before going into an assembly-line type of production." [65]

Nevertheless, WAAC expansion was a calculated risk that the Army's current situation made necessary. In spite of earlier optimistic estimates, the armed forces were now experiencing what historians called "the manpower crisis of the summer of

[59] Telg, TAG to CG Second Army at Memphis, Tenn., 22 May 42. AG 220.31 (5-23-42)EA, in G-1/15839-10 (7-8-42).

[60] Personal ltr, Jones to Tasker, 27 Jun 42. Folder T, Tasker files, 1942 WAAC files.

[61] Ltr, Faith to Dir WAAC, 24 Aug 42. SPWA 400.311 (8-24-42).

[62] Ltr, Tasker to Faith, 28 Jul 42. Tasker files, 1942 WAAC files.

[63] Memo, Lt Col James L. Tarr to CofAdm Serv, 13 Jun 42. G-1/15839-10 (7-8-42).

[64] *Conditions in the Three Women's Services.*

[65] Ltr, Mudgett to author, 19 Aug 48. OCMH.

1942."[66] The Army Ground Forces was already short more than 160,000 men. Only two days before General Marshall's decision the War Department had admitted the necessity either to train units understrength or to slow down the activation of new units, thus jeopardizing invasion plans. It was also being forced toward unpopular measures such as drafting eighteen-year-olds and fathers and cutting more deeply into defense industry and agriculture.[67]

On the other hand, the WAAC was concurrently rocking along with its modest prewar plan for about 12,000 women in the first year. Meanwhile, thousands of women were clamoring for admittance, and requisitions for over 80,000 of them had by this time been received from Army commands, of which at least 63,000 seemed suitable for employment. This did not include the needs of overseas theaters and many other domestic commands which had also shown interest.[68]

Thus, WAAC expansion seemed logical in spite of its practical difficulties. The Army itself was faced with the same expansion problems, and men were training with wooden guns, sleeping in tent camps, and wearing clothing often inappropriate to the climate, all with greater success than might have been expected. The WAAC move was nevertheless the greater risk, since men could be drafted to face these hardships while women would have to volunteer.

Plans for expansion were drawn up by the Services of Supply at hasty conferences on 14 and 15 June and submitted to the Chief of Staff on 17 June. The WAAC's ultimate strength was to be increased at least fivefold, from 12,000 to more than 63,000. Within a year, by May of 1943, there would be in the field 25,000 instead of the 12,000 previously planned for that date. Within two years, or by April of 1944, the entire 63,000 would be trained and at work. The divisions of the Services of Supply concerned believed it would be possible to manage this expansion successfully, and gave it their written concurrence.[69]

The plan was approved by the Chief of Staff, although within three months it, too, was to be discarded by the General Staff in favor of still greater expansion. This moment marked an important turning point in Corps history: up to this date, it was obvious that the WAAC had been forced upon an unwilling Army by a combination of circumstances; from this time forward, its employment was not only voluntary but was extended more rapidly than its original sponsors had contemplated.

As the first result of the plan, construction at Fort Des Moines was stepped up by 20 percent, and action was taken by condemnation to lease Des Moines hotels, garages, office buildings, college dormi-

[66] Kent Roberts Greenfield, Robert R. Palmer, and Bell I. Wiley, *The Organization of Ground Combat Troops*, in UNITED STATES ARMY IN WORLD WAR II (Washington, 1947), pp. 206–08.

[67] Remarks of Lt Gen Joseph T. McNarney, Min, Gen Council, 12 Oct 42.

[68] Memo, Brig Gen James E. Wharton, Dir Mil Pers SOS, for CofS, Attn: G–1, 17 Jun 42. G–1/15839–10.

Command	*Requisitioned*	*Approved*
Total	80,750	63,750
AAF, AWS	31,600	31,600
AAF, airfields	19,300	19,300
AAF, ferrying commands	1,100	1,100
Service Commands	6,450	6,450
Signal Corps	7,500	[a] 2,500
Antiaircraft commands	750	750
Weather Bureau	2,050	2,050
Post Exchanges	12,000	0

[a] Disapproved 5,000 intended for overseas.

[69] *Ibid.* Concurrences: SOS Tng Div, Reqmts Div, Pers Div, Opns Div, CofAdm Serv, and Dir WAAC; also Rctg Div AGO.

tories and classrooms, and even the local coliseum, so that specialist training might be given in the city, leaving more room for basic trainees at the post.[70] A tent camp was set up at Fort Des Moines for an additional 634 male cadre, who were intended to stay only four months, until Waacs could graduate and replace them. Action was not taken until 3 July to order these men in, and most arrived at Des Moines scarcely one jump ahead of the women whom they were to instruct.

It was now determined that the first officer candidate class of 440, which began on 20 July, would be followed by classes of 125 women weekly until at least 1,300 officers had been trained. Officer training time had already been cut from three months to six weeks. It had also been decided to start the first class of enrolled women soon after the first officer candidate class, without waiting for WAAC officers to graduate and assist in their training. Enrolled women would have only a brief four weeks of basic training, instead of the originally planned three months.[71]

Specialist schools—for cooks, clerks, and drivers—would be ready to function in August when the first enrolled women graduated from basic training. Thus it was believed that it would be possible as early as 9 November 1942 to begin sending three fully trained headquarters companies, with officers and cadre, to field stations each week, and to train and return to their station 300 Aircraft Warning Service Waacs weekly, beginning in September.[72]

The First WAAC Officer Candidate Class, 20 July–29 August 1942

By 20 July 1942, with the opening of the first officer candidate class,[73] the mounting interest of press and public reached a point of near-hysteria. The old post at Fort Des Moines swarmed with dignitaries, invited and uninvited, who braved the steamy midsummer Iowa weather in order to view history in the making. Every angle of the new scene was well documented by the press, while photographers invaded the barracks in search of the still-elusive WAAC underwear. Authorities were particularly alarmed to discover that a number of women reporters had resolved to participate in the first week of processing in order to reveal to the waiting public a woman's every sensation as she was converted into a soldier. Because it was not yet known whether the women's sensations would be printable, this request was refused, and the press was informed that the post was theirs for one day only. After that, a two-week news blackout was to be enforced while the Army and the women became acquainted in comparative privacy. This step seemed necessary to avoid complete disruption of training, but was later believed by some staff officers to have alienated certain writers and news agencies.[74]

Before crowds of interested Des Moines citizens the arriving Waacs were whisked

[70] Memo cited n. 63. Also Folder, Lease of Real Estate, 27 Jun 42, SPWA 481 (6-7-42).

[71] Memo, Preplanners for G-1, 2 Apr 42. Tasker Staybacks.

[72] (1) Memo, Brig Gen Idwal H. Edwards, G-3, for CofS, 3 Jul 42. WDGCT 291.9 WAAC, in CofS 291.9 WAAC. (2) 1st Ind, AG WAAC for Dir Mil Pers SOS through CofAdm Serv, 29 Jun 42. WA 320.2 (6-25-42). Copy in Tasker Staybacks.

[73] Dir WAAC for CofS, 4 Aug 42, 1st Ind, Exec Off CofAdm Serv, for Dir WAAC, 10 Aug 42. SPAAI 320 WAAC, in Dir Adm Serv ASF Sp Coll DRB AGO.

[74] If no other source is stated, descriptions of this first class are based upon: (1) accounts given by women participants to the author, and upon her own impressions as a member of Company 1. Other accounts differ only as to whether Company 1, 2, 3, or 4 was the best company. (2) Hist of 1st WAAC Tng Cen.

WAAC STAFF AND MEMBERS OF THE PRESS *at Fort Des Moines, Iowa, 20 July 1942.*

from incoming trains to waiting Army trucks. At the fort processing moved efficiently and included a brief physical check, an immediate meal, and assignment to one of the four companies. Each woman was assigned a bed, a wall locker, a footlocker, and a metal chair, and soon discovered that the Army had no bathrooms, only something two flights of stairs away called a latrine. Male company officers and cadre clung to their orderly and supply rooms, designating the first arrivals as barracks guides for the rest and hastily, if inaccurately, informing all comers that only married men with children had been given this assignment.

The issuance of uniforms was the main interest. Mobilized at the clothing warehouse were almost all of the saleswomen and fitters from department stores in Des Moines, and the Waacs were passed down an assembly line from one curtained booth to another. The underwear and foundation garments proved to be of excellent quality, although the rich mud-brown color of slips and panties was slightly disconcerting.

The morning was not an hour old before more was learned about women's uniforms than had been discovered in the past six months of research. It was immediately apparent that the articles shipped to Des Moines by the Philadelphia Quartermaster Depot bore little resemblance to the model that had been approved by the preplanners. The heavy cotton khaki skirts were cut as if to fit men's hips, and buckled and wrinkled across the stomach, so that even the slimmest Waac presented a potbellied appearance after she had sat down once. When Waacs walked or marched the skirts climbed well above the knee, unless a desperate grip on the skirt was substituted for the required arm swing. The hot, stiff cotton jackets, made of men's materials, draped as becomingly as a strait jacket and were of much the same texture. Shrieks of dismay arose as women tried on the WAAC caps, uncharitably christened "Hobby hats," which were not only unbecoming to many women but also cut the forehead and were easily warped, crushed, or soiled.

Furthermore, the Quartermaster size range had apparently been set up for a race of giants, and there were not enough small sizes in outer garments. As extensive alterations were necessary to cut down larger sizes, most of the women emerged with nothing to wear for a day or two except the brown-striped seersucker exercise suits—short button-front dresses worn over matching bloomers. However, the cameramen were diverted to a few women whose uniforms fitted without alteration; these were placed in the front row for platoon pictures or posed beside the post's one cannon.

The rest of the processing, carried on intermittently during the week, closely followed the Army pattern for men. There were simultaneous tetanus, typhoid, and smallpox shots; outside the dispensary stood an embarrassed Army sergeant with smelling salts, which actually proved little needed. The Articles of War were solemnly explained, although they applied to a Waac only in the improbable event that she should find herself "in the field." Lessons were given in bedmaking, care of clothing and equipment, the difference between *fall out* and *fall in,* the beginnings of close order drill, and the other Army orientation that for men was usually given at reception centers. Women were given the standard Army General Classification Test (AGCT), and their skills were classified.

On the fourth day, with processing completed and the women at last neatly uniformed in khaki skirts and shirts, Director Hobby addressed the group for the first time. She had been at the Fort since the first day, observing somewhat wistfully. Now she spoke words calculated to remove the military trees and reveal the forest, to give the women perspective on history and where they now stood in it:

May fourteenth is a date already written into the history books of tomorrow. . . . Long-established precedents of military tradition have given way to pressing need. Total war is, by definition, endlessly expansive. . . . You are the first women to serve. . . . Never forget it. . . .

You have just made the change from peacetime pursuits to wartime tasks—from the individualism of civilian life to the anonymity of mass military life. You have given up comfortable homes, highly paid positions, leisure. You have taken off silk and put on khaki. And all for essentially the same reason—you have a debt and a date. A debt to democracy, a date with destiny. . . .

You do not come into a Corps that has an established tradition. You must make your own. But in making your own, you do have one tradition—the integrity of all the brave American women of all time who have loved their country. You, as you gather here, are living history. On your shoulders will rest the military reputation and the civilian recognition of this Corps. I have no fear that any woman here will fail the standards of the Corps.

From now on you are soldiers, defending a free way of life. Your performance will set the standards of the Corps. You will live in the spotlight. Even though the lamps of experience are dim, few if any mistakes will be permitted you.

You are no longer individuals. You wear the uniform of the Army of the United States. Respect that uniform. Respect all that it stands for. Then the world will respect all that the Corps stands for. . . . Make the adjustment from civilian to military life without faltering and without complaint. . . .

In the final analysis, the only testament free people can give to the quality of freedom is the way in which they resist the forces that peril freedom.[75]

As one of the women later said, "It was only then that I really knew what I had done in enlisting. I looked around the roomful of women and suddenly had no more doubts that it was right. My feet stopped hurting, and the war and my place in it became very real."

The Training Course

From this time on the work began in earnest. The women were up at five-thirty or before in order to be neatly dressed for six o'clock reveille, a process that took most women somewhat longer than men. After making beds, cleaning, and picking up cigarette butts and photographers' flash bulbs in the area, the women marched to breakfast and then began classes, which lasted until five in the afternoon, with an intermission for lunch. After supper there was a required study hour and then a session devoted to the washing and pressing of uniforms and the shining of shoes.

The WAAC basic and officer candidate courses were identical with corresponding courses for men, except for the omission of combat subjects. Women studied military sanitation and first aid, military customs and courtesy, map reading, defense against chemical attack, defense against air attack, interior guard, company administration, supply, mess management, and other familiar subjects. A few courses were more or less adapted to women. The hygiene course was designed by the local hospital

[75] WDBPR Press Release, 23 Jul 42, revised copy, Remarks by Mrs. Hobby, Dir WAAC, to Students at WAAC Tng Cen.

personnel to apply to women's hygiene. The Army organization course contained vague references to WAAC organization, which no one even in Washington understood too well. Drill was without arms, and the physical training course was devised and conducted by a civilian consultant.[76]

The military courtesies course, which set out to permit no deviations for women, soon ran into the inescapable fact that there were socially accepted differences for women to which the Army must perforce conform. For example, women of some faiths could not remove their hats in church as was proper for men, and most women felt self-conscious about removing hats in such places as dining rooms and lobbies. Endless commotion was also caused when women attempted to open doors and show other required courtesies for men who outranked them. Army officers were, for the first time, confronted with a situation in which a man might be either an officer or a gentleman, but not both; there resulted much sprinting for doorways and leaping out of the wrong side of vehicles to avoid the eager assistance of a nearby Waac.

As the women learned about the Army, the Army also began to learn the many minor yet important items about the administration of women that could be learned only by experience. Oddly enough, these were not the expected difficulties that the planners had feared, for the women were agreeable about inspections and did not have hysterics in the barracks or faint at the discomforts of latrine ablutions. Instead, unique specific needs arose, more practical than psychological.

For example, a few laundry tubs and ironing boards had been provided, although men's barracks ordinarily did not have them. On the first day, the Waacs as one woman headed for these—a true WAAC pattern which was not to vary through hundreds of stations in the field—and it was discovered that the women considered the number of laundry tubs, drying racks, and irons inadequate. Women stood in line until late at night for a turn at the irons. Many Waacs were found to have brought their own irons, in numbers which so frequently blew out barracks fuses that their use had to be limited. The allowance of four khaki shirts and four skirts did not permit use of local laundries, which took a week, for the women considered that neat appearance in the sticky hot climate required at least one and often two clean shirts daily.

Lack of opportunity to launder and iron her shirt for the next day was to prove a greater morale hazard to a woman soldier than any lack of movies, camp shows, and pinball machines. The women's anxiety about the matter gave strong support to a fact which civilian psychologists had already suspected, and which was to prove a new problem to station commanders: that a woman's grooming was vitally connected not only with her morale but with her health and actually with her conduct.[77]

Similarly, the women felt that the Army provided too few brooms, too little scrubbing powder, and not enough dustcloths. The women also suffered somewhat, as did male recruits, from sleeplessness, sore feet, heat and humidity, and—somewhat more than men did—from the general lack of privacy.

Other expected difficulties with wom-

---

[76] For a full discussion of physical training and all other courses, see below, Ch. XXXII.

[77] See later chapters, particularly XXII and XXXI.

FIRST OFFICER CANDIDATE CLASS, *20 July–29 August 1942. Left page, top, reveille; middle, instruction in Military Customs and Courtesy; bottom, close order drill. Right page, top, classroom instruction; middle, physical training; bottom, chow line.*

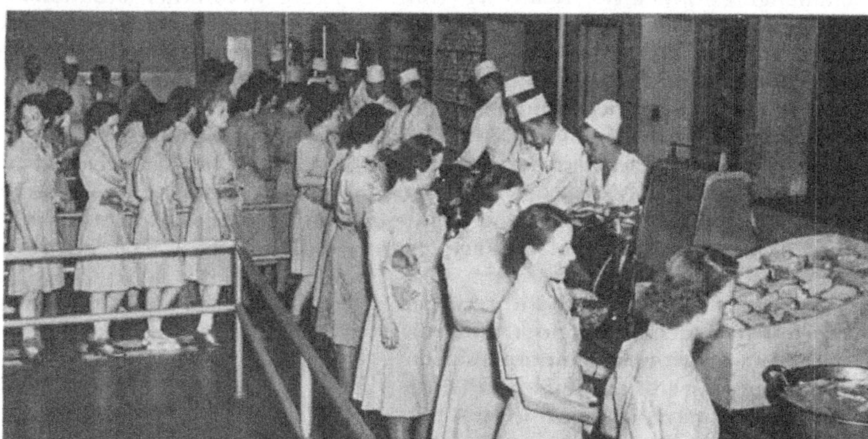

en's administration did not arise. Most Waacs were entranced with parades and bands and ceremonies of all sorts. They took well to drill, and after an exhausting day enthusiastic squads could be seen in the summer twilight, giving themselves extra drill practice. One company officer alleged, "They learn more in a day than my squads of men used to learn in a week." They were not disconcerted by the necessity for discipline and regimentation; their shoes were shined, uniforms spotless, beds tightly made, footlockers neat. Neither did they prove especially subject to quarrels, but appeared to be as good comrades as were the members of men's units.

On the other hand, the exciting publicity about the Corps and the steep competition for places had prepared women very little for the fact that the Army lacked wings and a halo. One disappointment was the caliber of the instruction at the training center. Ninety percent college women, the Waacs expected classes of challenging difficulty. Realizing this, WAAC Headquarters had sent Military Personnel Division a long account of the qualifications desired in the instructors assigned, saying, "The results to be obtained at the WAAC Training Center will depend in a large measure on the type of instructors which are detailed. . . . Time is too short to permit training these individuals after their arrival at Fort Des Moines." [78]

Colonel Faith felt that the resulting selection had been good, since the instructors were a "group of fine and sincere young men" who, although with little Army experience, were for the most part graduates of The Adjutant General's School or others of similar level.[79] Nevertheless, it became apparent to the students on the first day of classes that most of the instructors were not teachers, not public speakers, not college men, and sometimes without Army experience in the subjects they were teaching. It was a lucky young instructor who did not confront in every class some students with college training or years of teaching experience. The situation was similar to that which the WAVES were to encounter a few months later, of which the Navy noted, "The early indoctrination was largely farcical, with youthful male Reservists parroting material they had just received, to women their seniors . . . especially in teaching experience. Politeness and good humor on both sides made the situation tolerable." [80] The Waacs' most respected instructors were often the teachers of company administration, newly commissioned former sergeants who really knew the peculiarities of the morning report, although one of them startled his first class by welcoming them with, "Put dem hats on dose tables." [81]

There were other situations that the women solemnly noted for correction in the days when they would be running the Corps. All companies habitually marched out to mess at the same time, so that the last of the 440 stood in line for an hour. The aptitude tests were given at 7:00 P.M. of a stifling summer evening, immediately after a heavy meal. The summer uniforms were like nothing ever before worn by woman, with five sweat-soaked layers of cloth around the waist and three around

[78] 1st Ind, WAAC Hq for Dir Mil Pers SOS, 29 Jun 42. WA 320.2 (6-25-42), Tasker Staybacks.

[79] Ltr cited n. 55.

[80] WAVES Hist.

[81] Vol., General Information, SPWA 314.7, WAAC Hist File.

the neck.[82] The young officers did not seem to know how to put in their places the female apple polishers and classroom show-offs whose tactics were transparent to other women. The food was good and abundant but not always attractively prepared. These and similar occurrences gave the first hint of what was repeatedly demonstrated: that Waacs were often less tolerant of imperfection than men—a trait which some Army psychologists believed was due less to their sex than to the fact that all were volunteers, since male volunteers also frequently showed greater impatience than did draftees.[83]

This first class was nevertheless still too uncertain as to correct procedure, and too eager to please, to be very critical. General morale was at unprecedented heights. Women stood retreat with real appreciation of the ceremony. They trembled lest their company officers give them a gig. They crossed streets and ran out of doorways to practice their salute on everything that glistened, until the outnumbered male officers were forced to flee the front walks. No imperfection seemed of much importance when weighed against the general desire to justify the Army's admission of women to its ranks.

Graduation

On the eve of graduation there occurred one further test of the new soldiers' soldierliness: the release of the roster of class standing. Herein, it was discovered that young women who had dated their company officers had mysteriously soared in rating to outrank the older women who had led their classes in training.[84]

Since many of the company officers were ex-college boys of 19, any woman over 30 had evidently appeared antique. Katherine R. Goodwin, who later headed some 50,000 Army Service Forces Wacs, recalled: "Captain B. called me in and told me I was so old he didn't see how I could have made the journey to Des Moines, and that he couldn't imagine what use the Army could ever make of me, and perhaps I had better go home before I got any older."

Angry members rose in class to ask an explanation of the mathematics involved in rating, and an inquiry was launched by WAAC Headquarters. In general it was found that the honest tendency of the male officer was to consider a glamorous young woman, or a mannish loud-voiced one, the best material for leadership of women, whereas the women themselves had given highest ratings to women of mature judgment, and to those with a faculty for sweeping and polishing their share of the area. It was concluded that the problem would correct itself in later classes where women company officers had replaced men. However, the same situation was later to cause some difficulty in the field where men ordinarily selected and promoted WAC officers.

Graduation day, on 29 August, saw 436 new WAAC third officers, all with shining new gold bars but no WAAC insignia,

[82] Around waist: (1) heavy jersey panties, (2) sturdy girdle, (3) heavy jersey slip, (4) cotton shirt, (5) double 8.2 chino waistband of skirt; plus for dress uniform: (6) heavy cotton jacket, (7) double jacket belt. Around neck: (1) neckband of shirt (about four layers of cotton), (2) mohair tie; and for dress (3) jacket collar.

[83] Opinion of Maj Albert Preston, Jr., MC, head of Consultation Serv at Ft Des Moines. See below, Ch. XXXI.

[84] Rating was based one third each on grades, platoon rating, and company officer rating, but in the final results the last item was found to have outweighed the other two.

which The Quartermaster General was late in procuring. Mrs. Hobby spoke to the women, and Congresswoman Rogers told them: "You are soldiers and belong to America. Every hour must be your finest hour." [85] The Army was represented by Maj. Gen. James A. Ulio, The Adjutant General, and Maj. Gen. Frederick E. Uhl, Commanding General, Seventh Corps Area. General Marshall personally wrote out a congratulatory telegram for Director Hobby, saying, "Please act for me in welcoming them into the Army. This is only the beginning of a magnificent war service by the women of America."[86]

[85] *Congressional Record*, Vol. 88, No. 151, p. 7242, August 1942.

[86] (1) WDBPR Press Release, 8 Sep 42. (2) TWX, CofS for Mrs. Hobby at Ft Des Moines. WDCSA 291.2 (8-29-42).

CHAPTER IV

September–October 1942: Beginnings of Field Duties

Among the more important officer assignments were those of the eighteen carefully selected members of the first graduating class who arrived in WAAC Headquarters on 17 September 1942. They reported with a formality that delighted Regular Army staff members, having been particularly trained by Colonel Faith in how to knock, enter, and salute.[1] Third Officer proving unwieldy for direct address, the use of Lieutenant and Ma'am—instead of Sir, as WAVES officers were addressed—was adopted.[2]

For their own guidance, the Director and her advisers decided that WAAC Headquarters should be irreproachably military, with salutes rendered when entering the office of a superior, and with a command-type organization wherein no junior member of one section conferred with a junior member of another except through their respective Army section chiefs. The Director observed, "War Department officials will have to get used to the idea of treating Waacs as officers rather than as women."

The impeccably formal behavior of the Waacs did cause comment, not to say consternation, in War Department offices; it also was to impose considerable handicaps on staff operation, which did not reach maximum efficiency for many months, until the WAAC was part of the Army and could stop being military. The very junior WAAC members reported that few of them ever at this time saw the Director or had opportunity to influence Army policy. However, they grasped their assignments with rapidity and covered the scheduled three-month indoctrination course in three weeks.

Prominent among the eighteen members were Third Officers Helen H. Woods and Marjorie D. Fling, the former civilian assistants, who were now assigned, respectively, to Control Division and to Office Management. Third Officer Emily E. Davis, a former insurance company employee, was assigned to a position in the Executive Office. An especial point was made of securing a Negro Waac, Third Officer Harriet West, as one of those assigned to Personnel Division, in order that minority needs might not be overlooked in planning. Other offices receiving Waacs included Public Relations, Supply, and Plans and Training.

Possibly the most important assignment

---

[1] All references to staff opinions or meetings, unless otherwise specified, are from the WAAC Daily Journal. (See Bibliographical Note.)

[2] Office Memo 22, 15 Sep 42. SPWA 300. This directs that they not be called Third Officer except in correspondence, and never Miss nor Mrs.

WAAC OFFICERS ASSIGNED FOR DUTY IN WASHINGTON *sign the register, above, and receive explanation of Tables of Organization in Temporary M Building, below.*

was that of the Director's first WAAC aide, Third Officer Betty Bandel, formerly a reporter, music and drama critic, and woman's page editor for an Arizona newspaper. Her duties at once tended to expand beyond those of aide, for which in her own opinion she was the world's least qualified. She recalled later a moment in a London hotel when "the Director came to herself to find that *she* was sewing a button on *my* uniform; she pointed a finger at me and screamed, 'You're fired!' " Lieutenant Bandel came to be employed chiefly as a traveling deputy; on visits to field stations the two officers developed tactics whereby, if the Director was hopelessly immobilized by local dignitaries, the aide unobtrusively detached herself and explored the enlisted women's viewpoints.

New Graduating Classes

Meanwhile, a steady stream of new members, officer and enrolled, graduated weekly at Fort Des Moines. Most initial assignments were naturally to the training center. Enrolled women were shortly doing most of Fort Des Moines' office work, driving its vehicles, and staffing its messes. There was even a Waac bugler, and a band was being organized. WAAC officers at once took over much of the instruction: a former aircraft spotter taught the course in Identification of Enemy Aircraft; home economics teachers revised the mess management courses; a graduate chemist taught the course in Defense Against Chemical Attack. WAAC officers were also assigned to staff the incoming training companies, with the aid of Army "tactical advisers." No major responsibilities were turned over to women, but they were assigned as understudies to various section chiefs.

During these weeks Colonel Faith reported himself as harassed by all the ills known to similar growing institutions, and a few previously unknown. With thousands of new students scattered through officer candidate school, basic training, administrative school, motor transport school, and cooks and bakers school, the training center was still lacking in training manuals, film strips, balopticons, photographic equipment, and instructional materials of all kinds. For lack of typewriters, the administrative school could not train typists in Army methods; it was to be almost a year before a typing school could be opened. The motor transport school, unable to obtain vehicles, graduated women drivers who had driven an Army vehicle only once, and built its own demonstration engines from junk salvaged from nearby Army posts. The burden of cutting orders for hundreds of graduates was a heavy one, but teletypewriter service to Washington was not to be obtainable for eight months.

Colonel Faith's increasingly urgent reports of these matters were necessarily referred to the Services of Supply's Military Training Division, which delayed action while making further inquiry into the basis of his estimates and whether he had properly staggered schedules in order to reduce requirements.[3]

The most alarming shortage, however, was that of winter uniforms. For reasons unknown to the training center, The

[3] (1) Ltr, Comdt to Dir WAAC, 24 Aug 42, and Inds. SPWA 400.312 (8-24-42). (2) Min, Stf Conf, 14 Nov 42, WAAC Daily Journal. (3) Ltr, 1st WAAC Tng Cen to Dir WAAC, 15 Feb 43. SPWA 400.34 (3-21-42) (1) sec 3. (4) M/R, Stf Conf, 6 Jan 43. Folder, Memos Prior to 7 Jan 43 in WAC Planning, WAAC Planning files. (5) Memo, WAAC Hq for CofAdm Serv, 18 Jul 42. WA 483 (7-18-42).

Quartermaster General's specified shipment schedules were not being met. The supply of summer uniforms became erratic, while almost no winter clothing of any sort—underwear, coats, or uniforms—had been received. The Iowa weather proved un-co-operative, and an unseasonable cold spell struck in September, blanketing Fort Des Moines in snow. Temperatures stood near the freezing mark inside the newly constructed barracks and classrooms, known informally as Boomtown, where heating equipment had not yet been installed.

Within a few days the training center was plunged into a crisis of health—and, it was feared, of public relations—as respiratory disorders swept the student body. Hospitals filled suddenly, while continued instruction was made difficult by coughing and sneezing in the unheated classrooms.

At this point Director Hobby herself hastened to the training center. Putting on the summer raincoat the women were wearing, she immediately caught a bad cold, and, more practically, telephoned to higher headquarters in Washington in a manner that evoked a promise to expedite clothing shipments without delay. She also took the precaution of demanding several thousand enlisted men's overcoats from a neighboring station's surplus stock, and these were on their way before nightfall. They were to be worn all winter while action was being expedited in Washington.[4]

The men's overcoats proved excellent substitutes not only for WAAC coats but also for mittens and leggings, since they covered hands and feet and trailed on the ground. Unfortunately, Waacs found them so amusing in appearance that they could not forbear taking pictures, which they sent indiscriminately to friends and relatives, to the dismay of recruiters.

Close behind the Director came the Deputy Chief of Staff, Lt. Gen. Joseph T. McNarney, who verified her comments and immediately wrote to the Services of Supply: "Although the weather was cold, the Waacs were still wearing summer uniforms. They had been issued enlisted men's coats as an expedient. Please look into the matter of expediting winter uniforms."[5] Meanwhile, a brief Indian summer intervened to make summer uniforms again adequate.

In spite of supply difficulties, officers who inspected the training center in late September and early October reported that an excellent state of training had been attained. The WAAC graduates, both officer and enrolled, won unanimous praise. What seemed most remarkable to reviewing officers was the somewhat unremarkable fact that women could march and drill, a sight which melted the sternest critics. WAAC Headquarters' superior, General Grunert, Chief of Administrative Services, reported upon his return from Des Moines in early October:

> I was received at the entrance to the post by the band and an escort of honor which I complimented on its smart appearance, correct procedure, and excellent marching. . . . I was particularly impressed by the uniform smart appearance of all personnel, with the instructional setup, with the orderliness and arrangement of quarters, and with the punctilious saluting. . . . A late afternoon review was a gratifying sight to behold. The formations, evolutions, marching, and saluting compared favorably with the best I have seen anywhere. . . .
>
> I feel assured that the WAAC is off to an excellent start and that the commands re-

[4] From: (1) Persons at Des Moines at the time, including the author; (2) the Waac aide to the Director, 3d O Betty Bandel.

[5] Memo, DCofS for Somervell, 11 Oct 42. WDCSA 291.9 (9-11-42).

ceiving WAAC units will be agreeably surprised to find how much these units can contribute to the war effort.[6]

Also, in spite of the inconveniences of supply and climate, the women's morale remained high. Psychiatrists later remarked that the women seemed pleased with the idea that they were approximating the hardships of Army men. Whatever the cause, the trainees joked through their sniffles, laughed at their own appearance, and felt great pride in their new status and enthusiasm for the Corps' future. At night, wrapped in blankets, they sat on the floor in freezing unfurnished dayrooms to sing Army songs, including one written by a member of an early class which soon swept the training center:

All you fighting men, keep on fighting to win,
For the WAAC is in back of you;
If a plane you fly, keep it flying high
For the WAAC is in back of you.
Spread the news around that we're victory
bound,
And our hearts we pledge anew
That our flag shall wave o'er the home of the
brave,
And the WAAC is in back of you.[7]

The character and ability of the incoming recruits also continued to receive approval. By the end of September there were more than 3,000 Waacs at Des Moines, including 792 officers, with more arriving daily. Some 3,300 more had also been enrolled but could not yet be called to active duty for lack of training space.[8] The average ability of these recruits remained surprisingly higher than that of the Army or the civilian population. In the month of September, some 60 percent were found to be in the first two AGCT groups, I and II, having scores ordinarily required only of officers, and only 0.8 percent were in the fifth or lowest group.[9] Furthermore, every recruit possessed a civilian skill expected to be readily useful to the Army.

The Second WAAC Training Center

It had been known since summer that a second WAAC training center would be required to meet the expansion program directed by the General Staff. The only site offered the WAAC was that of Daytona Beach, Florida, where it was proposed to train women in leased city buildings, rather than to dispossess the occupants of any established military post. It was also believed that the climate would prevent suffering from the current lack of winter clothing for women.

However, soon after Colonel Faith first inspected the proposed facilities, Director Hobby appealed to the Services of Supply for any other location, stating that a military atmosphere and reasonable discipline could not be inculcated at any such establishment. The town itself was a resort area, with streets perpetually crowded with sail-

[6] Ltr, CofAdm Serv to Dir WAAC, 6 Oct 42. SPAAE 333.1. Such favorable impressions of Des Moines do not appear to have been exactly accidental. A staff note concerning other important visitors observes privately: "Des Moines will be prepared to . . . put on the same sort of show which has been arranged for General Grunert." Min, Stf Mtg, 6 Oct 42.

[7] By Ruby Jane Douglass. *WAAC Parodies*, pamphlet published by WAAC Publications Office, Ft. Des Moines, May 1943. Copy in OCMH.

[8] (1) Memo, Stat Br GS for CofS, 6 Oct 42. WDGSA 291.9. (2) Memo, Dir WAAC for CofAdm Serv, 28 Sep 42. SPWA 291.9 (9-28-42) E. Copy in SPWA 314.7 sec 3.

[9] Table III, in History of Military Training: WAAC/WAC Training, Army Service Forces, prepared by Maj. Lavinia L. Redd, WAC, 5 Feb 46. OCMH. Hereafter cited as ASF Hist of WAC Tng. An even higher average score in July was due to the relatively large number of officer candidates, but by September there were only 350 in this group.

ors, soldiers, and coastguardmen from nearby military stations. For barracks, it was proposed to house the Waac trainees in a number of scattered hotels and apartment houses, ranging in capacity from 37 to 600, as well as auto courts, inns, and villas; and a 6,000-woman tent camp was to be built in a cantonment area. For classrooms there were to be provided the Methodist, Baptist, and Presbyterian churches, a golf course, several garages and storerooms, the city auditorium, the Fifth Avenue Gown Shop, a theater, and other business houses. There were no recreation facilities other than the resort nightspots plentifully sprinkled among the buildings to be leased.[10]

In her appeal to the Services of Supply, Director Hobby asked instead for any regular Army post, preferably part of Fort Benjamin Harrison, which currently had vacant space.[11] This request was refused by the Services of Supply, the various facilities at Daytona were leased by the Army, and Colonel Faith was ordered to divide his staff and to activate the Second WAAC Training Center on 10 October, with recruits to arrive 1 December.

Both Army and WAAC cadre at Des Moines were divided as equally as possible, and efforts were made to duplicate courses of study, instructional materials, and the operation of staff sections. Over his written protest, Colonel Faith was assigned as commandant of the Second WAAC Training Center, with the promise that he could return to Des Moines in six months. A member of the Des Moines staff was promoted to be commandant at the First WAAC Training Center, but lasted only a few days after planning a series of measures considered discriminatory by the National Association for the Advancement of Colored People. A third commandant was then selected—Col. John Hoag, Field Artillery, who was to remain at the First and Fifth WAAC Training Centers for the next seven months.[12]

Aircraft Warning Service Units

The first enrolled women to reach the field were those of the Aircraft Warning Service, to which shipment of units began in early September. These units went out complete with officers and formally activated under Tables of Organization for operations and filter companies.[13] Within less than two months, nine operations companies and eighteen filter companies were in the field, located at First Fighter Command stations along the east coast—New York, Norfolk, Boston, Philadelphia, Portland, Albany, Harrisburg, Baltimore, Syracuse, Wilmington, Charleston, Jacksonville, and Miami.[14]

In preparing these units for the field, there became apparent in rapid succession the unexpected problems that these early WAAC field units were to face. Although a number of the units went to northern and eastern stations where winter clothing was already required for men, winter uniforms were not available at the training center, and the first units went

[10] (1) Min, Stf Mtg, 24 Aug 42. WAAC Daily Journal. (2) Ltr, Faith to Chief of Mil Hist, 18 May 50. OCMH.

[11] (1) Memo cited n. 8(2). (2) Min, Stf Mtgs, 24 Aug and 29 Sep 42. (3) Rpt, Daytona Beach, WAAC Housing files; (4) Memo, CG SOS for CofEngrs, 30 Nov 42. SPWA 600.1 (7-2-42) sec 1.

[12] (1) Memo, CofS for G-1, 15 Oct 42. WDCSA 291.9. (2) Interv with Actg Dir WAAC who received NAACP in Dir Hobby's absence, 9 Oct 48.

[13] T/O 35-37, WAAC Opns Co, AWS, 4 Jul 42; T/O 35-57, WAAC Filter Co, AWS, 4 Jul 42; Also series of Memos in AG 320.2 (6-15-42)(14) Const and Activ of WAAC Units.

[14] WDBPR Press Release, 26 Nov 42.

out in summer khaki. By October the supply of even this was exhausted, and whole units trained, graduated, and crossed the country by public carrier, costumed only in the short seersucker exercise dresses and bloomers. The long-curious citizens who lined the streets to see a Waac in uniform were reportedly thrown into a state which wavered between panic and hilarity. The problem soon became one of health as well as publicity, as Waacs in short-sleeved seersucker marched to work through the October weather of Maine and Massachusetts.[15]

It had been planned to have WAAC regional directors on hand to complete advance arrangements for the women, but this had not been possible because of the Services of Supply's continued delay in publishing authorization for these positions. In the total absence of published Army Regulations on the subject, the Air Forces' station commanders fended for the women as best they could in the unfamiliar matters of auxiliary supply, finance, and discipline. Luckily, few discharges, promotions, or other personnel actions were yet required. It was supposedly understood that the companies were assigned to the Services of Supply and attached to the Air Forces for duty; it became clear at any rate that the Air Forces were firmly attached to the units.

The women were assigned at once, a few as clerical help in the wing headquarters offices, but the majority to staff the filter boards and operations centers on which they had worked before enlistment. Here they received telephone calls from civilian spotters as to aircraft sighted, plotted the information on the boards, and traced the path of every aircraft to cross the area, for the benefit of liaison officers from the various services who identified them or sent up interceptors. Some filter boards had as many as fifteen or twenty positions, each filled by a woman wearing headphones and enduring endless boredom while waiting for the scarce telephone calls. Company commanders argued ceaselessly to keep the women assured of the importance of the work, since an unfilled position meant an uncovered area of several square miles into which an enemy plane might penetrate before detection.

The Air Forces soon reported that the boards and centers showed a marked increase in efficiency when freed from reliance upon the uncertain attendance of part-time volunteers. It also appeared that the use of Waacs might be more efficient than the use of soldiers in duties of such a monotonous, sedentary nature. Therefore, the Air Forces announced that Waacs would be requested to staff literally hundreds more Aircraft Warning Service stations to be established on both coasts and eventually inland. Other units were requested for regular air bases, and a new type of company, the WAAC Headquarters Company, AAF, was devised to fit the need. After less than three months' experience with the employment of Waacs, the Air Forces in November approached WAAC Headquarters with a discussion of the possibility of obtaining 540,000 more and of giving them full Army status.[16]

In spite of their efficiency, the units encountered a series of difficulties which em-

[15] Unless otherwise specified, statements in this section are from interviews with former Staff Directors, First and Third Service Commands, who had many of the units and who were charged with correcting the conditions described, which had occurred before their arrival.

[16] Memo, CofS for Gen Stratemeyer, 27 Nov 42. WDCSA 291.9.

FILTER BOARD, MITCHEL FIELD, NEW YORK. *The Waacs, wearing headphones, plotted the path of every aircraft that was called in by civilian aircraft spotters.*

phasized the need for regional directors to get to the field. Problems began when the Air Forces found themselves unable to recruit sufficient local women to fill the enlisted vacancies, although officers had been easily obtainable. To fill out the units, WAAC Headquarters was obliged to divert to the Aircraft Warning Service many women who had been recruited elsewhere for general service, although such women were seldom very happy about loss of the expected work on a military post. Since these nonlocal recruits could not live at home, government quarters had to be hastily provided. No sooner was this action under way than it was found that Army Regulations forbade any member of a unit to live at home if government quarters were provided. Many local women with husbands and young children had been enlisted on the promise of being able to live at home; the only solution in such cases was to offer immediate honorable discharge, which a number of newly trained women accepted.[17] Enlistment regulations were amended to prevent the acceptance of any more women with dependent children.

Since many of these centers were located in the poorer areas of large cities, the

[17] Ltr, WAAC Opns Co, Wilmington, N. C., to Wilmington Air Def Region, 19 Dec 42, and Inds. SPWA 620.

leased hotels sometimes proved to be, in the best phrasing of the matter, third-rate. Women living in such quarters were not only frequently insulted and accosted, but were at times endangered by drunken invaders who pounded on hotel doors or crawled up fire escapes and into windows. At several stations the surrounding slum areas were so dangerous that women were advised not to leave their quarters except in groups. The effect of such housing upon the women themselves was not serious, since they proved far from helpless to meet their own problems. One commander later recalled with pleasure, "When a man climbed the fire escape at the ——— hotel, he was bopped by a Waac at the top."

The effect upon civilian opinion was a different matter; combined with the appearance of the women's clothing, it appeared sufficient to bring the whole WAAC recruiting program to a halt in the cities concerned. Director Hobby eventually besought the Engineers to avoid acquiring WAAC housing previously used for "questionable purposes." [18] In a few cities, hotel accommodations were excellent. In Philadelphia, one of the best hotels was secured for WAAC housing, but was lost at once to the WAVES, since the Navy was able to pay more per enlisted person.[19]

At all stations proper food for the women was frequently lacking, since members had been expected to eat at home and units contained no mess personnel. On moving to government quarters, women were therefore given the Army ration allowance of about $1.25 a day—a sum that might have been adequate in some localities, but was found by those stationed in large cities to be scarcely enough to purchase one nutritious meal in city restaurants.[20]

The impact of enlisted men's opinion was another problem that had not been fully anticipated. One company commander reported:

> My women came of good local families, and as civilian volunteers had all been respectfully treated . . . now every man assumed that their uniform gave him the right to insult them. They were actually afraid to ride up in the elevators with the enlisted men because of the language they heard. . . . I went to the commanding officer but he didn't do anything; I heard that the men's actions were only reflections of his private remarks to his staff.[21]

Also, because of the Air Forces' inability to recruit women to staff the stations completely, it was necessary temporarily to retain part of the volunteer civilian workers. Many of these, although for various reasons not desiring to enlist, wished to take part in some form of patriotic activity, and were not pleased at the prospect of being eventually ousted by Waacs. Prominent women in several communities caused Congressional pressure to be placed upon the Air Forces to withdraw the Waacs. Where working groups were mixed, it quickly became clear that Waacs were in no position to compete with "free" labor which could quit at any time. Waacs were almost invariably given the

[18] Memo, WAAC Hq for Reqmts Div ASF, 18 Jul 43. SPWA 600.12 (1942).

[19] Passages on AWS, except as otherwise documented, are based on: (1) The WAC Program in the Army Air Forces, by Lt Col Betty Bandel, hereafter cited as AAF WAC Hist. (See Bibliographical Note.) (2) Intervs with WAAC Stf Dirs, 1st and 3d SvCs, and with company officers of Jacksonville, Philadelphia, Charleston, and Norfolk.

[20] Ltr, Stf Dir 1st SvC to Lt Betty Bandel, 2 Dec 42. SPWA 319.2 WAAC Stf Mtg, WAAC Planning files.

[21] Intervs with Maj Caroline Essex, formerly of Norfolk unit, Capt Caroline Varn and Maj Madge Williams, formerly of Jacksonville company, and Maj Mary Anderson, formerly in Charleston and Philadelphia.

graveyard shift, or the least desirable hours, while supervisory positions usually went to civilians and the least interesting work to the Waacs. A few civilian women were as insulting in their remarks as had been the enlisted men.

Under the circumstances, it was considered remarkable that few applications for discharge were received. The women remained in generally high spirits, although expressing considerable indignation against a vague WAAC Headquarters, which they regarded as all-powerful, because of the absence of published regulations, regional directors, necessary supplies, and attention to other obvious needs.

Appeals for Publication of Command Channels

The same months of September and October had seen a final struggle waged in WAAC Headquarters to get WAAC command channels published and WAAC regional directors sent out. In the months since the passage of the WAAC bill in May, the Services of Supply had repeatedly refused to take any such action. Its reasons for this stand were never to be made entirely clear. Only what staff members called "mysterious silence" had met their repeated attempts, throughout the summer, to clarify the power and responsibilities of WAAC Headquarters.

The Director's appeals had begun in July, even before the first Waacs had reached Des Moines. At this time she first began to be apprehensive about the forthcoming winter clothing shortage, as a result of the action of Requirements Division, Services of Supply, in refusing to approve signing of contracts for winter clothing. Although General Marshall himself had personally approved every item on the list, the Services of Supply took issue with some of his decisions. Appeals from both the Director and The Quartermaster General failed to get approval of contracts until winter was at hand and the September supply scandal unavoidable.

The Director believed this problem to be only one facet of the larger issue: the need for publication of some directive on the powers of her office and its relation to other Services of Supply staff sections. A parallel problem that she cited was the action of The Surgeon General in calling a meeting of civilian doctors, at which the WAAC was not represented, which formulated and came close to announcing to the public a plan for issuance of prophylactics and contraceptives to the women to be enlisted. This plan was suppressed only when Mrs. Hobby went directly to the Deputy Chief of Staff, General McNarney.

In August, after a series of such incidents, the Director appealed in writing to General Marshall to clarify her responsibility in relation to Services of Supply offices, alleging that the current lack of co-ordination was resulting in "serious jeopardy of the military and civilian acceptance of the whole idea of the Corps." [22] She recommended that:

> (a) all papers dealing directly and specifically with WAAC matters be routed through the Director WAAC for comment or concurrence.
>
> (b) . . . clarification [be made] as to final authority for decision in cases of non-concurrence. . . .
>
> (c) no meeting of civilian agencies or individuals be called . . . without prior approval of the Director.[23]

General Marshall did not receive the appeal addressed to him. The Services of

[22] Memo, Dir WAAC for CofS, 4 Aug 42. Copy (partial) in SGO files, SPMC 322.5-1 (Wash, D.C.)F; Copy in Adm Serv ASF Sp Coll DRB AGO.

[23] *Ibid.*

Supply returned it to Director Hobby in August with the statement that there would be no change in "established practice."[24]

The same reception was given to the Director's request that the Services of Supply approve publication of the WAAC command system with its authorization for regional directors. The necessary circular was prepared by WAAC Headquarters and forwarded for approval in midsummer, but no publication resulted. It was rumored that the Services of Supply was not in accord with either the circular or the WAAC Regulations on which it was based, although previously these had been approved by General Marshall. In August, without notice to WAAC Headquarters, the Services of Supply published a directive that contradicted these regulations. In a section of its new Organization Manual, it gave the commanding generals of service commands the authority, vested in the Director by WAAC Regulations, to transfer and assign WAAC personnel; this action could be taken without reference to WAAC Headquarters. The Manual also, without advance notice, removed the training center at Des Moines from under the Director's command and placed it under the Seventh Service Command.

This publication thoroughly puzzled the Director's staff, since by removing command from the Director it contradicted, but did not rescind, Circular 169 and WAAC Regulations. It came as a surprise to the Air Forces and Signal Corps, who liked SOS control even less than the system of command by WAAC regional directors. The Director's staff appealed to General Grunert for clarification, but was informed only that the Manual meant what it said.[25]

Director Hobby therefore requested her Army staff officers to explain the situation to her and to formulate some plan of operation that would comply with all directives. The senior staff member, Colonel William Pearson, had just been removed from the office after a dispute with The Adjutant General over the custody of WAAC files, which were also removed from the office, over its protests.[26] The new Executive, Colonel Tasker, stated that he was unable to devise any military method of reconciling the directives, since it appeared impossible for the Director to exercise those remaining portions of her command responsibility concerning discipline, discharge, and promotion while lacking that portion concerning assignment and transfer. Even to locate the women, if they were transferred by other agencies, it would be necessary to go through some four echelons to reach post headquarters companies and possibly seven to reach certain Air Forces units. The time factor rendered this system impossible, since evidently more than a month would elapse before any action could be secured through so many echelons. On the other hand, unless the cases were sent to WAAC Headquarters, it would never be possible to expand the Tentative Regulations, as had been intended, on the basis of experience.

The staff therefore informed the Direc-

---

[24] 1st Ind, SGO for CofAdm Serv, 8 Aug 42, to Memo cited n. 22. SPMCR.

[25] (1) SOS Orgn Manual, Aug 42, sec 405.04. (2) Memo, Dir WAAC for CofAdm Serv, 4 Aug 42. SPWA 300.3 (4-29-42) sec 2. (3) Memo, CofAdm Serv for Dir WAAC, 11 Aug 42. SPWA 300 (8-8-42).

[26] (1) Office Memo, 8 Aug 42. SPWA 300. (2) AGO Memo 111, 7 Aug 42. SPWA 314.7 (6-2-41)(1) sec 3. This states that all file papers on the WAAC, regardless of where prepared, will be sent to file in AGO Civilian Components Branch. Col Pearson had maintained that the WAAC needed its files in the same building for planning purposes.

tor that in coping with current emergencies it was still "violating all the above" by writing and telephoning directly to Colonel Faith. Staff members warned her that she was now in the position, feared since passage of the Rogers bill, of bearing legal responsibility without authority; the Services of Supply had evidently removed not only policy but command powers, while leaving her legally designated as the women's commander.[27]

New Expansion Plans

The WAAC staff's alarm over the Director's position was increased by the fact that, meanwhile, a series of directives from the General Staff had made the June expansion plans obsolete and posed problems of administration that would have been difficult to solve under even the clearest command system.

The first such directive called for overseas shipment. The Director had intended that no units be sent overseas for "the experimental period of a year," since supply and supervision would be proportionately more difficult overseas.[28] However, in August word was received from Europe that General Marshall, on tour, had promised an allotment of Waacs to General Eisenhower. The Joint Chiefs of Staff informed the Director briefly that "Eisenhower's headquarters is badly in need of clerical help and Waacs seem to be the best answer." [29] The Director at once dispatched Major Macy to England to discover how WAAC supply and its vague administration could be stretched so far. As the first few classes graduated at Des Moines, the training center was charged with organizing and equipping two overseas units, and much of the scarce winter equipment so far received was held for their use in case the momentarily expected shipment call should come from the General Staff. The Director was informed that dozens more overseas units would be wanted shortly.[30]

The problem of expansion overseas paled into insignificance when, on 15 September 1942, G-3 Division dropped a veritable blockbuster on WAAC Headquarters—a directive that immediate steps be taken to plan for an expansion of the Corps to a strength of well over a million members.

As the Chief of Staff had predicted a year earlier, the nation was now in the grip of a manpower shortage. Industry had already turned to womanpower as a supplement; in one aircraft company the employment of women had increased by 2,575 percent.[31] Not to be left behind in the developing competition, G-3 Division stated:

1. Tentative mobilization plans projected through 1946 indicate a necessity for a materially increased use of the Women's Army Auxiliary Corps in the overall structure of the Army.

2. The widespread character of current and projected operations is well known. One major and very serious effect of such a scheme of operations has been to force the design of an Army in which the ratio of personnel in service and overhead installations to personnel in combat units is undesirably high. As the Army further increases in size, it is evident that unless this ratio is reduced, and unless sources other than the able-bodied manpower of this nation are used to provide

[27] (1) Min cited n. 10(1). (2) Memo, PRO for Dir WAAC, 9 Sep 42. SPWA 300.4.

[28] Memo, Tasker for Dir WAAC, 21 May 42. Tasker Staybacks.

[29] Memo, Secy JCS for Somervell, 3 Aug 42. Dir Adm Serv ASF, Sp Coll, DRB AGO.

[30] Memo, CofOpns SOS for TAG, 20 Sep 42. SPOPU 320.2 (9-20-42).

[31] "The Margin is Womanpower," *Fortune*, January 1943.

a major portion of this service and overhead, we face a situation where we have mobilized a tremendous force which is strong everywhere except on the battlefield.

3. It is requested that your Division [G–1] cause a study to be initiated to determine all the occupations in the Army which can satisfactorily be filled by members of the WAAC. Such study should contemplate their use at home and abroad in practically every field except the actual handling of weapons. They should constitute the bulk of personnel assigned to overhead installations in the Zone of the Interior. Extensive use in defense command installations appears logical.

4. As a basis for an approach to the solution of the problem, it is recommended that an expansion of the Corps from its present authorized strength to an ultimate strength of 1,500,000 be contemplated.[32]

The earlier June plans for an increase of 500 percent—to an ultimate 63,000 members—were thus obliterated after only three months of existence. Complete new plans were obviously required: new schools, new recruiting quotas, new clothing procurement, and new legislation. Director Hobby was convinced that the new quota could not be filled by voluntary recruiting but would require the drafting of women.[33] Such plans, and such an expansion in size, were clearly impossible without immediate agreement upon, and publication of, a system of command for the WAAC.

General Marshall's Intervention

Simultaneously with the arrival of G–3's directive, General Marshall sent his staff officers to confer with the Director's staff, at which time they learned of the contradictory existing instructions on Corps administration and command.[34] On 17 September the Chief of Staff intervened to direct General Somervell to restore the Director's command powers. Even had the WAAC been legally subject to military law, General Marshall was of the opinion that it would be dangerous to assume that no new Army-wide policies would have to be set. He stated:

I find that the WAAC has been fitted into the SOS somewhat on the same basis as the Military Police. For a new organization, particularly one composed entirely of women, I think that this is not an effective arrangement. While units assigned to various localities and theaters must come under the local commanders, yet it would seem to be important for some time to maintain a rather direct relationship between the Director and these highly special organizations. There is too much that is entirely new and that demands a woman's point of view to decentralize to the extent that we do with Infantry, Cavalry, and Field Artillery, as well as other special units.[35]

For his guidance in publishing necessary instructions, General Marshall sent General Somervell a copy of the regulations that had been prepared earlier by the Director's staff.

At the same time, General Marshall informed General Somervell that WAAC Headquarters must be furnished with more assistance to carry the new planning load. He noted that Mrs. Hobby looked "rather overworked" and that, while her assistants "are doing a good job," she needed a high-ranking officer with experience and prestige in the War Depart-

[32] (1) Memo, G–3 for G–1, 15 Sep 42. WDGCT 291.9 WAAC (9–15–42). (2) Memo, Opns Div SOS for Dir Mil Pers SOS, 19 Sep 43. SPAOG 201, copy in AG 320.2 (6–15–42)(14) Const and Activ of WAAC Units, with Incl G–3 Memo, G–1 paper, and request for study.

[33] WAAC Daily Journal, Vol. I, particularly Min, Mtg, 14 Jan 43.

[34] (1) Memo, Tasker for Dir WAAC, 31 Aug 42, sub: Results of Confs with Col Faith, 29 and 30 Aug 42. SPWA 334.8 Conf (8–21–42). (2) Memo, Tasker for Young, 17 Sep 42. SPWA 300.

[35] Memo, CofS for Somervell, 17 Sep 42. WDCSA 291.9 (9–17–42).

ment "to get things going on the most efficient basis and to relieve her of many time-consuming details and also to obviate many delays." Such an officer, he stated, "might make an important contribution as a sort of military secretary during the formative period of the Corps." [36]

This second portion of the Chief of Staff's instructions was the first on which General Somervell took action. Directing that "immediate action be initiated to assist and aid the WAAC in the execution of its duties," [37] he sent inspectors from his Control Division to repeat the inspection they had just made in August, at which time little had been reported out of place except the furniture.[38] Control Division now noted that "the WAAC occupies an unusual and peculiar position . . . it is both within and without the military framework." However, they vetoed the Chief of Staff's suggestion for a "military secretary" and proposed instead an executive, who would be in the chain of command instead of an adviser, for which "a Regular U.S. Army Officer is essential . . . to guide the Director in Army methods." [39]

General Marshall had called for an officer "conversant with War Department procedure," but such an officer was not found available and the Director was assigned a retired Regular Army officer, Col. Thomas B. Catron, who had not served in the War Department in twelve years. Colonel Catron had, however, thirty-seven years' experience in the Regular Army, and was personally known to General Grunert's deputy, Brig. Gen. Madison Pearson, WAAC Headquarters' immediate superior.[40] The Services of Supply also regrouped the six major divisions of the Director's office into two—Personnel and Operations-Training. Both divisions were to be headed by newly assigned Regular Army officers since, Control Division noted, "The War Department and the Services of Supply must rely on the data furnished . . . . so it should be reliable." [41] WAAC Headquarters' Supply Division was abolished, since "it is apparent that the functions of a supply division are not of sufficient importance to justify a separate division." [42]

General Somervell approved the reorganization pattern, saying, "This is splendid. . . . With General Grunert personally assisting, I think you are on the right track for future progress." [43] Colonel Catron replaced Colonel Tasker as Executive; Colonel Tasker and other senior members of the former staff remained for a few more months, while new Regular Army officers gradually arrived. The junior Army members had already begun to depart as new contingents of WAAC officers reported to take over routine duties.[44]

Under Colonel Catron's direction, the office adopted the motto: "Simplify, Qualify, Consolidate." Colonel Catron also directed that the staff begin to use

[36] *Ibid.*

[37] Rpt, Col Betters, Contl Div SOS, 7 Oct 42. SPWA 020 (1943).

[38] Memo, Chief, Contl Div SOS for CofAdm Serv, 1 Aug 42. SPWA 317.7 (6-2-42) sec 3. The Director's Office then had 14 Army officers and 50 civilians. There were four divisions, corresponding to G-1, G-3, G-4, and OPD of a general staff.

[39] Rpt cited n. 37.

[40] Personnel Abstracts, WAC file, OCMH.

[41] There were also three small branches: Office Management, Control and Inspection, and Technical Information.

[42] Rpt cited n. 37.

[43] Memo, CG ASF for Dir WAAC, 7 Oct 42. File SPICY, in Folder, WAC Recruiting, ASF, Sp Coll, DRB AGO.

[44] (1) Intervs with members of the group. (2) Min, 24 Aug, 17 and 21 Sep 42. WAAC Daily Journal. (3) Memo, Macy for Dir WAAC, 24 Sep 42. SPWA 352.0. (4) Mimeographed Tng Program, 11 Sep 42. SPWA 352.0 (8-24-42).

both sides of the stationery, and that they write or wire the training center instead of telephoning.[45] The training center was restored, by General Somervell's direction, to WAAC command, under which it had in fact operated all summer.

Some four weeks elapsed before the publication of the system of WAAC command prescribed by General Marshall. Meanwhile, Director Hobby was not informed that the Chief of Staff had already decided the issue, and her staff continued in some confusion, attempting to devise another system which would be acceptable to General Somervell, for the command of the WAAC units, now already in the field. In one paper the Director offered to use a British system of command, in which the Director retained assignment authority only over company officers and cadre, thus allowing Army commanders to transfer other women, yet insuring that transfer would not be made to stations where barracks and cadre were not located. When this suggestion received no response, she proposed an alternate command system wherein she would gradually give up all command power to commanding generals of service commands, provided that they refrained from exercising this power until civilian auxiliary rules could be published on the basis of cases sent directly to WAAC Headquarters.[46] This proposal also received no reply. The Director begged General Grunert's office for some published system, stating, "If the War Department and the public are to hold the Director of the WAAC responsible for the proper administration of the Corps, the proper authority to accompany this responsibility is a natural and automatic requirement."[47]

By the second week in October, there were already a dozen Aircraft Warning Service companies in the field and some two hundred officers in recruiting stations and Army specialist schools. On 11 October General McNarney, General Marshall's Deputy, sent a second message to General Somervell:

> On a recent trip to the West Coast, including a stop at Des Moines, I gained the distinct impression that there exists no basic plan or schedule for the use of the WAAC. At two stations, Randolph Field and Fort Huachuca, they had been informed that they would receive WAAC companies, but they seemed very hazy as to the trades, skills, qualifications to be expected, where they should and could be used, together with a distinct lack of any indication that the use of Waacs should result in a nearly corresponding decrease in the number of men assigned to an activity. Station commanders scheduled to receive Waacs should be furnished, without delay, sufficient information to permit effective utilization, and directives for a corresponding decrease in men assigned. . . .[48]

It was not until 13 October 1942 that the Services of Supply published an explanatory statement to the field—Circular 344, the second important War Department publication in WAAC history. The circular at last made WAAC command channels clear to the field, although the channels themselves were complicated.

---

[45] Min, Stf Conf, 27 Oct 42.

[46] (1) Memo, Civ Components Br AGO for AG WAAC, 18 Jul 42, and reply, 10 Aug 42, office notes atchd. WA 300.3 (4-29-42) sec 2. (2) Memo, Tasker for Dir WAAC, 5 Oct 42. SPWA 314.7 (6-2-41)(1) sec 3. (3) WAAC Cirs 4 (10 Jul 42), 13 (14 Oct 42), and 16 (11 Dec 42). Cases are cited in Memo, Dir WAAC for CofAdm Serv, 3 Oct 42. SPWA 300 (10-3-42).

[47] (1) Memo cited n. 8(2). (2) Memo, WAAC Hq for CofAdm Serv, 29 Sep 42. SPWA 300.3 (4-29-42) sec 2. (3) Memo cited n. 46(3). (4) These were all recalled on 5 October 1942, when the Director first learned of General Marshall's action, by Memo, Dir WAAC for Contl Div SOS, 5 Oct 42, sub: Revision of WD Cir 169. SPWA 300.3 (4-29-42).

[48] Memo cited n. 5.

Three sets of channels were required in order to maintain the Corps' legal auxiliary status while allowing it to work for, and be supplied by, the Army.

Job performance was rated through the same normal channels of command as for soldiers and civilians at a station, and so offered no problems or causes of complaint from any station—Air Forces, Signal Corps, service command, or other. *Supply and routine post administration* were also the responsibility of the post commander, just as they were for men. However, above post level only service command stations could follow normal channels of supply, since WAAC clothing was stocked only at a few service command depots. Non-service command stations had to follow unfamiliar channels of requisitioning through the service command to get WAAC supplies. Finally, in *personnel matters,* the short WAAC command channel would operate, from the WAAC company commander to the WAAC regional director to Director Hobby. Personnel channels were specifically defined as including the disputed right of assignment and transfer, as well as promotion, discipline, and discharge, plus "all policies involving the welfare of members of the WAAC which, by virtue of the fact that they are women, differ from those prescribed for men."

Assignment of Regional Directors

Upon the moment of publication of Circular 344, nine WAAC regional directors departed for their posts in a race to remedy the Corps' field situation. These nine women had been chosen in early October and brought into WAAC Headquarters for assignment to the nine service commands as soon as the Services of Supply published authority for them. Of necessity the choice was made in ignorance of military aptitude although, as Director Hobby later noted, "Every one [service commander] said, 'You must put your very best officer on this.' "[49] The women chosen were all required to be over 35 years of age. They included two former business administration supervisors, a college teacher and a dean of women, a sales director, a government executive, a lawyer, and a housewife. In spite of the haste, the selection proved generally good, and included four women who were later to play a leading part in world-wide WAC administration—Third Officers Katherine R. Goodwin of the First Service Command, Jessie P. Rice of the Third, Westray B. Boyce of the Fourth, and Mary-Agnes Brown of the Eighth.[50]

The ten-day indoctrination of these women in WAAC Headquarters understandably proved something of a farce. Detailed auxiliary regulations and guides to WAAC needs had not yet been worked out, and could not be until more Waacs reached the field and until the regional directors themselves discovered their needs. Members of the group related that Colonel Catron told them, "No one can tell you what to do, because there's never been a job like this. No one knows what you're going to encounter when you get there." [51]

Circular 344 gave the job a difficult dual status: when handling matters pertaining to the "Personnel Channel" the regional director was responsible only to Director Hobby and signed herself "Director's Representative" (later, "WAAC Service Command Director"); but in all

[49] Speech at WAAC Stf Dirs Conf, 15-17 Jun 43. SPWA 337 (6-1-43).

[50] WDBPR Press Release, 28 Oct 42.

[51] Related to author by Lt. Col. Jessie P. Rice and other members of the group.

other matters such as supply she was only a staff adviser to the commanding general of the service command and could take action only through military channels, signing herself "Chief, WAAC Branch." [52] The latter position was roughly comparable to that of service command special staff officers such as the surgeon, finance officer, or quartermaster. However, these officers were chiefly colonels and were expected to be experts in one field only, while the Waac was a second lieutenant and was supposed to be an adviser in all fields—health, supply, housing, personnel, discipline, finance, public relations, and all others. If there were no WAAC Regulations to cover a situation, she must determine whether Army Regulations could legally apply, or whether it would be necessary to recommend new WAAC Regulations. In addition to her work in headquarters, she must visit every service command station where Waacs were assigned or expected, as a traveling trouble shooter to resolve difficulties and promote efficiency. She was also responsible for the command and well-being of Waacs at airfields, ports, Signal Corps stations, and all other exempted activities within the service command, most of which were already bristling at the thought of control through alien channels.

Because of the alarming responsibilities of the positions, some thought was given to the possibility of promoting all service command directors at once to the rank that was allotted to the position: assistant director or lieutenant colonel. After much debate, Director Hobby decided to postpone promotions until each woman's efficiency on the job was proven. However, Acting Director Helen H. Woods some days later secured permission to send an explanatory letter to each commanding general, noting: "We were conscious of the fact that we were sending out second lieutenants to the field to break ground, and to confer with major generals concerning their relative command powers." [53]

Only one major policy was fully impressed upon the regional directors before their departure: they must as far as possible act as if the Corps were a part of the Army; they must demonstrate that women needed no extra frills and comforts such as Army commanders might feel a gallant desire to provide.[54]

Field Action

The initial reception accorded these women upon their arrival at service commands varied from "all the co-operation a person could ask or need" down through various degrees of lesser warmth to "standoffish." [55] One fact became abundantly clear the moment each staff officer reported, and that was that most of the service commands—in fact, eight of the nine—had no intention whatever of complying with War Department Circular 344 insofar as it pertained to a WAAC command channel. Said one director, "When I arrived at this station and was presented to the Commanding General, he informed me, with some emphasis, that he was in charge of the Service Command and did not propose to tolerate any dual channels; that my loyalty was to him and to no one else." Others told much the same story.

[52] AGO Memo S635, 4 Dec 45. SPWA 314.7 (6-2-41) sec 3.

[53] Interv with Mrs. Woods, 9 Oct 48.

[54] Undated Draft of Dir WAAC speech to SvC mtg, Dec 42, in personal papers of Col Rice.

[55] Ltrs from SvC Dir, 30 Nov–9 Dec 42. SPWA 319.2 (12-2-42) WAAC Stf Mtg, WAAC Planning files.

LT. COL. HELEN H. WOODS (*then Third Officer*).

The four service command directors on the Atlantic seaboard plunged at once into the work of aiding the Aircraft Warning Service units to obtain supplies, warm uniforms, three meals a day, and safer quarters, as well as of providing a legal means for the personnel actions that began to be necessary. Particular difficulty was experienced in persuading non-service command stations to permit such efforts. Some measure of success was eventually obtained. For example, after repeated telephone calls and correspondence had failed to convince the Norfolk Air Defense Wing, Third Officer Jessie P. Rice of the Third Service Command paid a personal visit and reported a "long and pleasant" conversation with the commanding officer, during which he alleged that he had never seen War Department Circular 344, but agreed to abide by it and WAAC Regulations in the future.[56]

In stations at which units had not yet arrived, service command directors set themselves to prevent the hardships and poor publicity that unpreparedness had brought to the Aircraft Warning Service units. One surviving memorandum on "How to Visit the Field" offered strenuous advice:

1. Visit the commanding officer and explain the WAAC's status.

2. Visit the quartermaster and check with him methods of requisitioning WAAC uniforms and other supplies and equipment peculiar to females; where stocked, and so on.

3. Visit the post engineer and verify safety and proper construction of WAAC barracks, including plumbing, fire escapes, shades, and distance from men's barracks.

4. Visit the post surgeon and arrange for separate sick call and separate wards for women, different medical supplies, and the like.

5. Visit the finance officer and explain how the Auxiliary is paid.

6. Visit the post exchange officer and suggest toilet articles, sanitary supplies and other items to be stocked for women purchasers.

7. Visit the athletic and recreation officer and arrange for recreation suitable for women, and for admission to post theater, service club, and other activities.

8. Visit the provost marshal and discuss the Auxiliary disciplinary system and its wide difference from the Army system, the power of military police over Waacs, and related matters.

[56] (1) Ltr, Stf Dir 3d SvC to 3d O Betty Bandel, 1 Dec 42. SPWA 319.12. (2) Ltr, Stf Dir 3d SvC to Dir WAAC, 1 Jan 43, sub: Rpt on Trip to Norfolk, Va., 27 Dec 42. Same file.

9. Visit the post chaplain and discuss the special problems of a woman's adjustment to the Army.

10. Visit the post adjutant and advise that WAAC officers be allowed to use the officers' mess and club, and discuss other matters of administration.

11. Visit any other staff officers who might have special problems caused by the arrival of female personnel, such as the public relations officer, or those who might employ Waacs, such as the signal officer.[57]

Director Overseas

A few days before the regional directors went out, Director Hobby was obliged to entrust their further indoctrination and guidance to her staff, having received orders to accompany Mrs. Eleanor Roosevelt on an observation tour of the British women's services. Third Officer Helen Woods was left as Acting Director. On 19 October the Director and her aide, Third Officer Bandel, left for the British Isles, not to return until 11 November.[58]

At this time it appeared that the Corps was in fair condition to meet the strain of the oncoming expansion. As a result of General McNarney's intervention to expedite WAAC supply, every Waac at Fort Des Moines at last had a winter uniform and an overcoat, and supplies had been forwarded to those women who had left the training center without uniforms. As for the Second WAAC Training Center at Daytona Beach, a full conference of Services of Supply agencies assured Director Hobby before her departure that all necessary supplies and uniforms would reach it before the 1 December opening date. Also encouraging was the office's move to the newly completed Pentagon, where co-ordination with other Army offices was greatly simplified.[59]

The expansion program that had been mastered was, however, merely that which had originated in June. Still to be reckoned with was G–3's September plan for 1,500,000 Waacs.

[57] Table V–2, Vol. Gen Information, WAAC Hist files.

[58] Memo cited n. 5.

[59] (1) Min, Stf Conf, 27 Oct 42. (2) Weekly Rpt, WAAC Hq for CofAdm Serv, 1–7 Nov 42. SPWA 319.12. (3) Admin Memo 56, Hq SOS, 14 Oct 42. SPWA 310.2.

CHAPTER V

November 1942–January 1943: Plans for a Million Waacs

While Director Hobby was overseas, the necessary steps were taken by the Services of Supply to carry out the September expansion plan. To verify G–3's estimate of a need for 1,500,000 Waacs, The Adjutant General's Office was asked to make a list of all the Army jobs suitable for women. A committee of classification experts at once began the project, the most comprehensive study ever made of the outer limits to which replacement of men in the United States Army could theoretically be pushed.

The Adjutant General's Estimates

After pondering over each of the 628 military occupations listed by the Army, The Adjutant General's committee came to the conclusion that 406 were suitable for women and only 222 unsuitable. A job was deemed unsuitable if it involved combat, if it required considerable physical strength, or if the working conditions or environment were improper for women. Jobs requiring long training time were also eliminated. Typical of the jobs ruled unsuitable were brake inspector and railway operator (too heavy and requiring too much training time), switchboard installer (too heavy), laborer (too heavy), intelligence specialties (improper environment), and so on. All supervisory jobs were automatically declared unsuitable for women. The committee likewise, after some disagreement, ruled that for psychological reasons women should not be used in such jobs as classification specialist, personnel consultant, or psychological assistant, since in these jobs they might be called upon to classify recruits for combat duty, and it was felt that men would resent being classified as suitable for combat by individuals who were themselves noncombatants. The committee nevertheless admitted that, in an acute manpower shortage, women could actually be used in many of the 222 jobs thus ruled unsuitable.

The Adjutant General, using these lists, determined that in 1943 there would be 3,972,498 men in Army jobs appropriate for women—over half of the Army's strength at that date. In the interests of caution, this number was reduced to allow for the fact that some jobs, although of a suitable type, might be rendered unsuitable by their location or for some other reason. For instance, it would not be economical to maintain women at small stations where only a few could be used; it would not be wise to replace all of the supply clerks at any station, since work too heavy for women might occasionally occur; women could not conveniently do the office work in men's units. With such deductions, it was decided that women could fill 450,000 of the Army's 665,256 administrative and clerical jobs; 100,000 of the 834,600 motor transport jobs; 25,000 of the 121,000 radio operators' positions; 30,000 of 96,755 supply jobs. The total of women who could be used immediately was 750,000, and by the end of 1943 it was estimated that at least 1,323,400 could be used. Thus it was discovered that G-3 Division had, if anything, been conservative in estimating that 1,500,000 Waacs could be used by 1946.[1]

Estimates Based on Field Requisitions

Simultaneously with its request for The Adjutant General's theoretical estimates, the Services of Supply conducted a field survey to provide a practical check on the numbers usable by the Army. On 22 October, a few days after Director Hobby's departure, the Services of Supply informed its subordinate commands that further use of Waacs was contemplated, and asked for recommendations as to which jobs under their jurisdiction could be filled by WAAC personnel.[2]

Requisitions prompted by this letter were soon pouring into WAAC Headquarters, increasing those already arriving as a result of the earlier June expansion plans. They were listed in order of receipt, except for urgent cases, and forwarded for final approval in turn to the Chief of Administrative Services, the Chief of Operations, and G-3 Division of the War Department.

In order to determine which requests merited final approval, it was necessary to devise a number of new policies concerning the types of work and places of employment which would be authorized for Waacs. The number and variety of the requests went far beyond the four simple types of jobs originally authorized. The Air Forces requested Waacs for some thirty-one different technical and mechanical duties; the Signal Corps, Transportation Corps, Corps of Engineers, Ordnance Department, Chemical Warfare Service, and other agencies listed dozens of new types of work, many of which could not be filled by women except after Army technical training. There were also requests for new types of companies: Machine Records Units, Finance Department Companies, Base Post Offices, Mess Companies, and others for which no WAAC Table of Organization as yet existed. There were many requests for Waacs to replace civilians: as clerks in post exchanges, as laundresses, and in other non-Civil Service work. Several commands wanted mobile units for combat theaters, and the Signal Corps wanted these units mixed, with men and women in the same company.[3]

As a guide for determining which of these requests to honor, a most important policy had been recommended by Director Hobby before her departure: that Waacs not be allowed to replace civilians and that they be considered only for the 406 Army jobs included in The Adjutant

[1] (1) Memo, Classif & Repl Br AGO for Col Catron, 31 Oct 42; (2) Rpt, Classif & Repl Br AGO for WAAC Hqs, 25 Nov 42. SPWA 201.6 (10-2-42) (1) secs 1, 2.

[2] Memo, TAG for Chiefs of Servs SOS, 22 Oct 42. SPX 320.2 (10-17-42) PR-W-SPAAE. Also AGO Memo W635-2-42, 22 Oct 42.

[3] (1) Min of Stf Confs, Aug-Nov 42. Folder, Stf Confs, Hobby files. (2) Memo, MPD SOS for G-1, 4 Nov 42. SPGAM 320.2 WAAC (9-15-42). (3) Ltr, FA Sch to Dir WAAC, 25 Sep 42, and reply. AGF 320.2 WAAC, Binder 1, DRB AGO.

General's study. The Director stated that she had understood from the beginning that Waacs were to replace only enlisted men, except in the Aircraft Warning Service. General Grunert disagreed, recommending that stations be allowed to use Waacs in civilian jobs that they could not get civilians to accept.[4] G-1 Division upheld the Director, and the policy was established that *"First priority will be given to requests which result in WAAC personnel relieving enlisted men of the Army for combat duty."* [5] This was a policy of enormous importance to the Corps' entire future employment.

It was also decided that first priority was to be given to fixed installations in the United States and second priority to fixed installations in foreign theaters of operations. Mobile units were not, however, ruled out of consideration at some future date when all other needs were filled.[6]

No requests for Waacs to perform new or unusual military duties were disapproved, but it was obvious that actual provision of Waacs for such work would depend upon the establishment of new training schools and of types of units other than the standard Table of Organization company. Research was authorized to find some more flexible means of allotment which would permit Finance, Hospital, Mess, and other units.

The number of Waacs that could be used was greatly limited by a decision not to employ them in Washington, D. C., where as much as one third of the Navy women were eventually to be employed. This prohibition had originally applied to the Services of Supply only, having been announced by General Somervell at a staff meeting as an off-the-record policy for his Chiefs of Services, one of whom accidentally recorded it:

> Our policy can be only the policy that he has announced to us. That policy he particularly does not want committed to writing, because the minute that anybody gets ahold of it and puts it in the paper, there will be a mess. This is the policy . . . he does not want any of them here, except in the Office of the Headquarters of the WAAC, and certain ones in G-2, whom for some reason they prevailed upon him to have. Do you get it? We don't want anything said about it.[7]

However, a few weeks later, when it became clear that the War Department and the Air Forces were planning to bring Waacs into their headquarters, General Somervell recommended the policy in writing to the Chief of Staff, "in order that field agencies may have full benefit of the services of members of the WAAC." This policy was approved by the Chief of Staff and put into effect, with exceptions to be granted only by his own office for important secret work.[8] It was also generally agreeable to WAAC Headquarters, since assignments to Washington were not popular with enlisted women who wished to work on a military post. However, the policy caused added confusion in current planning, since agencies such as the Signal Corps, which had attempted to get WAAC officers to aid in planning for enlisted

[4] Memo, Dir WAAC for CofAdm Serv, 16 Sep 42, and Inds. SPWA 320.2 (9-5-42).

[5] D/F, G-1 to Dir WAAC through Adm Serv SOS, 1 Oct 42. AG/WDGAP 212.31 WAAC; 1st Ind, 2 Oct 42, Adm Serv SOS for Dir WAAC. SPAAE 320 (9-16-42).

[6] Memo cited n. 2.

[7] Tp Conv, Col H. L. P. King and Brig Gen James A. Code, Jr., OCSigO, 25 Sep 42, 11:05 A. M. SigC 320.2 WAC, 1942-45.

[8] (1) Memo, Somervell for CofS, 4 Oct 42, with approval 7 Oct 42. CofS 291.9. Also AGO Memo W635-1-42, 12 Oct 42. (2) Memo, Opns WAAC for CofAdm Serv, 28 Oct 42. SPWA 320.2 (10-28-42) O, in Dir Adm Serv ASF Sp Coll DRB AGO. (3) Memo, G-1 for CG AAF, 9 Dec 42. WDGAP 320.2 WAAC (12-9-42), in WDCSA 291.9 WAAC.

women, were obliged to relinquish them when informed that, "by order of General Somervell, no WAAC officers will be placed on duty in the space occupied by the Office of the Chief Signal Officer." [9]

Even with the various limitations, the results of the field survey generally upheld The Adjutant General's estimates. It was difficult to make an exact calculation of the constantly shifting requests, but the total, which had been 63,000 in midsummer and about 82,000 at the beginning of October, jumped to 386,267 toward the end of November and to 526,423 in December. These figures included only "actual requests . . . based on present tentative plans as to how and where Waacs can be used." [10] Thus, if more than 500,000 women could be assigned immediately, before overseas theaters and many other commands were heard from, it was clear that the G-3 estimate of 1,500,000 by 1946 had not been in excess of probable field requirements.

Proposals To Draft Women

Considerable change in planners' thinking resulted from the somewhat amazing revelation that, in a modern Army, possibly half of all "soldiers'" jobs were appropriate for women. With little natural limit thus existing upon the use of womanpower, major decisions were clearly required as to the Army's future policy: the number of women that it would attempt to employ in World War II, the rate at which it would take these, and the means by which it would attempt to get them.

The first and easiest solution to occur to most planners was that of drafting women. Although Director Hobby was not available for consultation, the Services of Supply on 4 November 1942 recommended to the War Department that, in furtherance of G-3's plans, legislation be sought to draft some 500,000 women yearly through regular induction facilities. According to SOS estimates, there were 25,605,179 women of ages 20 through 44 in the United States, of whom about 13,000,000 should be available for the war effort. Even if the WAAC required the present maximum Adjutant General's estimate of 1,300,000, this would constitute only 10 percent of the pool, leaving 90 percent for industry, farm, and other armed services. In recommending that the Army draft women, the Services of Supply noted:

> The nation is obligated to see that women will be placed to best advantage, not only from the standpoint of economical use, but to avoid any disastrous aftermath due to misplacement in the war effort, or by the employment of women who have definite obligation in other fields. . . . It is to the advantage of the nation that the best use be made of this pool of labor, and it certainly should not be frittered away in useless competition between competing agencies in our country. [11]

Unfortunately for the speed of WAAC planning, this proposal was so serious as to require extended War Department consideration and high-level negotiation with other agencies. It was realized that any such proposal from the Army would provoke considerable public outcry and Congressional opposition, and therefore could

[9] R&W Sheet, Col King to Aux C Sec through Deputy Chief and Exec Off, 24 Nov 42. SigC 320.2 WAC, 1942-45.

[10] (1) Memo, Dir WAAC for CofAdm Serv, 28 Sep 42. SPWA 291.9 (9-28-42) E, in SPWA 314.7 (6-2-41) sec 3. (2) Outline for speech "Role of the WAAC in the Service Command," December 1942; given to author by Lt Col Jess Rice.

[11] Stf Study, Dir Mil Pers SOS for G-1, 4 Nov 42. SPGAM 320.2 WAAC (9-15-42).

be publicly made, if at all, only with the full co-ordination of the sponsoring agencies. For this reason, as the winter of 1942 continued, no War Department reply as to the possibility of drafting women was received by either General Somervell or WAAC Headquarters.

The Weakness of the Auxiliary System

A second possibility which presented itself to planners was that of seeking legislation to give Waacs full military status. If a draft was later to be sought, such status was essential; even if it was not sought, there was already considerable evidence that no greatly expanded employment of womanpower was possible on an Auxiliary status.

In particular, it became clear during the autumn that most plans for overseas service must be canceled if the Auxiliary status continued. This was a serious blow, for the whole idea of the Corps had originally been thought of by the War Department as a means of preventing the confusion of civilian overseas service that occurred in World War I. During her visit to England, Director Hobby received from Lt. Gen. Dwight D. Eisenhower a request for immediate shipment of the two overseas companies then in training and for quick organization of a number of others. She also discovered—at that time top secret information—that the invasion of North Africa was imminent and that it was General Eisenhower's intent to take Waacs with him.

Director Hobby returned to the United States on 11 November, much impressed by the need for Waacs in General Eisenhower's headquarters, where she had observed a colonel typing a letter by the one-finger method. However, after conferring with her staff about current Comptroller General and Judge Advocate General decisions on the WAAC's legal status, her enthusiasm abated. It had recently been decided that Waacs were not eligible for the various benefits that protected men overseas—they could not, like the men, receive extra overseas pay; they were not eligible for government life insurance as men were, and would probably invalidate their civilian insurance by entering a war area; if sick or wounded they would not be entitled to veterans' hospitalization; if they were killed their parents could not collect the death gratuity; if captured by the enemy they could be treated as the enemy pleased instead of being entitled to the rights of prisoners of war. Under the circumstances, it appeared impossible for the Army to assume the responsibility of ordering women overseas; nor did the military situation yet require it, as the Director pointed out to the Chief of Staff:

> One of the primary reasons for the establishment of the WAAC was to replace men. Until necessity dictates otherwise, conditions in the theaters of operation indicate replacement in the United States.
>
> The Army classification system should be able to find among men the skills needed by the Army in the European Theater of Operations. Members of the WAAC can thus properly replace men found in the United States and release them for essential Army service overseas. . . .[12]

General Marshall therefore somewhat reluctantly postponed any consideration of immediate shipment overseas except for the two all-volunteer units already trained, saying: "They do not enjoy the same privileges that members of the Army do who

[12] Memo, Dir WAAC for CofS, 14 Nov 42. SPWA 320.2 sec 1.

become a prey to the hazards of ocean transport or bombing." [13]

In addition to negating plans for overseas shipment, it appeared possible that continued Auxiliary status would cause a rift with the Army Air Forces, which had been expected to employ more than 500,000 of the projected million-plus Waacs. Shortly after the Director's return from Europe, representatives of the Air Forces informed her that they could not employ Waacs on an Auxiliary status and would prefer to seek legislation for their own corps. General Marshall temporarily squelched this effort with a note to the Air Forces on 27 November:

> I believe Colonel Moore this morning took up with Mrs. Hobby the question of her attitude toward a separate women's organization for the Air Corps. I don't like the tone of this at all. I want to be told why they cannot train these women, why the present legal auxiliary status prevents such training.[14]

The attitude of the Signal Corps and other employing agencies continued to be very similar to that of the AAF.

It also began to appear that the Army would have difficulty in obtaining any large corps of women on an Auxiliary status in competition with the newly organized WAVES, who could offer women all military benefits—free mailing, government insurance, allotments to dependents, reinstatement rights to jobs, veterans' bonuses, and other advantages, none of which the Waacs were entitled to.

However, the granting of full Army status to women still appeared to many Army men as a drastic step, particularly in view of the fact that only a few units had as yet left the training centers. In late November it was finally decided to delay decision on both draft legislation and on Army status for the WAAC. The head of the Army's Legislative and Liaison Division, Brig. Gen. Wilton B. Persons, advised the General Council to postpone consideration of further legislation of any sort "until the Waacs have taken the field and demonstrated their usability." [15]

A few days later, Director Hobby submitted a substitute proposal believed to be more palatable to Congress: that of merely registering women in order to obtain accurate knowledge of the manpower potential of the nation according to numbers, skills, and other data. This recommendation was also approved by General Somervell and sent to the Chief of Staff on the first anniversary of Pearl Harbor.[16] However, it also was known to have small chance of immediate approval.

Proposals for Expansion on Auxiliary Status

There remained to planners the least desirable of the possible means of Corps expansion: to attempt to undertake G-3's directed program by voluntary recruiting and on an Auxiliary basis. A certain amount of action could be taken under existing legislation. It was therefore decided that, as a preliminary measure, the full 25,000 authorized by executive order should if possible be enrolled by the end of December 1942 instead of by the following June. Next, the Secretary of War petitioned the President to change the executive limitation from 25,000 to the full 150,000 authorized by Congress. The

[13] Memo, CofS for G-1, 3 Dec 42. CofS 291.9 WAAC.

[14] Memo, "G.C.M." for Gen Stratemeyer, 27 Nov 42. WDCSA 291.9 (11-27-42). Reference is to Col. Aubrey Moore, head of AAF's Allocations and Programs Division.

[15] Min, Gen Council, 23 Nov 42.

[16] Memo, Dir WAAC for CG SOS, 26 Nov 42. SPWA 324.71 (11-26-42) D, and Ind. Folder, WAC Recruiting, ASF Sp Coll DRB AGO.

President agreed and on 19 November 1942 raised the WAAC strength limit to 150,000. The Secretary of War indicated that the 150,000 women would be enrolled as fast as training facilities permitted, after which further legislation would probably be recommended.[17]

Meanwhile, recruiting quotas were opened as wide as training capacity would permit. From a beginning of 1,259 in July and 2,019 in August, the number of reported recruits had increased to 3,536 in September. In October and November, as Des Moines neared its maximum capacity, slightly over 4,000 women a month were enrolled. For December, estimated training capacity permitted advance quotas of approximately 6,000.[18]

In order to determine how fast the rest of the 150,000 could be enrolled, the General Staff called upon WAAC Headquarters for estimates as to the recruiting and training facilities that would be required to train 150,000 women by various possible dates, as well as greater numbers by later dates. The two WAAC training centers' combined peak capacity would plainly be insufficient to supply such numbers except over a period of years. The General Staff's desire was to take as many women as possible as rapidly as possible, since Army jobs were already waiting and since applicants might accept other war work if required to wait. Against this desire had to be balanced the practical question of where to get basic and specialist schools, of which the Army currently had none to spare. Compilation of a list of suitable schools was begun. It was also necessary to work out large flow charts to govern intake and output of proposed training centers and specialist schools, so that on a given date each would simultaneously produce exactly the required number of basics and specialists to make up exactly the number of WAAC companies departing for the field on that day, and for whom exactly the right amounts of housing and clothing were ready. Such charts had to be worked out for every possible total strength that the War Department desired to consider for every possible date.[19]

Opinions differed widely as to how quickly the WAAC should attempt to reach the 150,000 mark. Maj. Gen. Miller G. White, the new G–1 of the War Department, recommended to the General Staff in November that the Corps reach this number by July 1943—a deadline that would require at least 20,000 recruits a month for the intervening months, as well as several new training centers. An even more optimistic recommendation came from the new head of WAAC Headquarters' Recruiting Section, Capt. Harold A. Edlund, who advised a target of 50,000 a month, or 750,000 in 1943. Captain Edlund stated: "It may be that the goal outlined will appear to some as outside the realm of reason or possibility . . . but the reaching of a more modest goal is doubly assured if we bend every effort for the greater enrollment figure." On the contrary, General Surles, chief of the War Department Bureau of Public Relations, gave his opinion that not even 150,000

[17] (1) Ltr, SW to Bur of Budget, 12 Nov 42. WDGAP 320 WAAC (11–7–42), in WDCSA 291.9 WAAC. (2) Exec Order 9274, 19 Nov 42, in WD Bull 58, 28 Nov 42.

[18] (1) Weekly Rpt, WAAC Hq to CofAdm Serv, 8–16 Nov 42. SPWA 319.12 (11–9–42) sec 1. (2) Memo cited n. 10(1). (3) Table III, ASF Hist of WAC Tng. OCMH. (4) Although not reported fully at the time, actual nationwide enrollments as finally credited by The Adjutant General were: 1,259, July; 2,019, August; and 3,536, September.

[19] (1) Min, Stf Mtgs, 25 Sep, 6 Oct 42. (2) Flow Charts, WAAC Planning files, 1943.

could be obtained, and added: "We have about exhausted that group of women who are willing to volunteer to leave their homes." [20] This was also Director Hobby's opinion; after her return from Europe she expressed the opinion that not even 150,000 women could be recruited by voluntary means.[21]

The somewhat unfortunate effect of this division of opinion was that, for the remainder of 1942, the War Department was no more communicative upon the subject of WAAC 1943 goals than it had been upon the questions of legislation. As a staff member later observed:

> The Corps was placed in the position of a small businessman who overnight was told he must increase his business more than eight times, and to do it at once even before he knew what he was to produce, out of what materials, when or how the product was to be made, and with practically no organization to assist.[22]

The Winter Months

While it was realized that major decisions concerning womanpower could not be made easily in these, the darkest months of the war, their absence caused continued confusion in the new headquarters. At the same time, developments in the two training centers gave grounds for grave doubt as to whether any sort of expansion program could be supported.

In the second week of November, just before Director Hobby's return from England, urgent telegrams were received from the commandant at Fort Des Moines, pointing out that clothing shipments had again failed just as recruits were increased. The improvement caused by General McNarney's intervention in October had lasted scarcely a month.

Immediate investigation by WAAC Headquarters drew an alarming reply from The Quartermaster General: clothing promised to Fort Des Moines could not be delivered as scheduled. Also, no clothing could be sent to Daytona Beach by its opening date, as previously agreed by the Services of Supply. The Quartermaster General stated that his previous commitments had been contingent upon immediate letting of contracts and, for unexplained reasons, Requirements Division, Services of Supply, had again held up approval of the contracts for more than six weeks.[23]

On 11 November Director Hobby returned from England to find a recurrence of the September crisis at Des Moines: hundreds of recruits without winter clothing, and sickness rates again high. Colonel Hoag reported the entire supply of clothing now exhausted except for a little summer-weight underwear. The civilian clothing in which recruits arrived was far from adequate protection for four weeks of winter training in snow-covered Fort Des Moines. Most women had been instructed to bring only one outfit, and this, particularly in the case of women from the southern and southwestern states, frequently included nothing warmer than a dress, a light topcoat, and open-toed shoes.

---

[20] (1) Memo, CofS SOS for CofAdm Serv, 26 Nov 42. SPEX, in SPAA 320.2 WAAC, CofAdm Serv file, Sp Coll DRB AGO. (2) Surles statement in Min, Gen Council, 23 Nov 42. (3) Memo, Capt Edlund, Chief, WAAC Rctg Sec, for Dir WAAC, 22 Dec 42. Folder, Memos Prior to 7 Jan 43 in WAAC Planning, WAAC Planning files.

[21] Min of Speech, 31 Mar 43. SPWA 334.8 Conf, 1943.

[22] Miss B. Eugenia Lies, Civilian Consultant to Dir WAAC, The Overall WAAC Picture, (first draft), 28 Mar 43. Folder, Planning Serv Notes, WAAC Planning files, 1943 in SPWA 291.9.

[23] Weekly Rpts, WAAC Hq to CofAdm Serv, 1-7 Nov and 8-16 Nov 42. SPWA 319.12 (10-19-42) and (11-9-42) sec 1.

Units were now being shipped to the Aircraft Warning Service in the same civilian clothing in which the women had reported for training. It was known that trainees had written of these conditions to friends and relatives.[24]

Director Hobby's attempts at remedial action were somewhat delayed by the fact that, upon her return, she was immediately ordered to Fort Leavenworth, Kansas, to lecture to the Command and General Staff School on the proper administration of Waacs. She had brought back from England much valuable material on this subject, none of which had any effect on the situation. The fact was that the American WAAC was helplessly repeating every one of the British WAAC's mistakes, not because the WAAC did not know them and how to avoid them, but because their occurrence hinged on outside factors over which it had no control.[25] Nevertheless, three days after her return from England, the Director called in the two training center commandants, Colonels Faith and Hoag, for a meeting with Colonel Catron, Colonel Tasker, and herself. Here, in a stormy session, Colonel Faith alleged that the Services of Supply were neglecting the Waacs because they were not military personnel. He had particular reference to his experiences in trying to open the Second Training Center; for example:

Colonel Faith: I feel it is the function of this department [WAAC Headquarters] to get the War Department to understand that the Waacs are soldiers. It looks like I am going to have to open a training center at Daytona without anything. . . . I am not going to lie down because I do not have clothing, but I do sense the feeling that the War Department believes they have an Army, and some Waacs. . . .

Colonel Catron: No, the War Department feels that the Waacs ought to have what they need, but the clothing is not available. . . . I do not believe there is a tendency to say there is an Army and some Waacs; not in the Services of Supply. . . .

Lt. Colonel Tasker: I think there was that condition. . . .

[Colonel Faith goes on to describe how The Services of Supply refused him other equipment normally furnished to Army schools.]

Director Hobby: . . . Let me straighten out the position of this headquarters. You are given a job to do and a time schedule to do it on, and when your equipment does not come through I want to know about it. I will say another thing: you will find a lot of people with a bureaucratic frame of mind that would kill anything new unless you fought for it. It is my job to fight for it.[26]

A number of events therefore occurred on 21 November. On that day the Director stopped all further shipments of uniformless units to the Aircraft Warning Service. She simultaneously directed recruiters to warn recruits to bring enough warm civilian clothing to last for several months. She also requested and received from her Army staff a report on the action they had taken to date. This report showed not only that all Services of Supply staff divisions had concurred before expansion plans had been put into effect, but also that the Quartermaster General's Office had been informed by telephone in June and again in September as soon as WAAC Headquarters had itself learned of the plans. With this report in her office, and also General Madison Pearson, she therefore telephoned General Somervell's Chief of Staff, Maj. Gen. Wilhelm D. Styer.[27]

---

[24] (1) *Ibid.*, and Rpt of 15–21 Nov 42. (2) Account in the possession of Capt. Charlotte Dyer, made available to the author, 31 January 1946.

[25] Folder of British materials in 1942 WAAC files.

[26] Min, Tng Cen Comdts Conf, 14 Nov 42. Folder, Stf Confs, Hobby files.

[27] (1) Ltr, WAAC Hq to all SvCs, 21 Nov 42. SPWA 421 (11–19–42). Also Rpt, 15–21 Nov 42, cited n. 24(1). (2) Memo, Tasker for Dir WAAC, 21 Nov 42. WA 421 (11–20–42).

What was said was not recorded, but observers reported that it was sufficient to bring General Styer down the hall to WAAC Headquarters. Staff members were under the impression that Director Hobby mentioned that her own resignation would be necessary if the Services of Supply did not intend to give the WAAC the equipment for survival in the coming winter. In any event, General Styer was found to be not in agreement with the previous management of the WAAC supply program by his subordinates, which he attempted to correct by personal telephone calls on the spot. Emergency procurement of any available kind of cold-weather garments was authorized.[28]

The situation at Daytona Beach also was found to be critical. Before Director Hobby's return, a meeting had been held in General Grunert's office to consider whether to cancel the scheduled opening of the second training center, since it could not be supplied with uniforms. Lieutenant Woods personally made the decision that the Army's stated needs must be met, in spite of the inconveniences to trainees. The Second WAAC Training Center therefore opened on 1 December 1942 as scheduled. It was virtually without equipment; shoes had been sent, and 600 utility coats—scarcely enough for one week's trainees.[29]

Colonel Faith forwarded requests for improvement in the scattered housing and other facilities at Daytona Beach. A British report brought back by Director Hobby noted that to house women in scattered billets usually damaged health, administration, and the "amenities," and that "for young recruits undergoing basic training it is generally very unsuitable." [30] It was obvious that Waacs in Daytona Beach must either do without recreation or attend civilian night spots of doubtful respectability. Since a tent camp was being built to house some 6,000 of the trainees, Director Hobby asked in early December that it include the recreational facilities normally provided for an Army camp of 6,000 men—one theater, one service club, one library, and one post exchange. However, the Chief of Engineers objected on the grounds that critical building material would have been required, and the request was rejected by Requirements Division, Services of Supply. Colonel Faith also believed the commercial laundry facilities inadequate, but his requests for military supplements were refused by the Services of Supply.[31]

Colonel Faith's most serious concern, however, was not the lack of material facilities so much as the difficulty in getting competent Army cadre to augment the staff, which remained inadequate even after the original Army cadre was divided and the newly graduated Waacs assigned to all possible duties.[32] The greatest need was for highly competent key Army offi-

[28] (1) Interv with Col Woods, 9 Oct 48. (2) Rpt, 15–21 Nov 42, cited n. 24(1). (3) Memo, ACofS, Opns Div SOS, for QMG, 27 Nov 42. SPDDQ 420 Clothing, in SPWA 421 (11-27-42) 1942. (4) 1st Ind, Reqmts Div SOS to QMG, 3 Dec 42. SPRMD 421 (11-26-42), in WA 400.34.

[29] (1) Memo, OQMG for CG SOS, 10 Nov 42, with 1st Ind, 13 Nov 42. SPQXA 420 WAAC, in SPWA 400.34 (3-2-42) sec 1. (2) Ltr, WAAC Hq to Ft Des Moines, 19 Nov 42; and Ltr, WAAC Hq to 2d Tng Cen, 20 Nov 42. SPWA 421.

[30] *Conditions in the Three Women's Services.*

[31] (1) Memo, WAAC Hq for Chief, Reqmts Div SOS, 17 Dec 42; refused by 1st Ind, 6 Jan 43. WA 600.1 (1942). (2) Memo, WAAC Hq for OQMG, 2 Nov 42, and Inds. WA 486.3.

[32] Waacs were made chiefs of sections (Motor Transport, Officers Candidate, Basic, and Administrative) on 6 February 1943 at Fort Des Moines; dates for other centers not preserved. On 23 April 1943 at Fort Des Moines, Waacs became commanders of training regiments, chiefs of training sections, mess officers, public relations officers, and the like. On 24 April 1943 Waacs were designated Chief of Plans and Training, Assistant Commandant, and Chief Public Relations Officer at Fort Des Moines. MS Hist of 1st WAAC Tng Cen.

cers. The Director's requests for certain individuals by name had been refused, and General Grunert informed her that the WAAC would be assigned only officers unfit for duty with male troops. At the November commandants' meeting, Colonel Faith and Colonel Catron came to open dispute over this decision, with Colonel Faith stating: "I absolutely refuse any reduction in the quality of material. We cannot go any lower. We cannot allow any more culls. . . . I don't want a one of them. I cannot make it any stronger." To this Colonel Catron replied, "You will find lots of good material in the over troop age group. I think you are going to have to take it." Director Hobby added, "My own limited service in the War Department has taught me that it is difficult to get quality for the WAAC. It seems to be against the theory." [33]

The seriousness of the objections to the Daytona site was not to become fully evident for several months, but it was sufficiently clear to Director Hobby, when she saw the center for the first time in December, to cause her to demand that the already scheduled Third WAAC Training Center be located on an Army post. Accordingly, she requested assignment of Fort Oglethorpe, Georgia, just outside of Chattanooga, Tennessee. Fort Oglethorpe was in a central location, which would make it convenient to draw trainees from most of the United States; it was also in a temperate climate. Nonconcurrences on its use for Waacs were offered by the Provost Marshal General, who maintained an internment camp there, and by the Commanding General, Army Ground Forces, one of whose mechanized regiments would also be forced to move. These objections were overridden by the Chief of Staff himself and Fort Oglethorpe was designated the Third WAAC Training Center. Col. Hobart Brown was selected to command it, and its activation date was set for 1 January 1943, with recruits to report in February.[34]

In order to co-ordinate the work of the three training centers and the others that might be expected, it was decided to group them into a WAAC Training Command, which would form an intermediate echelon between the training centers and WAAC Headquarters.[35] Colonel Faith was chosen to command the new agency, and was promoted to the rank of brigadier general—an action that created a somewhat peculiar military situation, since the head of the WAAC Training Command was himself immediately under the command of Director Hobby, who wore colonel's eagles. Nevertheless, the rank of brigadier general was obviously quite modest for the commander of three large posts. General Grunert accordingly proposed that action be taken to change the Director's relative rank to that of brigadier general, and the Director immediately asked that equal rank be given Colonel Catron. General Somervell shelved both proposals and directed that no action be taken, pending decision on his proposal to draft women, which would necessarily place the WAAC in the Army. If the Corps should be given Army status, its Director and regional directors would become staff advisers instead of commanders, and an upward adjustment of

[33] Min, Stf Conf, 14 Nov 42. Folder, Stf Confs, Hobby files.

[34] (1) Memo, G-3 for CofS, 15 Dec 42. WDGCT 291.2 WAAC (12-8-42), in OCS 291.9 WAAC. (2) Weekly Rpts, WAAC Hq for CofAdm Serv, 7-12 Dec and 14-19 Dec 42. SPWA 319.12 (11-9-42) sec 1.

[35] Request for establishment approved 17 Nov 42 by CofAdm Serv, according to notes for speech, "Role of WAAC in the SvC."

BRIG. GEN. DON C. FAITH *shortly after his promotion and assignment to command the WAAC Training Command.*

the allotted grades of female leaders as the Corps expanded would not be obligatory.[36]

Shipment of the First Post Headquarters Companies

For some time disgruntled station commanders had complained about the repeated postponement of arrival dates originally given them for WAAC post headquarters companies. Many, ignorant of the fact that all available Waacs had gone to staff the Daytona Beach training center, began to repeat the rumor that the Corps was a failure and would never materialize in the field. One WAAC service command director noted, "I spent most of my time apologizing to my General for the peculiar behavior of WAAC Headquarters." [37]

The necessity for staffing the Third WAAC Training Center with cadre again set back the shipment schedule of post headquarters companies. By Christmas of 1942 only two Army stations had received their promised shipment—Fort Sam Houston, Texas, and Fort Huachuca, Arizona.[38] These units went out well equipped with winter uniforms at the expense of those remaining in training centers, in compliance with the Director's ban on shipment in civilian clothing. They were also more fortunate than the pioneer Aircraft Warning Service companies, in that careful preparations had been made for them by the WAAC service command directors.

The women as a whole were consumed with eagerness to reach genuine Army stations, and both morale and discipline were good. Mrs. Woods noted later, "In spite of the difficulties topside, the spirit of the women was remarkable, especially the fervor with which they accepted all sorts of unexplained difficulties. There sprang up among them an unexcelled *esprit de corps*." [39]

The arrival of the first Waacs at Fort Sam Houston on 17 December was described by a WAAC spectator, who wrote:

> The consensus of opinion at Fort Sam Houston was that it would be unfair and unwise to parade a company upon the day of its arrival by troop train. In fact the remark was made: "They will not make a good impression while disheveled." You would have been

[36] (1) Memo, ACofS Pers Div SOS for CofAdm Serv, 23 Jan 43, SPAP 322.5 WAAC (1-18-43), ref to Memo, CofAdm Serv for Gen Dalton, 13 Jan 43; (2) Memo, Dir WAAC for MPD SOS, 14 Jan 43, and Inds, SPWA 210. All in ASF Sp Coll DRB AGO.

[37] Intervs with Stf Dirs, 1st and 3d SvCs.

[38] Date of shipment given as 2 Dec 42 for Ft Huachuca units, in MS Hist of 1st WAAC Tng Cen; WDBPR Press Release, 19 Dec 43, gives 15 Dec as date of shipment and 17 Dec as date of arrival.

[39] Interv with Mrs. Woods, 9 Oct 48.

proud, however, if you had been standing there when they got off the train—their noses were powdered, their shoes were shined, they showed that they were well-trained and well-disciplined. Before the last ones had got off the train, the cannon was fired for retreat. Like one woman they came to attention and saluted. . . . The salutes so impressed the photographers that they stopped taking pictures and faced the flag. . . . I am sorry that no pictures were taken at that time.[40]

The other unit, composed of Negro enrolled women with Negro officers, was likewise well received at Fort Huachuca, and housed in new barracks with hairdressing facilities, recreation room, and athletic area. Upon arrival the women found their bunks already made up and their first meal prepared.[41]

Shipment of post headquarters companies to the seven other service commands was postponed indefinitely to permit the opening of new training centers, but the defrauded commands set up such protest that certain exceptions had to be made. Just before Christmas Director Hobby called a meeting at New Orleans of WAAC directors from the service commands, to explain the situation and to point out that delay at this stage would result in tremendously greater supply when all training centers came into full production. Nevertheless, all appealed so earnestly for at least one unit per command that Director Hobby promised to send it, as token to the service commands that Waacs really did exist. This was done, and during the last of December 1942 companies were placed on orders for shipment to stations in the remaining seven service commands. Very little could yet be told of what their assignments or job performance would be, although WAAC Headquarters eagerly awaited such news.[42]

In December also two experimental antiaircraft units reached the field, their "secret" station being actually Bolling Field, within sight of the Pentagon. The Antiaircraft Artillery, as directed by General Marshall, began the task of determining which of its duties could be performed by women and with what degree of success.[43]

First Overseas Shipment

Considerably more complications attended efforts to get the first two units overseas in December. These units had been ready for shipment since September, and were by this date trained to a hair-trigger state of tension while waiting for a shipment priority from England, which had not arrived. To ease the crowding at Des Moines, they were shipped to Daytona Beach, as soon as its cadre arrived, for another period of training.[44] Their morale was scarcely improved by the constant turnover of officers. Since no one yet knew what qualities a women leader needed, there was a process of trial and error by which several officers were successively relieved from duty because of in-

---

[40] Ltr, Lt Charlee Kelly, Rctg Off, to Capt Mary-Agnes Brown, WAAC Stf Dir 8th SvC, as quoted in Ltr, Capt Brown to Dir WAAC, 9 Jan 43. SPWA 319.1.

[41] (1) Ft Huachuca file, SPWA 291.21. For difficulties experienced by the first Negro Wac units, see Ch. XXX, below. (2) WAAC Daily Journal, Vol. 1, 30 Jan 43.

[42] (1) Notes of Dir's speech to CGs of SvCs, New Orleans, La., 17–18 Dec 42. SPWA 334.8 Conf (11–18–42). (2) Weekly Rpts, WAAC Hq for CofAdm Serv, 14–19 Dec 42 and 27 Dec 42–1 Jan 43. SPWA 319.12 (1942).

[43] AG Ltr, WD 320.2 WAAC (12–23–42) PR-W-WPOPU, 24 Dec 42, sub: Activation of 150th and 151st WAAC Tech Cos, and 1st Ind from MDW, 17 May 43. AG 320.2 WAAC (5–10–43).

[44] Ltr, WAAC Hq to Comdt 2d WAAC Tng Cen, 31 Oct 42, sub: 75th and 76th Cos, and two Inds. SPWA 320.2 sec 1.

ON THE JOB AT FORT SAM HOUSTON, TEXAS. *Motor pool attendant, above; dental assistant in the post dental school, below.*

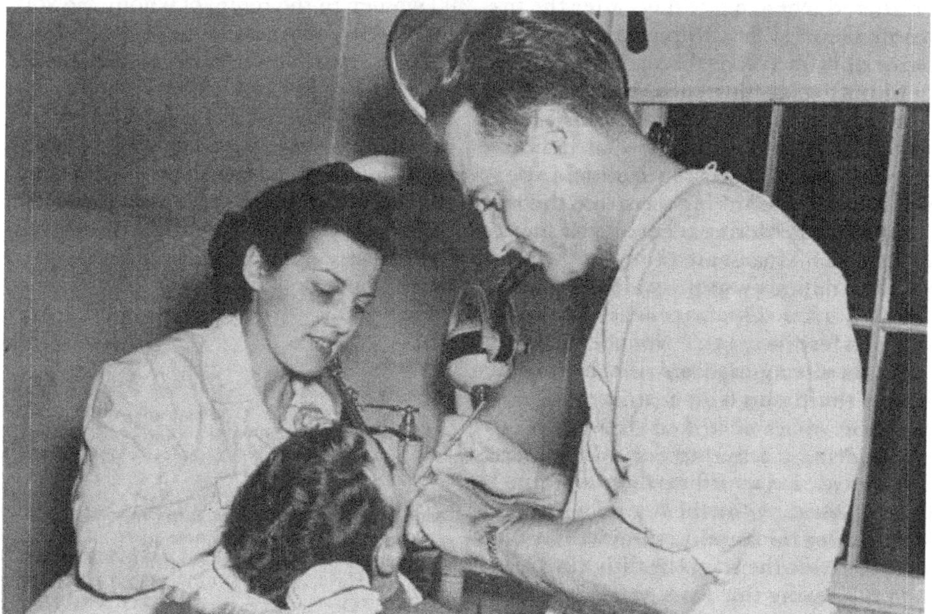

sufficient experience, lack of outstanding ability, or domineering traits and a habit of "treating adult women like high school freshmen." [45]

The prolonged delay could not be explained to the women, since even their destination, like all troop movements, was classified; in general, they attributed it to vacillation and bungling in WAAC Headquarters. Actually, shipment waited upon authority from the European theater, which had requested activation of the units but had not yet furnished shipment priority. At the same time, Operations Division of the General Staff informed WAAC Headquarters that shipment without a priority was impossible because of the shipping jams resulting from the invasion of North Africa in November.[46]

Disbandment of the units was being considered when, a few days after the invasion, General Eisenhower cabled for a company of WAAC typists and telephone operators for North Africa, and provided a high shipment priority. However, although the units earmarked for London contained the necessary specialists, Director Hobby felt unable to assume the responsibility of ordering them into the more active North African area without the protection of military status.

The danger was graphically demonstrated a few days later when, in answer to a cabled request from the European theater setting high air priority, the first five of the Waacs were flown to England—five officers qualified to act as executive secretaries. These officers were immediately placed by the theater on a ship for North Africa; General Eisenhower had in fact cabled for them on the fifth day of invasion. Nearing the coast, the ship carrying the Waacs was sunk by enemy action, with loss of all equipment. The five Waacs were fortunate enough to be rescued by a lifeboat and a British destroyer, but the scare was sufficient to emphasize the War Department's untenable position had the issue of hospitalization, capture, or death gratuity arisen publicly.

At this, Director Hobby herself flew to Daytona Beach and told the assembled women that their shipment must be either canceled or diverted to an unnamed but dangerous combat theater to which she would not order them against their wishes. She noted later that she reminded them of their lack of military status and its protection and "that I wanted them to appreciate the full significance of what I was saying." Then, in a speech which spectators called reminiscent of another Texan's at the Alamo, she called for volunteers to step forward to fill the unit. There were 300 women in the room, of whom 298 volunteered upon the instant. At this, Director Hobby was unable to continue speaking and hastily sought privacy in a broom closet.[47]

Half of this number were chosen, processed, and ordered to the staging area in December. The remainder were held until the European theater had three times more postponed its shipping priority, when they were disbanded and sent to other duties.[48]

[45] Documents dated 23 Oct 42 and 27 Nov 42. SPWA 320.2 sec 1.

[46] (1) D/F, OPD to G-1, 24 Nov 42. OPD 322.999 WAAC (11-14-42); (2) Cbls, London to AGWAR, 4416, 3 Nov 42, London to WAR, 5057, 19 Nov 42, and London to AGWAR, 5445, 4 Dec 42. All in SPWA 320.2 secs 1 and 4*a*.

[47] The two nonvolunteers had invalid parents. (1) Senate Com on Mil Affairs, 78th Cong, 1st sess, *Hearings on S 495*, 3 Feb 43. (2) Interv with Col Bandel.

[48] (1) Hist of WAAC-WAC in ETO, 1942-44, by Hist Sec, ETO. (Hereafter cited as ETO WAC Hist.) OCMH. (2) Ltr, WAAC Hq to Col Abbott, Hq ETO. 27 Jan 43. SPWA 320.2 sec 2.

FIRST AMERICAN WAACS EN ROUTE TO NORTH AFRICA. (*Photograph was taken in London, 30 November 1942.*)

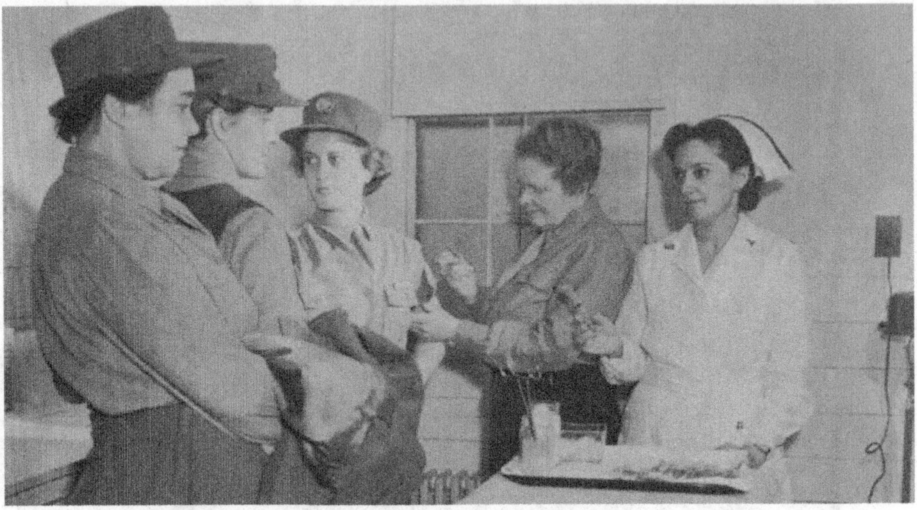

PROCESSING FOR OVERSEAS. *Members of the first WAAC unit to go overseas receive their immunization shots, above, and are inspected by the WAAC director, below.*

PROCESSING FOR OVERSEAS. *Waacs give each other a lift to waiting trucks, above, and, below, are helped over the rail and onto the ship's deck.*

End of 1942

Thus, by the end of 1942, the WAAC could boast two training centers open and a third being readied, twenty-seven Aircraft Warning Service units, nine service command companies shipped or ready to ship, two secret units with the Antiaircraft Artillery, and one on its way overseas. In late December Director Hobby reminded her staff that, in spite of the uncertainties and hardships of rapid expansion, "Our six months' achievement is really something to boast about." [49] Whereas she had originally promised the War Department 12,000 Waacs by July of 1943, the Corps had achieved that strength by December of 1942, with 12,767 enrolled. The year had witnessed a complete revolution in military thinking on the use of womanpower: in January, 12,000 Waacs had seemed too many; but in December, 1,200,000 seemed too few.[50]

The WAAC's first Christmas brought added praise from the Chief of Staff, in the name of the Army, for the Corps' spirit and contributions.[51] At Christmas also, anticipating the greater effort that would be required by the impending General Staff decision, the Director telegraphed her staff, "May God bless you and give you strength for the task ahead." [52]

Indecision as to Planning Goals

Strength of some sort was needed, for by January of 1943, planners feared that the WAAC could not survive unless the General Staff's decision as to its size and future status came quickly. It had now been almost five months since G-3 had called for 1,500,000 Waacs without setting definite goals. When these were not announced by the beginning of 1943, it was necessary to set an interim recruiting quota for January of about 10,000, based on existing training capacity. Reports indicated that this was being met, and that training centers were seriously overcrowded, yet by all evidence the Corps was still far from approaching the peak of General Staff expectations. It was learned informally that it was the General Staff's intent to authorize at once still another training center, the fourth. Again field shipment schedules were outdated and previous plans held up.[53]

At a meeting on 6 January, staff officers noted that, until it was known how big the Corps would be and how many new training centers would be required, it was impossible to set up shipping schedules or to tell Army stations when Waacs would arrive. The Quartermaster General and the Chief of Engineers were frantic for estimates on clothing and housing requirements which could not be given them.[54] Neither was it possible to publish detailed discharge and other regulations—to remove the burden of cases still coming to the office—since the Judge Advocate General desired to wait for a decision on the Corps' future status.[55] Staff members, in-

[49] Note on Memo, Mrs. Knight for Dir WAAC, 2 Dec 42. SPWA 300.6 (3-30-42) sec 2.

[50] M/R, Stf Conf, 6 Jan 43. Folder, Memos Prior to 7 Jan 43 in WAAC Planning, WAAC Planning files.

[51] Ltr, CofS to Mrs. Hobby, 19 Dec 42, in CofS personal file, made available to author in December 1945 by his secretary.

[52] Telg, Dir WAAC to Stf, 24 Dec 42. SPWA 335.15.

[53] Memo, Tasker for Exec WAAC, 1 Feb 43. SPWA 600.914 (2-1-43).

[54] Comment 2, WAAC Hq to CofAdm Serv, 5 Dec 42, to Memo, Dir Reqmts Div SOS for CofAdm Serv, 26 Nov 42. SPRMP 291.9 WAAC, DRB AGO.

[55] (1) ETO WAC Hist. (2) Ltr, WAAC Hq to Col Oscar B. Abbott, Hq ETO, 27 Jan 43. SPWA 320.2 sec 2.

terviewed later, commented, "No one seemed to know what he was doing." One of the ablest of the early staff members, Colonel Tasker, who had been replaced as Executive by Colonel Catron, now gave up in despair and transferred to the General Staff in spite of Director Hobby's protest that "his knowledge concerning the creation and growth of this activity is such that his reassignment would be a very real loss." [56] Similarly, First Officer Helen Woods, the first woman to serve as Acting Director, applied for transfer to a regional directorship as far from Washington as possible.[57]

By the end of the year it was obvious that the October office reorganization directed by General Somervell's inspectors was a failure, and Colonel Catron again reorganized the headquarters. Previous divisions were abolished and two new main halves were set up—Operating and Planning. It was thus hoped that the daily crises and emergencies of operation could be kept from disturbing the calmer atmosphere necessary to finely calculated schemes and schedules, while at the same time action on urgent field needs would not be delayed because staff members were busy preparing estimates for the General Staff. The Planning Service was headed for a few weeks by Colonel Tasker and later by a new arrival, Lt. Col. Robert F. Branch; the Operating Service was headed by another new arrival, Col. Howard F. Clark, a regular Army officer of many years' experience. For the more technical aspects of personnel planning and statistics, a civilian woman consultant, Miss B. Eugenia Lies, was also employed. The employment of a civilian appeared to be the Director's first move to overcome the rigid channelization which prevented her contact with junior staff members, since Miss Lies was given authority to ignore all office channels in securing information for the Director.[58]

This reorganization also was a failure almost from the beginning. Colonel Branch soon recommended that his large Planning Service be dissolved, saying, "It is a good deal like having a Diesel engine connected with a sewing machine." [59] With Planning and Operating Divisions both the same size and considering the same subjects, it proved virtually impossible for planners not to operate, or for operators not to plan; overlapping and disagreement resulted. Also, as Colonel Branch pointed out, it was useless for the WAAC to make any plans at all in the face of its dependence on other un-co-operative or dilatory agencies. He said:

> The opportunity scarcely exists for original or creative planning, since the problems of the Corps have been principally composed of situations after the fact. Action is therefore remedial in nature and choices are limited. Congressional action, War Department policy, and the attitude of other agencies are limitations which, regardless of their propriety or necessity, reduce the scope of planning work.[60]

Many badly needed improvements, which were never made or which were delayed, were ordered by Director Hobby during this period. Finer adjustments and accurate research were to be impossible until

[56] Personnel Abstracts. WAC files, OCMH.

[57] AAF Western Flying Tng Comd, Santa Ana, Calif.

[58] Memo, Plng Serv for Dir WAAC, 18 Mar 43, sub: Progress Rpt, 16 Feb–15 Mar 43. SPWA 319.1.

[59] Memo, Chief, WAAC Plng Serv, for Dir WAAC, 10 Apr 43. SPWA 020.

[60] Memo, Chief, WAAC Plng Serv for Dir WAAC, 15 Apr 43, sub: Progress Rpt, 16 Mar–15 Apr 43. SPWA 600.914.

a solution could be found to the basic problems of survival.[61]

By the end of January, when no directives were forthcoming from the General Staff, the WAAC planners, milling about helplessly among their schedules and flow charts, were in dispair. The office's Daily Journal for 1 February indicated: "Director asked General Grunert to obtain from the General Staff a definitive directive governing recruitment, training, and use of Waacs, particularly within remainder of fiscal year." As a result, General Grunert's office wrote the General Staff asking a quick decision on: "(1) how many Waacs we are to recruit and train by July, (2) where we can get training capacity, and (3) how we are to slow down recruiting if we do not [get the above]." [62]

[61] (1) WAAC Office Memo, 23 Feb 43. SPWA 300.6 (3-30-42) sec 2. (2) Folder, Stf Confs (1942), and WAAC Daily Journal, Vol. I (1943), in Hobby files. See also list of 24 Planning Service projects in WAAC Planning files, 1942.

[62] WAAC Daily Journal, dates cited.

CHAPTER VI

The Need for Military Status

The delay in the decision to launch the expansion program was partially attributable to debate on a controversial matter: whether to give full Army status and rank to women. The War Department's decision in this matter was never to be fully understood by many outside its planning group, representing as it did a complete reversal of the opinion of the previous year. For those inside the War Department, and for many station commanders employing WAAC companies, there actually was no room for question in the matter: the General Staff in fact was obliged to make the decision to grant Army status before it could authorize the expansion program. The evidence at hand overwhelmingly indicated that no great employment of woman power was even remotely possible under the Auxiliary system.

Administrative Handicaps of an Auxiliary

Almost from the moment the first Waacs reached them, Army field agencies had begun to bombard WAAC Headquarters with questions concerning which Army administrative procedures were applicable to the WAAC and which were illegal. This had once been considered a simple matter; it was now discovered that in almost all legal and technical matters it was far from simple, and that the problem was far more than that of dispensing information. The questions were such that it was frequently necessary to consult authorities ranging as high as the Secretary of War, the Comptroller General, and the Bureau of the Budget before a legal answer might be obtained.

Thus, on the universal question of whether Waacs might use the franking privilege when mailing their letters, The Adjutant General ruled that Waacs were "persons in the military service" for this purpose, but the Judge Advocate General ruled that they were not, and the privilege was rescinded after being widely used—a situation that distressed Waacs and caused many to realize for the first time that they were not soldiers or members of the military team.[1] As time passed, almost no two authorities agreed upon the occasions when Waacs could be considered "persons in the military service." To permit WAAC units to shop at post exchanges and share dividends, the Secretary of War was forced to declare them persons in the military service.[2] On the other hand, General Grunert, the Judge Advocate General, and the Veterans' Administration united to overrule Director Hobby's request for Na-

[1] (1) Memo, WAAC Hq for TAG, 21 May 42, and Ind. Published in WAAC Cir 2, 6 Jun 42. (2) Ltr, Postmaster Gen to SW, 17 Aug 42, and atchd decisions. WAAC Cir 9, 16 Sep 42, rescinded Cir 2.

[2] Ltr, Tng Gen to Dir WAAC, 31 Jul 42, and Inds. SPWA 331.3 (7-31-42). AR 30-2290 amended 24 Oct 42.

tional Service Life Insurance for Waacs, declaring:

> Such persons are not in the active service in the land or naval forces of the United States. . . . The WAAC as constituted under existing law is essentially a civilian group. The principle upon which war risk insurance is founded . . . has no application to those in civilian occupations.

G–1 Division believed this decision unfair, and attempted to process new legislation, but this was vetoed by the Bureau of the Budget.[3]

Likewise, the Judge Advocate General, General Grunert, and the Comptroller General agreed that Waacs were not "persons in the military service" for purposes of making allotments under the Servicemen's Dependents Allowance Act, nor could they even allot money from their own pay, since the act granted this privilege only to "members of the Army, Army Nurse Corps, contract surgeons, and civilians overseas." [4] On the other hand, Congress itself indicated that Waacs were "persons in the military service" under the Soldiers and Sailors Relief Act, so that The Adjutant General was able to issue Certificates of Military Service to protect members from litigation.[5] As for re-employment rights, Congress indicated that Waacs who left Civil Service jobs had soldiers' rights to reinstatement, but that those who left other jobs had not. When queried concerning state bonuses and other state benefits, The Judge Advocate General decided that each state would have to determine for itself whether Waacs were "persons in the military service" as interpreted by its laws.[6] In some matters of finance, the Chief of Finance ruled that Waacs were "members of the military forces of the United States," but the Chief of Transportation was unable to find similar authority to pay for mileage or for transportation of dependents.[7]

Such administrative annoyances were endless and apparently would continue to be so while the WAAC remained an auxiliary. It was necessary to refuse Army station commanders' requests to award Waacs the Good Conduct Medal, to issue them the command's sleeve patch, to appoint them as warrant officers; even the chaplains were obliged to secure a special amendment before they could give each Waac a free New Testament.[8] The final

[3] (1) Ltr, WAAC Tng Cen to WAAC Hq, 19 Jun 42, and Inds; (2) Memo, Dir WAAC for Adm Serv SOS, 14 Jul 42, WA 291.9; (3) D/F, CofAdm Serv to WAAC Hq 29 Jul 42, SPAAS 243 (7–6–42); (4) Ltr, VA to SW, 21 Jul 42; (5) Min, Gen Council, 28 Jul 42; (6) Ltr, Bur of Budget to SW, 1 Oct 42, and incl ltr, AG 291.9 (6–2–41). All in SPWA 314.7 (6–2–41)(1) sec 4.

[4] (1) TWX, WAAC Tng Cen to Dir WAAC, 3 Jul 42; (2) Memo, Dir WAAC for Adm Serv SOS, 6 Jul 42, with Ind, SPJGA 354.01; (3) WAAC Cir 4, 10 Jul 42; (4) Memo cited n. 30(3); (5) Ltr, Compt Gen to SW, 6 Aug 42, B–27558; (6) WAAC Cir 7, 29 Aug 42; (7) Min, Gen Council, 28 Jul 42; (8) Decision, JAG to CofAdm Serv, 4 Aug 42. SPJGA 354.01, 2d Ind. All in SPWA 314.7 (6–2–41)(1) sec 4.

[5] Memo, JAG for CofAdm Serv, 6 Aug 42. SPJGA 354.01, in WA 014.3.

[6] (1) Ltr, State of N. J. to TAG, 11 Jul 42, and reply; (2) Ltr, Los Angeles to JAG, 5 Oct 42; (3) Ltr, Induction Station, Portland, Ore., to Hq 9th SvC, 19 Apr 43; (4) Ltr, Pittsburgh to WAAC, 17 Oct 42, and Ind, SPJGA 354.01, in SPWA 315.7 (8–2–41)(1) sec 4. All in SPWA 230.57 (6–15–42).

[7] (1) 2d Ind, CofFin to San Antonio Avn Cadet Cen, 8 Dec 42. SPWA 012.33. (2) Memo, TC for Dir WAAC through CofFin, 4 Sep 42. SPTOT 531.2 (AR 30–925), in SPWA 300.3 (4–29–42) sec 2.

[8] (1) Memo, WAAC Hq for G–1, 24 May 43, sub: Good Conduct Medal. SPWA 200.6. Eventually granted. (2) Memo, Dir WAAC for CofAdm Serv, 21 Apr 43. SPWA 400.34 (3–21–42)(1) sec 3. T/E amended to permit sleeve patches. (3) Ltr, Ft Oglethorpe to O Dir WAAC, 8 Jul 43, sub: Appointment of Bandleaders as WOs, and Inds. SPWA 211 (11–13–43). (4) Ltr, Dir WAAC to CofChaplains, 20 Jul 43. SPWA 300.5, 1943. (5) Compilation for ready reference to all points made by 3d Officer Bernice Keplinger, WAAC Hq, 23 Nov 42. SPWA 314.8 (6–2–41)(1) sec 4.

word came from The Quartermaster General, who ruled that Waacs were not eligible for burial with a flag, military honors, or an escort to accompany the remains home. After conducting a Waac's funeral, an indignant American Legion post protested: "It would probably interest you to know that, at this particular Waac's funeral, she was given full military honors regardless of what her status is, even though she has passed on into another world." To this a harassed War Department made no reply, probably believing that it was difficult enough to determine a Waac's status in the present world without pursuing the matter further.[9]

There were minor points without number in which Waacs did not receive all of a soldier's privileges. Under the new military pay scale, the WAVES were able to offer recruits $50 a month, whereas, the Auxiliary legislation limited the WAAC to $21 a month. A particularly serious discrepancy was the matter of disability benefits and retirement. Soldiers and Waves who suffered disabling injuries or illnesses were entitled to lifetime pensions and hospitalization, but Waacs, under the Federal Employees' Compensation Act, were entitled only to a few dollars' compensation. Under the law, the Army could give a Waac hospitalization only until her disability was pronounced permanent, at which time she must be discharged and ordered to leave the military reservation. She had no right to a pension or to veterans' hospitalization.[10]

The Auxiliary Disciplinary System

Even had these lesser administrative difficulties not existed, there was another, important enough in itself to have invalidated the Auxiliary system, and not capable of solution by any course short of full Army status; this was the question of establishing a legally sound and equitable WAAC disciplinary system.

It had been known since passage of the legislation that Waacs, like other civilians, could not legally be made subject to court-martial and to the Articles of War unless they were "in the field." This was not even a matter of discretion; no citizen of the United States could legally be tried by military courts unless he had military status or unless he accompanied an army in the field. When not "in the field," the WAAC had its own code of conduct, which was extremely strict but which was limited to civilian punishments for infractions—fines, reprimands, restrictions, or discharge (white or blue); it could not legally subject members to court-martial, imprisonment, or dishonorable (yellow) discharge.[11]

During the autumn of 1943, Judge Advocate General decisions began to make it clear that Waacs could almost never be considered "in the field" unless they were overseas or in the unlikely event that they should accompany an army on maneuvers.[12] One station commander did actually try a Waac by court-martial on the grounds that his station on the Eastern seaboard was subject to attack and therefore "in the field," but by the time the

[9] Ltr, OQMG to 6th SvC, 21 Dec 42; also Ltr, N.J. American Legion Post 16 to WAAC Hq. SPWA 293 (12-10-42).

[10] Senate Com on Mil Affairs, 78 Cong, 1st sess, *Hearings on S 495*, 3 Feb 43, p. 18. For six months after discharge, pay and allowances continued if dischargee was hospitalized; no provision was made for hospital expenses in excess of WAAC pay scales. WD Memo 35-46, 24 Jan 46. WDWAC 705. Also Min, mtgs, Jun-Jul 42. WA 300.3 (4-29-42) sec 2.

[11] WAAC Regs, 1942 (Tentative), sec 47.

[12] Memo, M. C. M., N. K. B., C. E. F. for Dir WAAC, 15 Dec 42, sub: Interpretation of "In the Field." SPWA 016.2.

Judge Advocate General had considered the legality of the proceedings, the Waac had served her sentence and been discharged.[13] The Judge Advocate General decided that other stations, if they wished to take court-martial action against a Waac, must first get a legal decision as to whether the post was currently subject to attack, and that "military police would be liable for suit if restraint was found improper."[14]

At first there appeared to be little need for military discipline to hold Waacs in their jobs, but when recruiting quotas were raised standards fell somewhat and several WAAC "Awols" occurred. The military police occasionally apprehended a few and began to ask the correct procedure to follow in returning them. One service command inquired: Could the procedure in Army Regulations and the Articles of War be followed? Or were only members of the WAAC qualified to arrest a Waac? Was it legal to confine them until they could be returned?[15]

WAAC Headquarters at first replied that the regular procedure should be followed, except that the women should not be confined in the same guardhouses with men and should have WAAC escorts on the return trip. As WAAC Headquarters attempted to revise the regulations to this effect, it became evident that there was no legal way to provide for the return of a Waac to WAAC control; in the few cases shipped back so far, the expenditure of government funds had been illegal.[16]

Army staff members surmised that nothing could be done except to let the woman go and to mail her a blue discharge. To this the Judge Advocate General added that there was probably no legal foundation to the whole WAAC disciplinary system: any punishment meted out by WAAC company commanders might be unconstitutional, since it was not done by court-martial and therefore deprived the individual of civil rights without due process of law. Only admonition or reprimand would be safe. Restriction to quarters would be legal if it was not considered "confinement." Summary (blue) discharge, said the Judge Advocate General in an informal conference, was also probably "legal" although "without due process of law." This so startled WAAC Headquarters that it hastily drew up a directive to the field forbidding any further company discipline until the matter was settled. After consideration, this was not sent out; the Corps had gone too far to retreat, and since the Office of the Judge Advocate General itself had written the WAAC disciplinary regulations only a few months before, it was believed that some legal basis must exist.[17]

Accordingly, steps were taken to set up an exact procedure for trying offenses, so as to protect the individual. Also, after considerable debate, the authority of the military police was extended to apply to Waacs. This was somewhat risky in view of representations previously made to Congress that the women would be under virtually no more restraint than civilian employees except when "in the field." Therefore, to ward off Congressional criti-

[13] Folder, Discipline, files of Air WAC Div, Hq AAF, 1943.

[14] Memo cited n. 12.

[15] Memo, 7th SvC Internal Security Div for Gen Faith, 18 Dec 42, and Inds. SPWA 300.3 (4-29-42) sec 2.

[16] WD Memo W635-6-43, 1 May 43, sub: Apprehension of WAAC Absentees.

[17] (1) Memo, Maj James R. Gifford for Exec WAAC, 6 Jan 43. SPWA 300.3 (4-29-42) sec 2. (2) Memo, JAG for G-1, 7 Mar 42. JAG 354.01, in WA 300.3 (4-29-42) sec 1. (3) Two Memos, Stf for Dir, 15 Dec 42. SPWA 300.3 sec 1. (4) Weekly Rpt, WAAC Hq to CofAdm Serv, 14-19 Dec 42. SPWA 319.12.

cism, the provision was made that Waacs would not be placed in guardhouses. Such official extension of authority was needed quickly if WAAC Awols were to be returned at all, since some service commands had already forbidden their military police to take any responsibility in effecting apprehension of absentees.[18]

Even if absentees were returned to their company, there was no way of holding them. "Restriction to quarters" could obviously not be enforced on a determined absentee unless several shifts of guards were placed in the barracks.[19] As time went on, the situation became more than faintly ridiculous. Two young sisters named Lydia and Elvira wired Director Hobby:

> HAVE BEEN IN WAACS 3½ MONTHS. NOW AT HOME ON EMERGENCY FURLOUGH. HAVE NO INTENTION TO RETURN. CANNOT TAKE BEING IN CORPS. NERVOUS WRECK AND WILL LOSE OUR MIND IF NOT RELEASED. MA NEEDS US BOTH AT HOME AND CANNOT RETURN UNDER ANY CIRCUMSTANCES. PLEASE TAKE IMMEDIATE ACTION AND REPLY.[20]

This was probably the only occasion in history when a deserter directed a commanding officer to "take immediate action and reply." There were other cases even more annoying; for instance, four women who were picked up in Washington after repeated absences without leave. The report said: "The girls claim they know Regulations and nothing can be done to them to hold them if they want to go AWOL except discharge, which is what they want most."[21] In another case, a Waac from a large city alighted from the train at the dreary and isolated station which was to be her first assignment. Upon being informed that there were 149 Waacs at a nearby station behind sand dunes and brush, she remarked, "Well, there won't be 150"; she climbed back on the train and was never seen again.[22] Some new recruits even refused to comply with initial orders to active duty if they had changed their minds since taking the oath of enlistment.[23]

In other words, it was now clear that the Army had little more hold over Waacs than over civilian workers. The women could obviously leave the service at any time they desired, just as a civilian employee could. Under such conditions it would be risky to replace thousands of trained men on any very vital or secret work with an equal number of Waacs who might depart as readily as other civilians. The large women's corps which the War Department planned could scarcely be built on such shaky foundations.

Remedial Legislation

Long before the examples cited had come to light, the War Department had perceived that remedial legislation would be necessary. Some three months before, General Marshall had directed preparation of such legislation to give the WAAC the new pay scale beginning at $50 a month. Noting that the Services of Supply had taken no action on it in this period, the Chief of Staff expressed some annoy-

[18] (1) Memo, Dir WAAC for CofAdm Serv, 27 Mar 43; (2) Memo, WAAC Hq for JAG, 15 May 43, and Inds. SPWA 300.3 (1-7-43) sec 1. (3) Ltr, 2d SvC to WAAC Tng Cen, 16 Apr 43. SPWA 251.1.

[19] Ltr, SFPOE to WAAC Hq, sub: AWOL—WAAC. SPWA 251.1.

[20] Telg, Lydia and Elvira B. to Mrs. Hobby, 15 Jun 43. SPWA 251.1.

[21] Memo, Stf to 1st O Kerins, Field Insp Div, 23 Jun 43. SPWA 251.5.

[22] Remarks of WAAC Stf Dir, 9th SvC. Min, Stf Dirs Conf, Chicago, 15–17 Jun 43, SPWA 337 (6-1-43).

[23] Ltr, 7th SvC to Dir WAAC, 20 May 43. SPWA 300.3 (1-7-43) sec 2.

ance, in view of the WAVES' competition, and wrote: "It should be pressed to the limit." [24] In the same bill, it was asked that the confusing WAAC grades, officer and enlisted, be changed to correspond exactly to the Army's, except that general officer grades would not be open to women. Considerable Congressional objection resulted. The House first voted to leave the Director's and assistant directors' pay at the equivalent of major and captain, respectively, on the grounds that "the duties and responsibilities of these officers" did not merit higher pay. After some delay, the pay bill was finally approved intact on 26 October 1942.[25]

Other piecemeal legislation followed. On 23 July Congresswoman Edith Rogers had introduced a bill to give Waacs the benefits of National Service Life Insurance, but this got nowhere.[26] On 8 October Mrs. Rogers introduced a bill to grant Waacs hospitalization and domicilary care by the Veterans' Administration. Five months later, after numerous vicissitudes, this provision, plus burial benefits, was enacted into law.[27] In October, November, and again in January, bills were introduced to give free postage to Waacs, but these also got nowhere.[28] In October Mrs. Rogers introduced a bill to place the WAAC in the Army Reserve Corps.[29] This was intended to supplant the pay bill and other partial legislation by making pay and everything else equivalent. It did not pass, because General Marshall informally sent word to Congress that he believed the pay bill urgent and did not wish to delay it with a more controversial measure. General White reported:

> He [General Marshall] said that at that time the Corps was still in a formative period, we had no units actually organized and functioning with the Army [as distinguished from the Aircraft Warning Service], and under the circumstances he preferred that we go ahead as we were until we had gone far enough to know just how valuable this Corps was going to be to the Army.[30]

Introduction of WAC Bill

By the beginning of 1943 General Marshall had changed his mind. He informed the Director:

> Although the Corps is still in the formative period of organization, its members have convincingly demonstrated their ability to render a vital military service. The standards of discipline, training, and general efficiency are on the highest level and are a complete reassurance to the officials of the War Department as to the outstanding services which will be rendered by this organization.[31]

In January Mrs. Rogers worked with the Director's Office to draft a bill for Army status. This she introduced on 14 January 1943. The bill very simply stated that women might be enlisted and com-

[24] Memo, "G. C. M." to G-1. WDCSA 291.9 (8-31-42), in AG 291.9 (6-2-41).

[25] (1) Memo, G-1 through CofS to MPD SOS, 24 May 42. WDGAP 421 WAC, in CofS 291.9 WAAC. (2) Memo, Dir WAAC for Col Abbott, 20 May 42; (3) Ltr, Bur of Budget to SW, 2 Sep 42; (4) Memo, G-1 for CofS, 2 Sep 42; (5) Ltr, SW to Chmn Senate Com on Mil Affairs, 3 Sep 42; (6) S 2751, 77th Cong; (7) Senate Com on Mil Affairs, 77th Cong, 2d sess, *Hearings on S 2751*, 4 Sep 42; (8) Senate Calendar No. 1655, Rpt 1603; (9) HR 7539, 77th Cong, 10 Sep 42; (10) HR Rpt 2475, 21 Sep 42; (11) HR Rpt 2524, 12 Oct 42; (12) PL 761, 77th Cong, 26 Oct 42. All except (1) in AG 291.9 (6-2-41).

[26] HR 7435, 77th Cong, 2d sess, 23 Jul 42.

[27] (1) HR 7673, 77th Cong, 2d sess, 8 Oct 42. (2) HR 665, 78th Cong, 1st sess, 6 Jan 43. (3) Subcom of Senate Fin Com, 78th Cong, 1st sess, *Hearings on S 230*, 25 Feb-1 Mar 43. (4) HR 1749, 78th Cong, 1st sess. (5) PL 10, 78th Cong, 2d sess, 17 Mar 46.

[28] HR 7748 (27 Oct 42) and HR 7812 (24 Nov 42), 77th Cong, 2d sess; HR 844, 78th Cong, 1st sess, 7 Jan 43.

[29] HR 7718, 77th Cong, 2d sess, 16 Oct 42.

[30] Senate Com on Mil Affairs, 78th Cong, 1st sess, *Hearings on S 495*, 3 Feb 43, p. 4.

[31] Memo, 16 Dec 42. WDCSA 291.9 WAAC.

missioned in the Women's Army Auxiliary Corps, Army of the United States. A later version changed the proposed name to Women's Army Corps. No limitations were placed upon the Secretary of War's authority over the size of the Corps, rank, pay, or any other matter.[32]

The War Department's official attitude toward the new bill was not at once made clear, but on 1 February the Secretary of War announced his support, stating: "Although in the past the War Department has not advocated the establishment of the Corps as a part of the Army, experience has proved that the present arrangement will not be satisfactory." [33] On 3 February General White of G-1 Division told a Senate committee:

> We are now in the position of favoring that which we have heretofore in a sense opposed—not that we did not want it in the beginning. . . . We have more than 10,000 now actually organized into units and on duty in the field. . . . Wherever we have put them they have proved highly valuable.[34]

However, General White informed the committee that in his opinion a WAAC unit would not be able to replace an equal number of men, stating concerning the replacement: "Well, I do not think it will run quite one for one. I have always estimated that three women would release two men." [35]

By the date of the next hearings in early March, War Department support was considerably more enthusiastic. Reports from the first unit at Fort Sam Houston indicated that three women had replaced, not two men, but four. It was indicated that one woman could not replace one man in heavy work such as motor transport, but that women stenographers, typists, and switchboard operators could often replace two or more men apiece. Equally good reports also began to arrive from the North African unit and the antiaircraft experiment, accompanied by requisitions for more Waacs. Thereafter the War Department without hesitation directed every station receiving 150 Waacs to effect replacement of 150 enlisted men—a rate which was actually better than one for one, since part of the women had to be used for unit housekeeping and were not available to replace men.[36]

In an effort to secure speedy enactment of military status for women, General Marshall informed Congress in early March:

> In recent months great strides have been made in the organization. Sufficient numbers have now been trained, organized into appropriate units, and placed on duty in various activities to enable us to know that we are moving in the right direction. I am now certain that the women's organization will be of great value to the military service. . . .
>
> It is evident also that the operation of the women's organization will be simplified and its efficiency vastly improved if it is made an actual component of the Army. The personnel of the Army have definite, defined duties, are subjected to prescribed disciplines, and have clear-cut relations to each other and to the civilian population. These have been found necessary to the effective functioning of military establishments the world over. They are equally important in the WAAC, which in its present half-civilian, half-military status will be seriously handicapped. . . .
>
> Having the Corps with the Army, but not

---

[32] (1) Weekly Rpt, WAAC Hq for CofAdm Serv, 10-16 Jan 43. SPWA 319.12 (11-9-42) sec 1. (2) Identical bills HR 1188, 14 Jan 43, and S 495, 21 Jan 43, and later version, HR 1751, 8 Feb 43; 78th Cong, 1st sess.

[33] Ltr to Hon. Robert Rice Reynolds, Chmn, Senate Com on Mil Affairs, 1 Feb 42.

[34] Senate Com on Mil Affairs, 78th Cong, 1st sess, *Hearings on S 495*, 3 Feb 43, p. 4.

[35] *Ibid.*

[36] House Com on Mil Affairs, 78th Cong, 1st sess, *Hearings on S 495*, 9 Mar 43. Statement by Gen White.

in it, also results in inequalities and injustices to its members. . . . There is, finally, the important element of morale. Membership in the Army carries with it a natural and proper pride . . . for which service in an adjunct of the Army provides no satisfactory substitute.[37]

General White in his testimony was even more specific on the administrative annoyances of an auxiliary system. He said:

We have found that we run into administrative, disciplinary,[38] and command difficulties. . . . The Army Regulations are not applicable to the Waacs. The Army Regulations are contained in a set of books as long as this table. It took years to develop them. Now we are faced with having to develop almost a parallel set of regulations to govern another part of the Army. . . . There are little things that are constant irritants. They are the same difficulties we had in trying to operate the Army Specialists Corps—people not in the Army working with and doing the same thing people in the Army were doing, on a different basis and in a different status. There are questions of relative rank, command, administrative records, and so forth. It just simplifies the whole operation of the Corps to have them all alike. In other words, in the Army we want one category of people.[39]

No difficulty was encountered in the Senate, which passed the bill on 15 February 1943. The House of Representatives was another matter; its Committee on Military Affairs appeared hopelessly divided into two factions, which at public hearings fell to quarreling so vigorously that their chairman was obliged to remind them that such discussions should be taken up in executive session and not in the presence of Army witnesses.

One faction held with the congressman from Texas who said, "I spoke against it [putting the WAAC in the Army the previous year] four times from the floor. . . . Then we went ahead and organized the other women's auxiliaries as integral parts of the other services; I feel that we are going to have to do this now, no matter how good my argument was before." However, a representative from Indiana argued that draftees should receive soldiers' benefits, but Waacs and men who volunteered should not because they were not "forced into the Army." He also maintained that Waacs should not get soldiers' benefits because they did not go into front lines; this provoked an argument over whether General Eisenhower and the men in his fixed headquarters were entitled to military status. A congressman from Ohio joined in objecting to compensation for both husband and wife if both were in the Army and both were injured.

It was also feared that a Waac's husband would get a government allotment whether he needed it or not, since the law provided this for a soldier's wife whether she needed it or not. On this General White observed, "I am inclined to believe that the Comptroller General, who at one time recently ruled, in effect, that women were not persons, would rule that husbands are not wives." The representative from Ohio also failed to understand why the Secretary of War could not legally make the authority of military courts applicable to civilian women without giving them military status and benefits; this brought on another futile discussion of what "in the field" meant.[40]

It very shortly became clear that the

[37] Ltr to Hon. Andrew Jackson May, Chmn, House Com on Mil Affairs, 1 Mar 43.

[38] At the word *disciplinary,* Congressmen interrupted to inquire details, and discovered that General White was referring not to Waac misbehavior but to the legal and technical questions discussed above.

[39] *Hearings* cited n. 36.

[40] *Ibid.*

House of Representatives would not pass the bill without any number of hampering amendments concerning the size of the Corps, the top rank attainable, the types of duty permitted, and the benefits to be allowed. These amendments would have to be fought out with the Senate and a compromise reached, and when, if ever, the legislation would pass now became uncertain.

The first six months of 1943 were therefore, from the Corps' viewpoint, to be much like the uncertain period of the previous spring, when passage of the first WAAC bill was not sure and all plans had to be made subject to change without notice. However, in this case, the Corps was not merely a plan on paper; it was a living organization which grew from 12,767 women in December 1942 to 60,243 in June 1943, and which required detailed administration. Passage of the bill was to be expected in March, in April, in May. Director Hobby said, when it did not pass in April, "I am sure it will come in May as a birthday present to the Corps.[41] But it was not to come in May, or in early June.

Meanwhile, from January of 1943 onward, the War Department, insofar as legally possible, proceeded with plans for the expansion program, both in the United States and overseas, as if it were an assured fact that the Corps would be given military status.

[41] (1) Lecture by Comdt, Sch of WAC Pers Adm. OCMH. (2) Min, Gen Council, 5 and 26 Apr 43.

CHAPTER VII

Spring, 1943: Expansion and Decentralization

A few weeks after the decision to seek military status, the General Staff on 30 January and 6 February 1943 announced the long-awaited series of further decisions on the Corps' future—perhaps the most important staff action in the WAAC's existence.

The first was that a draft of women would not yet be sought: "Plans for the procurement of personnel will be based on the continuance of voluntary enlistment." [1] The proposal to draft women was still highly shocking to many individuals, and clearly would stand no chance of approval by Congress unless the military successes in North Africa should unexpectedly turn to disaster and the situation should approach that of Britain, which had perforce adopted national service laws. Events of a year later, when national service was finally proposed by other agencies, confirmed the present surmise that a draft proposal would merely have involved the Army in a bitter and ultimately futile legislative battle, without providing womanpower in time to aid in the current war.[2]

After making this decision the War Department faced a choice between attempting to secure a large Corps by other means, or abandoning G-3's idea and increasing the draft call on men. In a preliminary conference with G-1 and G-3 Divisions on 29 January, Director Hobby again informed them that in her opinion the WAAC could not recruit anything like a million women. Her frequently expressed view was: "I don't believe we are going to get even 150,000 volunteer women unless there is some move by the Federal Government requiring national registration or compulsory service." [3]

Expansion Program Decided

The important 6 February conference of General Staff representatives, which decided upon the expansion rate to be attempted, was headed by Maj. Gen. Idwal H. Edwards of G-3 Division, and included General Madison Pearson and other Services of Supply representatives, and both Director Hobby and Colonel Catron from WAAC Headquarters. There resulted a decision that appeared a moderate compromise to all concerned, although actu-

---

[1] (1) Memo, G-3 for TAG, 30 Jan 43. WDGCT 291.9 WAAC (1-30-43). (2) Ltr, TAG to CG SOS, 10 Feb 43. AG 320.2 WAAC (1-30-43) PR-W-WDGCT.

[2] See Austin-Wadsworth bill, HR 1742 and S 666, 78th Cong, 1st sess, 8 Feb 43; *Congressional Record,* Vol. 89.

[3] (1) WAAC Daily Journal, Vol. I, 29 Jan 43. (2) Speech, in Min, Conf of Tng Cen Comdts, Washington, D. C., 31 Mar 43. SPWA 334.8 Conf (3-29-43).

ally it was to have calamitous effects upon every phase of the Corps' future program.

To the conferees it was obvious that G–3's earlier optimistic calls for 1,500,000 Waacs could not be met by voluntary recruiting and that it would be necessary to set a far lower figure, while simultaneously increasing draft calls for men to make up the deficit. Director Hobby outlined the various recruiting difficulties that were closing in upon the WAAC, particularly the competition shortly to be expected from the Navy and the War Manpower Commission. The date at which these factors would be felt was uncertain, but she expressed the view that by June they would have cut off most if not all of the WAAC's intake.

On the other hand, it was evident that they had not yet had serious effect. Instead, the Army Recruiting Service had overrecruited every quota to date, and was obliged, for lack of training space, to turn away qualified applicants. At Des Moines 70 women slept in poorly ventilated stable-barracks intended to hold 50; at Daytona Beach, trainees were housed five to a tent and eight to a double hotel room. Desperate calls and letters from training authorities indicated that "Training Centers cannot stand such overloading."[4]

It was realized that to turn away applicants was probably to lose them. In order to insure next month's quotas, Army recruiting stations therefore often swore in surplus applicants and placed them on inactive duty until training space was available.

By February this trick had backfired, placing the Corps under additional public and Congressional pressure. Enrollees who had resigned from jobs, given up apartments, and given away civilian clothing were vociferously indignant at months of delay and frequently wrote their congressmen. Negro recruits on the list alleged that they were not being called up because of their race. The number of enrollees on inactive duty, awaiting orders to report, was by this time almost one fourth of Corps strength. Some authorities attributed the later difficulties of WAAC recruiting to the fact that the presence of thousands of recruits awaiting training in these early days convinced other prospects and the public that the Army actually had no need for Waacs.[5]

As a result, conferees decided that the sensible course was to make available to the WAAC enough additional training space to scoop up at once all inactive members and current applicants—a course that appeared doubly wise, since Army jobs were already waiting and applicants would probably accept other war work unless taken at once. It would then be appropriate to decide, on the basis of the national picture, how much farther it would be possible to go and whether further legislation should be sought.

Accordingly, the goal finally set was that which General White of G–1 Division had earlier recommended—150,000 by the end of June 1943, a mere one tenth of G–3's first total. Minutes of the meeting noted that "present training facilities could . . . provide for a total enrollment on June 28 of 122,234 . . . and an additional training capacity of approximately 10,000 . . . will permit a total enrollment by June 28 of 150,234." The matter of later expansion was "discussed, accepted in principle, and left for future detailed consideration, with the idea that

[4] (1) Min cited n. 3(2). (2) Weekly Rpt, WAAC Hq to CofAdm Serv, 14–20 Mar 43. SPWA 319.12.

[5] Contemporary Rpts, 1 Mar 43: Active—35,662 (21,925 in tng, 13,737 in fld); Inactive—8,022.

training capacity adequate for 150,000 by June 30 would be adequate for a greater number at the proper time." [6]

The greatest obstacle to the training of such a number was seen as the matter of supply, since, in both the previous recruiting step-ups of June and September, the Services of Supply had not been able to meet its commitments. However, there was placed in the record a schedule furnished to Colonel Catron by Requirements Division, Services of Supply, which indicated that in the new program all essential items for the numbers contemplated could be provided.[7]

The conference therefore directed, in language which in retrospect was to appear curious, that "a controlled recruiting of approximately 5,800 a week will be put into effect." Although this goal possibly appeared small by comparison with G–3's earlier ideas, actually the Corps was called upon to triple its size in four months. Corps strength in March would approach 50,000, of which only 14,000 would be trained or in the field, the rest being still in training or awaiting call to active duty. The new requirement meant that by 1 July there must be about 104,000 in the field and 46,000 in training. This was a voluntary recruiting effort unlike anything ever previously achieved by the U.S. Army in any comparable circumstances in its history.[8]

Within hours of the General Staff decision, the Corps was irrevocably committed to large-scale expansion. Recruiting quotas were at once doubled, with future tripling a necessity. Some 6,126 recruits had been admitted in December and 10,421 in January. With a fourth and fifth training center in prospect, the quota was now lifted to 18,000 for February and 27,000 for March. Thereafter it was expected to hold steady at around 33,000 monthly through June.[9]

The Services of Supply was already negotiating for a site for the Fourth WAAC Training Center. After conferences with the Chief of Engineers, it had recommended a part of Fort Devens, Massachusetts, which was excellently located to accommodate recruits from New York and New England. This site was shortly approved but, like the site for the Third, only after the General Staff overrode the nonconcurrence of the using command. It was directed that this center's staff of more than 1,000 be ordered in and be ready to open the training center in March.[10]

The site offered for the Fifth WAAC Training Center was less desirable; it consisted of three prisoner-of-war inclosures at Camps Polk and Ruston, in Louisiana, and Camp Monticello, Arkansas. This minimum-standards housing was available only because the Army had, at this date, taken few prisoners. Its use for Waacs was therefore proposed by G–3 Division, according to conference minutes:

> General Edwards stated that the housing situation had been carefully surveyed and that neither the Ground Forces, the Services

[6] Min, Mtg in Office of G–3, 6 Feb 43. Only copy found in file of Dir Mil Tng SOS, SPTR 337, DRB AGO.

[7] *Ibid.*

[8] All figures are from M/R, WAAC Overall Picture, 1 Mar 43. SPWA 314.7 (1–7–43)(1) sec 1. No two sets of estimates in the files ever quite agree; all are unimportant in that even the lowest estimate was never reached. WAAC Daily Journal, Vol. I, 6 Feb 43, indicated that these goals were known to the WAAC by that date.

[9] For recruiting data, see Ch. X, and Table 2, Appendix A, below.

[10] (1) Memo, Tasker for Exec WAAC, 1 Feb 43. SPWA 600.914 (2–1–43). (2) WD Cir 63, 1 Mar 43. (3) Memo, G–3 for SOS and AGF, 3 Feb 43, and Inds. Tab, Fort Devens, WA 600.1 (7–2–42)(1) sec 2, 1943 WAAC Housing files.

FIRST OFFICER ELIZABETH H. STRAYHORN, *of the Fifth WAAC Training Center, with Governor Sam Jones of Louisiana, a visitor on the post.*

of Supply, nor the Air Forces could make any further housing available without deferring the activation of military units. He stated, however, that there was a possibility that Prisoner of War housing, much of which is now completed and unoccupied, might be temporarily made available. . . . The total housing program for prisoners of war was in the neighborhood of 196,000 spaces, of which we had commitments with the British for only 150,000 spaces, while actual prisoners of war therein total only 6,600.[11]

The Provost Marshal General concurred in allowing Waacs to train there, provided they did not modify the group toilets and showers and other masculine-type plumbing facilities, and provided they would move out on thirty days' notice if German and Italian prisoners should need the place.[12]

The decision as to whether to accept this location was, according to Director Hobby, a difficult one. Its unattractive features were obvious, especially when compared to the WAVES and SPARS training schools at fashionable colleges. One recruiter asked, unnecessarily,

[11] Min cited n. 6.

[12] Memo, CofAdm Serv for TAG, 1 Mar 43. SPWA 314.7 (10-7-43)(1) sec 1.

whether the WAAC with its prisoner-of-war camps was not "suffering by contrast with the WAVES with Smith College, Holyoke, and Hunter?" [13] The Director therefore appealed to the Services of Supply for any other location, even city hotels, but was offered nothing else except a desert camp below sea level at Tulare, California. The WAAC therefore was obliged to accept the prison camps in order to comply with the directed expansion program.[14]

The three scattered branches of this training center, some one hundred miles apart, made administration difficult. The buildings were bare and rough inside and out, located in desolate sandy stockades. Colonel Hoag, who was ordered from Des Moines to be commandant, said: "It was necessary to place the moving-picture theaters of Ruston 'off-limits' until they were cleared of rats and other vermin. . . ." [15] General Marshall himself was concerned about the situation and, in a personal conference with Colonel Hoag, told him to push Special Services operations at this training center. However, the Eighth Service Command later turned down Colonel Hoag's request for these services on the grounds that it had had no instructions from the Army Service Forces.[16] Director Hobby, upon a later visit to the Fifth WAAC Training Center, said, "I know of no finer example of patriotism by Waacs anywhere than that which was shown by the women who worked and trained successfully at these camps." [17]

Expansion of WAAC Training

In accordance with General Staff directives, Brig. Gen. Don C. Faith's WAAC Training Command during the spring months underwent an expansion which overshadowed previous efforts. On 1 March the Fourth WAAC Training Center was activated at Fort Devens; on 15 March the Fifth WAAC Training Center was activated in the three prisoner-of-war camps, with recruits arriving a bare two weeks later. With these additions the five basic training centers had a capacity of 46,388. This meant that some 7,000 trained women—about 50 companies—could now be sent to the field weekly. There were 53 WAAC companies in the field at the end of February; by July, it was intended that there should be at least 375.[18]

In addition to the five basic training centers, several specialist schools were set up to relieve the training command of some of the load of specialist training. Most important of these were seven administrative specialist schools set up at women's colleges in the South and Southwest. These were under the jurisdiction of The Adjutant General, not General Faith, and were commanded by Army officers, with WAAC officers as instructors and company officers. Each had a capacity of about 600, and offered a six-week course designed to turn out trained administrative specialists.[19] Students of The Adjutant General's schools, like men, were given the rank of private first class, although graduates of identical courses at

[13] Miss B. Eugenia Lies, The Overall WAAC Picture (first draft), 28 Mar 43. SPWA 291.9, in Folder, Planning Serv Notes, WAAC Planning files.

[14] (1) Weekly Rpt, WAAC Hq to CofAdm Serv, 7–13 Feb 43. SPWA 319.12 (1942). (2) WAAC Daily Journal, Vol. I, 5 and 8 Feb 43.

[15] Ltr to Col Catron, 27 Apr 43. SPWA 291.1.

[16] (1) Statement by Comdt, Col Hoag, at mtg cited. (2) Ltrs, Col Hoag to Col Catron, 14 Apr 43, Tab, Ruston, and 3 May 43. SPWA 291.1.

[17] Lecture notes, Lt Col Jess Rice. WAC files, OCMH.

[18] M/R, 1 Mar 43, cited n. 8.

[19] WDBPR Press Release, 3 Mar 43.

WAAC training centers could not be so promoted, a circumstance that caused some discontent at the latter.

The new schools were shortly successful in removing a considerable load from the WAAC Training Command; on the other hand, inspectors soon noted a natural decrease in uniformity of operation, with some schools making no reports of personnel actions. There was also reported at some Army schools a growing need for a senior woman officer to correct a certain laxity of supervision of living conditions which was causing public comment unfavorable to recruiting.[20]

In addition, three Army schools to train WAAC radio operators and mechanics were set up at civilian schools in Missouri, New Jersey, and Pennsylvania. These had a capacity of about 300 each, and were under the jurisdiction of the Signal Corps.[21] In February, the Army Air Forces admitted 50 Waacs to the twelve-week photographic laboratory technician course at Lowry Field, Colorado, and others to later classes.[22]

From this time forward it became increasingly common to send Waacs to Army schools when their numbers did not justify separate schools for women. No particular difficulties were noted, and an ASF training inspector stated a few months later: "WAAC training, strangely enough, presents few problems. WAAC students at schools are eager to learn, and maintain discipline with less corrective action than men students in the same class." [23]

Delegation of Command Authority

Before the increased numbers of trainees could reach the field, hasty administrative changes were necessary. While the training system was in process of expansion, WAAC Headquarters therefore hurried to complete the delegation of its command authority to the field. It was realized that such decentralization was perhaps too early for full justice to individuals, since the scanty field experience to this date had not provided a basis upon which all types of Auxiliary regulations might be carefully formulated, and Army Regulations could not yet apply; nevertheless, the approaching expansion made the step necessary. Ready or not, it was obvious that in a Corps of 150,000 women WAAC Headquarters would no longer be able to process the thousands of individual transfers, discharges, and promotions. Already the length of time required to process discharge cases was such that, the field reported, "Many of these members are rapidly becoming mental cases due to the length of confinement awaiting authority for discharge." [24]

It was desired to realign the WAAC organization in a way that would, as nearly as possible, parallel the command system for men and leave the Corps prepared for conversion to Army status, although, as the Director noted, planners were limited to doing "only those things which can be done legally" under existing legislation.[25]

Delegation of command authority to the service command level had been under way for some time. Before the end of 1942

---

[20] Memo, CG WAAC Tng Comd for Dir WAAC, 24 May 43. SPWA 319.1.

[21] WDBPR Press Release, 14 Apr 43.

[22] WDBPR Press Release, 5 Feb 43.

[23] Min, Conf of CGs of SvCs, Chicago, Ill., 22–24 Jul 43.

[24] (1) Ltr, 1st Tng Cen to Dir WAAC, 28 Jan 43; (2) Ltr, WAAC Stf Dir 4th SvC to Dir WAAC, 18 Feb 43; (3) Ltr, TAG to Tng Cen Dirs, 27 Feb 43, SPX 291.9 WAAC PR–W. All in SPWA 300.3 (1–7–43) sec 1.

[25] Min, Stf Mtgs, WAAC Daily Journal, Vol. I.

service commands had been given power to send Waacs on temporary duty and detached service away from their stations, provided that enrolled women were sent only to stations where they could be housed with a WAAC company. Army stations were also given power to grant WAAC leaves and furloughs, with no restrictions except those that applied to men.

At the end of December that much-disputed power—assignment and transfer—was delegated to the WAAC service command directors with no reservations, since these officers now thoroughly understood War Department ideas on safe and appropriate locations. The wording of the circular shortly had to be changed. It at first read: "Orders authorizing assignment . . . will be issued by the commanding general of the Service Command upon the request of the WAAC Service Command Director." This was pronounced offensive to commanding generals, since it appeared to place them under the command of a WAAC officer. Accordingly the wording was changed to: "The WAAC Service Command Director will request the commanding general to issue orders. . . ."

In February authority to promote enrolled women was delegated to WAAC service command directors and company commanders, likewise the authority to demote, with certain safeguards to protect the individual's rights. In March, after some delay to get the approval of the War Department, detailed instructions were published to guide in the selection of women for attendance at officer candidate school; Army commands would receive quotas, appoint selection boards, and issue orders sending the selected applicants to the school.

Finally, an even more important power was delegated, that of discharge, which would now be ordered by the commanding general upon the recommendation of his WAAC director, unless he was in doubt, in which case he could send it to WAAC Headquarters. This delegation had required months of study to set up exact and just grounds for each of the WAAC's three types of discharge—honorable discharge, discharge without specification as to character, and summary discharge—which did not correspond exactly to the Army's three types because they could not legally include a dishonorable discharge.[26]

At the conclusion of this series of delegations of authority, in early April, WAAC Headquarters retained within itself no more power over operating functions than that which The Adjutant General retained over male personnel—initial assignment of newly trained units, promotion of officers, and appeal on doubtful cases. Of these it could not dispose until Army status allowed The Adjutant General's Office to assume them.

Inclusion in the Troop Basis

At the same time, the General Staff decisions concerning Corps expansion and status had committed the Army to a course implicit in the program: the more complete integration of womanpower into every area and command. On 8 February, two days after the expansion program was launched, General McNarney informed the General Staff of a further decision which represented a new era in War Department thinking on the employment of womanpower: the WAAC strength was to

[26] WAAC Cirs 14 (4 Nov 42), 17 (29 Dec 42), 5 (13 Mar 43), with note, in SPWA 300.3 (1-7-43) sec 1; WAAC Cir 5 (10 Feb 43), 8 (17 Mar 43), 10 (9 Apr 43).

count against the Army's Troop Basis. G–3 Division, in recommending this action, stated:

> Since the legislation was passed, conceptions as to the most profitable methods of employing members of the WAAC have undergone changes. . . . It is necessary to employ both soldiers and members of the WAAC only on those tasks which cannot otherwise be performed.[27]

General McNarney stated that the Army would voluntarily agree to draft one less man for every WAAC recruited, and that

> From now on the strength of the WAAC will be included as a part of the authorized troop basis and will form a part of the 7½ million Army.

On 22 February the General Council noted:

> General McNarney emphasized and reiterated again that the purpose of Waacs is to replace soldiers. When Waacs are assigned to a post, a corresponding reduction of enlisted men must be made immediately. G–1 will check on this.[28]

Accordingly, the War Department directed that Waacs be removed from all employment that did not result in the replacement of soldiers. The decision particularly affected the plans, which some agencies still had not abandoned, to employ Waacs in laundries, post exchanges, officers' clubs, and similar duties which had no allotment of soldiers. It was directed that WAAC Headquarters refuse all such requisitions.

More important, the General Staff decision required the removal of approximately thirty WAAC units currently assigned to the Aircraft Warning Service.[29]

The demise of the Aircraft Warning Service WAAC units caused no grief in WAAC Headquarters. Nothing but difficulty had resulted from the use of Waacs mixed with civilians in city stations, with the Waacs inevitably getting less desirable assignments and becoming convinced that they had replaced not a soldier but a debutante. Already the possibility of an air attack on the United States appeared remote. The Aircraft Warning Service, on the other hand, declared Waacs essential and refused to give up about half of the women; instead, it allotted military vacancies for small clerical detachments at various operations and filter centers. The rest of the women were reorganized into the newly created post headquarters companies, AAF, similar to service command companies, for shipment to various air bases to perform real military duties.[30]

The Services of Supply now suddenly became concerned with the fact, which had not previously troubled it, that all WAAC units were assigned to the Services of Supply and only attached to the using service. It was immediately clear that Waacs would have to be assigned to the using commands in order to count against their troop allotment and not against the Services of Supply's. This coincided admirably with the desires of the Air Forces, which had endured the service commands' administration of the city-stationed Aircraft Warning Service units but could not view with equanimity a Services of Supply invasion of actual air bases.

Of the approved requests now on file, some 57 percent were from the Army Air

[27] Memo, G–3 for CofS, 7 Jan 43. WDGCT 291.2 WAAC, in OCofS 291.9.

[28] Min, Gen Council, for dates cited.

[29] Transcript of Mtg, 15 Feb 43 (Col Catron, Gen Grunert, Gen Madison Pearson). Dir Adm Serv ASF Sp Coll DRB AGO.

[30] (1) See Memo and Ltr cited n. 1. (2) Weekly Rpt, WAAC Hq to CofAdm Serv, 17–30 Jan 43. SPWA 319.12 (1942). (3) AAF WAC Hist, p. 14. See Ch. XVI, below, on AAF.

Forces and the remaining 43 percent from the Services of Supply and the Ground Forces combined. Accordingly, a tentative division of the first 150,000 was planned on this basis, with the Air Forces receiving the largest slice for field duty, although the training center allotment continued to belong to the Services of Supply:

Command	*Strength*	*Percent*
Total	150,000	—
Trainees, SOS	20,000	—
Field duty	130,000	100
AAF	70,000	54
SOS	55,000	42
AGF	5,000	4

Although this division had neglected to take into consideration the future needs of overseas theaters and the War Department and was soon to require revision, it roughly determined the proportion in which shipments were to be made for the next few months.[31]

On 17 February it was directed that all Waacs would in the near future be assigned, not attached, to the using agency, which might be the Army Air Forces, Army Ground Forces, Services of Supply (shortly renamed Army Service Forces), or an overseas theater. The formal transfer from attachment to assignment was set for 1 May.[32]

Staff Directors Assigned to New Commands

If there had been risk involved in early delegation of discharge and other powers to the service commands, where were located experienced regional directors, it was recognized that real danger lay in the necessity for scattering Waacs through all other Army commands and overseas theaters, which had neither WAAC advisers nor experience in Auxiliary administration. Therefore, it was decided that, before the formal assignment of enlisted women on 1 May, a WAAC officer corresponding to the service command director must be assigned to each major command involved. These included an eventual fifteen Air Forces commands in the United States, two Ground Forces commands, and many of the Army Service Forces' administrative and technical services, as well as ports of embarkation, overseas theaters, and other miscellaneous commands. To designate all such WAAC advisers, the title WAAC Staff Director was devised, and the titles Service Command Director and Regional Director were dropped. All powers which had been previously delegated to a service command director were simultaneously made applicable to all WAAC staff directors.[33]

Immediate difficulty was experienced in finding qualified WAAC officers in sufficient numbers to fill these new and responsible positions. The best possibilities appeared to be those women with some experience as assistant service command directors, as senior recruiting officers, or in WAAC Headquarters. These officers, except for a few in headquarters, frequently proceeded hastily from the old assignment to the new without any such indoctrination as it had been possible to give the original nine service command directors.

Another problem in sending out these officers was posed by an unexpectedly

[31] (1) Transcript cited n. 29. (2) Memo, Dir WAAC for CofAdm Serv, 26 Jan 43. SPWA 320.2 (1-26-43)PS, in WAAC Planning files.

[32] (1) WAAC Daily Journal, Vol. I, 17 Feb 43. (2) WAAC Cir 11, 10 Mar 43. Since December, when there first appeared the possibility of unprecedented expansion into commands other than the service commands, WAAC third officers had been attached to the AAF, the Signal Corps, and The Adjutant General's Office, to assist those agencies in the necessary planning. WAAC SO 43, 5 Dec 42. SPWA 300.4 (8-18-42).

[33] *Ibid.*

quick transition from Auxiliary to Army rules of promotion. The system of requiring all officers to graduate from officer candidate school, with initial commissions only as second lieutenants, had, in Mrs. Hobby's opinion, proved sound and democratic in avoiding the choice of officers on political or personal grounds and in preventing the award of higher ranks to individuals who failed on the job. Nevertheless, as the WAAC Pre-Planners had recognized, the system was tenable only if women who had proved capable were advanced to ranks that would provide a normal spread of grades for Corps administration. In line with this policy, WAAC Headquarters had in December promoted service command directors and battalion commanders to the rank of first officer, or captain. It was planned to repeat the process in April to give equal rank to the new staff directors being sent out, to make experienced key officers majors, and company commanders captains.

However, when the Director proposed this move, she was informed by the Army Service Forces that the Army time-in-grade now applied to the WAAC, in anticipation of Army status later in the year. The Director protested that such abrupt application of the Army promotion system was impossible since the Corps had not initially followed the Army system of direct commissions:

> The Army of the United States, faced with a similar necessity for providing officers for a rapidly expanding army, has commissioned many men directly from civilian life, determining their grades both by the nature of the job they have been asked to undertake and by the nature of their civilian experience and training.

If the WAAC had done the same, she noted that in February the WAAC with its current strength of over 30,000 members would have had, at least, on the Army ratio, 183 lieutenant colonels, 329 majors, and 848 captains; actually it had no lieutenant colonels, no majors, and only 82 captains. While no one desired these maximum numbers in a new organization, the Director suggested that staff directors and unit commanders be advanced to the rank to which their age, civilian experience, and military duties would have entitled them under the Army system of direct commissions. Otherwise, she noted, "The responsibilities devolving upon unit commanders and staff directors will not be adequately recognized, and they will be at a disadvantage in dealing with other officers of like responsibility but higher rank." Even under Army promotion policy, she noted that in similar emergencies West Point graduates and Air Corps men had been granted exceptions to the required time-in-grade.

However, the Army Service Forces' final decision was that the Army time-in-grade must thenceforth be observed; even women occupying Air Corps positions must observe the longer time-in-grade of the Army instead of the shorter time of men in Air Corps jobs. General Grunert noted: "Notwithstanding the contemplated expansion of the WAAC, there appears to be little justification for such rapid promotion." One concession was made, in which some promotions to captaincies were permitted in April, but none to field grade. As a result, the new staff directors went out as captains or lieutenants; by the end of the war, very few had reached the originally allotted rank of lieutenant colonel, and many company commanders were not yet captains. For promotions to the grade of major, only two exceptions were made during the remainder of the WAAC's existence, the first of these being for the officer sent to fill, in the Army Air Forces, a position comparable to Director Hobby's in the Service Forces.[34]

Major Commands Receive Waacs

Army Air Forces

At this point it was evident that the Air Forces' share of the WAAC field units would be almost double that of all other commands combined. It was currently scheduled to receive 253 field companies as against the Service Forces' 120 and the Ground Forces' 7. The receptive Air Forces attitude toward the employment of women was due largely to the announced policy of Gen. Henry H. Arnold, Commanding General, AAF, who repeatedly wrote field commanders that efficiency required the employment of Waacs to the widest extent possible, to make up manpower deficiencies.[35]

The Air Forces Waacs began arriving in March, accompanied by staff directors for the various subordinate air commands. The first unit, fifty-seven WAAC enrolled women and two officers, arrived at Jefferson Barracks, Missouri, on 3 March 1943 to work for the AAF's Map Chart Division. On 22 March the AAF's first two WAAC post headquarters companies arrived at Chanute and Scott Fields. Twenty-three more WAAC units arrived at air bases in April; by the end of the summer there were 171 air bases which had WAAC personnel as part of the permanent party.[36]

The choice of a woman to head this organization and to serve on General Arnold's staff was not announced until May. Although the importance of the position was second only to Director Hobby's, and in fact equal in echelon as long as the Director remained assigned to the Army Service Forces, the top rank which could be allotted was that of lieutenant colonel, in view of the legislative restrictions which limited the Corps to one colonel. The position was given the title of Air WAAC Officer, and was set up as parallel to that of the Air Quartermaster, the Air Engineer, and other such staff officers.

The woman chosen for this position was the Director's former aide and current acting deputy, First Officer Betty Bandel, who upon assuming this duty was promoted to be the Corps' first field director, or major; she was to remain the WAC's second-ranking officer throughout Director Hobby's tenure of office. The new Air WAAC Officer was at this time thirty years old; as the Director's aide she had accompanied Mrs. Hobby to Europe and on most field trips and thus had a firsthand knowledge of the WAAC situation. As Acting Deputy Director since February, she had written many of the Corps' more important staff papers and had played an active part in co-ordinating the decentralization plans with ranking officers of other agencies.[37]

Army Ground Forces

Since the Army Ground Forces ordinarily trained on ASF-serviced posts and needed overhead only at a few schools, it had been allotted only about 5,000 Waacs of the first 150,000 expected to be re-

[34] Memo, WAAC Hq for CofS through CofAdm Serv, 18 Feb 43, and atchd corresp. SPWA 314.7 (6-2-41)(1) sec 3 (1941).

[35] See Ch. XVI, below, for details.

[36] AAF WAC Hist, p. 17.

[37] WDBPR Press Release, 4 May 43.

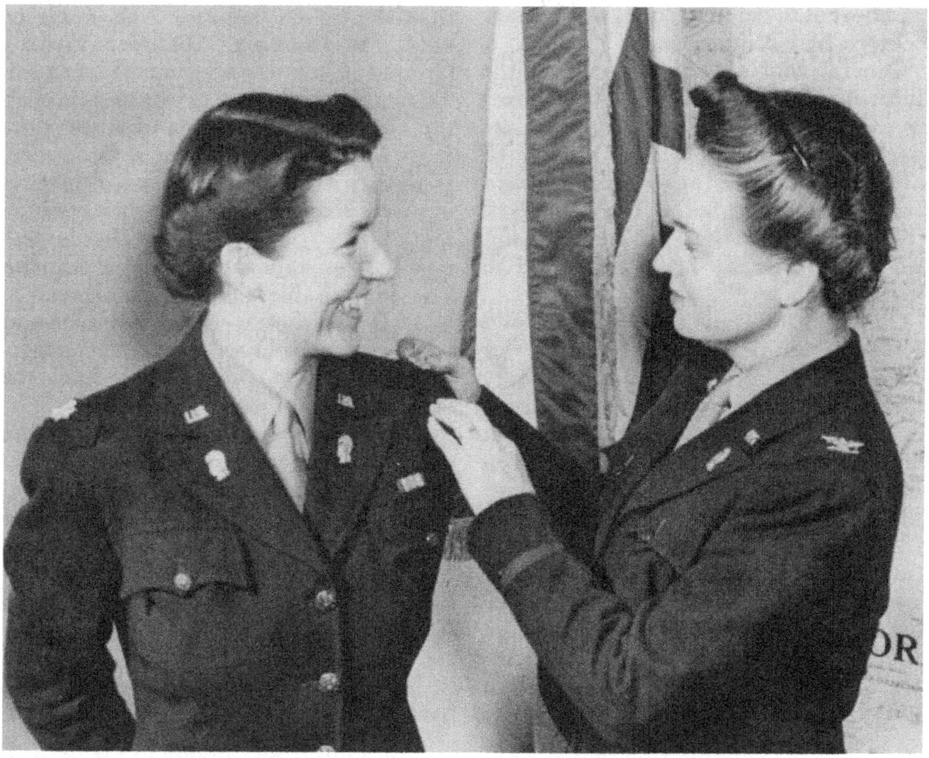

MAJ. BETTY BANDEL *receives her leaves from Colonel Hobby.*

cruited.[38] Comment within its headquarters was unfavorable to integration of even this number, and AGF Plans Section stated:

> In view of the educational, occupational, and physical training of the average American woman, it is anticipated that it would be extremely difficult to adapt them to military duties. . . . With the exception of a very limited number of assignments . . . there is no reasonable field for utilization of women in the military structure.[39]

Other comments added: "WAAC activities should not be expanded if it can be avoided . . . definitely opposed to coed organization."[40] Accordingly, Lt. Gen. Lesley J. McNair, Commanding General, AGF, wrote the War Department that "for the present, the WAAC should be continued only in limited size," and should not be placed in the Army: "It seems premature to anticipate manpower shortages

[38] Ltr, TAG to CG AGF, 1 Apr 43. AG 320.2 WAC (3-12-43) PR-W-WDGAP.

[39] The WAC in Army Ground Forces in World War II (hereafter cited as AGF WAC Hist), Vol. I, p. 4. This study was prepared by the AGF WAC Officer, Lt. Col. Emily E. Davis. (See Bibliographical Note.) Folder, WAC, AGF Sp Coll DRB AGO.

[40] *Ibid.*, comment of G-1 AGF.

so critical as to demand the extraordinary measures here contemplated."[41]

General McNair objected also to the allotment of 5,000, which had been based by the General Staff on proportionate overhead strength of male personnel, without his comment or consent; he asked that the allotment be set back to 3,600. This was done by the War Department, and the remaining 1,400 reallotted to overseas theaters.[42]

Under the impression that Waacs would always be administered by the Service Forces, the Ground Forces had already consented to receive some 1,700 Waacs at six different stations. The pioneer unit went to the Second Army at Memphis on 26 March 1943, and in April the others followed in rapid succession to Camp Hood and Forts Benning, Sill, Riley, and Knox. These installations followed their own inclinations, some allowing service commands to administer the Waacs, and some prematurely employing AGF channels to WAAC Headquarters. Within AGF headquarters, all WAAC matters were referred to G-3 Section rather than to G-1, under the impression that WAAC problems chiefly concerned units rather than personnel. The Ground Forces declined to accept the assignment of a WAAC staff director or any WAAC officer in its headquarters.[43]

Army Service Forces

Within the third of the major domestic commands, the Army Service Forces, the number of the service command staff directors was increased by the addition of other staff directors for the ASF technical services. This action was highly pleasing to most such services, which had long resented the interference of service command directors and had repeatedly asked that their own command channels be used. For example, Aberdeen Proving Ground objected to sending officer candidate applications and discharges through the Third Service Command, since Aberdeen was under the Chief of Ordnance. Inasmuch as many of the technical services were engaged in highly secret research and other activities, most felt that inspection by a member of another command was inadvisable. The appointment of WAAC staff directors for Ordnance and other services ordinarily ended such objections, and made command channels the same as those for men.[44]

Employment of Waacs among these technical services now promised to become rather extensive. The Transportation Corps placed a senior WAAC staff director in its Washington headquarters, and others in the various ports where Waacs were to be stationed in numbers—New York, Hampton Roads, New Orleans, San Francisco, and Seattle. Since most ports were in labor-short areas, they were already hard-hit by the loss of general service men.

The Chemical Warfare Service likewise foresaw a future lack of military personnel for permanent domestic installations, and desired to obtain Waacs whose research, once begun, would not be interrupted by withdrawal for combat duty. A WAAC staff director was therefore requested by the Chemical Warfare Service in March of 1943, and its first Waac company arrived in April.

[41] Memo, CG AGF for CofS, Attn: G-3, 1 Jan 43. GNDCG 320.2 WAAC (12-7-42). Full text in appendix to *Ibid.*

[42] 1st Ind to AG Memo, sub: Utilization of WAAC Pers in AGF. GNGCT 320.2/5 WAAC (4-1-43). Copy in App. VI of AGF WAC Hist.

[43] AGF WAC Hist, pp. 12-13.

[44] (1) Memo, WAAC Hq for TAG, 5 Jun 43; (2) Ltr, Aberdeen Proving Ground to WAAC Hq, 7 Jul 43, and reply. SPWA 300.3 (1-7-43) sec 2.

The Corps of Engineers now also requisitioned Waacs for an important project at Oak Ridge, Tennessee, so secret that its character could not be revealed.

The Signal Corps secured adoption of a plan to place in the WAAC some 8,000 civilian women, known as WIRES (Women in Radio and Electrical Service), whom it was training in radio, telephone, and other communications work; a staff director had already been accepted.

Medical Department requests also came in for the first time on 13 March, and totaled 20,869 or from 30 to 50 percent of enlisted personnel in hospitals of 500 beds or more. The requests were for women trained in various medical skills and also for clerks, drivers, and mechanics, plus 4,000 orderlies and others for low-grade work around hospitals. No staff director was accepted.[45]

Overseas Theaters

The decentralization of authority applied even to overseas theaters, since approaching legislation would give the Waacs the military status that would make further shipment morally justifiable. Requests from the North African and European theaters were by now so extensive that a War Department decision on priority was necessary. Director Hobby favored priority to the zone of the interior, but theaters argued that skilled Waacs had proved able to replace from two to three men apiece in North Africa, and that, in view of the difficulty of maintaining personnel overseas, theaters should get the most competent. Both General Somervell and General Marshall inclined to this view, and finally informed the Director that overseas theaters' requests would have priority.[46]

The staff director for the European theater, Capt. Anna W. Wilson, arrived in England on 13 April 1943 to plan for later shipments. No firm shipment priorities had yet arrived, and the several requisitions for Waacs were by this time so confused and overlapping that it was impossible to tell whether they concerned the same or separate units. Captain Wilson found that the theater had so little information about the Corps' mission and capabilities that specific requests were difficult; when this was supplied, the Eighth Air Force in the theater requested a WAAC battalion, and supplied a firm shipping priority. The first unit for England therefore began to be organized, and was to reach the theater in the summer shortly after the Corps obtained military status. Other earlier shipments to the North African theater were also authorized.[47]

Readiness for Increase in Numbers

By late spring of 1943 the WAAC organization was thus fully decentralized and well prepared, as far as mere skeletal organization was concerned, for Army status and for expansion to any number. Its staff advisers were in almost every important Army command, learning its needs and peculiarities; its command powers were so delegated that shift to Army status would scarcely be noticed; its training system was fully staffed and capable of accommodating large numbers of recruits. The administrative network was ready. Had no other factor been involved, expansion to a million or more upon this frame appeared easily possible.

[45] See section on each service in Ch. XVIII, below.

[46] Office Memo, WAAC Hq Opns Br for Exec WAAC, 29 May 43. SPWA 320.2 sec 4*a*.

[47] See Chs. XIX and XX, below.

CHAPTER VIII

Stresses of Rapid Build-up: Personnel and Training

In spite of its excellent framework, the expansion program was hardly launched before it became evident what stresses such a rapid build up would carry with it. In some respects these had been anticipated, since they were comparable to those which the Army as a whole had recently encountered; but there remained one major point of difference: the women's services were dependent upon voluntary recruiting. As the spring of 1943 wore on, it became clear that stresses easily supportable in themselves might nevertheless have an exaggerated effect upon public opinion and the supply of recruits. It was gradually recognized that any attempt at quick expansion by voluntary recruiting carried within itself the seeds of self-defeat.

Training Center Confusion

From the viewpoint of the WAAC training centers, the first six months of 1943 had all the aspects of chaos. As successive new centers were staffed, commandants and key officers seldom stayed more than a few months in any location; experienced instructors, cadre, supply officers, classification officers, all moved to open new centers before projects could be completed or reports made. As new field companies had to be formed, some basic companies were reported to have lost their cadre several times in the four-week basic training. A recruit who trained at Camp Polk wrote:

> We changed officers so often that no one knew who was looking after us. We seldom had the same instructor or classroom two days in a row, and some instructors must have just taken over classes because they knew nothing about them. We didn't learn a thing. We never even had a retreat parade the whole time, and I made a fool of myself during one when I got to Fort Riley later. In our last class the workmen walked in and took out the furniture and just left us standing there.[1]

Field complaints indicated that training authorities sometimes shipped units without advance notice to the station; service records and classification cards at times did not accompany unit movement but arrived as much as two weeks late, incomplete and incorrect, after assignments had been made without benefit of records. Stations protested when Waacs arrived after cross-country trips in dirty coaches without lights or sanitary facilities and with no one responsible in charge.[2] At one station

[1] Interv with Miss Dorothy Pat Costello, former WAC sergeant, 4 Aug 48.

[2] (1) Ltr, 2d SvC to Dir WAAC, 9 Mar 43, sub: 42d WAAC Post Hq Co. SPWA 319.1. (2) Ltr, Col Howard Clark, Chief, Opns Serv WAAC Hq to 2d Tng Cen, 20 Mar 43. SPWA 319.1-os. (3) Ltr, 2d SvC to Dir WAAC, 1 Dec 42, and Inds. WA 513 (11-26-42) Transportation.

the unit was in the field ten days before an officer arrived. The staff director reported, "Undoubtedly the timing went wrong somewhere, but it created an unfortunate impression of WAAC organization." [3]

There were also more lasting inconveniences to stations receiving units. Enrolled women of low ability or questionable character, who once had been weeded out at training centers, now were undetectable in the general confusion and were the subject of complaint when they arrived at field stations. Company officers and cadre for outgoing units were lumped together from whatever was available at the moment, without much consideration of such factors as relative age and civilian experience, although these were often the margin between success and failure in groups of women destined to spend the rest of the war together in one installation. Untried cadre were promoted by training centers to the grades specified in the Table of Organization, often leaving units in the field saddled with incompetents who had too much rank to be reassigned. Officers were sent out as company commanders who had little or no experience or aptitude for the work; Maj. Betty Bandel noted, concerning the Air Forces, "Our greatest problem is our company commanders. I make a special plea that at least one experienced officer go out with any new unit to the field." [4]

Field stations frequently complained that neither WAAC officers nor enrollees arriving from training centers knew anything about WAAC rules and regulations, and did not have copies of them. Some company commanders considerably embarrassed their staff directors by informing station commanders that the Army had no jurisdiction whatever over WAAC units. Others had little idea of real Army life in the field, and attempted to inculcate a state of discipline and daily routine suitable only for a brief and strenuous basic course, but impossible to sustain in a working unit.[5]

Commissioning of Unqualified Officers

Among early casualties of the expansion program were the original perfectionist plans for avoiding British mistakes in officer selection. As soon as the expansion program was directed, it was evident that the original carefully selected 1,300 officers would be insufficient to staff the new units. Since officer candidate training was weeks longer than basic training, it became necessary to begin large officer candidate classes before the women to be commanded were even recruited. Unavoidably, almost all came from the training centers, since most companies in the field were still in transit or had not had time to set up boards and select officer candidates, although it was well known that many of the best-qualified women had been placed in early field companies, while those in the training centers had not yet had opportunity to demonstrate leadership under actual field conditions.

As available personnel in training centers was combed over again and again for successive classes, quality fell rapidly. Daytona Beach complained that it had to furnish four large groups of officer candi-

[3] Ltr, WAAC Stf Dir WFTC to Exec WAAC, 3 Jun 43. SPWA 319.1.

[4] Comment of Air WAAC Off, Min, Stf Dirs Conf, Chicago, 15–17 Jun 43. SPWA 337 (6–1–43). Also see: (1) other comments in Min; (2) Memo, Consultant (Miss Lies) for Dir WAAC, 6 May 43. Plng Proj 6, WAAC Planning files, 1943.

[5] (1) Rpt of 2d SvC, Min cited n. 4. (2) Interv with Lt Col Katherine R. Goodwin, former Stf Dir 1st SvC, 31 May 46. (3) Statement by Lt Col Jessie P. Rice, former Stf Dir 3d SvC, in speech at Purdue Univ., Apr 45. (4) Min, Stf Dirs Conf, New York, 1–3 Dec 43. SPWA 337 (11–10–43).

dates from its cadre alone, before trainees arrived, and that while the first group was good, the last three were extremely doubtful. Board members at Des Moines reported, horrified, that they had given the lowest possible scores to certain applicants who were judged impossible material because of poor appearance, manners, education, and possibly morals, yet to make up the required quota they were directed by the WAAC Training Command to go to the bottom of the list and accept every applicant on it.

General Faith, in directing this action, noted that the WAAC had no choice if it was to meet the expansion plans. He therefore ordered boards to be realistic and admit some candidates who would make good second lieutenants although they would never be able to advance higher: "The ability to advance to high rank, which was very important in the original selection of officer candidates, has become of less importance as the total number of officers already commissioned has increased." [6] He directed boards under his Training Command not to reject anyone finally, since the ones rejected one week might be better than those available the next. General Faith suggested to the Director that she order officer candidate boards in the field to follow this system, but she did not concur, and instead published a field directive which established a selection system much like the Army's.[7]

Even if good candidates were sent by posts or training centers, they did not invariably receive commissions. During the hectic months of expansion the WAAC Officer Candidate School came to share with men's schools a problem of which Army inspectors noted:

> In many [Army officer candidate] schools, performance on the drill field was used as a means of separating those with 'leadership' qualities from those lacking such characteristics . . . a loud voice and a general air of confidence indicated that 'leadership' was probably satisfactory.[8]

Numerous field complaints were received on this score; for example, the Fifth Service Command protested the "washing out" of three of its candidates who, in spite of leadership ability demonstrated in the field, had been rejected for "no voice and command," "could not drill," and "no pep and enthusiasm." [9] Similarly, the Transportation Corps protested concerning thirteen women who had all "washed out for voice and command, nothing else." [10] After frequent complaints, Director Hobby sent to the Training Command for transcripts of "murder board" hearings, which convinced her that too much emphasis was placed on youth and physical contour, and too little upon character and past accomplishments. At a time when field stations were pleading for mature officers, boards had informed applicants that an age as advanced as 35 years, if apparent, was disqualifying.[11]

---

[6] Ltr, CG WAAC Tng Comd to Dir WAAC, 3 Feb 43, and pencil notes thereon with Dir's nonconcurrence. SPWA 291.9 (2-3-43).

[7] (1) Ltr, Dir Tng, 2d WAAC Tng Cen to Dir WAAC, 8 Mar 43. SPWA 319.1. (2) Rpt, 2d O Ann Danovsky, *et al*, to 2d O M. E. Treadwell, Des Moines Office Dir Tng. Confirmed by intervs with Maj Helen Hanson, head of OCS and Dir Tng, 3d Tng Cen. (3) WAAC Cir 8, 17 Mar 43, Sec VIII. Circular had been requested a month earlier, Memo for CofAdm Serv, 19 Feb 43. SPWA 300.2 (1-7-43) sec 1. (4) Speech by Dir Hobby, 9 Jun 43, Min, Conf of Offs to Supervise OC Bds. SPWA 334.9 (6-10-43).

[8] ASF Hist of Military Training: Officer Candidate Training, 1941-45. (Hereafter cited as ASF Hist of OCT.) OCMH.

[9] Ltr, 5th SvC to Dir WAAC, 27 Jul 43. SPKEB 341, in SPWA 341 R (1942).

[10] Min cited n. 4, especially rpts by San Francisco POE and by Dir Hobby.

[11] *Ibid.;* also interv with Maj Patricia Lee Chance, Exec WAC, 1945.

Her discovery of the collapse of standards at training centers came too late. In the few brief months before the end of the summer, half of all WAAC officers were women without previous experience in jobs more responsible than those of clerks, typists, stenographers, and secretaries—including sixteen actresses, fourteen chorus girls, fourteen waitresses, and varying numbers of beauticians, charwomen, laborers, chauffeurs, and housekeepers, as well as one undertaker. Education had also declined from earlier standards. About 31 percent of WAAC officers now had only high school education or less—about 5 percent had less—while another 28 percent had not completed college; only 41 percent were college graduates. Still worse, 30 percent of officers were less than twenty-five years old.

Upon learning these facts, WAAC Headquarters took action to set up another campaign to get more officer candidates from civilian life, but before this could well come into operation, the Corps had reached almost the top officer strength it was to be permitted. The WAAC by the end of its first year had over 5,800 officers, who were, with the later failure of the expansion plan, to be enough for the rest of the war. Only a few hundred more were to be commissioned in the remaining years, so that many of the unsuitable officers so hastily commissioned did not remain second lieutenants, as General Faith had expected, but were pushed up gradually as the Corps expanded. Embarrassing as this deficiency was to everyone concerned, the WAAC had experienced no more problem in this respect than the Army; remarkably enough, even this lapse had not brought the average of WAAC officer education below that of Army officers. However, even a minority of unsuitable or rank-happy officers were always to be extremely conspicuous in a women's service.[12]

Unsuitable Mentors

The expansion program also made it impossible to carry out Director Hobby's recent order that male officers in training centers be replaced by Waacs, the most unsuitable to be the first weeded out. The "man-woman factor," unprecedented in Army schools, afforded opportunities for maladministration which all too frequently were found to have been realized. General Faith later estimated that male offenders under his command totaled not more than 5 percent, and that these were immediately relieved from duty if discovered in misconduct. Nevertheless, scattered reports continued to be received from all training centers of cases in which women's advancement was allegedly based on matters other than merit.

The gravest situation in this respect was that at Daytona Beach, which was investigated at Director Hobby's request by operatives from the Military Intelligence Service in Washington as well as by the Fourth Service Command's inspector general. These inspectors' reports noted that certain of the Army officers in highly responsible positions were drinking and misconducting themselves with Waacs and were promoting to positions of leadership only those who offered them favors in return. Members of the Army staff, when written statements were taken, officially accused each other of "running around"

[12] (1) Memo, Capt Ruby E. Herman, WAAC Hq, for Dir WAAC, 12 Nov 43, with incl statistics, 12 Nov 43, made as of 15 Sep 43. Unnumbered folder, 1943 WAAC files. (2) ASF Hist of OCT, 1941–45, pp. 15–16, 74. OCMH.

and "carrying on" with Waacs, and of using force on enlisted women while drunk.[13] A Waac visitor described a resulting situation of "sycophancy" by senior WAAC officers and of "obsequious flattery" among junior ones. When a WAAC inspector arrived, she learned that women were in many cases not assigned in accordance with real skills and abilities.[14]

In view of the fact that many of the officers assigned to WAAC training centers had been pronounced "culls" by the commandant when first assigned, it was perhaps remarkable that more serious situations did not develop in all of the various training schools. Concerning the common practice of assigning second-rate officers to staff women's schools, the Bureau of Naval Personnel observed that in the WAVES' program "it was a penny-wise-pound-foolish policy not to detail the handful of first-rate men it would have taken to have established the whole program on the best possible basis."[15]

Observers noted that these unsuitable mentors often had an unfortunate effect upon newly commissioned, inexperienced, and eager-to-please female officers. Through a process of what Army psychiatrists called "mimesis," or unconscious imitation, as well as through a conscious desire to secure advancement, women copied supervisors who were not in themselves models of Army leadership. Imitation was noted not only in standards of morals and public behavior, but in language and attitude toward trainees. The senior WAAC officer at Des Moines, beset by critics at a staff directors' meeting, admitted, "I have been told that some of our officers are very rough in their language. I think it is due to some of our Army officers at the fort."[16] It was some time later that Director Hobby formulated the difficulty and told a group of women officers:

> It seems to me lately that I have seen too many women officers who are hard. The last thing we want to accomplish is to masculinize a great group of women. . . . I think that [harshness] often comes from the officer's own insecurity. When she is at a loss how to handle a problem, she relies on what she thinks would be the Army attitude. Actually it is not an Army attitude among leaders. Only insecure persons revert to harshness.[17]

The "Nightmare" of Basic Training

In these months many recruits used the word *nightmare* to describe basic training, although at earlier and later periods the same course was pronounced "inspiring." One recruit, an Army wife, noted that this definition was literal: "For months after I left there, I used to wake at night crying, dreaming I was back at Des Moines."[18]

The difficulty, insofar as it could be diagnosed, appeared to lie chiefly in an overabrupt and deliberately harsh introduction to Army life. The early high state of training had been produced when recruits' ideas of membership in a military team had been more nearly met, with plentiful ceremonies and instruction in Army tradition which produced a lasting pride in military status regardless of what hardships followed. This system continued to be followed, by deliberate intent, at the

[13] (1) Memo, Dir WAAC for TIG, 13 Mar 43, sub: Request for Investigation, and atchd corresp. SPWA 330.14, Folder, Daytona Beach, Dir Adm Serv ASF Sp Coll DRB AGO. (2) Memo, IG 4th SvC for CG 4th SvC, 9 May 43, atchd as Exhibit B to Rpt, CIC to MIS, 19 Jul 43, sub: Origin of Rumors Concerning WAAC, G-2 files MID 322.12 WAAC. (3) Also see Ch. XI, below.

[14] Memo, 1st O Anna W. Wilson for Chief, WAAC Plng Serv, 11 Mar 43. SPWA 319.1 (3-11-43).

[15] WAVES Hist.

[16] Min cited n. 4.

[17] Speech at Stf Dirs Conf, New York, 1-3 Dec 43. Min cited n. 5(4).

[18] Diary of an Auxiliary, winter 1942-43, made available to the author in a personal interview, 31 January 1946.

WAVES' training centers, with the first weeks devoted to inculcating a knowledge of Navy tradition and ceremony, and with only fully trained women graduates instead of recruits being used for the more menial tasks about the school.

Gradual adjustment seemed even more necessary for female recruits than for male draftees, since the women were apt to be more uncertain about their ability to "take it," and had been subjected to the sales talks of recruiters who accented the glamorous side of Army life. At a later period these facts were to be recognized by WAAC training authorities, but during the winter and spring of 1943 there was reported, at most training centers, an initial emphasis on military "hazing" which, as one authority later noted, "tried to belittle a recruit's ideals, laugh at her patriotism, and make her a soldier in the first five minutes." [19]

Thus, one recruit's diary noted what other accounts confirmed as a typical experience at Fort Des Moines: "I was sent out to scrub one of the offices on my first day, before I got my uniform." Emphasis on fatigue duties continued to take precedence over military indoctrination:

> We seldom got to attend classes regularly, as we were pulled off for all kinds of duties: we did KP at our own mess, the officers' mess, and the consolidated mess; we scrubbed classroom floors and the theater daily; we cleaned offices, orderly rooms, dayrooms, storerooms, cleaned the outside of buildings and washed the windows and the white pillars, though in zero weather. We hated to miss those classes as there was so much to learn in only four weeks.[20]

The absence of a uniform during much of the training period was also keenly felt, though to a less extent than the fact that "the men who drilled us and marched us through the snow had complete warm outfits." The fact that almost all recruits at one time or another during training contracted colds and coughs, sometimes influenza or bronchitis, did little to increase their endurance. Even more lamented was the lack of drill and ceremony, which was prevented at some centers by the winter weather and lack of warm clothing, and at others by lack of space, by scattered housing, or by the necessity for pulling out half-trained women to meet the shipment schedule.

Under these conditions, some women suffered breakdowns and had to be discharged before receiving a field assignment, in such numbers as later to cause investigation by The Surgeon General. For the majority, it was evident that surprisingly little permanent damage was done. The time of basic training was brief, the hoped-for military ceremonies were eventually encountered on Army stations, and in looking back from the vantage point of years, trainees were if anything inclined to boast about how they "took it." Morale was reported in even the most confused months of the expansion program as "unbelievably good," and "the women were excited about their work, determined to succeed, and felt that the worst would be over if they could only get to the field and to work." [21]

The women's generally good-humored reaction was fairly well expressed in their favorite songs, which grew up during these months: [22]

> WAAC Days, WAAC Days,
> Dear old break-your-back days. . . .

---

[19] Speech by Lt Col Jess Rice, "Development of the WAC," at Sch of WAAC Pers Adm, Apr 45, notes in WAC files, OCMH. Also see bulky studies in SPWA 331.1 (3-1-43), now in file of Dir Pers ASF.

[20] Diary cited n. 18.

[21] *Ibid.*

[22] Pamphlet, WAAC Parodies. The only song not from the pamphlet is "AR 35-1440," most of which is unprintable.

KP DUTY, *Daytona Beach, Florida. Note seersucker exercise dress.*

When you come to the end of a perfect day,
And you sit alone with your gigs. . . .

K-K-K-Ka-P, beautiful KP,
You're the only Army job that I abhor.
When the moon shines over the mess hall,
I'll be mopping up the K-K-K-Kitchen floor.

Troublesome rules and regulations were memorized to the tune of ditties such as:

> AR 35-1440,
> Deals with matters very naughty.

Running a close second to the earlier favorite, "The WAAC Is In Back of You," was a new original at Des Moines, entitled the "G. I. Song": [23]

Once her Mommie made her bed,
Cleaned her clothes and buttered her bread,
And her favorite dress was red—
Oh me, Oh my, that ain't G.I.

Hats and shoes and skirts don't fit,
Your girdle bunches when you sit,
Come on, rookie, you can't quit—
Just heave a sigh, and be G.I. . . .

In the Mess Hall she now stands
Buried 'neath the pots and pans
Getting pretty dishpan hands,
Oh me, Oh my, gotta be G.I.

Then she came to camp one day,
Quickly learned the WAACKIE way,
Underwear cafe au lait—
Oh me, Oh my, strictly G.I.

Winter, summer, spring or fall
Should you try to end it all
You can't die until sick call
You see, if you die, you gotta die—G.I.

However, even among such lighthearted versions of school days, the final song was:

> We're in the Staging Area
> And we soon will go away
> We've finished all our basic
> Glory be and happy day
>
> Glory Glory we are staging
> Glory Glory we are staging
> Glory Glory we are staging
> Before we travel on.

The more damaging aspect of the dislocations to basic training during the months of expansion appeared to have been the effect upon recruiting. In the letters written to friends and relatives during the first week or weeks, there was noted, at worst, a frightened disillusionment or frantic desire to escape from what at first glance seemed a trap and, at best, a tendency to boast about the unspeakable hardships the writer had survived. The Army as a whole was not unfamiliar with such a phenomenon, which had little effect upon draft boards. British women's services had noted a less favorable public reaction, and that "The large majority adapt themselves in time . . . but a certain number, when writing home, may easily exaggerate. . . ." [24]

Just how great a part the brief "nightmare" of basic training had played in the later failure of the recruiting program was never to be easy to determine. General Ulio, The Adjutant General, ascribed some importance, in a staff study made during these months, to "people writing home about . . . conditions." [25] Recruiters attributed damage especially to publicity on the prisoner-of-war camps, in which no male soldiers had been reported as required to train. Later Gallup surveys also showed that many acceptable recruits were allegedly deterred from enlistment by the belief that they lacked the necessary strength to survive the reputed rigors and hazings of basic training.[26]

It came to be the general belief within WAAC Headquarters that the echelon of the Training Command had been in itself

[23] Words and music by Lt. June Morhman.

[24] *Conditions in the Three Women's Services.*

[25] See Ulio rpt, Ch. X, below.

[26] See Gallup rpt, Ch. XIII, below.

one of the less happy results of the expansion program. The Acting Deputy Director noted:

> We were completely cut off from the training centers and could not even communicate with them; instead, we should have had direct telephonic communication all winter in order to find out their difficulties and expedite corrective action in supply, officer selection and training, and other matters.[27]

In March, Director Hobby petitioned the Services of Supply to abolish the echelon of the Training Command and restore its functions to her office, as well as General Faith in person, for its better assistance and information:

> The Training Plans Division [of WAAC Headquarters] has no authority and receives problems for decision which it is unable to give. In addition, it is uninformed on training activities and programs. . . . Since the functions now being performed by the WAAC Training Command are properly headquarters functions, it is recommended that . . . the functions of the Training Command be absorbed as rapidly as possible in the appropriate organizational units of WAAC Headquarters.[28]

This request was disapproved by the Services of Supply, but some increase in effective liaison was achieved by moving the Training Command headquarters from Daytona Beach to Martinsburg, West Virginia—a location as near to Washington as possible, and one removed from the scene of the scandal concerning maladministration at the Daytona Beach training center. At the same time, General Faith was given additional duty as head of a new Training and Field Inspection Division in the Director's office, where he now spent part of his time. It was hoped that the field inspection function would provide training centers with assistance and some guide to field needs.[29]

Personnel Problems in Headquarters

Its loss of control over the training program was only one of the difficulties that expansion had brought to WAAC Headquarters. The office still suffered from chronic understaffing and overwork; since January it had been reported that "the pressing common problem throughout Headquarters is a shortage of competent dependable civilian help. Officers are doing clerical work." [30]

The Director, while supplying thousands of Waacs to the Army, was in the position of being unable to get a requested allotment of eleven for herself, because of the Services of Supply's rule against bringing enlisted Waacs to Washington. Finally, a few enlisted women were brought in on temporary duty, which prevented their promotion, and were soon also suffering from overwork, while the headquarters feared to let them go lest shipment orders remain untyped and "a tremendous backlog accumulate at Training Centers."

Letters concerning individual Waacs were referred unanswered to their posts, and letters of inquiry from Army stations were answered without being seen by the Director or top authorities—a system that provoked much unfavorable comment from the public and the field, and resulted in vague or inaccurate routine replies from junior officers to inquiries which were really matters of policy. The volume of mail was clocked at an average of 406

[27] Interv with Lt Col Betty Bandel, 5 Oct 45.

[28] Memo, CofS ASF for Dirs of Stf Divs and Chiefs, Sup and Adm Svs, 1 Mar 43; Dir WAAC's reply, 19 Mar 43. SPWA 331.1. (3-1-43), located and read in Office Dir Pers ASF.

[29] TWX, Faith to Dir WAAC, 29 Jun 43. SPWA 314.7 (1-7-43)(1) sec 1.

[30] Memo, 1st O Anne Alinder for Contl Div WAAC, 6 Jan 43. WAAC files.

pieces daily. The stenographic pool in April had a ten-day backlog and was obliged to reject all but urgent work, while some 1,200 papers remained unfiled. In April, the headquarters petitioned the Service Forces for an increase of civilian personnel from 56 to 77, stating: "Necessary work is being delayed or left undone with the result that the WAAC program is being seriously impaired." [31]

The problems of an expanding organization did not receive too much sympathy from the Army Service Forces, which had passed its own critical period and was at the moment engaged in a drive to reduce personnel. On 1 March WAAC Headquarters, in common with all ASF offices, was required to report on steps taken to abolish nonessential functions and personnel. The Director protested that reductions could not safely be made at present, but that in a few months, if legislation passed, "it will be possible for WAAC Headquarters to avail itself of existing Army machinery for handling certain functions, which will permit dropping activities which parallel functions done elsewhere in the War Department." In this, she was overruled by General Styer, who indorsed the papers back with a demand for immediate reduction of the office force, stating: "You will submit a further report . . . [which] will list . . . the number of military and civilian personnel released." [32]

The Director took the occasion to point out that the January 1943 reorganization of her office by General Somervell's Control Division, like that of October 1942, was unworkable. She indicated that there were friction and confusion among her Army section chiefs, with different section chiefs claiming the same projects. She therefore asked that authority be given her to redefine duties and regroup the sections concerned according to her own ideas of office organization. There was also some discontent among the WAAC officers who made up the junior working personnel, in that male officers still retained key jobs and policy-making powers and did not allow WAAC officers to say anything in staff meetings that had not been written out and approved in advance, or otherwise to give the Director "stories of what is going wrong." [33]

In late April, Control Division, ASF, partially acceded to the Director's request and again revised the WAAC Headquarters organization—the third revision in seven months—changing the wording of the ASF Manual accordingly. A personnel increase was also briefly granted, but nullified three weeks later, and a further cut ordered.

The headquarters situation, during the entire time when the expansion program ran its course, therefore remained generally critical. The only prospect of relief and improved efficiency was Army status, which did not come until the hope of recruiting success was ended.[34]

[31] (1) Memo, Dir WAAC for IPD ASF, 2 Apr 43, sub: Civ Pers, and atchd. SPWA 230.14 (4-2-43). (2) Memo, Dir Adm Div for Chief, Opng Serv WAAC, 31 Mar 43; Memo, Dir Adm Div for Dir of Opng Serv, 22 Jan 43. SPWA 311.1.

[32] Three studies: (1) Ltr, ASF to all Divs, SvCs, etc., 1 Mar 43; Dir WAAC reply, 20 Mar 43; 1st Ind, CofS ASF, 10 Apr 43. (2) Memo cited n. 28, and 1st Ind, CofS ASF, 14 Apr 43. (3) Memo, CofS ASF for CofServs and SvCs, 1 Mar 43; Dir WAAC reply, 30 Mar 43; 1st Ind, CofS ASF, 19 Apr 43; 2d Ind, Dir WAAC, 28 Apr 43. All in SPWA 333.1 (3-1-43), now in Dir Pers ASF Sp Coll DRB AGO.

[33] (1) WAAC SO 78, 22 Apr 43. SPWA 300.4 (8-18-42), 1942. (2) Interv with Maj Elizabeth C. Smith and Maj Irene Galloway, 11 Oct 45. (3) Interv cited n. 27.

[34] ASF Cir 30, 15 May 43; also SOS Manual M301, Jul 43, par 202.03.

Shipment of T/O Units

In the field, possibly the most serious publicly known personnel problem caused by the expansion program was in the matter of job assignments. Again, the difficulty was not the expansion itself, but the fact that it was required before the end of the Auxiliary system.

The separate WAAC Table of Organization—which could not yet legally be merged with military job allotments—had proved in itself a disaster for the Corps. Although large Army Tables of Organization had vacancies for almost any variety of skill, the small 150-woman T/O had positions for only a few general skills—chiefly office workers and chauffeurs. As recruiting restrictions were eased, recruits were accepted who had other and more varied skills. However, when a WAAC was recruited with some rare and badly needed skill, such as teacher of Braille, she could not be assigned as such because the WAAC allotment did not call for it. The WAAC T/O could not be amended to add the job without requiring every WAAC unit in the world to have a teacher of Braille.

Under this system, the temptation for training centers to misclassify women was great. The teacher of Braille sat waiting assignment which could never be given her; a WAAC company could not be shipped until a mail clerk was found to complete its T/O; the teacher of Braille overnight became a mail clerk. Still worse, a Table of Organization could not be amended by a post commander; when the WAAC unit reached a station, the station was helpless to remedy the misclassification and was obliged to assign the Braille expert as a mail clerk, even though its hospital was badly in need of her skill and had a military vacancy for it. "Many of our stations faced this problem," reported Staff Director Jess Rice of the Third Service Command.

> The WAAC Table of Organization called for six housekeepers, so they fired the maids in the guest houses and put the Waacs to scrubbing floors, even though they might be doctors of philosophy, teachers, or technicians, who had been classified as housekeepers just to get them out of the Training Center.[35]

If training center classification teams rebelled at the dishonesty of falsifying such a woman's records and classified her properly, she could not be assigned at all. Training centers with such scruples gradually accumulated a number of women with more or less odd skills—dental hygienists, X-ray technicians, translators, key-punch operators, opticians, dietitians, and many others. These, although badly needed to fill military vacancies, could not be assigned in the WAAC T/O unless classified as unskilled. Training centers reported that these women were in a state of very low morale: many advised civilian friends that the Army could not use their skills; and some nervous individuals actually became psychopathic cases, after months of delay and malassignment, and had to be discharged.[36]

Malassignment of skilled workers was, of course, well known to Army men in times of rapid expansion. The effect of malassignment on a woman, however, was often greater, because she could not receive a combat assignment which might in her own eyes justify the waste of a special skill. The only reported cases in which malassignment caused no morale damage to Waacs were those instances in which a

[35] Speech, "Hist of WAC," at WAC Pers Adm Sch, Purdue Univ. OCMH.

[36] (1) Memo, Classif and Asgmt Sec WAAC Hq for Sp Consultant to Dir, 7 Jun 43. SPWA 201.6 (6-7-43). (2) Rpt of WAAC Stf Dirs, 2d, 3d, and 4th Tng Cens. Min cited n. 4.

clerical or sedentary skill was ignored in favor of a field-type job, which most women were found to prefer.

The whole WAAC training system was soon distorted to meet the rigid requirements of the separate WAAC T/O. Exact calculations were made as to how many cooks, clerks, and chauffeurs would be needed in a certain number of companies, and the WAAC specialist schools were geared to turn out this number. Once the school quota was set, it became an inexorable demand which had to be met weekly regardless of the qualifications of the new recruits who presented themselves that week. Thus, if in one week there were large numbers of typists, badly needed in the field, they must be divided into three parts, one for cooks and bakers school, one for motor transport school, and one for administrative school. If, a week later, the recruits were all skilled mechanics, they must nevertheless also be divided among the three schools.

General Faith admitted later, "We arbitrarily tried to maintain a balance. Women who hated to cook were sent to Cooks School and women who couldn't spell were sent to Administrative School." [37] One Air Forces representative noted:

> We don't mean to complain of the classification at the Training Centers, but we received a unit of 76 from one of the centers and all 76 were supply clerks. When we interviewed . . . we found we had college graduates who had years of experience in hospitals or other jobs, and they had been sent to administrative schools to become supply clerks.[38]

Expert stenographers had been trained as cooks, while former cotton pickers were graduates of Administrative Specialists School. Women had been sent to radio school who had not only hated radio work but also had medical or clerical skills.[39] Airfields that requisitioned weather observers got former waitresses classified as "potential weather observers." [40]

The War Department had for some time realized that the T/O system, while appropriate for rifle companies and other identical combat units, was a poor means of supplying either men or women to zone of the interior stations, no two of which had exactly the same needs. The Bulk Allotment system, prepared during these same months, promised relief from the distortion of both WAAC and Army skills to fit an inflexible Table of Organization.

Both WAAC and Army allotments were eventually to specify only the numbers and grades of individuals sent to most noncombat commands, leaving their duties to be determined and requisitioned according to the needs of each station. Although WAAC Headquarters worked all spring to hasten the application of this system to the WAAC, such did not become possible until May. In late March Army commands were warned to prepare for the new system. For units already shipped, they were expected to prepare Tables of Allotment representing the women's real skills; for units yet to come, any desired assortment of skills could be requested. On 1 May WAAC T/O units were formally inactivated, and Army commands were on their own in determining WAAC jobs within the limits of the total Bulk Allotments.[41]

---

[37] Speech. Min cited n. 4.

[38] Speech of WAAC Stf Dir AAF Tng Comd. Min cited n. 4.

[39] Rpt of 2d SvC. Min cited n. 4.

[40] Rpt of Insp, Great Falls AAB, Montana, 20 Aug 43. SPWA 333.1.

[41] (1) WD Memo W635-3-43, 31 Mar 43. (2) Ltr, TAG to CG ASF (AGF, AAF), 1 Apr 43. AG 320.2 WAAC (3-12-43) PR-W-WDGAP.

Replacement of Soldiers

Even the Bulk Allotment system did not solve another problem that resulted from Auxiliary status, whether under a Table of Organization or a more flexible Table of Allotment: under either system it proved difficult to determine whether a Waac had really replaced a man. The General Staff sent out repeated injunctions that Waacs must count against the Troop Basis and replace men one-for-one; a series of increasingly stringent War Department directives required a full report by skill and rank of men shipped, to be tallied against the number of Waacs received. Only thirty days—fifteen in the Army Service Forces—were allowed for the replacement, a shorter time than customary with male replacements, on the theory that a Waac could learn a job faster.

Nevertheless, it became increasingly evident that few men were actually being sent away, and might not be until the WAAC became part of the Army and its allotments could be merged with the military. Through a technicality of paper work, minor but basic, it was found that most Army posts were able to retain both Waacs and men. Each station had an allotment of military personnel and an allotment of civilian personnel; when Waacs arrived it had a third type, since each company came complete with its own Table of Organization or Allotment under which its members could be promoted and paid. Whether Waacs filled military or civilian jobs, all three allotments remained intact. All efforts, therefore, to get many posts to send a soldier away failed except on paper. A Waac might actually be given a man's desk and see the man depart, only later to find him at a new desk a few doors down the hall, in work for which the installation often had a perfectly legitimate allotment. If the man actually departed, the station had a vacancy in its military allotment, and promptly requisitioned another man. So long as Waacs were not legally recognized as military personnel, and could not fill vacancies in military allotments, the General Staff directives concerning the Troop Basis remained administratively impossible to enforce.[42]

The apparently simple solution of reducing a station's military allotment for each WAAC allotment, although directed by the General Staff, was also administratively unworkable. A station's allotments, whether in the form of Tables of Organization or the later Bulk Allotments, were complicated affairs which could not be altered at station level. To reduce the allotment, it would be necessary to republish it each time a new WAAC unit arrived, or each time a Waac was promoted from one grade to another. The time lag in this process was such that, reports indicated, most stations would not apply to higher echelons for republication of their military allotments, but clung to both WAAC and Army personnel. Only approaching military status could entirely solve this problem, by making one military allotment apply to both men and women.

[42] (1) WD Memo W635-2-43, 10 Feb 43. (2) WD Memo W635-3-43, 31 Mar 43. (3) WD Memo W635-4-43, 31 Mar 43. (4) WD Memo W635-5-43, 26 Apr 43. (5) ASF Cir 39, 11 Jun 43.

CHAPTER IX

Stresses of Rapid Build-up: Supply and the WAAC Uniform

As was to be expected from past Army experiences, the difficulties of expansion were also felt immediately in the matter of supply. For the WAAC, supply problems had been more or less chronic since the Corps' establishment, with crises following each new expansion plan, none more than temporarily remedied.

Shortages of Clothing

Of these supply problems, the one most apparent to the public was the Corps' inability to issue even one set of military outer garments to recruits—a situation that rarely pertained to male draftees. As the Corps neared the end of its first winter, fully half of the women in some training centers still went through their entire training without uniforms, while at others only summer clothing was available. Fort Dix, New Jersey, for example, was surprised to see descending from a train—in the midst of a March snowstorm—the entire 42d WAAC Post Headquarters Company from Daytona Beach, dressed in summer cottons. The women were promptly restricted to barracks and put to bed to keep warm, while Fort Dix forwarded comments derogatory to the intelligence of WAAC authorities in training centers.[1] One west coast airfield, expecting a company, received a single untidy-looking Waac whose only clothing was the begrimed civilian outfit she had worn from her home to the training center and throughout basic training. After acid comment by telephone to Washington, one uniform was obtained. Other Waacs arrived with one or two shirts at desert airfields where the daily temperature averaged 110 degrees. At other times stations complained that women had been given their complete clothing issue by a process of issuing them grotesquely ill-fitting garments and, still worse, wrong-sized shoes, with the assurance that "you can change them when you get to your permanent station," even though the stations obviously had as yet no maintenance supplies of WAAC clothing.[2]

[1] (1) TWX, Comdt 1st WAAC Tng Cen to Dir WAAC, 24 Dec 42. WA 421 (11-20-42) Status. (2) Ltr, 2d SvC to Dir WAAC, 9 Mar 43, sub: 42d WAAC Post Hq Co. SPWA 319.1.

[2] (1) AAF WAC Hist, p. 39. (2) Memo, Dir WAC for QMG, 30 Aug 43. SPWA 400.34 (3-21-42)(1) sec 4 (1942). (3) Ltrs, WAAC Hq to field, 24 Feb 43, and replies; Memo, Lt Ryan for Plng Serv, 3 Apr 43. Same file, sec 3*a*.

Such failures were generally attributed by the public, and by WAAC trainees, to the numbers involved in the expansion program, or to the difficulty of procuring women's garments in an organization accustomed to dealing only with men's. The latter view was at times held by WAAC Headquarters itself, which noted in one criticism of The Quartermaster General's action: "Two manufacturers of women's wear said . . . any one of fifty stores in the country supplied many more women with many more items." [3] However, upon later analysis, shortages—as distinguished from defects in style—were seldom found to be due to Quartermaster inexperience or inability to let contracts for the relatively small numbers of women in the Army, but rather to simple failure to let WAAC contracts to manufacturers in time. This in turn was caused by the repeated refusal of Requirements Division, Services of Supply, to approve The Quartermaster General's proposals without lengthy delay and debate over possible discrimination in favor of women. At other times failures resulted, in The Quartermaster General's opinion, from the Services of Supply's action in approving G-3's expansion plans without the concurrence of supply agencies.

Difficulties in this respect could be traced back to the moment, a few days after the passage of WAAC legislation, when WAAC Headquarters and The Quartermaster General completed and sent to Requirements Division for approval the Table of Basic Allowances for winter clothing. Its approval was believed routine, since General Marshall had personally viewed and approved every listed item at the time of the WAAC's establishment. Nevertheless, approval was not forthcoming, and upon inquiry it was found that Requirements Division took issue with the fact that women's military clothing, like their civilian clothing, was scheduled to cost more than men's. A Waac's outfit would cost $177.45 as against $102.33 for a man's. In its efforts to equalize the total costs, Requirements Division noted that, whereas men received only one overcoat, women were scheduled to get two—the heavy winter coat like the men's and the light waterproof utility coat instead of the men's field jacket. Requirements Division at once determined to save $15.00 per woman by deleting the utility coat.

Director Hobby, when informed, pointed out that Requirements Division's action had been taken in ignorance of the previous months of planning on the WAAC uniform, in which the whole outfit had been keyed to the two coats. Thus, women's winter uniforms and underwear were of a lighter weight than men's, and women were not given either wool shirts or field jackets. The heavy overcoat was too warm for wear in spring and fall, and she felt that sickness would result if women faced those seasons with only cotton shirts and light covert cloth uniforms. The Quartermaster General sided with the Director, pointing out that no saving would result from the deletion, since one utility coat and one overcoat would wear as long as two of the more expensive overcoats, while a utility coat would protect the overcoat from rough wear in rain, mud, and motor convoys. The Chief of Administrative Services, General Grunert, likewise backed the Director in her arguments.[4] Requirements Division nevertheless re-

[3] MS Draft, The WAAC—Its Mission, 6 Apr 43. Folder, Plng Serv Notes, WAAC Planning files.

[4] Memo, CofAdm Serv for Reqmts Div SOS, 25 Jun 42. Same file.

fused to restore the item: "Statement as to possible increase in sickness may be questioned. On a cold winter day, it is a common sight to see women dressed in light weight clothing, including silk stockings and light shoes, while men are clothed in heavy wool garments." [5]

For five weeks the matter remained deadlocked. The Philadelphia Depot telephoned to state, "The girls will freeze this winter," [6] and wrote, "The time has passed when they may elect the fabric they desire and must take that which is available." [7] Director Hobby twice more protested the refusal to approve her requests.

At the end of July, with Waacs already at Des Moines, The Quartermaster General's Office (OQMG) had still received no procurement authority for either coat, and in desperation telephoned Director Hobby to ask that she put "a lot of pressure" on Requirements Division. This feat she believed to be militarily impossible since Requirements Division was on a higher SOS echelon than her office and was headed by a major general. The Director stated, "It's going to be a tragedy," to which the OQMG representative added: "And we don't want to be responsible for it. I know damn well that we will be blamed for it if the things aren't available. And these people that sit up there and can't make up their minds are going to crucify us, and I don't like it. . . . God Almighty, they are grown people!" [8]

What happened next was not recorded, but two days later Requirements Division, SOS, agreed to permit the procurement of both coats. There was some indication that the matter had not previously reached a high echelon in the Requirements Division.[9] At all events, it was admitted that the delay in letting contracts was already irreparable; [10] there shortly ensued the September and October supply debacle at Des Moines, with its accompanying high sick rates and adverse publicity.

Nevertheless, when increased expansion plans were adopted in September of 1942, the same situation arose again. In explaining his inability to supply uniforms to either Des Moines or Daytona Beach in November and December, as promised, The Quartermaster General stated that Requirements Division had again delayed letting of uniform contracts for six weeks, for unexplained reasons. In addition, it proved impossible for The Quartermaster General to get from the Services of Supply authority to procure emergency cold-weather items to make up the deficit. The climate at Des Moines by this time had proved such that, even had full winter uniforms been available, additional arctic-type clothing would have been needed for outdoor workers and for all who had to march or walk across the extensive post. Temperatures of from zero to 20 degrees below were not uncommon, and a foot or more of snow often covered the parade ground.

On 3 November 1942, the newly appointed commandant, Colonel Hoag, wired: DUE TO SEVERE CLIMATIC CONDI-

---

[5] (1) Memo, Dir WAAC for Reqmts Div SOS, 25 Jun 42; (2) Memo, Reqmts Div SOS for Dir WAAC, 2 Jul 42. SPWA 400.34 (3-21-42)(1) sec 1.

[6] Memo, OQMG for Dir WAAC, 20 Sep 42. SPQRP 421-C. Copy in SPWA 421 (9-30-42).

[7] Risch, *Wardrobe for Women of the Army*, p. 67.

[8] Tp Conv, Col Hugh B. Hester, OQMG, and Dir WAAC, 20 Jul 42. SPWA 400.34 (3-21-42)(1) sec 1.

[9] Memo, Reqmts Div SOS for Dir WAAC, 2 Aug 42. SPWA 400.34 (1943). The memorandum says only that the Director, Requirements Division, SOS directed its Allowance Branch, the contending party, to change its stand.

[10] Risch, *Wardrobe for Women of the Army*, pp. 70, 82-85.

TIONS, FOLLOWING ITEMS WAAC CLOTHING URGENTLY NEEDED TO PROPERLY PROTECT AND SAFEGUARD HEALTH OF THIS COMMAND. He asked for over 15,000 each of four-buckle arctic overshoes, nurses' lambskin-lined mittens, and wool-lined trousers, and concluded: IT IS IMPERATIVE THAT THE SUPPLY OF THESE BE EXPEDITED.[11] A few days later Colonel Hoag appealed again by letter, asking for wool socks, caps, shirts, and trousers, also mittens, wool underwear, and overshoes; for drivers, he asked for flannel-lined coveralls to replace the cotton ones.[12]

In Director Hobby's absence in England during this month, Colonel Catron immediately appealed to Distribution Division, Services of Supply, for orders to ship emergency cold-weather items. Some debate then took place between Distribution Division and Requirements Division. Distribution Division felt that action on the radio request should not be delayed but that the whole problem must be approached more deliberately:

> With reference to the broad problem posed by basic radio, this division has noted numerous requests initiated, piecemeal, by the Women's Army Auxiliary Corps for the adoption of additional items of clothing. . . . It is recommended that caution be observed in authorizing any special procurement to assure the adequacy of items hurriedly requested to fill the requirement for which intended. This division has no information as to the adequacy of the trousers desired, nor as to the suitability of the Arctic overshoes requested for wear with shoes issued to the WAAC.[13]

As a result, no decisive action was taken except to schedule a conference for 25 November, some two weeks away.

The Quartermaster General also was unable to take action, and in fact was in a state of rebellion against the repeated expansion directives his office had been receiving. It was his opinion that the War Department supply agencies that agreed to G-3's expansion program—particularly Requirements Division—had not co-ordinated the matter fully enough with their operating agencies.

OQMG, which had requested twelve months' notice of expansion plans, now was lucky to get twelve minutes, so rapidly did the WAAC estimates go up. The Quartermaster history of women's clothing stated: "Army Service Forces did its planning at much too short range, without admitting the necessity for stock-piling cloth. . . ." Admittedly OQMG was kept fully informed, often by telephone, of the latest developments, but no sooner did it complete plans and let contracts than the figures were again raised. The figures furnished The Quartermaster General had increased within a few months from a March 1942 estimate of 12,000 in a year to later estimates of 53,000 for 1943; then 113,000; then 150,000; then half a million, although the last figure was not used for procurement purposes. As a Quartermaster historian later remarked, "The consternation this created can be imagined." [14]

Accordingly, an OQMG council was held in November of 1942, at which it was stated that no increased supply of winter uniforms could be procured and no cold-weather or arctic equipment, unless the office was given twelve months from date of notice if cloth must be bought, or six months if the items could be procured ready-made. This was the time customarily allowed for the procurement of men's

---

[11] TWX, Comdt 1st WAAC Tng Cen to Dir WAAC, 3 Nov 42. SPWA 420.

[12] Ltr, Col Hoag to Dir WAAC, 9 Nov 42. SPQ 420, in SPWA 400.34 (3-21-42)(1) sec 1.

[13] Covering Memo and Inds to TWX cited n. 11.

[14] Risch, *Wardrobe for Women of the Army*, pp. 5, 6.

clothing and equipment. This recommendation was approved by the Services of Supply's Chief of Staff, General Styer, in ignorance of the fact that the women were already at Des Moines.[15]

Upon Director Hobby's return from England in mid-November and her threatened resignation, General Styer shortly reversed this policy and directed The Quartermaster General to get 2,000 winter uniforms to Des Moines at once and 1,000 a week thereafter, regardless of his twelve months' notice. General Styer's immediate subordinate, the assistant chief of staff for operations—Maj. Gen. LeRoy Lutes—then directed The Quartermaster General to procure at once 2,500 wool-lined trousers, 2,500 wool shirts, and 2,000 pairs of gloves for WAAC motor transport personnel. General Lutes' immediate subordinate, the director of Requirements Division—Brig. Gen. Walter A. Wood, Jr.—also directed the Quartermaster Corps to procure 17,700 wool caps and wool gloves and 2,574 more overshoes, and to take action to standardize necessary articles for winter wear.[16]

To achieve this increase without diverting standard material from men's uniforms, the Philadelphia Depot was obliged to gather up various odd lots of different colored woolen materials which were then in the depot, and to have them redyed to the enlisted shade. These inevitably came out in a wide range of nonstandard colors. One branch of OQMG directed that some skirts, some jackets, and some caps be made in each shade, so that a woman might receive a complete matching outfit even if she did not match her neighbors. Since this would have caused administrative difficulty for the depot, another branch of OQMG canceled this order without the knowledge of the branch that had issued it. As a result, skirts and jackets were of different shades and it was impossible to match them. Observed a Quartermaster historian: "The most grotesque combination was that of the chocolate-brown barathea and the mustard shade olive-drab." [17] From the viewpoint of the Recruiting Service, these uniforms were worse than the shortages, but they were at least warm.

To avoid such hasty procurement in the future, The Quartermaster General again recommended that his office be given twelve months' notice on any further increases, saying:

> Experience indicates that it has required approximately twelve months from the initiation of procurement of cloth required for WAAC uniforms until the finished article is ready for issue.

WAAC Headquarters again professed its helplessness to give such notice:

> Since the passage of the bill creating the WAAC, four (4) distinct expansion plans have been ordered by higher authority. In each instance the plan was ordered into operation immediately, and The Quartermaster General was given the proposed figures without delay.

Accordingly, the Services of Supply was forced in December of 1942 to give The Quartermaster General procurement authorization in advance of the General Staff decision on the Corps' future size, stating that the clothing could be used for main-

[15] Memo, CofS SOS for CofAdm Serv, 19 Nov 42. SPEX, SPWA 421 (11-19-42).

[16] (1) Interv with Lt Col Helen Woods, 9 Oct 48. (2) Weekly Rpt, WAAC Hq to CofAdm Serv, 15-21 Nov 42. SPWA 319.12 (11-9-42) sec 1. (3) Memo, ACofS Opns SOS for QMG, 27 Nov 42. SPDDQ 420 Clothing, in SPWA 421 (11-27-42). (4) 1st Ind, Reqmts Div SOS (Gen Wood) to QMG, 3 Dec 42. SPRMD 421 (11-26-42), in WA 400.34.

[17] Risch, *Wardrobe for Women of the Army*, p. 51, ns. 95, 97.

tenance stocks if the expected draft of women did not occur. The enormous amounts of 400,000 by the end of 1943 and 750,000 by June of 1944 were authorized, in the expectation that the plan for a draft law to obtain those numbers would be approved. In the light of future events, it was perhaps fortunate that the paper was recalled, and that it was decided to await the General Staff decision on draft plans.[18]

In spite of General Styer's intervention in November, there could be no immediate improvement in the supply of uniforms since the procurement process had been long delayed. On Christmas Eve Colonel Hoag again telegraphed from Des Moines, reporting many items exhausted, all stocks low. At Daytona Beach, which opened in December, only summer uniforms were available for most of the winter. Trainees were necessarily shipped out in summer khaki, regardless of destination. The Third WAAC Training Center, at Fort Oglethorpe, opening in January, reported that many women went through their entire training without uniforms; at the same time, the rate of respiratory disorders soared to twice that of men in the area. Graduates were at first held idle in staging areas. The previous shipment of women in civilian clothes to the Aircraft Warning Service had had such an adverse effect upon recruiting that Director Hobby had issued a ban on further shipments not properly uniformed. Overcrowded conditions in the staging companies soon forced repeal of this order, and a report from Fort Oglethorpe noted that "Fifty percent of personnel departing this station during the week of March 21–28 will not be uniformed." [19]

In early March WAAC Headquarters was informed that it was impossible for The Quartermaster General to meet the supply schedule furnished by Requirements Division at the time of G–3's February decisions on Corps' size. As a result, the Fourth WAAC Training Center was scheduled to open in the still-wintry Massachusetts weather without any clothing supplies whatever.

At this, the Director sent her chief of operations, Colonel Clark, to the training centers to determine the actual situation. Upon his return Colonel Clark directed sharp criticism toward the Army Service Forces' supply agencies. There had been minor administrative faults at the training centers, Colonel Clark reported, such as occasional failure to establish close co-operation with depots or to utilize stock fully. Thus, the Atlanta Quartermaster Depot had shipped items to Daytona Beach that were needed at Fort Oglethorpe, and vice versa, and Quartermaster had refused to exchange them:

> He objected strenuously on the grounds that such action would confuse his records. My reaction to that is that records can always be straightened out but a death from pneumonia due to improper clothing can hardly be laughed off.

In spite of such occasional administrative faults, Colonel Clark concluded emphatically that fully 80 percent of the difficulties could be traced to insufficiency of production rather than of administration. He recommended that proper authority assure that individual contractors were on time, that surplus men's material "known to be available in abundance" be used,

[18] Memo, Dir Reqmts Div SOS for CofAdm Serv, 26 Nov 42, and Inds dated 5, 21, 27 Dec 42 and 12 Jan 43. SPRMP 291.9 WAAC, in Dir Adm Serv ASF Sp Coll DRB AGO.

[19] (1) Weekly Rpt, WAAC Hq to CofAdm Serv, 14–24 Mar 43. SPWA 319.12. (2) Ltr, 3d Tng Cen to Dir WAAC, 20 Mar 43. WA 421 (1942).

"that a determined inquiry be made at once to ascertain why certain articles are, and have been, habitually lacking over a period of months, and that steps be taken to remedy this deficiency." [20]

In reply to Colonel Clark's report, The Quartermaster General stated that WAAC supply had been "expedited in every manner"; "contractors are well ahead"; "this office, in supervising and directing the procurement of WAAC clothing through the procuring depot, has taken every action possible to eliminate delinquencies on current contracts." [21] This view was supported by General Madison Pearson, Deputy Chief of Administrative Services, SOS, who wrote, "The report of action taken by The Quartermaster General indicates that everything possible is being done to expedite the manufacture and shipment of clothing." [22]

Finally, representatives of WAAC Headquarters, in a meeting at OQMG, secured promise of makeshift action to supply the Fourth Training Center. A sufficient quantity of Civilian Conservation Corps spruce-green mackinaws were made available; trainees were issued Army officers' serge shirts in lieu of winter jackets; and 20,000 more off-shade skirts were procured, also shoes and raincoats. The Fifth Training Center, at Camps Ruston and Polk in Louisiana and Monticello in Arkansas, could fortunately get along with summer khaki. The Quartermaster General, said a WAAC representative, was "most sympathetic and cooperative."

The Quartermaster General now decided that the matter of supplying the WAAC warranted the full-time specialized attention of one officer, and a WAAC officer was assigned to his office to work on the problem. Such an assignment had not been possible previously because of General Somervell's order that no WAAC officers would be assigned to any office within his headquarters other than that of the Director WAAC.[23]

If the supply of uniforms to training centers was insufficient, that to field stations was virtually nonexistent. Army supply officers at these stations often were uncertain what the WAAC supply channel was or what depots stocked WAAC clothing, and they had no copies of the WAAC T/BA, T/E, and T/O, which were apparently not available at training centers or in the field. Even if requisitions were properly submitted, it took over a month to get them filled from the two depots—Kansas City and Atlanta—that stocked WAAC uniforms. Sometimes the requisitions were never filled, and the wait for shoes was often two or three months. As late as 15 March a service command which had administered WAAC units for seven months was told that there was still no schedule set up for maintenance allowances for Waacs at posts, camps, and stations, and that none could be set up until sufficient quantities of clothing were available for new recruits.[24]

Appearance of the Uniform

In addition to exaggerating shortages, the expansion program had permanently

[20] Memo, Chief, Opns Serv WAAC, for Exec WAAC, 6 Mar 43. SPWA 421 (11-20-42) 1942.

[21] Memo, QMG for WAAC Hq, 10 Mar 43. SPWA 421 (1942).

[22] Memo, DCofAdm Serv for Dir WAAC, 13 Mar 43. Same file.

[23] Memo, Chief, Opns Serv WAAC for Dir WAAC, 23 Mar 43. WA 421 (1942). For Somervell order, see pp. 94-95, above.

[24] (1) Ltr, OQMG to field, 23 Sep 42. SPQX 420. (2) Ltr, 8 Dec 42, sub: same. SPWA 421 (5-16-42) sec 1. (3) Memo, 1st Tng Cen for WAAC Hq, 29 Oct 42. SPWA 421. (4) Ltr, 1st SvC to Dir WAAC, 15 Mar 43. SPWA 400.34 (3-21-42)(1) sec 3.

affected the appearance of the WAAC uniform, by making correction of early mistakes impossible before mass production was required. For every amateur theorist who supposed that any other cause had damaged WAAC recruiting, there were ten who put the blame on the appearance of the uniform. Some nine out of ten of the unsolicited letters of advice received by the WAAC went so far as to place sole responsibility upon this one factor.

Actually, OQMG had been well aware from the moment uniforms were issued that many items of clothing needed modification, and plans for such modification were being made when the need for immediate mass production intervened. The Quartermaster General, soon after the opening of the First WAAC Training Center, sent to Des Moines an expert committee which included a representative of the Philadelphia Depot and a civilian consultant, Miss Dorothy Shaver, vice-president of Lord & Taylor. These catalogued various complaints, and Colonel Faith reported others. All garments were cut with wide collars and narrow hips, as for men; skirts, shirts, and jackets were for this reason generally ill fitting, uncomfortable, and unbecoming to the average woman. Hats were out of shape before they were issued; raincoats leaked at every seam in even a light shower; seams of hems were sewed down so that they could not be easily raised or lowered, and some garments had no hems at all. The suspenders on girdles were too short and pulled runs in stockings, as the War Production Board had allotted insufficient elastic.[25]

The attempt to discover the reason for the poor appearance of what had been a basically sound uniform design revealed that the Philadelphia Depot had never made a graded designers' model of the uniform, but had somehow got hold of a rough pattern cut by a manufacturer to estimate the cloth needed. This the depot had henceforth called "the master pattern," and from it each manufacturer had developed a set of sized patterns of his own. Those manufacturers who received contracts were all in the men's-wear industry, since, The Quartermaster General reported, "The manufacturers of women's clothing were not able to handle the production of WAAC uniforms at the prices which the Philadelphia Depot was willing to pay." Moreover, winter uniforms had already been made upon the defective patterns used for the summer ones.

For the poor results obtained, the Philadelphia Depot blamed The Quartermaster General for failing to consult it earlier, thus giving it insufficient time for development of patterns. Further developmental work was therefore relinquished to the depot so that it might not be hampered. A new jacket was developed, with a better-fitting collar. It had no belt, as the women had shown a tendency to pull the belt too tight. Unfortunately, the depot, in omitting the belt, respaced the jacket buttons so that the bottom one fell quite low; it was now impossible for a Waac to sit down without unbuttoning the lowest jacket button or suffering what Director Hobby described as an "unsightly gap." Since large quantities had been acquired, The Quartermaster General refused to correct the spacing. The new jacket, as turned out by the men's garment contractors, was also too flat across the bust and still too stiff and awkward.

[25] (1) Rpt, 20 Oct 42. SPWA 421. (2) Ltr, Faith to Dir WAAC, 12 Aug 42. SPWA 421 (8-22-42). (3) Risch, *Wardrobe for Women of the Army*, Ch. IV. The following paragraphs, unless otherwise stated, are based on Risch, especially pages 40-47 and 61-63.

WAAC BAND, *Fort Des Moines.*

The skirt pattern also was modified to round the hipline somewhat and to prevent wrinkling and rolling. The summer material was too stiff to permit insertion of a pleat, even if War Production Board rules would have allowed it. The 8.2-ounce cotton twill used in men's uniforms had admittedly proved totally unsatisfactory for women's; OQMG now tried to devise some other suitable summer material which would withstand laundering in Quartermaster laundries, but various cotton-rayon combinations all proved doubtful. Seersucker was dropped from consideration because it would have required a new uniform design and because of a shortage of production facilities, although the WAVES, SPARS, and women Marines all found an adequate supply for summer uniforms. Finally, it was decided that the khaki uniforms must be retained.

Until the spring of 1943, both the Director WAAC and The Quartermaster General found themselves in the position of helpless bystanders, able only to make suggestions, which were accepted by the Philadelphia Depot but never seemed to appear in the finished garments. In May, OQMG discovered the cause—the depot was frequently ignoring its recommendations, and for six months had not made all the requested corrections, nor had it ever placed the pattern in the hands of a good tailor of women's clothing, which it had been directed to do.

The Quartermaster General now took back some control over developmental work and, at the request of WAAC Headquarters, called in representatives of four leading clothing firms and expert designers of women's patterns.[26] These immediately discarded the Army sizes, which had been set up in long, regular, and short, as

[26] M/R, 19 May 43, Min of Mtg, and atchd Memo, ASF for OQMG, sub: Summer Uniforms, WAAC. SPWA 421 (1942).

for men. Now at last patterns in the usual women's sizes were developed—half sizes, misses', women's, and long sizes. These proved better, but the Quartermaster's Storage and Distribution Division insisted that the old size designations and stock numbers be used on the new patterns, to avoid complicating its storage and issue system, which was geared to handle only short, regular, and long.

Even with improved patterns, the jackets still remained stiff and unbecoming, and it was not until some months later that OQMG detected the trouble: the Philadelphia Depot was still writing specifications, which manufacturers were required to follow, that called for the use of the same heavy construction with canvasses and interlinings of the weights used in men's uniforms.

In late 1943 and 1944 a satisfactory uniform pattern was to be gradually developed, but its production by this time was, as The Quartermaster General said, now merely a matter of academic interest, as large stocks of the earlier patterns had been ordered in an effort to meet the scheduled expansion program, and no new procurement could be authorized until these were worn out.

At first sight of the original ill-fitting uniforms, Director Hobby in July 1942 had recommended to Requirements Division, SOS, that Waacs going to the field or on leave be allowed to purchase individually tailored white dress uniforms, at their own expense, as were male Army officers, Army nurses, and, later, the Navy women's services. It was believed that such a measure would eliminate the worst of the general criticism encountered earlier, since the dress uniform could be worn for public ceremonies and appearances.

Requirements Division disapproved this request: "No useful purpose can be served in connection with efforts toward a successful prosecution of the war." [27] Three subsequent requests by the Director to General Marshall were stopped by the Services of Supply from reaching him, and the WAAC was not authorized a dress uniform until later entry into the Army automatically gave it to women officers.

The appearance of the WAAC cap was also much criticized. It was worked over by experts—the visor was made twice as thick and heavy, to prevent buckling; the side and top stiffening was increased; a new type of lining was devised. Still the cap remained almost impossible to clean and block. Director Hobby in January 1943 finally gave up the idea of a distinctive cap and requested a garrison (overseas) cap for all but dress wear. The Director pointed out that winter and summer garrison caps at a cost of 71 cents and 39 cents respectively would eliminate the need for issue of extra visored caps at $2.32 and $1.99 apiece. This request was refused by the Services of Supply on the grounds that it would increase the number of issued items. Although men got both dress and service caps, it was at the time believed that the War Department was shortly to cease issuing the garrison cap to men.[28]

The WAAC shirtwaist was also the subject of dispute, and more than one Army man reported, quite accurately, "I never yet saw a WAAC whose shirt collar fit her." Commercial producers of women's shirts had always sized them by bust sizes rather than by collar size, since civilian

[27] (1) Memo, OQMG for G-1, 18 Jul 42, with Inds and covering Memos; (2) Memo, Dir WAAC for CofAdm Serv, 24 Aug 42, and Inds. SPQRE 421, in SPWA 421 (7-16-42).

[28] Memo, WAAC Hq for CofAdm Serv, 20 Jan 43, and Inds. SPWA 421.

women's blouses seldom had a standing collar and tie and therefore did not require exact collar measurements. Men's shirts, on the other hand, were sized by neck and sleeve measurements.

The Quartermaster General now perceived that women's military shirts should have been sized like men's, but the shirt contracts had been let to manufacturers of women's garments, who were unable to produce men's-type shirts with a standing collarband and graded neck and sleeves. To cancel their contracts and let new ones would have caused a serious delay; it was therefore decided to retain the convertible collar and the original sizing system. Several sleeve lengths for each size were recommended, but the Quartermaster Storage and Distribution Division would not concur, because of the "extra work this would entail" in storing and shipping several types of shirts.

One difficulty was solved, however: The Quartermaster General discovered that the Philadelphia Depot had again been guilty of furnishing manufacturers a basic pattern cut by makers of men's shirts. A new pattern was devised with a collar that was smaller than the previous one and a bust that was larger, and this proved fairly satisfactory. Again, large purchases of the defective pattern had been made and had to be issued throughout 1943 and 1944, and complaints about the shirt never ceased to come in.

The problem of providing satisfactory government-issue shoes for women had from the beginning been difficult to solve so as to please everyone. The Director's Office made surveys of the women's opinions as to various types of shoes and found that there was a dissatisfied minority for any type. Medical opinion also was divided. One school of thought recommended a flat-heeled shoe like the men's for all purposes, pointing out that most orthopedic patients in WAAC training center hospitals suffered from sprains and fractures caused by marching or doing heavy work in the medium-heeled oxfords. Other experts said that, on the contrary, a sudden change to flat heels for constant wear would cause even greater foot trouble and fallen arches for the majority of American women. The dispute was resolved by the adoption of the nurses' high-laced flat-heeled field boot for outdoor work and marching and the service oxford for office and dress. Even for this limited use, the field boot had to be broken in by an elaborate routine involving heavy wool anklets and ten-minute daily wearings. The Director therefore recommended that the women get one pair of these field shoes in addition to the two pairs of ordinary shoes.

This request was refused by the Army Service Forces: men got only two pairs, both of course flat-heeled, so women could have only two—one of oxfords and one of field shoes. The Director appealed again, pointing out that in such case a woman could not let her service oxfords be sent away for repair, since repair of women's shoes proved to be always slower than that of men's and ordinarily took from two weeks to a month, a prohibitive length of time to be without shoes for office wear. Finally, after the Director wrote a memorandum on the subject to General Somervell, the ASF agreed to issue Waacs three pairs of shoes—two of oxfords and one of field shoes for office workers, one of oxfords and two of field shoes for outdoor workers.[29]

[29] (1) Request, Dir WAAC to OQMG, for field shoes, 4 Sep 42; (2) Rpt on test of 500 WAAC oxfords, 14 Nov 42; (3) Ltr, Surg, Ft Oglethorpe, to Dir

YOU'LL JUST HAVE TO GET USED TO LOW-HEELED SHOES!

The WAAC utility bag had likewise proved unsatisfactory. The cheap imitation leather adopted by the economy-minded Quartermaster General almost at once cracked and peeled. The Director then presented him with a leather bag designed by Richard Koret, and requested its adoption. Mourned a Quartermaster historian: "It was apparently immaterial that the bag was expensive and that it used calfskin leather, a critical material."[30] This sample was delivered to the Philadelphia Depot, which immediately let the contract to such low-priced manufacturers that the resulting bag was unrecognizable and quickly wore out. Finally, OQMG announced with pleasure that it had located a quantity of seal leather, tanned goatskin, and genuine water buffalo. These were durable and noncritical materials, and an excellent utility bag resulted.

The Navy avoided most clothing difficulties by giving its Waves a clothing allowance with which to purchase uniforms. Various commercial firms produced the uniforms, apparently without any of the Philadelphia Depot's misfortunes, and sold them through department stores which gave expert fitting. Shoes and other personal items could thus be chosen from standard models whose fit suited the individual. The Army was unable to adopt this system, since to provide a money allowance in lieu of clothing would have required Congressional action to amend existing Army legislation. Army women eventually benefited in that they received maintenance of the uniform by free replacement of worn-out items, whereas the Waves' monetary maintenance allowance of $12.50 every three months was reputedly scarcely enough to keep them in stockings, much less in new uniforms, shoes, and underwear.[31]

The Need for New Types of Work Clothing

As Waacs began, under the expansion program, to do various kinds of Army jobs which had not been anticipated by the WAAC Pre-Planners, a need arose for new types of work clothing. Of the four types of jobs which had been originally authorized—clerks, drivers, cooks, and telephone operators—the first and last could be performed in the A uniform. Accordingly, the only work uniforms authorized for Waacs were white dresses for cooks and coveralls for drivers to put on over their uniforms while making motor repairs. These work uniforms were issued only to the few specialists concerned. The majority of Waacs when shipped to the field had only the A uniform plus two of the seersucker exercise dresses, which were worn to protect the service uniform during physical training, kitchen police, and barracks fatigue duties.

With the expansion of the Corps, Waacs soon were working in hospital wards and laboratories, driving light trucks as well as staff cars, and being assigned as full-time mechanics, welders, pier checkers, messengers, and gas pump attendants. In the Air Forces many worked "on the line" in aircraft maintenance and in other jobs that required them to climb in and out of aircraft and up and down control towers. Neither the A uniform nor the fatigue

WAAC, 23 Feb 43, and Inds; (4) Ltr, WAAC Hq to Des Moines, 3 Apr 43; (5) Ltr, WAAC Hq to Reqmts Div ASF, 10 Jun 43, with Inds; (6) Memo, Col Hobby for Gen Somervell, 9 Jul 43; (7) Change to T/E 21, Sep 43; (8) Ltrs, Field Stas to Dir WAAC, various dates, atchd to above. All in SPWA 421.3 Shoes (9-4-42) 1943.

[30] Risch, *Wardrobe for Women of the Army*, p. 73.

[31] Memo, Asst Exec WAC for Dir WAC, 14 Mar 44, sub: Uniform Issue of Marines, WAVES, and SPARS. WDWAC 421.

dress was appropriate. The A uniform was easily ruined by grease or medicines, and its tight skirt made tower-climbing either impossible or immodest; the exercise dress was even skimpier, being well above the knee and unsuitable for appearance outside the WAAC area.

One of the most important reported work needs was for some form of trousers or culottes, requests for which poured in from the field. The Military District of Washington, after a study of the needs of Waacs in its motor pool, requested a culotte skirt for drivers. The regular skirt had proven

> slightly embarrassing at times to the wearer, particularly in climbing over tail-gates into trucks, or standing on a rack about 3½ feet high and leaning over washing the tops of passenger cars. . . . The coverall currently provided has been found unsuitable and unsightly for wear when driving staff cars and the regulation skirt too short and tight.

WAAC Headquarters concurred in this request and asked the development of a divided skirt for drivers.

However, Requirements Division, Army Service Forces, refused the request on the grounds that the Director herself had rejected culottes at the time the uniform was decided upon. The Director thereupon acknowledged her mistake but renewed the request, saying, "In March 1942 there were no Waacs on duty in the field and therefore no recommendations or decision could be based on actual experience." This question went as far as the Acting Chief of Staff, Army Service Forces, who supported Requirements Division's view and refused the request.[32]

The demand for a work garment continued too strong to stem, and later in the spring of 1943 Director Hobby again appealed to the Army Service Forces: "It has become increasingly evident as larger numbers of WAAC units are assigned to the field . . . that a trousered garment for exercise, fatigue, and other heavy work is vitally necessary."[33] Since Requirements Division would not concur in the design and procurement of slacks or culottes for Waacs, Director Hobby requested, as the only other alternative, that the herringbone twill coveralls be issued to all Waacs, instead of only to drivers.

This action was agreeable to Requirements Division, but, since men had only one type of fatigue clothes, the seersucker exercise dress was deleted from the authorized issue except for use in the training center. Instead of two exercise dresses, each Waac assigned to the field was now issued one coverall. WAAC Headquarters requested two coveralls, but Requirements Division decided that one would be adequate and that Waacs in active work could later get two by turning in one skirt and one shirt. The Des Moines Motor Transport School attempted to get three or four coveralls for student drivers, so that coveralls could be sent to the ten-day Quartermaster laundries. The Quartermaster General's Office concurred in this, but Requirements Division disapproved it "in view of the requirements of the using arms and services and the production facilities of industry."[34] The later training centers for a time had difficulty in getting even the allotted number, and drivers at Daytona Beach wore men's Class B blue denim.[35]

---

[32] Ltr, MDW Motor Cen to CG MDW, 30 Jun 43, with Inds. SPWA 400.34 (3-21-43)(1) sec 4 (1942).

[33] Memo, WAAC Hq for Reqmts Div ASF, 21 May 43, sub: Trousers, HBT. Same file. Also Ltr, Camp Campbell, Ky., to 5th SvC, 20 Mar 43, and 16 Inds. Same file, sec 3.

[34] Ltr, Ft Des Moines MT Sch to CO 1st WAAC Tng Cen, 22 Sep 42. Same file, sec 1.

[35] Ltr, 2d WAAC Tng Cen to Dir WAAC, 18 Jun 43. Same file, sec 4.

PHYSICAL TRAINING *at an Army Air Forces Training Command base in 1943.*

The adoption of the coveralls did not solve the need for cold-weather work garments, for which requests continued to come in from both Army and Air Forces stations. Whereas a man might comfortably wear cotton coveralls over his long winter underwear and wool trousers, a Waac had nothing underneath but her winter panties, only one fourth as heavy as men's winter underwear. In March, what OQMG believed to be women's wool shirts and trousers were issued to drivers only, but these proved greatly oversized and unsuitable.

In late April of 1943, after the Waacs had somehow survived the first winter, The Quartermaster General sent a committee to investigate the situation at Fort Des Moines and make recommendations on winter clothing. The committee recommended that all Waacs in cold climates, regardless of job, get warmer panties and vests (50 percent wool instead of the current 25 percent) and also wool shirts and knee-length wool stockings. All outdoor workers should also get long wool drawers and long-sleeved undershirts, trousers with an inner wool liner and outer windproof cover, field jackets with liners and covers, leggings, and wool caps. Director Hobby concurred, pointing out that much training time had been lost at Fort Des Moines and Fort Devens for lack of warm clothing. However, Requirements Division did not favorably consider the field jacket, leggings, and 50 percent wool underwear for women, or the wool waist, trousers, and liner for any except drivers.[36]

Warm-weather clothing was equally deficient, according to reports from hospitals and stations in semitropical climates. These, finding the heavy A uniform totally unsuited to a hot climate or to hospital ward work, had fallen into the admittedly undesirable practice of allowing Waacs to wear the short seersucker exercise dress for many types of jobs, even desk jobs in headquarters where they were fully visible to employees and visitors. The dress was also worn for kitchen police and barracks duties where the long-sleeved coverall was too hot. Just as the decision was rendered to substitute coveralls for exercise dresses, many southern stations were writing to plead for four or more of the exercise dresses instead of two.

The Army Air Forces' large Training Command, which at one time utilized almost one sixth of the entire WAAC personnel, was particularly hard-hit, since the majority of its airfields were along the southern border in Florida, Louisiana, Texas, and the southwestern desert. A study prepared by this command pointed out that at many fields the temperature never went below 90° or 100° for months at a time and often rose as high as 135° F. Enlisted women were obliged to wear two shirts a day and wash them at night, since there were no laundries on the fields. The Air Forces study concluded that Waacs must be immediately issued at least five short-sleeved shirts in addition to their regular ones, three exercise suits instead of the two then issued, and four pairs of cotton anklets instead of the three wool ones then issued. In reply, the AAF received what it considered an inexplicable decision that no additional clothing would be issued and that even the two exercise suits would be taken away and replaced by a heavy coverall.[37]

In May of 1943, three months before the refusal of the Air Forces' request, the Director had made a similar recommendation: that the Quartermaster Corps redesign the exercise dress into a seersucker summer uniform like that of the WAVES and Women Marines. The Quartermaster General concurred and visualized the development of one standard short-sleeved work dress, longer and better fitted than the exercise dress, which could be worn by cooks, hospital workers, women in tropical climates, and any others who needed such a dress. The Quartermaster Corps began to study this problem in 1943 but no summer dress for universal on-duty wear was produced during World War II.

The lack of such a dress was felt most strongly by hospitals, where WAAC workers had no uniform comparable to the nurses' white or seersucker uniforms. Some Waacs wore seersucker fatigue dresses if they had received any before issue was discontinued. This practice was pronounced unsanitary, since the same dress was worn for physical training, recreation, kitchen police, and barracks fatigue duties, and with only two dresses a Waac could not

[36] Ltr, Comdt to OQMG, 22 Apr 43. Same file, sec 3*a*.

[37] Ltr, AAF CFTC, Randolph Field, Tex., with Ind, Hq AAF Tng Comd, Ft Worth, Tex., to Dir WAC, 20 Aug 43. Same file, sec 4.

use the hospital laundry but either wore her dress a week unwashed or washed one herself every night. Even so, the two exercise dresses were more suitable than the one coverall that replaced them.[38]

The Quartermaster Corps, on being informed of the problem, was agreeable to issuing hospital Waacs the WAAC cook-baker dress, a neat cool wrap-around white garment which could be dyed any desired pastel shade to distinguish Waacs from nurses. However, The Surgeon General's Office objected: "Fatigue clothes are worn by enlisted men of the medical department and there should be no exception for members of the WAAC stationed at hospitals." Waacs, this office said, could wear either their coveralls or surgical gowns. Requirements Division, ASF, concurred with The Surgeon General's Office and overruled the Quartermaster Corps. They held firm in this ruling all summer even after stations in the field pointed out that enlisted men on duty in clinics and laboratories often did get four pairs of cook's white trousers and six white coats. The surgical gown shortly proved most unsatisfactory for women working in men's wards: it was too long, frequently dragging on the floor. Worse, it was open all the way down the back and a uniform was not customarily worn underneath in warm weather.[39]

With hundreds of similar field requests on file, the Army Service Forces held unswervingly to its policy that no additions or changes would be made to the originally authorized uniform either in type or amount. Fort Bragg, North Carolina, asked increased issue of dresses and aprons for cooks and bakers because of the intense summer heat, the coal ranges used, and the ten-day laundry service. This was refused by The Quartermaster General's Office on the grounds that the material was needed instead for grain bags and civilian clothing. Recruiting offices in Virginia asked that all stockings issued them be rayon instead of half the issue in cotton, which produced a bad public reaction. The Director approved this request and asked that all Waacs assigned outside Army posts be issued rayon stockings only, but was refused by The Quartermaster General on the grounds that rayon did not wear as long as cotton stockings.

Director Hobby asked that, if Waacs could not have a garrison cap, they be issued two summer hats, since the current issue of one made cleaning and blocking impossible. No immediate action was taken on this request; OQMG had on hand only 262,000 extra summer caps and feared that a sudden influx of recruits might cause a shortage. In only one minor case did the Army Service Forces grant a request for change: late in August of 1943 it authorized WAAC recruiters to receive an extra summer jacket, skirt, and cap, after it was demonstrated by a year in the field that recruiters had to wear the full uniform daily and could not wear fatigue clothing while their one jacket and cap were being cleaned.[40]

The trend of Requirements Division's policy was in fact in the opposite direction—the systematic deletion of all items of the original authorization that men did

[38] Ltr, CO Hosp Unit, Ft Oglethorpe to Dir WAAC, 10 Apr 43. Same file, sec 3.

[39] (1) Memo, SGO for QMG, 6 May 43, with Inds; (2) Ltr, Randolph Field, Tex., to Dir WAC, 20 Aug 43. Same file, sec 4.

[40] (1) Ltr, Ft Bragg to 4th SvC, 28 May 43; (2) Ltr, Richmond R&I Dist to Dist R&I Off, 23 Aug 43, and Inds; (3) Memo, ASF for Reqmts Div ASF, 30 Aug 43; (4) Memo, Reqmts Div ASF for OQMG. Same file, sec 4.

not wear. These included summer and winter pajamas, galoshes, handkerchiefs, dress shields, athletic shoes, and summer and winter bathrobes.[41]

Next, a Quartermaster committee "made personal investigations at Fort Des Moines" and found that only 25 percent of the women were wearing the issue girdles, others preferring to wear none or to purchase a favorite brand commercially.[42] The Quartermaster Corps felt that Waacs should be given, in lieu of the issue, money to purchase the brassières and girdles of their choice, and WAAC Headquarters concurred. However, this plan would have required amendatory legislation and, after various unsuccessful attempts to draw up the legislation in general terms, the Army's Legislative and Liaison Division retreated in disorder. One colonel in this division recommended that the proper physical appearance of Waacs be attained by exercise and good posture, and not by the use of "surgical contraptions." If the government was to issue brassières and girdles to women, he wrote, then "such devices could well be considered for the officers and enlisted men." [43] The issue of girdles and brassières was therefore stopped. Several months later, when such items became very scarce on the commercial market, the Director requested that the issue be resumed, but this request was refused by The Quartermaster General.

Public Reaction to Unsuitable Uniforms

By this time it was unfortunately impossible to conceal from the American public the fact that the WAAC uniform and supply program had become a tragicomedy of errors. Gallup polls showed that eligible prospects for enlistment rated the WAAC uniform last in attractiveness, after the Marines', Waves', and Spars' uniforms. The citizenry at large was inspired with a desire to be helpful, and letters poured in from housewives, designers, Congressmen, soldiers, and other interested bystanders. All writers were firmly convinced that adoption of their ideas would immediately end the WAAC's recruiting difficulties.[44]

Many ideas, particularly those from Army stations and Waacs in the field, were practical and sound, but required major revisions, which could not have been made during the expansion period without greatly delaying contracts, and which could not be made afterward in view of the necessity of wearing out huge stocks of already-procured items. The same public that condemned the Waacs' appearance was not ready to support expensive measures such as discarding thousands of dollars worth of unsightly and ill-fitting garments. When recruiting advertisements pointed out that Waacs received certain fine-quality items worth a total of $250 (although actually costing the government half that) irate readers wrote to ask why the taxpayers' money was spent in this fashion. Later, Congressional committees were to investigate the alleged expenses.[45]

[41] (1)Memo, Distrib Div SOS for Reqmts Div SOS, and Ind. SPDDQ 421 Shoes, in same file, sec 1. (2) Memo, WAAC Hq for CofAdm Serv, 23 Nov 42. SPWA 400.34.

[42] Rpt cited n. 25(1).

[43] (1) Memo, Col E. J. Walsh, L&LD, for G-1, 28 Oct 42. SPWA 422.2 (9-16-42) Foundation Garments. (2) Memo, QMG for USW, 16 Sep 42, sub: Proposed Amendment to PL 554. Folder, WAC Recruiting, ASF Sp Coll DRB AGO.

[44] See entire SPWA 400 files, as well as: (1) Memo, Dir WAC for OQMG, 20 Aug 43. SPWA 400.34 (3-21-42). (2) Ltr, Miss Erma Reece, Kansas City, Mo., to Mrs. Hobby, 10 Jul 43. SPWA 421. (3) Ltr, Mr. A. J. Drugan to the President of the U.S., 9 Oct 42. SPWA 421. (4) Comments in Min, Rctg Conf, Chicago, 7-8 May 43, SPWA 337. (5) For Gallup polls, see Ch. XV, below.

[45] *Ibid.*

The final solution of the WAAC uniform problems was impossible at this time. It waited upon the day when the Director's Office was removed from the Army Service Forces and when The Quartermaster General had organized a specialist group for the development of Wacs' and nurses' clothing. Meanwhile, the extent to which the state of the uniform had damaged recruiting in the intervening months was always to remain a matter for speculation. WAAC authorities actually attributed less importance to this factor than did the general public. Although many women informed recruiters that this was their reason for not enlisting, and although many men used this argument to discourage women from enlisting, later events were to indicate that this excuse was nothing more than a red herring offered by individuals who would have disapproved of enlistment in any case.

Director Hobby offered scant sympathy to those who wrote her that their patriotic desire to enlist was impeded only by the uniform. To one such person she personally dictated and signed a reply: "Since the uniform is of so much importance to you in making your decision to join one of the women's services, I suggest that you select the service which, in your opinion, has the most attractive uniform." [46] Significantly, the complainers did not do so. The WAVES, with their smart Mainbocher-designed uniforms, never recruited the numbers of women that the WAAC and WAC did, and the weekly intake of the WAVES, SPARS, and Women Marines fluctuated constantly in close proportion to that of the WAAC.[47] The cause of recruiting difficulty, for any women's service, was to be proved more complex than any mere reaction to the appearance of a uniform.

[46] Ltr, Miss Elizabeth Williams to Mrs. Hobby, 4 May 44, and atchd reply. WDWAC 095.

[47] See following chapters.

CHAPTER X

Stresses of Rapid Build-up: Recruiting

The Army Recruiting Service, to which had been delegated the recruitment of women, inherited a discouraging tradition: from Revolutionary days onward the Army had never been able to recruit the number of men needed to fill its ranks in a major war. The Continental Congress had been unable to persuade each colony to contribute its fair share of men, and General George Washington had lamented, "We should not have found ourselves this spring so weak as to be insulted by 5,000 men . . .; we should not have been for the greatest part of the war inferior to the enemy . . . for want of force which the country was completely able to afford." [1] In the War of 1812 and during the Indian wars, even the meager forces authorized by Congress could not be obtained by recruiting. Both the North and the South, during the Civil War, had to resort to conscription before the war ended.

With the first World War, it became an established tradition that the nation would draft soldiers if they were needed. Both the public and the Army accepted the fact that a volunteer system would never persuade the needed number of men to enlist in a national emergency. In the interval of peace between the two World Wars, Congress whittled the enlisted strength of the Regular Army down to 118,750 men—less than the total number of women who were eventually to be recruited for the WAC.[2] The Army Recruiting Service therefore was geared to bring in only a modest number of men, and this during the depression years when surplus manpower was abundant. Even before the nation entered World War II Congress again provided for conscription, so that the Army had never been obliged to perfect a high-grade sales organization. After enactment of Selective Service legislation, only limited personnel and facilities remained available to the Recruiting Service, which was devoted chiefly to Aviation Cadet recruiting and other special projects.

It was thus a severely handicapped organization that undertook WAAC recruiting upon the Corps' formation. In the opinion of the Army officers later assigned to salvage the program, the Army Recruiting Service was placed at a great disadvantage in being forced to attempt a feat entirely outside its experience: to bring in larger numbers of women than it had ever

[1] *Military Policy of the United States* (U. S. Military Academy, West Point, 1944), with foreword by Col. Herman Beukema, Prof., USMA, p. 8.

[2] *Ibid.* The top WAC strength at any one time was just below 100,000; total number recruited was 143,435. See Tables 1 and 2, Appendix A. See also Ch. XXXIV, below.

obtained of men, and this in a nation that indorsed the idea that "women's place is in the home." The 1943 chief of the WAAC Recruiting Section, Captain Edlund, said, "The Army had never had a real recruiting setup. . . . It was never a sales office in any sense of the word, for in peacetime they were always able to get the quota they were after, without actually recruiting them." [3]

The Navy, in recruiting Waves, recognized the difficulty by giving direct commissions in high ranks to well-known advertising experts and authorities on sales methods. The enlistment of Waves was taken entirely out of the hands of the agencies that enlisted men and given to the Office of Naval Officer Procurement, in consideration of the fact that the recruitment of women was more selective than the drafting of men. In the Army, the Army Service Forces chose to use the personnel already assigned to its Recruiting Service, and The Adjutant General, General Ulio, stated that no intensive recruiting effort for women was deemed "necessary or advisable." [4]

The Army Recruiting Service responded well enough as long as no strain was put upon it. WAAC Headquarters furnished it with assorted instructions and pamphlets, and the numbers of applicants at first so far exceeded training space that the selection of well-qualified women offered no great problem. It was directed that recruits be thoroughly screened by mental and physical tests and by inquiries to their references and to police courts and schools. Women were also required to prove their occupational claims, and to be interviewed finally by a WAAC officer who had the power of administrative rejection. Even so, recruiting efficiency was far from 100 percent, for early reports showed that recruiters were losing sales in more than two thirds of all cases in which women were interested enough to approach them.

Staff directors also noted that recruiters frequently gave out untrue information, or took other action not in the best interests of the Corps.[5] About mid-November of 1942 Colonel Catron reported to General Grunert that there were some signs of what was called "careless enrollment." [6] Each applicant was at this time the subject of a Provost Marshal General check, although her enlistment could be completed pending its receipt, and an increasing number of derogatory reports began to be received which necessitated the discharge of women after enrollment. Particularly embarrassing was one case in which a recruit's unsuitability had not been detected in time, and, although discharged, she was thereafter able to bill herself as "The Original WAAC Strip-Teaser." A study was initiated to check such cases, and the office also took up with The Surgeon General the number of physically unfit individuals being accepted. Colonel Catron urged The Adjutant General's Office, which conducted WAAC recruiting, to increase its efforts both in numbers and quality.

---

[3] Memo, Conf of Offs to Supervise OC Bds, 9–10 Jun 43. SPWA 334.8 Conf 1943.

[4] Memo, TAG for CG ASF, 5 Apr 43. AG 341 (4–5–43) PR-I, in SPWA 341.

[5] (1) Ltr, TAG to all Corps Areas, 4 Jul 42; Additions: TWXs, TAG to all Corps Areas, 24 Jul and 12 Aug 42; also Ltr, WAAC Hq to Corps Areas, 16 Jul 42. SPX 291.9, in SPWA 341. (2) Weekly Rpt, WAAC Hq to CofAdm Serv, 7–12 Dec 42. SPWA 319.12. (3) Ltr, Stf Dir 2d SvC to Betty Bandel. Folder, WAAC Stf Mtg, SPWA 319.2 (12–2–42), WAAC Planning files. Also Rpts in Folder, Stf Dirs Conf, Chicago, 15–17 Jun 43, SPWA 337 (6–1–43).

[6] Weekly Rpt, WAAC Hq to CofAdm Serv, 8–16 Nov and 28 Nov 42. SPWA 319.12 (11–9–42) sec 1.

RECRUITING OFFICERS OF SIXTH SERVICE COMMAND *at a meeting to help promote enrollment in the WAAC, December 1942.*

Danger Signals

When the expansion program required a tenfold increase in quotas, The Adjutant General's Office began to put more pressure on its Recruiting Service. A booklet was circulated entitled *A Plan for Increasing the Rate of Enrollment in the WAAC.* In order not to disturb Army command prerogatives, the actual conduct of recruiting was left to the service commands, of which Captain Edlund observed, "In the Service Commands we had nine separate recruiting campaigns." [7] In January of 1943 all service commands received a visit from a liaison party consisting of representatives of The Adjutant General and WAAC Headquarters, accompanied by advertising experts from N. W. Ayer & Son, the advertising agency that held the contract for WAAC recruiting. Advertising matter was placed in seven magazines and in many newspapers; several WAAC posters were printed and distributed by The Adjutant General's Recruiting Publicity Bureau in New York City; information pamphlets of several types were also provided. Through the Office of War Information, a radio campaign was obtained.[8]

In spite of these efforts, certain danger signals began to be apparent. One was the adverse attitude of the men of the armed forces. In early January of 1943 the unfavorable nature of soldier opinion had become so unmistakable that Director Hobby

[7] Memo cited n. 3.

[8] (1) Memo cited n. 4. (2) Memo, Col Charles A. Easterbrook for Dir WAAC, 1 Feb 43. SPWA 334.8 Rctg (12-28-42) 1942.

requested the Services of Supply to make a formal survey.[9] This was done at once by Special Services Division, and the results, available in March, were more unfavorable than even the worst expectations. When asked, "If you had a sister 21 years old or older, would you like to see her join the WAAC or not?" the reply was:

No	40 percent
Yes	25 percent
Undecided	35 percent

The percentages were almost identical when men were asked whether they would advise a girl friend to join—only one fourth would. Those who said *no* gave as their reasons: "Women are more help in industry, defense work, farm"; "better off at home"; "Army no place for a woman"; "too close contact with soldiers"; "too hard a life"; "don't like Army life and wouldn't want her to be in anything similar."

When asked, "How do you think a young woman without family responsibilities can serve her country best?" over half designated war industry; others preferred farm or "government work." Only from 10 to 20 percent of different groups named the WAAC; only from 1 to 2 percent the WAVES. These opinions were offered although most had never seen a Waac. Almost none had any idea what the Waacs were to do; about 25 percent said "work for the government," and as much as 13 percent in some groups said "combat duty."

General Somervell expressed considerable alarm at the report and wrote, "This not only creates a severe drag on WAAC recruiting, but also creates an unnecessarily lengthy adjustment period when WAAC personnel reports for duty in the field." He directed that a high-priority film be made for the purpose of influencing soldier opinion, also a handbook to provide soldier information, and that any other media available to the Army concentrate on the problem. General Somervell also, upon the request of General Ulio, sent a letter to the commanding generals of service commands, stating, "Intimations have come to me that our men in the Army have not been too sympathetic to the idea of a women's auxiliary . . . ," and urging them to take steps to change the minds of their subordinates.

By early spring, however, the unfavorable attitude was still growing. General Madison Pearson commented on "the wave of letters coming from service men urging their wives, sisters, or girl friends not to join the WAAC." Director Hobby added that the WAAC had received fairly good publicity from press and radio, and only adverse comments on enlistment from men of the armed forces; she added, "WAAC Headquarters had been aware for some time of the attitude of the Army to the WAAC."[10]

Also ominous was the attitude of the newly created War Manpower Commission, which in December of 1942 had been given jurisdiction over the Selective Service system. Since one of the Commission's primary missions was to secure 4,000,000 women for civilian industry, its policies toward the womanpower needs of the armed forces from the beginning caused them some concern. The War Manpower Commission had immediately ended Army-Navy recruiting for men within draft age,

[9] Weekly Rpt, WAAC Hq to CofAdm Serv, 4–9 Jan 43. SPWA 319.12.

[10] (1) Memo, CG SOS for Dir Sp Serv Div SOS, 19 Mar 43. SPAP 320.2 WAAC. (2) Memo, TAG for CG ASF, 25 Mar 43. Vol. I, Gen Policy, WAAC Hist files. (3) Memo, DCofAdm Serv for CofS ASF, 5 Apr 43. SPWA 341 (4–5–43). (4) Memo, Dir WAAC for CG ASF, 7 Apr 43. SPWA 341.

limiting the armed forces to such manpower supply as could be obtained through authorized draft quotas. It was the Army's opinion that the Commission did not possess powers so to limit female recruiting, since in this case the armed forces' needs were not supplied by a draft. Nevertheless, in preliminary negotiations with the Commission over the extent of its influence, the suggestion was made that the WAAC restrict its recruiting to Group IV areas—the most limited and unpromising part of the nation's manpower area. It was also hinted that the WAAC should not recruit in occupational fields in short supply in industry—namely, the clerical, stenographic, and other fields most needed by the Army. This action followed the standard world-wide pattern: in almost every nation in which womanpower had become important, there had existed a stage in which military needs were considered secondary to those of industry. In at least one country the pressure of industry upon the national control group had been such that an army had been ordered to release to factories the women it had already recruited and trained.[11]

For its first months the WAAC had had no restrictions upon the sources from which it might accept recruits, although almost every civilian concern expressed the opinion that its womanpower needs were more important than those of the Army. A typical protest was one forwarded to a United States senator by a state manufacturers association, which said:

> A great deal of training is needed to produce a cotton mill operator. . . . Don't you think these women are more needed in the cotton mills than in the WAACs? Can you advise us of any steps this mill can take to prevent these girls from leaving their positions to join the WAAC? [12]

Early objections also came from government agencies, such as the General Accounting Office and the Federal Bureau of Investigation, which asked that the WAAC refuse to accept their female employees.

At this time Director Hobby still clung to what she felt to be the "only tenable position":

> So long as enrollment in this Corps is voluntary, and [women] citizens of this country are free to choose whether or not they shall make a war effort and if so in what manner, the Corps cannot refuse to enroll those who desire to join its ranks if they are otherwise eligible.[13]

On 6 February 1943, when the General Staff conference met to decide the WAAC's future recruiting quotas, the imminence of the Army's defeat in this encounter was perhaps not fully realized. However, within three months tremendous pressure from government and industry was to compel the armed forces to agree to sweeping limitations on women's enlistment: no women federal employees to be accepted without a release from their agencies; no women from "war industries" to be accepted without a written release from their employers or a statement of compelling reasons; no agricultural workers to be accepted at all.[14]

The Director informed the General Staff that recruiting prospects, excellent in 1942, had by this date deteriorated badly. For its first six months the Army had en-

[11] New Zealand. Memos, Alice R. Randolph (Civilian Consultant) to Dir WAAC, 1 Jan 43, and 4 Jan 43. WAAC Planning files.

[12] Ltr, 5 Jan 43, and atchd corresp. SPWA 201.6.

[13] Ltr, Atty Gen to SW, 15 Sep 42, and atchd replies. WA 044.2 FBI. Also Ltr, GAO to 1st Lt Henry Lee Munson, 5 Oct 42. SPWA 324.72 (1943).

[14] Ltr, TAG to all SvCs, 26 Feb 43. AG 341 WAAC (2-25-43) PR-I-A, in SPAP 341 WAAC (2-26-43).

countered little competition from the other armed services. However, in 1943 the WAVES, SPARS, and Women Marines began to offer numerous advantages to the better-qualified women: military status, direct commissions for college women, smart uniforms individually purchased, schools in attractive locations, and, in the case of the WAVES, technical assignments and a policy of permitting social association with officers. Also set up about this time was the government-financed U.S. Cadet Nurse Corps for civilian trainees, which gave attractive uniforms, free professional training, and a student's salary greater than that of a Waac private.

Disputes Over Lowering Standards

To overcome these obstacles, The Adjutant General took a step which, for the remainder of the Corps' existence, was to remain a most controversial issue. He lowered acceptance standards and simplified recruiting procedure in order to meet the increased quotas. Recruiters were permitted to enroll applicants before results of the serological test were obtained and to discharge them from the Reserve if the test proved positive. Presentation of a preliminary health certificate from a civilian doctor was no longer required, since this put the applicant to some expense and was supposedly duplicated by the Army examination. Letters of recommendation were no longer required, except in doubtful cases, and the WAAC questionnaire was eliminated, as well as the statements on occupational training. An aptitude test score of only 50 was made acceptable, and no educational requirement was included. Instead of the previous Provost Marshal General investigation there was substituted a check by the local office of the Retail Credit Company, Inc., optional with the recruiter.[15]

The place of WAAC Headquarters in this recruiting picture was not clear. Captain Edlund said later:

> Back in November we had a recruiting section here in Headquarters which consisted of two people. We were called into The Adjutant General's Office and the only thing we had to say about it was not in regard to the way the job should be done. . . . As early as January we were convinced that the recruiting campaign would not be successful . . . we fought for all the things we felt were missing. The quality slipped badly.[16]

On the other hand, General Ulio said, "Close liaison has been maintained at all times with WAAC Headquarters"; and also, "All important changes in recruiting activities affecting the WAAC receive the comment or concurrence of WAAC Headquarters before issue." [17]

In spite of these drastic measures, the recruiting campaign gradually began to disappoint its sponsors. In January of 1943 the picture was still rosy—the higher quota of 10,000 was actually exceeded. In February the quota was raised to 18,000; most of those on the waiting list were called to active duty and all others who presented themselves were accepted immediately. Nevertheless, at the end of the month it was discovered that the quota had not been met in spite of the lowered standards:

[15] (1) Ltr, TAG to all SvCs, 14 Jan 43. SPX 341 WAAC (1-14-43) PR I, in SPWA 341 (7-4-42) 1942. (2) Ltr, TAG to all SvCs, 26 Feb 43, AG 341 WAAC (2-25-43) PR-I-A, in SPAP 341 WAAC (2-26-43). (3) Historical Monograph, The Loyalty Investigations Programs, prepared by the Office of the Provost Marshal General, p. 17 and Tab 38A. OCMH.

[16] Memo cited n. 3.

[17] Memo, TAG for CG ASF, 25 Mar 43. Vol. Gen Policy, WAAC Hist files, 1943.

only 12,270 had been enlisted. This figure, had recruiters only known it, was to be tops for all time. The bubble had burst. The quota was raised to 27,000 for March—the number needed to fill the new training centers—and recruiters were urged to make new efforts. This quota was not met and the take for March fell below that of February—11,464.[18]

There was mounting evidence that the campaign was not recruiting from wider and more promising areas, but instead was merely scooping up the undesirables who had been previously rejected. One veteran WAAC recruiter described the situation:

> In 1942 we had crowds of applicants and could pick and choose. By early 1943 these were well picked over, but the quota went up while the applications went down. Many of those whom we now called to duty were those who were acceptable but not quite the best. By March these were almost gone, and we were faced with a choice of accepting women we had previously rejected or of leaving our quotas unfilled.[19]

AGCT Standards

Within a few weeks, the results of lowering standards were so serious as to cause Director Hobby to appeal to General Somervell to overrule The Adjutant General. In the matter of AGCT standards, records preserved at Fort Des Moines illustrated the situation. In September and October of 1942 Army General Classification Tests given recruits at this training center showed that less than 10 percent fell into the lowest two grades, IV and V, 30 percent in Grade III, or average, while 60 percent were in Grades I and II, the two highest grades, in which Army officer candidates were required to score.[20] Even in December, only 12 percent were in the two lowest groups, as against the Army expectancy of 32 percent. However, during the first months of 1943 standards plummeted, and over 40 percent of all recruits received at Des Moines were in Grades IV and V. Similar records from other training centers were not preserved, but it was the opinion of authorities at Daytona Beach and Fort Oglethorpe, which drew from the Fourth and Eighth Service Commands, that they had an even higher percentage of Grades IV and V recruits. Almost half of all WAAC recruits were now falling in a group so low that their usefulness for any sort of skilled noncombat task seemed doubtful.[21]

Records at Fort Des Moines also showed a similar drop in the degree of education attained—in April a large number of recruits were found to have had only grade school education or less. Likewise, the degree of skill attained in civilian experience declined; of a typical group of women at Des Moines, about 70 percent were classified as semiskilled, unskilled, domestic service, or laborer. Not all of these records were available to Director Hobby when she began to make protests in March, but the trend was unmistakable.[22]

By the end of March about 1,000 of these women, about half of whom were white and half Negro, were waiting idly in the training centers, where they constituted a problem of morale and sometimes of discipline. It became clear now, if it had

[18] See Table 2, Appendix A.

[19] Intervs with Maj Jean Melin, then of New Jersey Rctg Dist.

[20]
Grade I	130 or above
Grade II	110–129
Grade III	90–109 average; 100 median
Grade IV	70–89
Grade V	69 or below

[21] ASF Hist of WAC Tng, Ch. II, Table III and footnote. OCMH.

[22] *Ibid.*, Tables IV, V.

not been before, that the Army literally had no military assignment for unskilled and unintelligent women. These could not be assigned to replace enlisted men of like category because such men were ordinarily assigned to heavy work which women could not perform. They could not be trained for more than a few types of military duties.

Colonel Branch, head of the WAAC Planning Service, suggested that they be formed into motor transport, mess, and laundry companies, but the idea proved unsuccessful. Laundry training was not attempted because no military jobs could have been filled. Cooks and Bakers School received many and failed to graduate most, pointing out that such women almost never made good cooks and could pass neither the cooks nor the bakers course; they were not even cleanly and reliable in lesser duties. The Motor Transport School was directed to take large numbers. When the women could not pass the driver aptitude test, they were assigned to the course without it; when they could not pass the course, the course was changed to eliminate all technical subjects. Even with these concessions the school was able to convert few into qualified drivers.

Next, attempts were made to train the women as ward orderlies in hospitals but were abandoned when it was found that the women could not replace men of like category because of the heavy lifting involved. The women obviously were useless for all except the most simple and repetitious duties in factories or the home. To enlist them was not only to deprive industry of needed factory workers but also to burden the Army's military manpower allotments with women who could not carry their share of the work.

Later, opportunity schools and special training units were to be attempted, but inasmuch as the women's deficiency was ordinarily mental and not educational, training authorities pronounced the time spent on them to be more than their subsequent usefulness warranted. Medical authorities also discovered that the rate of disability discharge among these women was about four times as high as among skilled workers. Psychiatrists noted that the continued failure to succeed on Army jobs provoked neurotic conditions and disciplinary infractions among the subjects.[23]

In addition to the low-grade personnel retained at training centers, thousands of others were placed in companies on various pretexts and shipped to the field. Their impact on field stations was unfortunate and threatened to destroy the Corps' early reputation of one-for-one replacement of men. An alarmed staff director reported that, at a certain Arizona airfield, 152 women had arrived, of which 53 were in Grades IV and V; 86 had scores below 99. At a neighboring field in California, 34 of 154 were in Grades IV and V, and 78 were below 99 in score. While such percentages were still within the Army average, they were not good enough for a headquarters company such as the WAAC provided. The staff director commented:

> It would seem that no company going into the field should be burdened with such a large number of auxiliaries in the two lowest brackets. It makes an unfortunate impression of the Corps when the station contains so many women of low-grade mentality, which so often means low-grade behavior.[24]

She reported that all had been assigned to "grease monkey" work where it appeared

[23] For full discussion see Chs. XXXI and XXXII, below.

[24] Ltr, 1st O Helen Woods, WAAC Stf Dir AAF WFTC, to Col Catron, 3 Jun 43. SPWA 319.1.

doubtful that they could ever replace men, one for one. The Second WAAC Training Center was called upon for an explanation, since it had shipped the offending companies. Daytona Beach replied: "The only thing significant about the fact that these companies came from Daytona is that we have had the misfortune of having these people sent to us to train. There have been sent to us 3,576 Grade IV and V people." [25]

At Director Hobby's direction, training centers compiled detailed reports of all women in Grade V, giving the station of their enlistment. These she forwarded to General Ulio with the request that stations be reprimanded and individual recruiters penalized for all such women accepted, since they obviously could not have passed the WAAC intelligence tests which stations were supposedly giving. Since, in one week, three fourths of the number came from two stations in the Eighth Service Command, such corrective action appeared warranted. However, The Adjutant General did not reprimand the stations, but instead asked them to forward the test scores of the individuals concerned; these, somewhat naturally, were reported as passable, and no further action was taken.[26]

Medical Standards

As soon as the requirement for a civilian doctor's examination was dropped it quickly became clear that physical and psychiatric standards were falling alarmingly. Director Hobby gathered from training centers a list of the worst cases, including 30 women who were pregnant when accepted, 52 cases of psychoneurosis, 39 of menopausal syndrome, and miscellaneous other disorders such as fibroid uterus, rheumatic heart disease, chronic cystitis, duodenal ulcer, severe bronchial asthma, hyperthyroidism, arthritis, epilepsy, atrophy of left leg and thigh, and dementia praecox. Also, two babies had been born to new enrollees before they could be discharged—one only six weeks after the date of the mother's enrollment examination.[27]

Director Hobby forwarded these lists to The Surgeon General with names of the stations from which each had been recruited, with a request for corrective action in the case of the medical examiners concerned. The Surgeon General, however, refused to take any such action on the grounds that an Army medical examiner could not have been expected to detect any of these cases, except that of the advanced pregnancy, under the induction examination used. This examination included serological tests, chest X rays, urinalysis, palpation of abdomen, and tests of blood pressure, vision, and hearing. A pelvic examination had not been required and neither had a complete social history, since these were not required for men's induction examinations. Without them, The Surgeon General pointed out, many illnesses could not always be diagnosed.[28]

Fort Oglethorpe authorities complained that Army medical examiners seemed to think that the training center was an induction center like the Army's and that a thorough physical examination would be given there. WAAC Headquarters welcomed this, saying, "This confirms the

[25] *Ibid.*, 3d Ind.

[26] Memo, WAAC Hq for TAG, 16 Mar 43. SPWA 210.11 (2-17-43).

[27] Memo, WAAC Hq for CofAdm Serv, 11 May 43. SPWA 702.

[28] *Ibid.*, 2d Ind. SPMCS 322.5-1, SGO.

Division's contention that publication of a circular is necessary to clarify standards." [29]

Returns from the field indicated that the chief difficulty was not willful negligence, but either the absence of an adequate authorized induction examination for women or medical examiners' ignorance of proper standards of such an examination. Thus, early in 1943, rejection rates for mental diseases ranged from a low of 0.6 per thousand in the First Service Command to a high of 30.6 per thousand in the Sixth Service Command; either far more women in Chicago were deranged than in Boston, or the medical standards differed widely in the two service commands.[30] Evidence indicated that more stations were too lax than too strict: of all disability discharges granted until June, 1943, one half of the women discharged had been in service less than four months and three fourths less than five and a half months, which strongly suggested that the disabilities had existed before enlistment. Furthermore, analysis showed that 73.7 percent of all disability discharges were for two reasons only—gynecological and neuropsychiatric. This appeared to some to be rather strong evidence that medical examiners were more or less efficiently screening out the disabilities common to men and women, but were baffled by those peculiar to women.[31]

There were a few WAAC staff directors who felt that some examining psychiatrists, in the absence of guidance from The Surgeon General, were methodically screening out the more stable applicants. In one extreme case, a director noted that the local psychiatrist "maintained that the only way to determine a woman's stability was to require her to walk into his office naked, and to sit down and answer his questions in this condition." He then, in order to test her emotional balance, asked her, "How often during the past month have you had intercourse with a soldier or sailor?" The staff director added, "We lost a number of nice young prospects who never came back after he interviewed them. I could just imagine what they told their parents about the purpose of the Women's Army Corps." [32] There also continued to be a disconcerting number of unwise acceptances, such as the woman who reported to Fort Oglethorpe and nine days later informed authorities: "I am the Duchess of Windsor." [33]

A company officer at Daytona Beach found that the parents of one of her recruits had removed her from an insane asylum to enlist her; when the officer called the girl and her parents to the orderly room and presented them with her discharge, the parents wept, while the girl, it was related, "leaped over and bit her mother." [34]

To remedy such matters, Director Hobby urged The Surgeon General's Office to set up a specialist group to determine and publish detailed guides to the medical examinations required for women

[29] (1) Extracts from a large file of Tng Cen Rpts on WDAGO Form 40 (C.D.D.). SPWA 220.8 (9–30–43), 1943. (2) Ltr, 3d Tng Cen to Dir WAAC, 14 Apr 43; also atchd Memo, WAAC Pers Div to Dir WAAC. SPWA 314.7 (1–7–43)(1) sec 1.

[30] (1) Quarterly Rpt, Exam of Candidates for WAAC, Nov 42–Jan 43, SGO, Jan 43. WDWAC 702. (2) Memo cited n. 27. (3) Memo, SG for Dir WAC, 15 Sep 43. WDWAC 370.01.

[31] Memo, Med Stat Div SGO for Dir WAAC, 18 Jun 43. SPWA 220.8 (6–18–43).

[32] Interv with Lt Col Katherine Goodwin, ASF WAC Off (then Stf Dir 1st SvC), 8 Oct 45.

[33] Rpt of 3d WAC Tng Cen, Min, Stf Dirs Conf, New York, 1–3 Dec 43. SPWA 337 (11–10–43).

[34] Interv with Lt Col Mary Weems Fullbright (former CO of Tng Co.), Apr 51.

and to their later medical care. In May, she brought the matter to the personal attention of Generals Marshall and Somervell. In return she received from both the assurance that they would "personally OK procedure and it [would] soon straighten out."

The action of these authorities was unrecorded, but The Surgeon General shortly went on record that "there are problems of health peculiar to women." He therefore yielded in the matter of specialization and, a year after the WAAC's establishment, appointed the office's first Consultant for Women's Health and Welfare—Maj. Margaret D. Craighill, formerly dean of the Women's Medical College of Pennsylvania. Major Craighill became responsible for recommendations concerning the health of 60,000 Waacs and 30,000 nurses, and for visits to them at stations both in the United States and overseas. Director Hobby stated, "This is our first ray of hope. Now we have real co-operation from The Surgeon General. Major Craighill has been given the whole problem." Major Craighill at once set herself to secure published standards for proper gynecological and psychiatric screening of applicants and for other medical problems.[35]

Moral Standards

Even more distressing was the lowering of moral standards, although the relative number of women concerned was not great. Some lowering of behavior standards was inherent in the acceptance of women of low ability or of neurotics and other unstable persons. In addition, women were now accepted whose character and past record were questionable.

Early in November 1942 the Office of the Director had pointed out to The Adjutant General the dangers of attempting to apply to women the moral standards used for men. The Director's Office protested that the check on male inductees was handled so as to admit all but vicious criminals, chronic offenders, and habitual drunkards. The Director believed higher standards necessary for a volunteer noncombat corps of women and, in all cases referred to her, refused to accept women who had been convicted of any criminal offense.[36]

Until January, when references and other time-consuming checks were discontinued except for doubtful cases, a woman might possibly be screened out by reports received in this way, unless she applied in a state where she had no record, but after January this safeguard was absent. Recruiters were further handicapped in smaller towns by the fact that a traveling recruiting team, spending only a day or two in the town, might unknowingly accept a presentable-looking woman whose unsavory record was known to everyone in town except the recruiters. For men, such incidents were prevented by the fact that Selective Service boards gathered case histories, but the Director of Selective Service refused to extend this work to WAAC recruits.

One WAAC officer enlisted a woman, about whom she was a little doubtful, only after the local police chief assured her that the woman had no police record. The

[35] (1) Memo, Exec SGO for Dir WAC, 15 Aug 44. SPMC/DD 322.5, in WDWAC 720. Appointment of women doctors in the Medical Corps had been authorized by PL 38, April 1943; those already in the WAAC were transferred to the Medical Department. (2) Speech by Dir WAAC. Min, Stf Dirs Conf, Chicago, 15-17 Jun 43, SPWA 337 (6-1-43).

[36] Memo, WAAC Hq for Apmt & Ind Br AGO, 11 Nov 42. AG 000.51 WAAC PR-W, in SPWA 000.

woman was subsequently discovered to be a well-known prostitute and drunkard with a long police record. When the police chief was asked for an explanation, he replied frankly, "Well, I thought it would get her off our hands and probably do her good too." [37]

To prevent such occurrences, the Director and The Adjutant General joined in a petition to General Grunert that no woman be accepted without investigation and clearance by the Provost Marshal General's local investigators. The Director stated, "In the present stage of development, with the quality of material under close [public] scrutiny, . . . this office is of the strong opinion that all recruits should be checked . . . for the next 60 days." However, the Chief of Administrative Services and the Provost Marshal General opposed the idea, since such a check was not made for men. Instead, General Grunert informed the Director that, as a concession to the publicity problem, a spot check would be made of one in every ten recruits.[38]

Inexpert Sales Methods

The appearance of recruiting stations and the manners of recruiters also, in the Director's opinion, frightened off some applicants. To check on this suspicion, she secured authorization for a civilian agency to make an impartial survey.[39] This agency sent women employees to recruiting stations in all parts of the United States. Some sat for as much as forty minutes without attention; others were told that they were not needed and were given no informational pamphlets.[40] Still others were unable to find the Army recruiting stations, which were usually in inaccessible locations, although in searching for the station one agent found a Navy station where she was most politely treated and showered with literature and enthusiastic accounts of life in the WAVES. One who did find an Army station reported, "We had to go through about 100 selectees, whistling and making friendly remarks." [41] Another gave an account of her experience as follows:

> Q: As you entered the recruiting station, what was the general reaction you got?
> A: Rather unfavorable, unattractively furnished, and clouded with cigar smoke. Man in uniform looking out the window.
>
> Q: What sort of reception did this person give you?
> A: Turned around, gave a brief look at me, hollered "Sergeant!," and started looking out the window again.
>
> Q: Was the recruiter a man or a woman?
> A: A man . . . I felt that he thought the campaign was of little importance. Here is a verbatim report of our conversation.
>
> Applicant: I just want to ask some questions about the WAAC.
> Sergeant: Are you over 21 and have a birth certificate some place at home?
> Applicant: Yes.
> Sergeant: OK. There isn't anything you could ask me that isn't in there. (Hands over a pamphlet.) Take it home and when you make up your mind, come back.[42]

It was the agency's verdict that any business concern selling products to women would shortly go bankrupt if it hid its shops in places where women would not or did not go, or if it employed untrained salespeople who had no knowledge of the

[37] Co Comdrs Forum on Discipline. Files of Sch of WAC Pers Adm, OCMH.

[38] Weekly Rpt, WAAC Hq to CofAdm Serv, 14-21 Feb 43. SPWA 319.12 (11-9-42) sec 1 (1942).

[39] Young & Rubicam.

[40] Statement by Maj Rice, Min cited n. 33.

[41] Rpt, Young & Rubicam. Min, 6th SvC Rctg Conf, Chicago, 7-8 May 43, SPWA 337.

[42] *Ibid.*

product or any desire to sell it. It was discovered that, as a result of these sales methods, only one out of every ten women interested enough to find their own way to a recruiting station ever filed an application.[43]

The Director pointed out that surroundings which were acceptable to male draftees were not always acceptable to women volunteers. For example, male inductees who must spend the night near a station were ordinarily housed in the cheapest hotel available. Some WAAC applicants reported that they had to spend the night in questionable establishments. This point was not one on which The Adjutant General cared to make a stand for identical treatment, and a directive went out:

> In contracting for lodgings for WAAC applicants for enrollment, the regulation . . . which requires the lowest available rate per man per day will be interpreted to mean the lowest obtainable rate per woman per day consistent with the proper standards of cleanliness and decency and the personal safeguards to which women are entitled.[44]

Although WAAC officers and enlisted women had at this time been assigned to the Army Recruiting Service for several months, protests had come in from almost every service command, saying that they had little voice in the selection of applicants. Not every station was criticized in this respect, but the condition was reportedly widespread. The Second Service Command's report on this matter was the most complete. It asserted that WAAC officers had been coldly received by the recruiting stations; they received orders from enlisted men; civilian women were used to interview applicants while Waacs sat idle. Army officers determined the disposition of WAAC mail without consulting the Waacs; they determined rejections; they made requests for waivers after the Waacs had declared the applicant not qualified. The Staff Director, Second Service Command, noted: "It is felt that the special knowledge, skill, and training of WAAC officers assigned to recruiting in the Second Service Command are not being fully utilized." She recommended that their position be defined or that they be transferred from the Recruiting Service to other jobs.[45]

Reports from other service commands corroborated this report. In the opinion of the WAAC Staff Director, First Service Command, the difficulty was due to lack of orientation of the Army Recruiting Service in the field. "They had no idea what a WAAC Recruiter was supposed to do," she said. "When the first two officers arrived at a certain station, the officer in charge asked, 'Can you type?' and when they said no, he instructed them to attend night school and learn so that they could be assigned as office typists. Apparently it had never occurred to these officers that women could be expected to plan and conduct a recruiting campaign or to determine the merits of enlistees." [46] The first WAAC recruiters in the Fifth Service Command noted, "All the station commanders forced us to take in women with scores of 40; they just changed the score." [47] On the other hand, some WAAC recruiters alleged that they never permitted the enrollment of unsuitable applicants, al-

---

[43] Memo, Dir WAAC for CG ASF, 5 Apr 43. SPWA 341 (4-5-43)C.

[44] Ltr, TAG to all SvCs, 22 Jan 43. SPX WAAC (12-17-42) PR-I, in WA 513 (11-26-42), 1942.

[45] Rpt, WAAC Stf Dir to Dir Pers 2d SvC; copy in Ltr, 1st O Cora Webb Bass to Col Catron, 11 Feb 43. SPWA 319.1.

[46] Interv cited n. 32.

[47] Interv with Maj Marion Lichty, National Sel Serv Hq (former Rctg Off), 5th SvC, 4 Mar 46.

though they had to defy senior officers when quotas were not met.[48]

Transfer of Responsibility to WAAC Headquarters

By the end of March, less than two months after the expansion program was put into effect and before the worst of the results were felt, Director Hobby became convinced that the Army Recruiting Service's methods and failure to uphold standards were wrecking the Corps' mission. At a staff meeting on 25 March she directed her Planning Service to "present factual and irrefutable evidence to higher authority that will accomplish the immediate transfer of responsibility for the WAAC Recruiting Program from the Office of The Adjutant General to Headquarters WAAC." [49] On 5 April the Director submitted to General Somervell a detailed WAAC recruiting plan to raise standards and a request for control of acceptances.[50]

The Adjutant General countered by offering a radical concession. He stated that a separate specialist group was needed to recruit women and proposed entirely to separate the Army Recruiting Service and the WAAC Recruiting Service, having a "Service Command WAAC Recruiting Officer" directly responsible to the commanding general, and WAAC recruiters responsible only to her and with full control of the acceptance of Waacs.[51] He listed what he considered the chief causes of the difficulties, as concurred in by the Director:

1. The uncertainty of the proposed legislation. [The public had apparently, until the WAC Bill received publicity, never before realized that the WAAC, unlike other women's services, had not been admitted to full military rights and benefits.]
2. The unsatisfactory setup at some of our service command headquarters and subordinate recruiting installations.
3. The disappointment and dissatisfaction of enrollees writing home and describing conditions.
4. The lack of proper outfitting at the time of enrollment.
5. Inadequate housing facilities to take care of more enrollees.
6. Lack of appreciation of the importance of the WAAC objective to the Army.
7. Competition with industry.
8. The fact that the recruiting service has been staffed largely with men who are not suited for and do not understand the psychology of recruiting women.[52]

To remedy these conditions, General Ulio proposed extensive reforms: better locations for recruiting stations; only Waacs allowed to process applicants; solicitation of help from women's clubs; follow-up inspection by his office. He planned to increase recruiting personnel, make a Hollywood film, conduct a Special Services campaign to change Army men's opinions, and call on all Army chaplains to assist. He proposed also to relocate some recruiting stations, increase the advertising campaign, and ask all Waacs to urge at least two friends to join.

More important, he proposed to lower the age limit to 19, lower the physical standards, and eliminate the aptitude test for high school graduates.[53]

---

[48] Maj Jean Melin, WAAC Stf Dir, AGF Repl and Sch Comd, formerly Rctg Off in New Jersey.

[49] Cited as authority in Memo, CofPlng Serv for Dir WAAC, 5 Apr 43, Tab A to Progress Rpt, 16 Mar–15 Apr 43. SPWA 600.14 (4–15–43).

[50] Memo cited n. 43 Also, for Director's list of deterrents, see Memo, Dir WAAC for CG ASF through CofAdm Serv, 31 Mar 43. SPWA 291.9 PS, under Planning Serv Notes, 1943 WAAC files.

[51] Rough Draft by buck slip, Col Vance L. Sailor, R&I Sec AGO, to Capt Edlund, WAAC Hq, 13 Mar 43. SPWA 341.

[52] Memo, TAG to CG SOS, 25 Mar 43. Vol. Gen Policy, WAAC Hist files.

[53] Memo cited n. 4.

General Ulio's plan was endorsed for the Chief of Administrative Services by his deputy, General Madison Pearson. General Pearson suggested that the age limit be lowered to 18, the physical standards be made less rigid, and an additional mental test or two be made available for those who failed the present one.[54]

The Adjutant General's proposals reached General Somervell on the same day that Director Hobby sent her request for transfer of control, 5 April 1943.[55] Called upon to comment, the Director gave the opinion that further lowering of standards would be a dangerous matter. She noted that what she had proposed was a major matter of policy concerning higher standards, whereas The Adjutant General's program of administrative detail would be highly appropriate for whatever policy was adopted.[56]

There was no reasonable compromise between the two points of view, since British experience had already strongly indicated that a women's corps dependent upon recruiting could not contain both types of workers: if low-grade personnel constituted the majority, office workers of good character and acceptable skill would not volunteer. Only selective service for women had eventually made it possible for the British forces to obtain adequate numbers of both types of workers in the same corps. The choice posed was thus whether the WAAC should continue its original mission of supplying the Army with those skills that were scarce among men—chiefly clerical and communications—or should relinquish this to invade the fields of heavy physical labor in which women were inefficient replacements for men, requiring possibly two-for-one replacement, but in which large numbers of low-grade women could easily be recruited.

General Grunert, after a flying trip to Daytona Beach, changed his office's previous stand and supported the Director's view, stating: "Enlistment of too many low grade factory, store, and restaurant workers . . . is a problem. WAAC clerical assignments predominate and this calls for class. We had better slow down and get quality than . . . speed up for quantity."[57] He therefore withdrew previous objections to better screening methods, and advocated that Provost Marshal checks be restored for all recruits.[58]

On the other hand, there was a considerable body of opinion within the Army Service Forces which held that the Chief of Staff's original policy should be reversed and low-grade women sought. This view was stated a little later in the summer by one of General Somervell's consultants in recommending that steps be taken to replace Director Hobby:

> The techniques appropriate to the original conception of the WAAC as a small highly-qualified group of women enrolled for more or less morale and propaganda purposes are completely inappropriate for a Corps of 385,000 to take over all of the dirty work in the Zone of the Interior. . . .
>
> This means drastic revision of the present policy insisting on the high standard of intel-

[54] Memo, DCofAdm Serv for CofS ASF, 5 Apr 43. SPWA 341 (4-5-43).

[55] The Director had discussed her plan on 3 April with both The Adjutant General's Office and the Chief of Administrative Services; she had held it for two days for their comments and incorporated in it the only suggestion, a minor one, received from them. The Adjutant General had not co-ordinated his paper with hers; neither had the Chief of Administrative Services. Memo, Gen Somervell for Dir WAAC, 6 Apr 43. SPEX, in SPWA 341 (4-5-43).

[56] Memo, Dir WAAC for CG ASF, 7 Apr 43. SPWA 341 (4-5-43).

[57] Folder, Daytona Beach, Dir Adm Serv ASF Sp Coll DRB AGO.

[58] Instead, a Retail Credit Company check was extended to all recruits, for which $2.25 per name was paid. For later types of checks, see Monograph cited n. 15(3).

ligence, dignity, and educational and social background. . . . Public Relations and Recruiting policies of WAAC Headquarters . . . [place] an overemphasis on high standards and dignity, which . . . are principally and directly responsible for decreasing WAAC enrollments.

The root of the whole matter is that Mrs. Hobby feels that deterioration of standards would result from any mass recruiting technique, yet, whether she admits it or not, she has to have 385,000 Waacs by 1 July 44. . . . Consideration [should] be given to the appointment of a new WAAC Director capable of mass organization. . . .[59]

One day after its receipt the Director's request for control of recruiting was approved. Some explanation of the decision was made in a note by the WAAC Planning Service: "We understand that the Director secured the concurrence of General Marshall to raise the enlistment requirements to insure better quality."[60] Thus, as in the crisis of the previous September when he had intervened to provide for more cautious administration, the Chief of Staff's personal decision now for a second time determined the Corps' future path and mission. For the next months the goal was to be quality, if necessary at the price of quantity—recruits with clerical, communications, and technical skills or the ability to learn them, rather than unskilled laborers.

The Adjutant General was directed to give the WAAC full co-operation and every practical assistance in its new responsibility. He shortly published a memorandum asking that all communications on recruiting be sent to the Director, and withdrew from any further part in operating details. The Chief of Administrative Services, as directed by General Somervell, notified the field of the new plan, inclosing copies of Captain Edlund's somewhat frank interoffice memorandum on the deficiencies of the Recruiting Service. This rendered Captain Edlund *persona non grata* with so many Army recruiters that within a few months he was replaced by a WAAC officer. At the moment he continued as Chief, Recruiting Section, and guided the campaign which was to follow.[61]

The Restoration of Standards

Director Hobby's first act upon taking control of recruiting was, on 7 April, to restore the enrollment standards. A minimum of two years in high school and a score of 60 on the aptitude test were now required. While high, these standards were still below those of the WAVES, which required high school graduation. At once the number of recruits fell from 11,464 in March to 6,472 in April, and in May, when the new standards were in force all month, to only 4,064. In June the requirements were raised again, to a score of 70 plus two years of high school, or 80 without high school; in this month only 3,304 recruits were enlisted. By July classification officers at headquarters and training centers were able to report that "problems in making assignments have disappeared due to the present higher enrollment requirements." Records from Fort Des Moines showed that the number of recruits in Grades IV and V fell at once from 40 percent in April to 12 percent in July and 6 percent in October. In November 1943,

[59] Memo, "K. J." for Gen Somervell and Brig Gen Clinton F. Robinson, 12 Jul 43. Folder, WAAC Rctg, ASF Sp Coll DRB AGO.

[60] Memo cited n. 49.

[61] (1) 1st Ind, ACofS ASF to Dir WAAC, 8 Apr 43, to Memo cited n. 54. (2) ASF Memo S635-4-43, 14 May 43, signed by Gen Ulio. (3) Buck slip, Capt Edlund to Col Catron, May 43. SPWA 341 (4-5-43). (4) Ltr, CofAdm Serv to all SvCs, 10 Apr 43. SPAAS 341, incl Memo, 31 Mar 43, cited n. 50, in SPAP 341 WAAC (2-26-43).

after the WAAC was placed in the Army and The Adjutant General resumed operation, the figure jumped from 6 percent to 20 percent, which was to remain its average for the rest of the WAC's existence. The number of women with less than high school education fell from 36 percent in April to 20 percent in July.[62]

The expansion program had not so much collapsed as never begun. Numbers obtained monthly were now approximately the same as before the campaign began. The influx of the unqualified had lasted only a brief two months and had been checked before it lowered the Corps' average standards too far, but it was clear that the expansion program could not continue unless a real recruiting campaign could be devised that would increase numbers without lowering standards.[63]

The Advertising Contract

General Ulio was also overruled in the matter of the advertising contract, for which he favored the current contractor, N. W. Ayer & Son. This agency had adopted the campaign slogan, "Release A Man For Combat," and had given it wide publicity. The slogan had proved unfortunate in several ways. In view of the prevalence of punsters in the population, it was discovered that it would be wiser to use the expression "Replace A Man" instead of "Release" or "Relieve." Worse, as the "phony war" vanished and combat became more real to the American public, the slogan appealed to no one: Army men in clerical jobs did not particularly appreciate being replaced for combat; mothers did not wish a daughter to enlist if this would send a son to his death; and a woman whose husband or sweetheart was killed overseas did not like to think that but for her or some other woman he would have been safe in a desk job.

Also, the advertising agency, in handling the WAAC account as part of the Army account, had not given particular consideration to women's psychology in the matter of advertising appeal, location of recruiting stations, training of recruiters, or other matters.

General Somervell's Control Division, after investigation, therefore recommended that the account be canceled because of a decision to revise the entire campaign and not for any particular fault. Accordingly, eleven of the nation's leading advertising agencies were invited to a meeting and asked to suggest campaign plans. On the basis of results, Control Division, ASF, and Control Division, Adjutant General's Office, awarded the contract to Young & Rubicam, called "the largest advertising agency in the country," and one which handled campaigns for many women's products and also had an unusually large number of women writers on its staff.[64] Young & Rubicam immediately began to make surveys, for which they employed Dr. George Gallup and a staff of female investigators, to determine the correct psychological approach. Every effort was made to suppress the theme of "Release A Man For Combat," but it was found that the slogan had sunk deep into the public mind and was by this time so widespread that for months every new re-

---

[62] (1) Memo, Capt Edlund for TAG, 7 Apr 43; (2) Memo, Dir Hobby for DCofS for SvCs, 6 Jun 43; (3) Memo, Lies, Special Consultant, for Dir WAC, 8 Jul 43. SPWA 341. (4) Tables III, IV, V, ASF Hist of WAC Tng, Des Moines figures only. (5) See Table 2, Appendix A.

[63] For a description of the average Waac, see Ch. XI, pp. 192-93, below.

[64] Rpt, Lt Col Paul V. Betters and Maj McGovern, Contl Div ASF; see Weekly Rpt, WAAC Hq to CofAdm Serv, 7-13 Mar 43 SPWA 319.12.

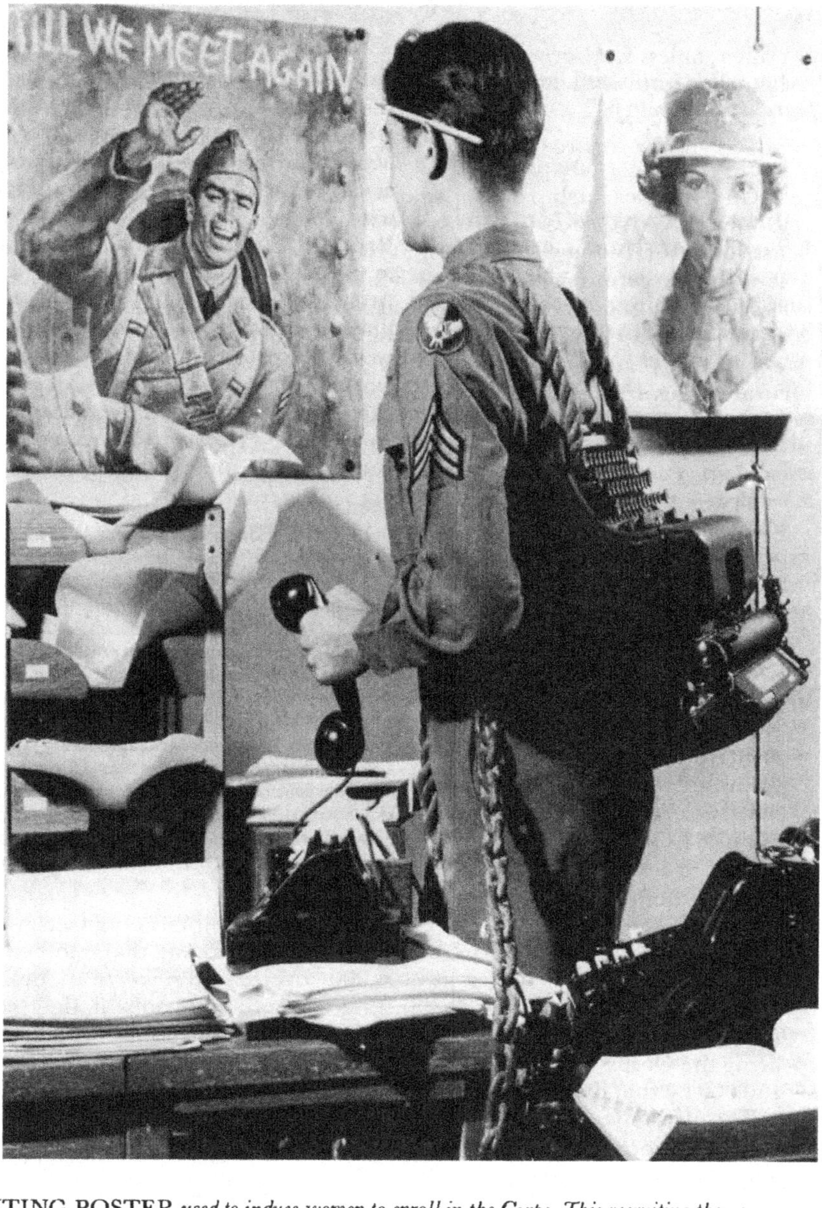

RECRUITING POSTER *used to induce women to enroll in the Corps. This recruiting theme was later discarded.*

cruiter, unless forewarned, enthusiastically launched forth with it.[65]

The Revised Campaign

Under the leadership of Maj. Harold A. Edlund, WAAC Headquarters with the assistance of Young & Rubicam at once launched a recruiting campaign which for desperate thoroughness had not been surpassed in Army history. The success of the expansion plan and the Corps' whole future hung upon the question of whether well-qualified American women could be induced to volunteer in the numbers scheduled.

Major Edlund, in civilian life a sales expert, believed that they could, but he did not underestimate the problem, which, he said, was believed by sales experts to be "one of the toughest selling jobs in the country today." He stated that recruiters who approached the problem casually had no idea of "the difficulties of persuading women of the sort we need in the WAAC to give up their homes and security and to voluntarily take on a way of life completely foreign to them." The WAAC recruiters, he said, were intuitively aware of the factors involved, but lacked sales training and the necessary authority. Nevertheless, Major Edlund's confidence in the infallibility of scientific sales technique was such that he predicted that 100,000 women could be recruited in the three months remaining before 1 July 1943, to bring the Corps' strength to the required 150,000 by that date.[66]

The first step of the giant campaign was the collection of accurate data, and a quick Gallup poll was made of a scientifically selected nationwide cross section of eligibles and the parents of eligibles. The results were admittedly shocking even to those closest to the WAAC program. Until this time it had been believed that if the women of the nation knew the Army's need for their services they would respond at once. Publicity had therefore been centered around statements by President Roosevelt, General Marshall, and others of the nation's leaders, pointing out the Army's need for personnel. Now it was discovered that the public was perfectly well aware of the Army's need; an overwhelming majority (86 percent) replied "Yes, we know the need; we know the Army is calling for more." But women did not respond, according to the Gallup survey, for five reasons:

1. *Apathy:* "Sure, it's important, but let someone else do it."
2. *Fear of Army life:* Many women feared they would not be equal to the reputed rigors of training and regimentation.
3. *Misunderstanding as to the jobs Waacs did:* The majority were most familiar with the least attractive types of work—cooking, laundry, scrubbing. Thirty-one percent said the Waacs' main job was kitchen police.
4. *Attitude of relatives and friends:* Those considering joining were almost unanimously discouraged by parents and male advisers.
5. *Army attitude:* Almost all eligible women had heard, or had been told by soldier friends, that the Army was opposed to the WAAC.[67]

Other interesting facts were uncovered. It was found that more eligibles feared overseas service than desired it. Some would join only if they could be near home; others, only if they could travel.

[65] (1) Ltr, SW to Pres., N. W. Ayer & Son, Inc., 7 May 43. WDGAP 095. (2) Speech of Maj Edlund at the mtg. (3) TAG's opinion in Memos cited ns. 4, 17. (4) Memo, WAC Hq to RPB, New York, 16 Oct 43. SPWA 000.7. (1943). (5) Ltr, Dir WAC to CSigO, 2 Jul 43. SPWA 411.36 (4-10-43).

[66] Incl to Memo cited n. 43.

[67] Ltr, Young & Rubicam, Inc., to Maj Jess Rice, WAC Hq, 1 Oct 43. SPWA 000.7. (2) See also statement cited n. 40.

Half the women believed that a Waac could not marry an Army man; many others believed that a Waac was not allowed to marry at all, or to have dates, or to use cosmetics.

The new campaign was shaped by these discoveries. If apathy was the trouble, the war must be made real to women. If women feared a strange new life, every detail of the life must be publicized so that it was familiar and not frightening. If they had mistaken ideas about the Corps, these must be corrected: they must be told that Waacs did interesting work as well as menial, that Waacs dated, married, and used cosmetics. If families and Army men disapproved, some approach must be found to change their attitude. The keynote of the campaign, in short, would be to make familiar to everyone the life and work of a Waac, and especially the good side of that life.

Accordingly, Young & Rubicam hastily prepared advertisements that pictured the interesting jobs Waacs did; they devised quizzes with many small pictures and questions; and, believing that women were inveterate coupon clippers, they inserted small coupons that could be mailed to recruiting stations for free literature. At the same time there was assembled an array of supporting talent that dwarfed previous efforts. Among others, the Writers' War Board pledged co-operation—Rex Stout, Clifton Fadiman, Paul Gallico, Margaret Lee Runbeck, Sarah Elizabeth Rodgers, Katherine Brush, Laura Z. Hobson—each to write articles to reach a different type of national magazine and audience. Ten writers of this board were flown to Fort Oglethorpe and other stations to gather material.

Famous photographers were asked to take pictures for picture magazines. In addition to stories, many magazines agreed to use a WAAC cover girl—the July *American,* the August *Cosmopolitan* and others. The May *Reader's Digest* had a story; *Life* used pictures sent from North Africa; the August *Harper's Bazaar* showed WAAC physical training in its least terrifying aspects; a June *Saturday Evening Post* began a five-part serial. Through the co-operation of the Office of War Information, radio time was obtained, including plugs on shows by Bob Hope, Kate Smith, and others. Daily papers got news stories and pictures, and Sunday papers carried paid advertisements. Motion pictures also were contributed, by both Hollywood and the Signal Corps. All of these various media were co-ordinated through WAAC Headquarters in order to maintain the same theme throughout.[68]

To co-ordinate recruiters' efforts with the advertising campaign, there set out from Washington a flying task force which, traveling by plane, train, and subway, achieved the apparently impossible feat of a two-day conference with each of the nine Service Commands in sixteen days.[69] These included representatives of Young & Rubicam to describe the campaign, of General Motors to initiate expert sales training methods, and of The Adjutant General's Recruiting Publicity Bureau to display their newest wares of posters and booklets.

The conferences were generally well received, although, said Major Edlund, "The Army Personnel connected with recruiting didn't quite understand what this was all about and perhaps resented it a little

[68] (1) Ltr, Hq 2d SvC to Dir WAAC, 15 May 43. SPWA 334 with Min of mtg. Also, Min cited n. 41.

[69] The First, Second, and Third Service Commands met together in New York. Exhibit A to Memo, WAAC Hq for CofAdm Serv, 24 Apr 43. SPWA 341 (4-5-43).

bit. We had to make sure that they realized we were perfectly friendly and honest about the whole thing and that the main purpose was to help them do the job they had been asked to do with insufficient help."[70] Instead of scolding about unfilled quotas, Edlund congratulated them on the 60,000 women already recruited. "In Canada they have been going much longer and have less. In England they had to come to conscription." Even the Waves, he said, were having difficulty in getting the relatively small number they were then aiming for.

Major Edlund had made the decision not to separate the Army Recruiting Service and the WAAC Recruiting Section, although events were to prove that it would have been wiser to adopt at once General Ulio's plan of removing all authority from the regular Recruiting Service, since such action eventually proved necessary.

Major Edlund's keynote, later deplored by veteran WAC recruiters, was, "Let's turn the bright side of the apple toward the general public." He proposed that pictures and releases all feature the most interesting jobs and the best-looking Waacs, and that Waacs not be pictured driving heavy vehicles, or in fatigue uniforms, or doing strenuous calisthenics, or even marching, since this frightened away many desirable prospects. This approach was supplemented by Young & Rubicam's advertisements with captions such as,

> "We're the luckiest girls in the world and we know it."
>
> "I joined to serve my country and am having the time of my life."
>
> "I felt pretty important when that tailor fitted that swank new uniform to me."

Such advertisements, while calculated to counteract previous misunderstandings, were also sufficient to bring on howls of derision from veteran Waacs, and to call forth complaints from later recruits who enlisted while under their spell.

The Cleveland Plan

Time was now working against the salesmen. The various surveys, the preparation of advertisements, and the flying tour had been accomplished in the unprecedented space of little over a month. Nevertheless, only six weeks remained to get the scheduled 100,000 women by 1 July. Major Edlund therefore determined upon a saturation technique of salesmanship that would literally hunt out and carry off potential recruits in a house-by-house canvass in major cities. He immediately began a demonstration campaign in a typical city. Cleveland, Ohio, was selected for the trial, and WAAC recruiters from every service command gathered to assist and observe.[71] The basic idea was that WAAC sales could be achieved only by personal interview, and not by rallies, bonfires, and previous tactics, of which Edlund said, "Everything that we could see being done [before] was a big rally or parade—a lot of talk mostly along patriotic lines and 80 percent of the people listening were not eligible to join."[72]

The Fifth Service Command ran the Cleveland Campaign, although Major Edlund had desired to do it. Theoretically such a campaign should have been preceded by a softening-up barrage of newspaper and radio publicity, but Cleveland's

[70] Speech by Maj. Edlund. Min cited n. 35(2).

[71] Cleveland was chosen because it was a representative large city, 52 percent of homes had telephones, and the Cleveland Advertising Club offered to help.

[72] Speech cited n. 70.

publicity media were busy "trying to buy a ship or something," and WAAC recruiters were obliged to land without protection. Since there were not enough WAAC recruiters in the nation to conduct a door-to-door survey in a city of this size, the Cleveland Plan called for the use of any civilian agency that had many women members in a city-wide organization. The American Red Cross was deemed the oldest and best organized, but when asked to assist, its officials replied that the Red Cross was "interested only in essential war work"; later they added that their charter did not permit help. The Office of Civilian Defense (OCD) therefore volunteered to assist, although its Block Plan was newly organized and its leaders inexperienced.[73]

The WAAC recruiters held meetings with the OCD sector leaders and supplied them with literature and questionnaires, which were passed on to block leaders. These leaders were not to attempt to enroll women, but only to locate eligibles. The final report stated that "block workers were most co-operative. Reports were that many women at home seemed to be waiting for someone to ask them to do something in the war effort. . . . The majority attitude represented complacency." These block leaders were soon able to furnish recruiters with the name and location of each eligible woman in the district.

WAAC enlisted women next went to work at a battery of telephones, trying to persuade each eligible to grant an interview to recruiting officers. These Waacs were especially trained by the Ohio Bell Telephone Company. They were instructed not to try to recruit, but merely to get an interview; when and if an interview was obtained, WAAC recruiters called on the prospect. The Realsilk Hosiery Company donated expert training in the technique of polite interviewing. All who still appeared eligible were urged to enlist, and those who seemed undecided were placed on a follow-up list.

The Cleveland Plan, supported by the nationwide publicity campaign, seemed the most intensive recruiting effort that it was possible for any organization to make. When the drive was finished and the score counted, it stood revealed as a miserable—almost an incredible—failure. The score was: [74]

Families contacted	73,589
Questionnaires completed	36,151
Considered eligible from questionnaire	12,886
Found eligible on contact	8,253
Signed applications	427
Enrolled	168

To achieve even the mere 100,000 female recruits that the Army needed immediately, recruiters would, if this rate held true, have to contact 43,800,000 families, although the nation was estimated to have only about 37,000,000 families. Interviews and sales talks would have to be held with almost 8,000,000 eligibles before 100,000 could be recruited. There were probably not that many qualified women in the United States.[75]

The Cleveland method, refined and improved by experience, was caught up and used all over the nation. Improvement in recruiting machinery continued. All summer the great radio and press campaign played upon the public. More and better-trained recruiters were thrown into

[73] This and the following are from a mimeographed booklet, WAAC Recruiting Plan, Cleveland, Ohio, June 1–10, 1943. SPWA 341.

[74] Memo, Asst Chief, WAC Rctg Serv, for Exec WAC, sub: Résumé of Rpt from 5th SvC, 4 Aug 43. SPWA 341 (1942).

[75] The Bureau of the Census in a Statistical Abstract estimated 37,100,000 families in 1944, in a civilian population of 126,606,000.

the battle. And still, amazingly, the number of qualified applicants fell. Finally, in August, the exhausted recruiters suspended their efforts to await completion of the conversion to Army status—the final hope. In this month only 839 recruits were enrolled—not enough to make up for normal attrition.

At last the full extent of public opposition stood revealed to the WAAC recruiters—an impenetrable wall against which the methods of supersalesmanship and expert recruiting techniques broke and fell ineffectually. It was not as yet fully understood why this wall had sprung up and solidified after the initial favorable public reaction, but it was all too clear that it had now done so.

CHAPTER XI

The "Slander Campaign"

British experience in two world wars had indicated that any women's service soon after organization would become the target of slanderous charges, which would lower morale, alarm parents, and make it impossible to secure a large corps except by drafting women. However, the WAAC had appeared to be relatively free of such charges during its first few months. It was therefore prematurely hoped that the American public had, since the early attacks on the Army Nurse Corps, outgrown the use of moral charges as a means of opposition to women in public life.

Record of the WAAC's First Year

This hope was sustained by the fact that the Corps' record, as it reached the milestone of its first birthday in May of 1943, remained good beyond even the highest expectations. Recruiting, training, and supply difficulties had had little if any effect upon the efficiency of WAAC units in the field. The War Department, which had supported military status in advance of proof of WAAC efficiency, now found its action justified by all field reports. From North Africa, Gen. Dwight D. Eisenhower's headquarters expressed enthusiasm for the performance of the first WAAC company and forwarded requests for hundreds more enlisted women, without whom, it was declared, it was "literally impossible to conduct effective administration." The theater requested that these units "be given earliest priority . . . and shipped at expense of ground force replacements." [1] Equally important for public opinion, both health and discipline in North Africa had been good, and there had been only one pregnancy, that of a married woman.

From the secret antiaircraft artillery experiment, which was concluded about this time, came an official report that Waacs could fill more than half of the jobs in AAA units, and were "superior to men" in many operations requiring dexterity.[2] War Department files abounded in reports of sudden conversion, such as that of the post commander who informed Col. Frank U. McCoskrie that Waacs would be sent to his post only "over my dead body," but who, a few months after their arrival, was discovered to be not only alive but writing to Des Moines for two more companies.[3] From other station commanders came similar indorsement.[4] The commanding general of a port of embarkation wrote, "I am greatly impressed with their discipline, intelligence, efficiency, and devotion to

[1] (1) ETO WAAC Hist, p. 7. (2) Cbl W3105, 20 Jun 43. WAAC 320.2 sec 5.

[2] Memo, G-3 for CofS, 7 Jul 43. WDGCT 291.9 in CofS file and in MDW.

[3] Related by Col. McCoskrie of a post commander in the Sixth Service Command.

[4] (1) Project 24, WAAC Planning files (1943). (2) Army Fact Sheet, Women in Uniform, 10 May 43, WDBPR. WAAC Hist files. (3) Dorothy D. Bromley, *New York Herald Tribune*, July 11, 1943.

duty. They have raised the standard of discipline of the command." Similar comment was received from the Signal Corps, Air Forces, Adjutant General's Department, and service commands, respectively: "Proved its value in hundreds of departments"; "Their work is splendid"; "looking forward to receiving more Waacs"; "highest type of intelligence and aptitude"; and so on.[5]

An Army Service Forces inspector in May of 1943 reported, after a visit to Waacs in the field:

> My impression in general of WAAC personnel is:
>
> 1. They are doing a fine job in the training center. . . .
>
> 2. They are performing the jobs to which assigned . . . in an excellent manner.
>
> 3. The officers and enlisted personnel with whom the Waacs work are more than satisfied with the efficiency and manner in which the women perform their tasks, as well as their attention to military courtesy.
>
> 4. The WAAC personnel is happy and doing a fine job.
>
> 5. The conduct of the WAAC personnel both on the job and after working hours is satisfactory. . . . The using people heartily endorse the use of Waacs and want to know when they are going to get more.[6]

Remarks of service commanders at a conference in July of 1943 likewise indicated that they were unanimously satisfied with their WAAC personnel and had experienced no noteworthy difficulties except in recruiting.[7]

A newspaper commented in July, near the first anniversary of the Des Moines school:

> The life history of the Waacs reads like the proverbial American success story. At their inception they were offered a chance to make good at only a handful of noncombatant jobs then considered suitable for women. Within little more than a year they had proved so effective that the Army now urgently asks that their ranks be increased to 600,000.[8]

In the summer of 1942 only four jobs had been authorized for Waacs; in the summer of 1943 Waacs were already filling 155 different Army jobs, and the number was increasing daily. In 1942 it had not been supposed that the WAAC's range of usefulness would require assignment other than to the Army Service Forces; in mid-1943 Waacs were already assigned to every major Army command in the United States and to two active overseas theaters; they were stationed in forty-four of the forty-eight states. In 1942 the admission of women even to auxiliary status appeared risky; in 1943 the War Department entertained no further doubts about the wisdom of full integration of women into the Army.[9]

General Somervell noted, in commending General Faith, "The excellent discipline, military courtesy, and appearance of the Women's Army Corps . . . are equalled by few and surpassed by no other group in the Armed Services." [10]

Statistical records in May of 1943 showed that, in spite of the brief lapse in recruiting standards, Waacs still surpassed in qualifications both the civilian average and that of Army men. The "average" Waac at this time was a mature woman, 25 to 27 years old, healthy, single, and without dependents. She was a high school graduate with some clerical experience. According to information derived from

[5] Quotations in Ltr, Dir WAAC to Edith Nourse Rogers, 26 May 43. SPWA 000.7.

[6] Rpt, Lt Col Hal P. Crane, GSC, to MPD ASF, 12 May 43. SPWA 319.1.

[7] Min of Conf, CGs of SvCs, Chicago, 22–24 Jul 43.

[8] *Washington Post*, July 10, 1943.

[9] Mimeo Sheets, 10 Jul 43, 142 Spec Jobs and 406 Suitable Jobs, and 9 Jul 43, 155 WAAC Jobs and Over 400 Others Suitable.

[10] Memo, Somervell to WD Decorations Bd, 15 Jan 44, sub: Award of DSM to Brig Gen Don C. Faith. Personnel Abstracts, WAC files, OCMH.

psychiatrists' interviews, she was inspired to enter the WAAC chiefly by a desire to do war work of a more active and responsible nature than was generally possible to a woman in civilian life. Of the small number of her eligible American sisters who were similarly inspired to the extent of getting an application blank, she was the one in three who completed it, after which she had survived tests and interviews that eliminated half of her fellow applicants. Before shipment to her job in the field, she had cost the Army only four weeks of basic training but no specialist training, which had not, for the average woman, proved necessary to successful assignment. On the men's AGCT test, she made an average score of 109. At her field station, she accepted assignment to routine clerical work although she would have liked to drive a truck. She had not yet been promoted, and her rank remained that of auxiliary (private), at $50 a month.[11] She constituted a permanent and reliable type of employee, not being subject to transfer to combat duty, nor to the usual causes of turnover in civilian personnel.

From the public viewpoint the Corps' moral record was even more important than its job efficiency. Statistical records indicated that, even with the brief lapse in standards during February and March, enlisted women's morality exceeded the civilian average. The WAAC rate of venereal disease was almost zero; many WAAC units had not experienced a single case, while training centers generally encountered only cases that had been undetected by faulty enlistment examinations. Even including cases existing before enlistment, the incidence in the WAAC was far below that of either the Army or of women in civilian life. As for pregnancy among unmarried women, the rate in the WAAC was about one fifth that among women in civilian life. This record was even better than that reported in 1942 by the British women's services, which was itself better than that of British civilian women.[12]

However, by May of 1943 it was already known within the War Department that the American corps, in spite of its actual record, was not to escape the traditional fate of slanderous attack, which became familiarly known to the Department's investigators as the "Slander Campaign," sometimes also called the "Whispering Campaign" or "Rumor Campaign." The slander campaign was, as its name implied, an onslaught of gossip, jokes, slander, and obscenity about the WAAC, which swept along the Eastern seaboard in the spring of 1943, penetrated to many other sections of the country, and finally broke into the open and was recognized in June, after which the WAAC and the Army engaged it in a battle that lasted all summer and well into the next year before it was even partially subdued.

Publicity Machinery

It was WAAC Headquarters' belief at the time that full and early publicity on the record of the WAAC's first year might have prevented what followed. However, during the period when the slander campaign took shape and gained momentum, it was the policy of the War Department

[11] Unless otherwise stated, all statistics in this and the following section are from two manila-bound typed booklets: (1) Info on WAAC, the Director's personal book of statistics, in the 1943 WAAC files, and (2) WAAC Pers Statistics, 6 Sep 43, 1943 WAAC files. See Table 3, Appendix A, for a breakdown of WAAC statistics.

[12] *Conditions in the Three Women's Services*, pp. 49–52.

Bureau of Public Relations to permit no specialized attention to the WAAC. There was no central agency charged with securing and releasing the true record and statistics of the Corps' first year; news stories on WAAC life were generally limited to those which news media secured and presented for clearance. The Bureau's Radio, Press, Pictorial, and Publications Branches all handled WAAC news releases separately and without co-ordination of policy.

General Marshall and others evidently supposed that there existed some central publicity group for the support of recruiting; as late as January of 1944 General Marshall addressed a memorandum to the "WAC Recruiting Section, Bureau of Public Relations,"although there was actually no such agency.[13]

To make up for the lack of a publicity campaign from the Bureau of Public Relations, the Director attempted to bring enough WAAC public relations personnel into her own office to supply good material to news media and guide them toward desired policies. For this purpose, she was allowed a small Office of Technical Information (OTI), such as other administrative services had, chiefly designed to check releases for technical inaccuracies but not to promote publicity. In the WAAC's first months the Director requested a larger allotment for this office, on the grounds that the WAAC must recruit personnel and was the object of more "extraordinary public interest" than other administrative services. The Services of Supply refused this request on the grounds that men's organizations, such as the Chemical Warfare Service, did not have a larger allotment.

Even had more personnel been allotted it, the WAAC OTI was not allowed by the Bureau of Public Relations to handle publicity. In April of 1943, the OTI appealed to the bureau for permission to contact editors directly to get more accurate and positive articles in magazines and journals, since the "personnel shortage" in the bureau had apparently made it impossible for that agency to achieve the desired results. In reply, the Bureau of Public Relations published a directive that all inquiries concerning the WAAC received by WAAC Headquarters would be referred to the Bureau of Public Relations, and that inquiries received by the bureau would be handled by it without reference to the WAAC except by telephone.

The extent to which the WAAC OTI was allowed to influence the bureau's decisions was limited; in June, General Surles reminded the Director by personal letter that when his Review Branch asked any OTI for comment, it desired only views on security and accuracy, and not opinions on method of presentation or tone.[14] Projects originated by the WAAC were frowned upon; thus, when the Director desired to use the WAAC Band on a radio program, the bureau vetoed the idea on the grounds that "publicity is moving along very well and the orderly procedure of it should not be disturbed too often by special appearances." [15]

Under this system, press comment on the WAAC varied. Many newspapers faithfully and favorably reported all that was furnished them, although this was not plentiful enough to build public knowl-

[13] Memo, CofS for WAC Rctg Sec WDBPR, 26 Jan 44. CofS 324.5 WAC.

[14] (1) Memo, Dir Pers SOS for CofAdm Serv, 20 Nov 42. SPGAO 322.5 WAAC, in SPWA 020. (2) Memo, WAAC OTI for WDBPR, 21 Apr 43. Vol. Policy, WAAC Hist file, SPWA 314.7.

[15] Memo, "TS" for Dir WAAC, 15 Aug 42. SPWA 311.23.

edge and acceptance to a degree that would insulate the WAAC against the later rumors. Certain anti-Administration newspapers from the beginning made the WAAC the subject of caustic comment, evidently regarding the Corps as a New Deal creation. Even in the most favorable press, there was a natural tendency for "news" to consist of those items amusing or spectacular enough to reach reporters directly, such as stories headlined, STORK PAYS VISIT TO WAAC NINE DAYS AFTER ENLISTMENT,[16] or, ARE WOMEN PERSONS? DEBATED BY HOUSE VETERANS COMMITTEE.[17] There was also a tendency for headlines to include the word WAAC in reporting all accidents, murders, suicides, and family troubles involving a member of the Corps; a headline that might more properly have read ARMY OFFICER TRIED FOR BIGAMY became, instead, WAC BRIDAL BRINGS TRIAL.[18] Even when friendly newspapers arranged for visits of their own reporters and photographers or foreign correspondents, the results frequently tended toward the coy or frivolous rather than a serious emphasis on actual WAAC jobs.[19]

Also, from the first, cartoonists had found the WAAC amusing and had contributed caricatures which ranged from light humor to emphasis on anatomical detail. Of these, a training center commandant protested to WAAC Headquarters: "There seems to be no restraint on funny papers and cartoons with the WAAC as subject matter." He was informed that "legally there is nothing that can be done to restrain cartoonists from caricaturing the WAAC." [20]

The more extreme actually were seldom commercial cartoons, but were more often soldier products from camp newspapers, which held with rather monotonous lack of originality to the idea that the best way of ridiculing a woman was to exaggerate those portions of her figure that differed from the masculine version. Although peculiarly masculine garments were not considered funny, the mere depiction of a brassière, empty or otherwise, was alone enough to seem comic to cartoonists.[21]

Certain clergymen had also published their protest against the Corps as an improper place for young Christian women. Of these, the Catholic chaplain at Fort Des Moines wrote:

> It is unfortunate that the few clergymen who have sounded off against the Corps get so much publicity, while the thousands I know who are in favor of it get little or no publicity. . . . I have had to have three masses every Sunday—about 1,400 Waacs each Sunday.[22]

Attacks by Private Letter and Gossip

These relatively minor cases of poor published and broadcast publicity or ill-chosen humor bore little resemblance in degree of virulence to the slander campaign which followed, and which had an entirely different character. The first manifestations of a deeper change in public opinion came early in 1943, when there began to be evidences of more vicious attacks on the Corps spread by word-of-mouth gossip and by private letter. Some of these were merely the "nut" letters that any organization might expect; quite often these included miscellaneous charges of

[16] Washington *Times-Herald*, May 1, 1943.

[17] Washington *Evening Star*, February 3, 1943.

[18] Philadelphia *Evening Bulletin*, August 8, 1945.

[19] For WAAC clippings, see Vol. IV, WAAC Hist file, SPWA 314.7 (1942–43).

[20] Ltr, Comdt 5th WAAC Tng Cen to Dir WAAC, 3 May 43, with atchd comment. SPWA 319.1.

[21] Malvina Lindsay, in *Washington Post*, July 17, 1948.

[22] Ltr, Chaplain Urban J. Baer, Ft Des Moines, to Dir WAAC, 9 Dec 42. SPWA 319.12.

WHAT MAKES YOU THINK THE WAAC IS COMING TO THIS CAMP?

vice in the armed forces, or attacks on "the people in the White House" for permitting card playing and drinking by soldiers.[23]

Some letters were more dangerous than their character warranted. A typical example was a letter from an Army nurse to an Arkansas radio evangelist, which began, "I am a Christian and a member of the . . . Church and hate sin as bad as anyone." She then alleged that Waacs at training centers were lined up naked for men medical officers to inspect, that no sheets were used on examining tables, and that medical officers showed Waacs pictures of naked men and of men sitting on toilets. She concluded, "Christ loves these girls and I know he does not like for them to have to line up naked and it is embarrassing for our girls every month. Please send me your book *The Truth About the Mark of the Beast,* also *Satan's Children.*"

The radio evangelist naturally became indignant and wrote his senator, saying that he intended to warn Arkansas parents and, "I am not going to permit Arkansas to become a Socialist State under the New Deal." He also sent copies of the letter to the governor, the Secretary of War, and three editors.[24] Upon the senator's request, an Army inspector general flew to Arkansas, launched a full-scale investigation, and discovered that none of the allegations was true and that the nurse in question was currently hospitalized with a diagnosis of mild psychoneurosis; she had also written similar letters about the Army Nurse Corps.

This, of course, was merely one of hundreds of such letters; the WAAC seemed to be a favorite target of mentally unbalanced persons. Unfortunately, although disproof for such allegations was readily available, it did not always reach all who had heard the charges.

Other scattered reports indicated that the WAAC was also encountering gossip and animosity from more responsible elements of the population. This was particularly true around all training centers, where numbers of Waacs were so great as seriously to inconvenience civilian users of streetcars, shops, and beauty parlors. An Army Service Forces inspector noted that, while Des Moines merchants and civic leaders had offered much co-operation, certain citizens were displeased, and added:

> It seems that any dislike of WAAC personnel is not caused by disorderly or promiscuous conduct, but rather by the fact that the WAAC personnel, in the strength superimposed upon a city the size of Des Moines, makes it appear that they take over the town at such times as they are free, particularly Saturday and Sunday, which causes some amount of inconvenience. . . .[25]

Investigation at Des Moines by the Army's Military Intelligence Service was unable to discover any basis for such dislike except "resentment on the part of local citizens . . . to the presence of strangers who they feel are usurping the old settlers in restaurants, stores, theaters, and hotels."[26] The Director of Intelligence, Seventh Service Command, after investigating Fort Des Moines, said "It was conclusively determined during the course of the investigation that the morals of the members of the Corps are exceptionally and surprisingly good."[27] He

[23] Photostat of anonymous Ltr to SW, 9 Jun 43. MID 322.12 WAAC (6-9-43) (6-11-43)

[24] (1) Ltr, Lt S., ANC, to Rev. James MacKrell, Little Rock, Ark., 18 Jan 44, and atchd. Hobby files. (2) Memo, CG ASF for TIG, 2 Feb 44, with 3d Ind SGO to TIG, 9 Mar 44. SGO Hist Div Cl file 322.5-1.

[25] Rpt cited n. 6.

[26] Memo, Chief CIC, MIS, for Maj Gen George V. Strong (G-2), 19 Jul 43. G-2 MID 322.12 WAAC.

[27] *Ibid.*, Des Moines Sec.

pointed out, however, that one example of WAAC misconduct in a bar would create much more gossip than identical conduct by scores of local civilian women.

As WAAC companies spread to the field, similar local animosity was sometimes noted near these units. For example, USO facilities and Stage Door Canteens at times discouraged or prohibited attendance by Waacs, defending this action by saying that Waacs broke rules by going outdoors with men between dances, which hostesses could not do. Waacs felt that they were not welcomed simply because USO hostesses were not interested in raising the morale of military personnel except of the marriageable variety.[28]

The Civilian Uniforms

Another variety of gossip began to plague the WAAC perhaps more than any other in the spring of 1943. It arose from the fact that the public began to attribute to the WAAC certain misconduct which, upon investigation, proved to be that of civilian women in near-military uniforms. For example, an Army recruiting officer in Louisiana reported that Waacs, probably on leave from the Fifth Training Center, were drinking heavily in Shreveport bars and taking men to their hotel rooms. Investigation by the provost marshal revealed that the women in question were indeed conducting themselves as stated; they were not Waacs but "a group of women ordnance workers wearing a uniform identical with that of the WAAC except for insignia." [29] The Women Ordnance Workers, better known as WOWS, were civilian employees of Army Ordnance; although the majority of such employees did not wear uniforms or misconduct themselves publicly, a certain number caused rumors in all parts of the country. For instance, Ninth Service Command and AAF authorities in California both protested to the Director that some of the Wows were drinking to excess and engaging in barroom brawls, and being mistaken for Waacs. Investigation disclosed that the Wows at Stockton Ordnance Depot were wearing khaki shirts and skirts, garrison caps, and enlisted men's or officers' insignia.

The ordnance depot, after the investigation, ordered removal of Army insignia, but allowed retention of the uniform, which was optional with such workers all over the nation. Their winter uniform was of olive-drab elastique, like a WAAC officer's except for patch pockets and garrison caps; they also wore depot sleeve patches and miniature shields and ordnance insignia. Being civilian workers, they were under no restrictions as to conduct, hours, or neatness.

The same ordnance depot, like hundreds of other Army installations, also employed civilian women drivers who were allowed to wear khaki shirts, slacks, and garrison caps, with sleeve patch and shield insignia. The same region in California also had a civilian volunteer group, the Women's Ambulance and Defense Corps of America, which had a khaki uniform with Army insignia of rank and the letters WADC on a sleeve patch; these women, however, were forbidden by their bylaws to drink in uniform.

[28] (1) Rpt, Stf Dir 4th AF, 2d SvC, and others, Min, Stf Dirs Conf, Chicago, 15-17 Jun 43. SPWA 337 (6-1-43). (2) Ltr, Dir WAC to all Stf Dirs, 13 Mar 43, incl Ltr from Mrs. Eliot Cross. SPWA 250.1 (3-12-43).

[29] Ltr, 8th SvC to Dir WAAC, 3 Apr 43, incl comment of Maj Warner Bishop, Ft Humbug, Shreveport, La. SPWA 250.1. WAAC Headquarters, suspicious of this address, investigated and found Fort Humbug a legitimate Army district.

The Ninth Service Command investigators, after study of uniforms of the WOWS, women drivers, and WADC, concluded that the public was undoubtedly taking all of them for Waacs, unless they investigated insignia and button design closely.[30]

This was only the beginning of what seemed to be an attempt by every woman in America to get herself into a military uniform without the inconvenience of subjecting herself to WAAC discipline. Civilian clerical workers in many Army offices bought officers' "pink" skirts and olive-drab jackets with gold buttons. The Civil Air Patrol women were authorized by the AAF, without clearance from WAAC Headquarters, to wear WAAC uniforms with red braid and silver buttons. Even the WAAC "Hobby Hat" was not sacred; secretaries at Valley Forge Military Academy wore a close copy, with a uniform almost indistinguishable from a WAAC officer's. When WAAC authorities protested this to the academy, they were informed by the professor of military science and tactics that the uniform was not at all similar since the skirt had a pleat and the buttons had the academy crest and not the WAAC eagle. At Fort Devens a soldier's wife was found wearing a WAAC uniform with gold U.S. Army buttons and her husband's insignia; she said the uniform was one that "they sold in Philadelphia" to girls whose husbands were in the service.[31]

Eastern stores advertised a "junior WAAC uniform" in sizes up through 14, guaranteed to be "an exact copy of the real WAAC uniform." The Quartermaster General informed the Director that its sale was not illegal, although its wearing by an adult might be, depending on circumstances.[32] A New York manufacturer supplied dress shops all over the country with a uniform quite similar to the Waacs', advising them:

> STAKE YOUR CLAIM. There is a vast new field open for you in selling to the army of Women Volunteer Workers—Air Raid Wardens—Minute Men—Canteen Workers—USO and scores of others. ALL DOING THEIR PART AND ALL WANTING TO DRESS THE PART.[33]

It was doubly annoying to WAAC Headquarters that these concerns were able to get olive-drab and khaki cloth in the early months when The Quartermaster General was still unable to get it for WAAC uniforms.

These department store uniforms were bought and worn not only by volunteer workers but by scores of organized prostitutes in Eastern cities. Staff Director Jess Rice of the Third Service Command, working with the provost marshal, gathered and forwarded to WAAC Headquarters irrefutable evidence of this practice. The streetwalkers, known as Victory Girls, were discovered in Harrisburg, Newport News, Baltimore, and other cities, wearing uniforms of material and cut very similar to the WAAC's. One was apprehended by military police while trying to buy a furlough-rate railroad ticket. Another was discovered when authorities checked a report that a WAAC officer was drunk in a disreputable Harrisburg hotel. The proprietress of a Baltimore clothing

[30] Ltr, Dir WAAC to 9th SvC, 4 Jun 43, sub: Unauthorized Wearing of Uniform, and Ind. SPWA 250.1.

[31] (1) Ltr, TAG to CG AAF, 10 Aug 43, sub: Authorization of WAAC Uniform for CAP Pers. SPWA 421. (2) Ltr, 4th SvC to 3d SvC, 26 May 43, sub: Valley Forge Mil Academy, Wayne, Pa., and Ind. SPWA 250.1 (4-3-43). (3) 1st Ind, WAAC Stf Dir to Dir WAAC, 11 Aug 43, to Memo, Chief of PM Br 3d SvC for Capt Rice, 11 Aug 43. SPWA 250.1 (4-3-43).

[32] Ltr, WAAC Hq to Legal Div OQMG, 14 May 43, and Ind. SPWA 421.

[33] Photo adv, Evelyn Alden Fashions, 134 W. 37th St., New York, incl with Ind cited n. 31(3).

store claimed that she was doing a good business in sale of these uniforms and "desired to know if she was doing wrong in selling them." [34]

At the Hampton Roads Port of Embarkation, the camp followers, dressed in khaki skirts and shirts, were so bold as to wait outside the gate, claiming that they were Waacs and picking up soldiers as they left the port. Here, the commanding general ordered the enrolled women to pin their insignia on the collars of their cotton shirts, at that time contrary to uniform regulations, so that they could be distinguished from prostitutes in khaki shirts and skirts.[35]

Gathering together all these examples, Director Hobby requested the Army Service Forces to amend Army regulations so as to forbid the civilian use of WAAC uniforms and insignia, or that of any insignia, buttons, and clothing which very closely resembled the WAAC items. She also asked that the Quartermaster Corps and Army Exchange Service co-operate in discouraging manufacturers of these uniforms. Since the Army controlled the manufacture and sale of all material used in commercial manufacture of Army officers' uniforms, it could, by withholding material, force most manufacturers to cease wasting it in the production of nonmilitary uniforms.

However, the Army Service Forces rejected the Director's idea, saying that it would be too difficult to enforce and that the Quartermaster Corps and the Exchange Service already co-operated with manufacturers of male officers' uniforms, so that no new instructions to them were necessary. The ASF added "Due to the fact that the WAAC is a comparatively new corps, the casual and uninformed observer is apt to believe that every woman in o. d. uniform is a Waac. It is believed that this situation will be overcome in due course." [36]

The Quartermaster General soon afterward authorized the lend-leasing of 5,000 WAAC winter uniforms, left surplus by the collapse of recruiting, to the French "WAAC" in North Africa—at that time a part-native corps without military organization. The Director shortly received a flood of derogatory letters from soldiers, in parts of North Africa where there were no Waacs, who nevertheless alleged that Waacs were disreputable in both appearance and conduct, wore earrings and bobby sox with uniforms, also long hair down their backs, and obeyed no military commands.[37]

The Organized Rumors

By late spring, even before the Cleveland Plan was launched, rumors were more widespread, more consistent, more vicious, and the tempo of their occurrence had quickened. Director Hobby and War Department officials now began to suspect that Axis agents had taken over the sporadic stories and were systematically spreading certain definite rumors in an attempt to discredit and wreck the WAAC and thus impede the Army's mobilization. This theory was supported by the fact that the onset of the more vicious rumors followed immediately after the Congressional hearings in March of 1943, in which Army

[34] Memo cited n. 31(3).

[35] Rpt, Stf Dir HRPOE. Min, Stf Dirs Conf, Chicago, 15–17 Jun 43. SPWA 337 (6-1-43).

[36] (1) 2d Ind, 19 Aug 43, to Memo cited n. 31(3). (2) D/F, MPD ASF for Dir WAAC, 1 Sep 43, atchd to same.

[37] 2d Ind, OQMG for Reqmts Div ASF, to Memo, Dir WAAC for Reqmts Div ASF, 16 Nov 43, sub: Redesigning of WAC Cap. SPWA 421.

leaders had asked Army status for the WAAC and had emphasized that thousands more women were sought to allow more men to be sent to strengthen the fighting front. The Waacs already obtained, it was said, would release for combat a number of men equal to that which had just defeated the Germans in North Africa.

As early as 18 May 1943, Director Hobby wrote to General Grunert that "there have been many indications of an organized whispering campaign directed against the WAAC" and asked investigation. The Army Service Forces sent the request for an investigation to G–2 Division, General Staff, which in turn sent it to the Federal Bureau of Investigation, claiming that it was out of the Army's jurisdiction.[38]

By early June, the situation was so far out of control that G–2 Division reversed its stand and requested the FBI to allow it to act. In a letter to the FBI, G–2 Division said:

> Consequent to the formation of such a women's auxiliary to any of the military services, a certain amount of indecent humor was to be expected. However, the inevitable so-called humor first has been supplemented and subsequently has been replaced by a circulation of plainly vicious rumors . . . what appears to be a concerted campaign has assumed such proportions as seriously to affect morale and recruiting.[39]

Supporting evidence for this view was plentiful. An identical rumor appeared almost simultaneously in New York, Washington, Kansas City, Minneapolis, and other cities, to the effect that large numbers of pregnant Waacs were being returned from overseas. Camp Lee, Virginia, was swept suddenly by the report that any soldier seen dating a Waac would be seized by Army authorities and given medical treatment. A widely repeated rumor circulated at Hampton Roads to the effect that 90 percent of Waacs had been found to be prostitutes, 40 percent of them pregnant. In the Sixth Service Command an apparently organized rumor appeared in many localities to the effect that Army physicians examining WAAC applicants rejected all virgins. In Philadelphia a "War Department Circular" with obscene anatomical "specifications" was reproduced and widely circulated, finally being found even in the foxholes of New Guinea. From Florida there came numerous identical stories that Waacs openly solicited men and engaged in sex acts in public places.[40]

One favorite theme for these organized rumors was that Waacs were issued prophylactics or were required to take such items with them when they left the barracks, so that they could fulfill the "morale purposes" for which the Army had really recruited them. It was this story that finally brought the whole slander campaign into the open. Until this time, Army and WAAC authorities had felt it wiser to ignore all rumors, since to deny any publicly would merely have given them greater circulation. However, on 8 June 1943, the charge that Waacs were issued prophylactics was made in a nationally syndicated column, "Capitol Stuff," in the McCormick chain of newspapers, by a columnist

[38] (1) Memo, Dir WAAC for Dir of Adm ASF, 18 May 43, with incl evidence; (2) 1st Ind, ASF for G–2, 22 May 43; (3) 3d Ind, G–2 for ASF. G–2 MID 322.12 WAC (5–18–43).

[39] Ltr, G–2 to FBI, 9 Jun 43. G–2 MID 322.12 WAAC (6–11–43).

[40] (1) Weekly Intelligence Bull, MID, 11 Jun 43. G–2 MID 322.12 WAAC. (2) Memo, AAF Hq for Dir WAC, 5 Apr 45. WDWAC 330.14. (3) Exhibit D, Rpt, CIC to MIS, 19 Jul 43, sub: Origin of Rumors Concerning WAAC. G–2 MID 322.12 WAAC (6–11–43).

THE PRESIDENT AND MRS. FRANKLIN D. ROOSEVELT *visit the Waacs in the spring of 1943. Above, Mr. Roosevelt reviews the troops at Fort Oglethorpe, Georgia. Below, Mrs. Roosevelt with Director Hobby and Col. John A. Hoag, commandant of First WAAC Training Center at Fort Des Moines.*

who had continuously opposed Administration measures. It was noted that in the weeks preceding the appearance of the column Mrs. Roosevelt had visited the Waacs at Des Moines and the President himself had reviewed those at Fort Oglethorpe, after which he had informed the press that "those of us who have seen the work they are doing . . . have only admiration and respect for the spirit, the dignity, and the courage they have shown." [41]

On 8 June the column stated:

> Contraceptives and prophylactic equipment will be furnished to members of the WAAC, according to a supersecret agreement reached by high-ranking officers of the War Department and the WAAC Chieftain, Mrs. William Pettus Hobby. . . . It was a victory for the New Deal ladies. . . . Mrs. Roosevelt wants all the young ladies to have the same overseas rights as their brothers and fathers.[42]

There was actually no truth in the statement. The Army did provide free prophylactic equipment for men, and it appeared theoretically possible that some station in the field might have attempted to apply the same rule to women. However, the most thorough investigation by Army operatives from G–2 Division failed to produce any such evidence. These operatives reported:

> There is apparently no factual basis for the . . . charge that contraceptives and prophylactics are issued to WAAC personnel. It is indicated that these articles are not even generally purchased in Post Exchanges and drug stores by individuals in the WAAC; in all cases of recorded sales the purchasers have been married women.[43]

It was G–2's opinion that the "supersecret" document referred to was a War Department printed pamphlet for the WAAC, *Sex Hygiene,* which prescribed six lectures to be given WAAC officers and officer candidates, to equip them to give their women a suitably modest version of the Army's required hygiene course for men.[44] This was, however, an unsensational document, part of the routine training course, which prescribed standard subjects no more radical than those given in high schools and colleges—feminine anatomy and physiology, the nature and dangers of venereal disease, and the facts about menstruation and menopause—and which said nothing whatever about the issue of contraceptives. Its wording had in fact been carefully reviewed by the Director, and its presentation limited to trained WAAC officers, because of the British experience:

> Exaggerated rumors appear to have gathered about hygiene lectures in the forces. . . . The mental reaction of a girl unaccustomed to attributing precise meanings to words, and bewildered by the impact of new and unfamiliar terms, must be kept in mind.[45]

G–2 described the pamphlet as "an excellent, frank, and wholesome manual . . . [which] counsels continence." [46] It definitely did not authorize any issue of contraceptives, and did not even tell the women what they were or how to use

[41] (1) *Des Moines Register,* February 15, 1943, and AP release, February 17, 1943. (2) Press Release, 15 May 43, signed Stephen Early. Envelope, Roosevelt, F. D., Statements by, WDBPR file of WAC Gp. (3) Text on new application form for enlistment, 31 May 43. SPWA 062.1.

[42] Washington *Times-Herald,* June 8, 1943.

[43] Memo, G–2 for Dir WAAC, 21 Aug 43, sub: Origin of Rumors Concerning Waacs. G–2 MID 322.12 WAAC (6–11–43).

[44] WD Pamphlet 35–1, 27 May 43, *Sex Hygiene Course, Officers and Officer Candidates, WAAC.*

[45] *Conditions in the Three Women's Services.*

[46] Memo, Chief CIC for Gen Strong (G–2), 23 Jun 43, sub: Prelim Rpt on WAAC. MID 322.12 WAAC (6–11–43).

them. This, just published on 27 May 1943, could have been the "agreement" mentioned in the 8 June column, although it was not "supersecret" or even secret or confidential but merely restricted to military personnel. Save for this, G-2 found that no directives on the subject had ever been issued by WAAC Headquarters except one letter in May, in answer to an inquiry, which said definitely that Waacs would not be given even so much as instruction in the use of prophylactics, much less the prophylactics themselves.[47]

War Department Denial

The War Department thus was in an excellent position to force the columnist to retract his statement. The question was whether it would be wise to lend the affair the dignity of a formal War Department denial. Many of the Director's advisers counseled against it, pointing out that persons would read the denial who had never read the attack, and that the present distress, anger, and humiliation experienced by the Director and all other Waacs would in time be forgotten. On the other hand, the shock which the column had caused the Waacs and their families was so great that an immediate denial seemed necessary to preserve the faith and self-respect of the Corps. An Army officer described a typical reaction in the WAAC company on his post:

> It raised hell with that company. Long distance calls from parents began to come in, telling the girls to come home. The younger girls all came in crying, asking if this disgrace was what they had been asked to join the Army for. The older ones were just bitter that such lies could be printed. It took all the pride and enthusiasm for the Army right out of them.[48]

An enlisted woman described the same reaction; she said:

> I went home on leave to tell my family it wasn't true. When I went through the streets, I held up my head because I imagined everybody was talking about me, but when I was at last safe inside our front door, I couldn't say a word to them, I was so humiliated—I just burst out crying, and my people ran and put their arms around me and cried with me. I couldn't understand how my eagerness to serve our country could have brought such shame on us all.[49]

Director Hobby herself, when she gathered her staff to tell them what had happened, broke down and was unable to continue speaking. The severity of the reaction of all Waacs, in whatever ranks, could be explained only by the fact that all, at this date, were the early pioneers whose enlistment had been motivated by a perhaps impractical idealism, intense enough to sustain them through the supply and training problems of the first winter, but too intense to receive such a gross attack with the indifference its inaccuracy merited.

Director Hobby therefore made the decision to reassure the women and their parents by public denials, whatever the effect on newspaper readers. Such denials were thereupon immediately made by the President and Mrs. Roosevelt, by Secretary of War Henry L. Stimson, by General Somervell of the Army Service Forces, by members of Congress, and by Director Hobby and other WAAC officers. The President told his press conference that it was a "deliberate newspaper job" and that the reporter had merely taken orders

[47] *Ibid.*, with incl Ltr, WAAC Hq to CG 2d SvC, 14 May 43. SPWA 726 (5-6-43).

[48] Interv with Joseph A. Bourdow, 9 Aug 48, then of Ft Washington AGO Sch.

[49] Sgt Amelia Madrak, Hist Div SSUSA, secy of author, 1947.

"from the top."[50] Mrs. Roosevelt in her press conference on 8 June stated that rumors about misconduct among Waacs in North Africa were Nazi propaganda, and that "Americans fall for Axis-inspired propaganda like children."[51] Naturally, she said, the Germans were interested in discrediting an organization that released so many men for the fighting front.[52]

Secretary Stimson's denial was the most publicized, and did in fact reach many people, especially on the west coast, who had never heard of the columnist or the rumors:

> Sinister rumors aimed at destroying the reputation of the Waacs are absolutely and completely false. Anything which would interfere with their recruiting or destroy the reputation of the Corps, and by so doing interfere with increase in the combat strength of our Army, would be of value to the enemy. The repetition of any unfounded rumor . . . is actually an aid to the enemy.

He pointed out that reflection on the WAAC was reflection on the whole of American womanhood, and that to malign the nation's women could easily destroy the morale of men at the front.[53]

In addition, General Somervell told a Congressional committee that the rumors were spread by a person sympathetic to the Axis and that the Waacs were "your and my daughters and sisters" and entitled to respect.[54] Representative Edith Rogers told Congress that "nothing would please Hitler more" than to discredit Waacs and American women. Representative Mary Norton said, "Loose talk concerning our women in the Armed Services cannot be less than Nazi-inspired."[55] Director Hobby told reporters that there was "absolutely no foundation of truth in the statement."

Under the barrage, the columnist was forced to retract his statement, which he did, although protesting that his information came from an "intelligent and trustworthy" official who swore that "his eyes had passed over" the alleged secret paper.[56] Nevertheless, three years later, religious publications were still to be found reprinting the story, and actually attributing the columnist's lines to Director Hobby. Director Hobby's picture was labeled "Astounding Degeneracy," and one article continued, "Mrs. William Pettus Hobby, chieftain of the WAC, says, 'Contraceptives and prophylactics will be furnished to members of the WAC according to a super-secret agreement reached by high-ranking officers of the War Department. . . .' "[57]

Investigation by Intelligence Service

In June, a full-scale investigation of possible Axis influence in the rumors was launched by the Army's Military Intelli-

[50] *Washington Post*, June 30, 1943.

[51] *Washington Daily News*, June 9, 1943.

[52] *PM* (New York), June 10, 1943.

[53] *Washington Post*, June 11, 1943.

[54] *Ibid.*, June 16, 1943.

[55] Both in the Washington *Evening Star*, June 10, 1943.

[56] "Capitol Stuff," June 10, 1943. The columnist later insisted he was vindicated when, in 1945, *Time* alleged the Army had declared surplus the remains of a stock of contraceptive jelly. He did not check far enough to discover that such jelly, if it existed, would have been sold in any Army PX, where Waacs formed an almost microscopic fraction of female customers, who included soldiers' wives and other authorized purchasers. In fact, upon investigation, Army officers were unable to locate any such stock of contraceptive jelly as reported by *Time*. Transcript of Tp Conv, Maj Chance and Mr. Marples, WDBPR. WDWAC 331.1 (1945–46).

[57] (1) *The Evangelical Beacon*, Vol. XIV, No. 22, February 27, 1945, incl to Ltr, AAF PDC to Air WAC Div, 20 Sep 45. WDWAC 330.14. (2) Ltr, Mrs. F. D. Roosevelt to Dir WAC, 25 Oct 44, incl Ltr from voter re religious magazine of alleged 1,500,000 weekly circulation, *The Missionary Worker*. WDWAC 330.4.

gence Service. The Federal Bureau of Investigation gave the Army full permission, in this one instance, to investigate persons who might otherwise fall under the investigative jurisdiction of the FBI.[58]

A more exhaustive investigation could scarcely have been made than that which Military Intelligence now undertook. First, Army agents covered sections of the nation near large WAAC installations, such as training centers. At Des Moines, for example, more than 250 interviews were held with a cross section of the local citizens, in an effort to expose the source of the rumors. Next, specific stories all over the nation were tracked down to their beginnings, each involving a large file of notes and records.[59]

The report reached a conclusion far less pleasant than the theory of Nazi activity:

> There is no positive evidence that rumors concerning the morality of WAAC personnel are Axis-inspired. There is some evidence that the Axis-controlled radio has followed a line of rumors already widely circulated by . . . *Army personnel, Navy personnel, Coast Guard personnel, business men, women, factory workers and others.* Most . . . have completely American backgrounds.[60]

Evidence indicated that in most cases the obscene stories had been originated by men of the armed forces at about the time of the change from the "phony war" to real combat in North Africa, and of Congressional publicity on the thousands of women sought to release men for combat. From Army and Navy men, the stories had spread rapidly, first to their wives and women friends, and thence to the whole population. Eventually the rumor spreaders included, according to G-2 secret files:

> (1) *Army personnel:* "Army officers and men who resent members of the WAAC . . . who have obtained equal or higher rank than themselves." "Men who fear they will be replaced by Waacs." "Male military personnel who are sometimes inclined to resent usurpation of their long-established monopoly." "Soldiers who had never dated Waacs . . . [or] had trouble getting dates."
>
> (2) *Soldiers' Wives:* "Officers' wives over bridge tables." "Women whose husbands are shipped overseas."
>
> (3) *Jealous Civilian Women:* "Local girls and women who resent having the Waacs around." "Younger to middle-aged women who deplore the extra competition." "Women who ordinarily participate in community enterprise and who are losing publicity as a result of women in uniform."
>
> (4) *Gossips:* "Thoughtless gossiping men and women." "Men [who] like to tell off-color stories."
>
> (5) *Fanatics:* "Those who cannot get used to women being any place except the home." "Those whose rabid political convictions cause them erroneously to see in the WAAC another New Deal creation."
>
> (6) *Waacs:* "Disgruntled and discharged Waacs."[61]

A typical example of an unfounded rumor that spread in the standard pattern was the case involving a Midwestern city and the surrounding area. In this city, agents reported, enlisted men and officers in bars had begun the rumors with statements of which the more printable included, "Waacs are a bunch of tramps"; "All Waacs have round heels"; and "Waacs are nothing but prostitutes." In small towns nearby, officers' wives soon

[58] Memo cited n. 43. Also, Ltr, J. Edgar Hoover to Col Leslie R. Forney, MIS, 30 Jun 43. G-2 MID 322.12 WAAC (6-11-43).

[59] Specific references will be given in each case in following paragraphs. The entire file of the summer's investigation is found in G-2 files, MID 322.12 WAAC.

[60] Memo cited n. 43. Author's italics.

[61] (1) Memo, Chief CIC, MIS, for Gen Strong, 13 Aug 43, sub: Closing Rpt on Investigation of Rumors Concerning WAAC. MID 322.12 WAAC. (2) Memo cited n. 26. (3) Memo cited n. 46.

afterward stated to friends that Waacs were really taken into the service to take care of the sex problems of soldiers. An Army chaplain then advised Waacs not to re-enlist in the WAC. The rumors spread to Protestant and Catholic ministers in the vicinity, who then urged Waacs to get out of the service. These were investigated by the FBI for sedition in urging desertion.

Agents were not able to verify even one case of pregnancy of an unmarried Waac in that area, or any other notorious misconduct.[62]

In New England, especially in the Fort Devens area, agents found that a number of stories about mass pregnancy, venereal disease, and immorality had originated with military personnel. The G–2 report added:

> Military personnel, commissioned and enlisted, were found to be a prolific fountainhead of WAAC rumors. Soldiers who had never dated Waacs, and consequently didn't know whether the stories were true, accepted the tales as gospel. Army nurses are allegedly jealous of the Waacs because the latter are promoted more rapidly, receive more publicity, and encroach on the nurses' dating territory. The wives of men replaced by Waacs are said to be angry because their husbands are sent overseas as a result of the Waacs supplanting them. No subversive intent is apparent in either case.

Soldiers in the Fort Devens area were credited by investigators with originating the rumor that "fantastic" numbers of pregnant Waacs had been sent back to Lovell General Hospital from North Africa. Agents descended on that hospital's records "without prearrangement" and reported, "No record of a pregnant Waac was found." In fact, no Waacs pregnant or otherwise had ever been returned from North Africa to Lovell General Hospital. Another Fort Devens' rumor among military personnel was that the WAAC venereal disease rate was skyrocketing. When 6,000 women were examined, only 11 cases were discovered, 8 of them having existed before enlistment and having been undetected by entrance examinations. This, agents said, was a rate which was "less than *any* civilian community." A third rumor in New England was that Waacs were officially advised to utilize contraceptives. Agents interviewed hundreds of Waacs and were unable to find even one who had ever been so advised; the Catholic chaplain also asserted that no Waacs in the area had ever to his knowledge been given such advice.[63]

In the Fort Des Moines area, where Waacs had been longest, rumors were not so vicious; the attitude of male military personnel, originally quite hostile, had reportedly upon closer acquaintance taken a marked change for the better. The Director of Intelligence, Seventh Service Command, concluded, "It is obvious that the Corps members, by force of their own composite opinion, do much to enforce proper conduct and freedom from even the appearance of evil on the part of other members."[64]

Agents at Fort Des Moines reported that no soldier could be found who had ever had sexual intercourse with a Waac; in fact, most had trouble getting a date. Sales of contraceptives at local drug stores had not gone up. Waacs drank less in public than civilian women, and it was found that "merchants agree Waacs are more courteous and patient, meet obligations more readily." All interviewed who knew Waacs placed the sexual morality of the

[62] Memo cited n. 61(1).
[63] *Ibid.*, Ft Devens area sec.
[64] Memo cited n. 26, Ft Devens area sec.

average member as higher than that of the average civilian girl. Nevertheless, a relatively few instances of drunkenness and misconduct had made a proportionately greater impression on local citizens. Since Waacs originally did not have to wear uniforms off duty, local citizens also showed a tendency to consider any drunken woman a disguised Waac, until training center authorities forbade wearing of civilian dress in the Des Moines area.[65]

Near the Fifth Training Center, a Capt. Charles S. wrote a Waac friend that he had heard there were 165 "pregnated" Waacs in one month at Camp Polk, and that before members went out on passes they were required to show that they had contraceptives with them. When asked for proof by Military Intelligence, Captain S. was, agents said, "much chagrined and embarrassed."[66]

In the area around Fort Oglethorpe, Georgia, agents reported evidence that rumors had been spread by men who resented WAAC rank or who thought they would be shipped overseas as soon as enough Waacs could be recruited to replace them. Local girls and women had picked up the rumors; one admitted to agents that she had spread stories which were not true but that, "I just get tired of seeing them around." A rumor about 100 pregnancies was traced to the local WCTU. Again, statistics failed to support the charges. Of 14,000 women trained or processed since the training center opened, three had been hospitalized for drunkenness, and eight had venereal disease, which agents described as a "negligible" percentage of cases compared to civilian rates. Hotel owners and the director of the Chamber of Commerce said Waacs conducted themselves better than civilian girls.[67]

Investigation at Daytona Beach

The most extensive training center investigation was conducted at Daytona Beach, where the Second WAAC Training Center and General Faith's headquarters had operated in the midst of a resort city. In January Director Hobby had renewed her attempts to get the women out of the city area, and in March the Services of Supply approved construction of classrooms and one theater but again refused to approve construction of recreational buildings.[68]

As a result, large numbers of recreation-seeking Waacs descended on the city nightly, provoking much civilian resentment and some of the most serious allegations encountered by investigators. At first these objections concerned only the crowding and food consumption by Waacs: one winter resident, who described himself as "a lover of the locality and its facilities as an adorable resort," wrote his senator to denounce WAAC service as "a grand vacation at Government expense . . . the Army group monopolizes our few sizeable restaurants to the detriment of civilians." Investigators found that most such complaints came from about 15 percent of the local residents described as "well-to-do property owners . . . the

[65] Memo cited n. 26. Also (1) M/R, 10 Jul 43, Summary of Investigation, sub: Origin of Rumors Concerning WAAC. G-2 MID 322.12 (6-11-43). (2) Ltr, Col Hoag to Dir WAAC, 11 Jan 43, sub: Wearing of Uniform. SPWA 300.3 (12-7-43) sec 1 (Cont). (3) WAAC Cir 2, 28 Jan 43. (4) Ltr, WAAC Hq to Tng Cens, 24 Feb 43, sub: Wearing of Uniform. SPWA 300.3.

[66] Memo cited n. 61(1), 8th SvC sec.

[67] (1) Memo cited n. 26, Ft Oglethorpe Sec. (2) Ltr, MID 4th SvC to MID (Washington, D. C.), 5 Jul 43, sub: Origin of Rumors Concerning WAAC. G-2 MID 322.12 WAAC (6-11-43).

[68] Ltr, 2d Tng Cen to Dir WAAC, 8 Jan 43, and Inds. WA 600.1 (7-2-42)(1) sec 2 (1942).

elderly conservative type."[69] Another civilian criticism alleged that WAAC messes wasted food so that their garbage cans were filled to overflowing while Florida civilians went hungry under the administration's food rationing system. Instead, investigators found that the disposable garbage rate in WAAC messes was only half that of the Army rate; the local garbage collector, when interviewed, stated that what he found in WAAC garbage cans was far less than that from "civilian sources."[70]

Rumors nevertheless grew more vicious. It was said that WAAC trainees drank too much; that they picked up men in streets and bars; that they were registered with men in every hotel and auto court, or had sexual relations under trees and bushes in public parks; that there was a nearby military hospital filled to overflowing with maternity and venereal disease cases. Finally, it was seriously stated that Waacs were touring in groups seizing and raping sailors and Coast Guardsmen.[71]

No evidence to support such statements could be found by military intelligence operatives, or by independent investigations by the Fourth Service Command, or by Colonel Clark of WAAC Headquarters.[72] The alleged government maternity home was nonexistent; only 18 pregnancies had been discovered, 16 of them among married women; inspectors could get locally only hearsay and gossip but "no single piece of correspondence which would indicate any tangible item."[73] Although the center had almost 10,000 trainees, the military police report for a typical Saturday night revealed a total of only 11 delinquencies:

2 kissing and embracing in public
1 no hat on
2 injured in auto accident
1 without identification card
1 walking with officer on street
2 found intoxicated
1 AWOL returned
1 "retrieved from Halifax River in an intoxicated condition"
—
11 Total

This was accounted a remarkable record in view of the fact that, among the thousands of women at the Second Training Center in this first week of May, 1943, many were the mental, moral, physical, and psychological problems that had been accepted in such large numbers before the restoration of recruiting standards. However, it was admitted that rumors would naturally spread through the local civilians if even one Waac out of 10,000 had to be retrieved from the Halifax River every Saturday night.[74]

The inspectors and the local authorities

[69] (1) Ltr, Mr. Fred Huntress to Senator Claude Pepper, 3 Mar 43; (2) Memo, Gen Grunert for Dir WAAC, 23 Mar 43. Folder, Daytona Beach, Dir Adm Serv ASF Sp Coll DRB AGO.

[70] For civilian complaints: (1) Memos cited n. 61. (2) File of Dir Adm Serv, with investigations, cited n. 72.

[71] Memos cited n. 61.

[72] (1) For MID investigation, see Memos cited n. 61. (2) For Col Clark's, see Memo, Col Clark, Chief Opns Div WAAC Hq, for Dir WAAC, 27 May 43. SPWA 319.1. For others, see (3) Ltr, Mr. Harry Hopkins to Brig Gen Frederick H. Osborn, 1 Mar 43; (4) Memo, Dir WAAC for TIG, 13 Mar 43, incl Ltrs from Senator Pepper and Gen White, SPWA 330.14 (3-8-43)E; (5) Memo, Exec 2d WAAC Tng Cen for Comdt, 13 Mar 43; (6) Memo, Gen Faith for Dir WAAC, 18 Mar 43, SPWA 250.1; (7) TWX, Gen Grunert to DCofAdm Serv ASF, 19 Mar 43, WA-212-119-Govt Z; (8) Memo, DCofAdm Serv for Dir Reqmts Div ASF, 20 Mar 43, with approval by Reqmts Div, 23 Mar 43, SPAAS 353.8; (9) Memo, CofAdm Serv ASF for Dir WAAC, 23 Mar 43; (10) Various Ltrs from White House, Congress, citizens, *et al.*, atchd to above. Folder, Daytona Beach, Dir Adm Serv ASF Sp Coll DRB AGO.

[73] Rpt, CIC 4th SvC to MIS, 19 Jul 43, sub: Origin of Rumors. G-2 MID 322.12 WAAC.

[74] *Ibid.*

unanimously blamed the location and the lack of recreational facilities. The area was described as the week-end mecca of soldiers and sailors from surrounding military and naval stations. An Army investigator observed, "From Saturday noon until midnight Sunday, Daytona Beach takes on an atmosphere of a large coeducational institution at which the home team has just won an important football game." [75] One officer reported, "It was a crazy idea to try to set up a Military Training Center in a place like this where the girls live in hotels and are surrounded by a carnival atmosphere." [76]

The Fourth Service Command inspector general recommended more recreation facilities so that trainees would not need to roam the city at night. The military intelligence operative agreed: "It is the opinion of this officer that recreational facilities are inadequate." [77] Ninety percent of the women had, he said, stayed in their quarters and suffered low morale, while only about 10 percent were seen in the city, but this 10 percent totaled a thousand women. General Grunert, who investigated in person, immediately telegraphed his office:

> RECREATION FACILITIES AVAILABLE TO WAAC TRAINING CENTER HERE AT DAYTONA BEACH ARE SO GROSSLY INADEQUATE AS TO MAKE IT NECESSARY TO RECONSIDER THE QUESTION OF PROVIDING TWO SERVICE CLUBS. . . . PRESENT USE OF INADEQUATE CIVILIAN FACILITIES HAS RESULTED AND WILL CONTINUE TO RESULT IN COMPLAINTS AND RUMORS AS TO DRINKING AND IMMORALITY AND DEPLETION OF CIVILIAN FOOD SUPPLY. . . . TAKE UP THIS MATTER WITH COMMANDING GENERAL ARMY SERVICE FORCES AT ONCE FOR EARLY ACTION. [78]

Now, belatedly, Requirements Division, ASF, reversed its earlier disapprovals and approved Director Hobby's four-month-old recommendation for two service clubs, but the time required to bring them into operation was considerable, and in any event the damage was irretrievable. [79]

In addition, inspectors blamed some rumors at Daytona Beach on the conduct of female dischargees who remained about town "conducting a campaign against the WAAC." [80] General Faith noted that, in the weeks before recruiting standards were restored, about one half of 1 percent of the women sent him had previous records which warranted their immediate discharge, and that in addition another 4 percent were guilty of "unseemly conduct . . . specifically, drinking, boisterousness, and petting in public parks and on benches." The remaining 95 percent he believed to be of exemplary conduct and discipline, a good statistical average for any civilian community. General Faith stated, "I am convinced that faulty recruiting is the primary cause of the conditions described." [81]

No matter how promptly such women were discharged, civilian gossip had opportunity to multiply their numbers and to confuse the conduct of dischargees with that of trainees. WAAC military police attempted to take uniforms from dischargees who were wearing them for a purpose other than the official one of return to the place of enlistment, but in one such case a woman discharged for neurosis promptly ran screaming into the street disrobed, causing even worse public comment.

---

[75] Memo, Asst IG 4th SvC for CG 4th SvC, 9 May 43, atchd as Exhibit B to Rpt cited n. 73.

[76] Post Engr, in Rpt cited n. 73.

[77] Rpt cited n. 73.

[78] TWX cited n. 72(7).

[79] Memos cited n. 72(8) and 72(9).

[80] WAAC Cir 11, 1 Oct 42, and correspondence in SPWA 420 (8-25-42), 1942. Army Regulations shortly superseded this circular.

[81] Memo cited n. 72(6).

Colonel Clark, who investigated for WAAC Headquarters, said:

> Being considerably alarmed by rumors . . . I took the trouble to ascertain from the hospital records the true facts with regard to pregnancies and venereal diseases. These figures are statistical and incontrovertible. . . . It will be seen at a glance that the experience of this center in this respect has been unusually fortunate. . . . It is extremely unfortunate that we have no recourse against these scurrilous and slanderous charges.[82]

The Army did have one recourse, and that was to abandon all city property—a move also dictated by the shrinking enrollment. At this, a local newspaper columnist cried:

> It may be that the present scare that the War Department would forego Daytona Beach . . . will close the traps of some of the scandal mongers. . . . Had it not been for the coming of the Waacs, Daytona Beach would by now have been a ghost town.[83]

In July, the Waacs gave up the leased city buildings and withdrew within the cantonment area, and the local radio station broadcast, "Cool Daytona Beach can once again accommodate thousands of summer visitors."[84]

Other Investigations

It was from overseas, where most soldiers had not as yet seen a Waac, that the worst opposition came. The Office of Censorship ran a sample tabulation and reported that, of intercepts of soldier mail which mentioned the WAAC, 84 percent expressed disfavor and most advised a woman not to join; some threatened to jilt or divorce her, as the case might be, if she did join.[85]

The only Waacs then overseas were some 200 in North Africa and about 600 bound for England, but, according to the soldiers' reports, each must have been shipped home pregnant several times. A counterintelligence operative, sitting in a restaurant next to an Army major, heard him say loudly that the Army had sent "a whole boat-load" home and that it was "difficult to keep Waacs at any station over there for any length of time without fully two-thirds of them becoming pregnant."[86] A civilian just returned from North Africa cautioned his women friends against joining the WAAC because of the low moral character of nurses and Waacs, but when pressed for facts was unable to furnish the names of any specific persons or places, and admitted he had no direct knowledge of the matter.[87]

One favorite rumor was that General Eisenhower had said that he didn't want Waacs "dumped on his command area." Foreign correspondents interviewed General Eisenhower and found that he had not only requested Waacs but said he would get British servicewomen if American ones were not sent. One reporter discovered that Waacs in North Africa lived in a convent under strict discipline, rose early to catch a 6:00 A.M. military bus to town, worked hard all day, securing enthusiastic recommendations from supervisors, and retired to the convent for an 8:00 P.M. curfew. He therefore failed to see how their night life could be very exciting.[88]

---

[82] Memo cited n. 72(2).

[83] *Daytona Beach Observer,* May 22, 1943.

[84] July broadcast over Daytona Beach stations. Exhibit N, Rpt cited n. 73.

[85] Memo cited n. 46.

[86] Memo, CID agent for Chief CIC, 16 Jul 43, sub: Dissemination of Rumors. MID 322.12 WAAC (6-11-43).

[87] Memo cited n. 61(1).

[88] John Lardner's syndicated column, June 10, 1943.

Among all the comments reported in later samplings by the censors, who of course had no power to delete them, there were many praising the Corps and its work, but not one reported instance in which a man advised a woman to enlist.[89] Typical soldier comments, from many units and from widely separated parts of the world, were:

Wife of one of the men in my company joined the Wacs. She simply wrote and told him that, tired of living off the fat of the land, she had enlisted. With no further ado, the man wrote his father's attorney to institute divorce action . . . and is at present engaged in cutting off her allotments, changing insurance beneficiaries. . . .

You join the WAVES or WAC and you are automatically a prostitute in my opinion.

I think they are great organizations, but I don't want any wife, or future wife, of mine joining them.

Are you going in the Wacs, Mother? If you did or do, I will disown you.

Velva, please don't join the Wacs. I have good reasons for not wanting you to. I persuaded my sister not to. Some day I'll tell you why.

You ask me to tell you what I think of the Wacs and Waves with the idea of you joining in mind. Darling, that sort of puts me on the spot. If the idea of you joining were not involved, I would say that they have proven a proud, worthwhile part of our armed forces. . . . But from the standpoint of you joining is something else again . . . very emphatically I do not want you to join.

I think it is best that he and Edith are separating, because after she gets out of the service she won't be worth a dime. . . . I would not have a girl or wife if she was in the service even if she was made of gold.

Any service woman—Wac, Wave, Spar, Nurse, Red Cross—isn't respected.

I told my Sis if she ever joined I would put her out of the house and I really meant it. So if you ever join I will be finished with you too and I mean it.

I think it is enough to say that I am not raising my daughters up to be Wacs.

It's no damn good, Sis, and I for one would be very unhappy if you joined them. . . . Why can't these Gals just stay home and be their own sweet little self, instead of being patriotic?

I just spoke to a couple of officers one of whose wife was so smart she joined and then told him. To give you an idea of what officers think . . . he told his wife to get a divorce. The other officer had to really tell off his daughter before she got such ideas out of her head.

Darling for my sake don't join them. I can't write my reasons because the censors won't let it through.

Honey don't ever worry your poor head about joining the Wacs for we went over all that once before, Ha! (Remember, over my dead body. Ha! Ha!) You are going to stay at home.

I would rather we never seen each other for 20 years than to have you join the Wacs. For gosh sakes stay a good girl (civilian) and I'm not just kidding either.

Ruth asked about the Wacs. The idea is noble but the widespread attitude of the public is narrow and bad. So I definitely don't recommend it.

I don't want you to have a thing to do with them. Because they are the biggest houres (I hope this gets through the censor.) Lousey, boy, they are lousey, and maybe you think my blood don't boil and bubble. . . . God, I'd disown anybody who would join.

The practice of women in the Army, still in my mind, is a glorified form of cheap bolshevism.

Lets not say no more about the auxiliaries for the idea makes me mad in the first place

[89] Exception: one comment was found in which, toward the end of the war, a man stated that it would provide a woman friend with GI benefits for a few weeks' service before the Corps was demobilized. All soldier comments are from (1) NATO AG file 319.1 Morale, Vols. II–V; (2) Morale Evaluation Survey in Women's Services, GHQ AFPAC, 16 Jun 45, in possession of former Staff Director, AFPAC.

and I don't want you galavanting around over the country for another thing. You will be better off chopping cotton. . . .

Your letter shocked me so and it was not appreciated by no means. If you join the Wacs, you and I are through for good, and I'll stop all allotments and everything.

I remember distinctly telling you before that your joining the WAC was one thing I would not tolerate. I told you distinctly that I didn't want to discuss it. Why do you persist? . . .

The service is no place for a woman. A woman's place is in the home.

About joining the Wacs the answer is still NO. If they really need service women let them draft some of the pigs that are running around loose in every town.

When you asked me the first time whether you could join the Wacs I refused and I meant that for all time. I want to come home to the girl I remember.

I cannot put this on paper how I feel and I am ashamed to tell my fellow officers. She cannot even consider herself as my wife from now on. I am stopping all allotments to her and am breaking off all contacts with her. Why she did such a thing to me I cannot understand. My heart is broken.

Get that damn divorce. I don't want no damn WAC for a wife.

In the last year of the war a censor summarized the situation:

Comments on the WAC continue to reflect much credit on their ability. Their military bearing and adaptability to military life bring praise . . . [but] there is no indication of a lessening of U.S. troops' opposition to friends and relatives joining. WAC individuals indicate their disillusionment.[90]

In the spreading of rumors, the Waacs themselves were far from guiltless. Letters written home by disgruntled individuals frequently contained remarks about food, living conditions, and forced association with low and unsympathetic characters. For example, a young woman just arrived at a new station wrote her father:

Dear Pop:

Please don't show this letter to Mom. This place is like a concentration camp and the C. O. and the First Sergeant have driven one girl to suicide and three to going AWOL. The girl who went AWOL last night was a fine, clean girl, and when she refused to submit to her Boss he threatened to court-martial her on a trumped-up charge. A boy in Co. W. killed himself last night, I saw the note he left with my own eyes. Two Chaplains have gone AWOL this week and can't be found. Please don't worry. I'll stick it out as best I can, but it is leaving its mark.

Somewhat naturally, her father sent this letter hastily to his congressman, who sent it to the Secretary of War, saying, "I have known her all of her life and have always found her to be honest, reliable, and trustworthy." A formal investigation was immediately held, at which it was discovered that none of the allegations was true: "Complainant stated that she addressed the letter to her father while in a mood of depression following her arrival at Camp D———. She acknowledged that the statements in her letter were based upon hearsay."[91]

The faulty recruiting techniques of the past months also resulted in receipt by civilians of long, incoherent letters from mentally disturbed individuals concerning persecution by officers, general misconduct among all officers and noncoms, or at least those who got promotions, and unwarranted punishments for the writer. Most alleged that they were sending copies to newspapers and friends. Investigation

[90] File cited n. 89(1).

[91] Ltr, Adm Asst to SW to Hon Michael Kirwan, 25 Sep 45; Ltr, Hon Michael Kirwan to SW, 6 Jul 45; Ltr, [WAC, Camp Davis], 6 Jul 45. CofS 324.5 WAC.

of such cases ordinarily disclosed only that the writer "had great difficulty in adjusting to the military service," [92] or had just been discharged.

With the approach of the conversion to the Women's Army Corps, many women deemed undesirable members of the company were refused re-enlistment by their commanding officers; these included neurotics, behavior problems, and other troublemakers. Recruiters came to dread the return of one of these rejects to her community, knowing that she would probably spread ridicule and falsehoods about the Corps.[93]

The Question of Axis Influence

Army intelligence agents at this point began to concentrate upon tracking down individual stories with, originally, the hope that Axis agents would be found at the source. In only one case was there any such Axis connection: this was the rumor about the return of thousands of pregnant Waacs from North Africa. When this was painfully traced to its beginning, it was found to be based on an actual incident in which the first three Waacs were shipped home, two sick, one pregnant, the latter having been for some years the lawful wife of an Army officer who had spent a leave with her a few months before.

The Axis radio station DEBUNK then broadcast in English to North Africa the modest amplification that 20 Waacs had been returned for pregnancy. This broadcast, however, was less effective than that of a Coast Guard lieutenant who told friends at the Capitol in Washington that he was a member of an armed guard which was required on a ship bringing back 150 pregnant Waacs "to keep some of the women from jumping overboard." On a train from Washington to New York, a Navy commander gave the number as 300. By the time the figure was given in Minneapolis by a Navy enlisted man, it had reached 250,000. When questioned by the Office of Naval Intelligence, the Coast Guard lieutenant said, "It was spoken in jest." He was, the Navy said, then "admonished." The Navy commander had the misfortune to relate his story on the train to a man whose daughter was a Waac; this man took his name and wrote the Secretary of War. The Secretary of War asked the Secretary of the Navy for action. The commander then wrote the Waac's father, "The private conversation in which we engaged informally seems to have been seized on by you as the basis of some far-extending investigation, in which I do not intend to become involved." The War Department was not informed whether any action had been taken against the commander.[94]

No other cases showed even a trace of Axis influence. A typical case was that of an insurance salesman in Philadelphia who wrote an Army officer that WAAC conduct was deplorable. When military agents appeared in his office and pressed for details, he alleged that Waacs had at various times accosted him and his friends, entered his hotel room, and showed him contraceptives in their purses. He refused to give names of the men friends who he

[92] Ltr, Pvt Ruth F. Bolger to Dir WAC, 23 Jun 44, incl, with investigation, in Memo, Air WAC Off for Dep Dir WAC, 18 Oct 44. WDWAC 330.14.

[93] Office Memo, 1st O Virginia Martin, Rctg Sec WAAC Hq, for Exec WAAC, 24 Jul 43. SPWA 319.1.

[94] (1) Memo, G-2 for Dir WAAC [WAC], 12 Aug 43, sub: Origin of Rumors Concerning WAAC. G-2 MID 322.12 WAAC. (2) Memo cited n. 61(1), sec re Lt Dix W. P., Coast Guard, and Comdr Neil B. W., USN. Also, same file, Ltr, John W., Jersey City, N. J., to SW, 30 Jul 43, with atchd Ltr, Comdr W. to Mr. W. (3) Ltr, SW to SN, 28 Jul 43. CofS files.

said were present when the alleged offenses occurred. His neighbors were asked for opinions as to his reliability and described him as "not reliable," "know-it-all," "a blowhard." He was not an Axis agent.[95]

Miss Helen A., a businesswoman in New York City, wrote Army authorities concerning the disgraceful situation at the ——— Hospital of Jersey City, where, she alleged, there were 50 pregnant Waacs. When questioned by intelligence officers, she admitted that she had heard that story from a friend, Miss W. Miss W. said she had the information from another friend who was having a baby at the hospital and heard two nurses in conversation outside her door. However, Miss W. finally admitted that the real source was a conversation she had overheard between two unidentified women sitting at the next table in a restaurant. The hospital submitted four affidavits to show that no Waac, pregnant or otherwise, had ever been there. Confronted with this evidence, Miss A. reportedly said that she "regretted writing the letter." None of the individuals concerned were Axis agents.[96]

In South Dakota, a civilian woman who had been assisting recruiters told them that she could no longer help recruit women unless they took steps to see that no more Waacs were sent overseas "to be tempted." She had heard that 5,000 had just been sent home from England for pregnancy. Recruiters told her that she must be wrong, since there were not that many Waacs in England, but she replied that she had the facts from her son, Capt. Harvey H., a medical officer in England. Recruiters reported this to the Director, who promptly wrote to the European theater requesting that they call to Captain H.'s attention "the implications of such statements on American womanhood." Scarcely was Captain H. threatened with court-martial in England than his father telephoned from South Dakota to tell the Director that the whole thing had been a misunderstanding. None of the family were Axis agents.[97]

One of the lengthiest investigations concerned a suspected Nazi in New Jersey, Hugo S., recipient of the Iron Cross and former Bund member. This individual, working at his bench in a war plant, observed to fellow employees that 500 pregnant Waacs had just arrived in New York from overseas. A fellow worker, whose daughter was a Waac stationed overseas, promptly knocked him down and reported him. Intelligence agents made every effort to prove that Mr. S. was acting as a Nazi agent in repeating this rumor, but were unable to do so. Several weeks before, Representative Daniel S. had risen in the Massachusetts legislature to make the same charge. He alleged that the story was told him by an Army medical officer whom he refused to name, and the representative was not a Nazi agent.[98]

Even when a rumor could be traced to its originator, there was no legal means of punishing loyal American citizens for gossip. An Axis agent, if proven such, could have been jailed had he spread identical stories with intent to hinder mobilization. Members of the armed forces could be punished only if it could be proved that they used their position to

[95] Memo, MIS for Dir MI, 3d SvC, 16 Jun 43. MID 322.12 WAAC (6-16-43) (6-11-43).

[96] Ltr, 2d SvC Hq to G-2, 22 Jul 43, sub: Origin of Rumors Concerning WAAC. MID 322.12 WAAC (6-11-43). Sec re Miss Helen A. and Miss W., New York.

[97] (1) Memo, Dir WAC for CG AAF, 6 Dec 44, sub: Rumors. WDWAC 330.11. (2) Ltr, ETO Hq to ETO, 17 Mar 45, sub: Rpt of Investigation. AG 333.5 MPMGA, in WDWAC 330.14.

[98] Memo cited n. 61(1).

spread false official information detrimental to the war effort. Director Hobby reluctantly recommended such action to Waacs who reported, for example, that an instructor in an Army school had told his class that Waacs were "a prostitute Army," that an officer's wife had told everyone in a beauty shop that large numbers of Waacs were pregnant, and other such cases. Observing, "It is a bitter thing that this had to happen in the Army family," Director Hobby advised that in every such case the offender's name be obtained and a report made for investigation of the charges.[99]

The conclusion that Axis agents were not primarily involved was supported, after the end of the war, by captured German intelligence files. These indicated that German agents had collected newspaper clippings and other public announcements, and were well aware of the Corps' strength and the plans to recruit more womanpower, as well as of the plans to place the WAAC in the Army. However, there was no indication that the Germans had placed secret agents on this project, as they had on others; apparently they were content with information gleaned from news and radio sources. If any orders had been given to spread rumors, they were not recorded.[100]

Ineffectiveness of Denials

It proved all but impossible for a Waac to defend herself or the Corps against the various rumors. For example, a young recruiter, in her speeches to civic groups, attempted to spike the rumors by telling the absolute truth: not that all Waacs were angels, but that only one Waac in the area had ever been picked up by military police; only one out of many drank heavily; the local police chief reported "very little trouble"; only nineteen women of thousands trained had ever been discharged for pregnancy and she "understood" that all nineteen were married. Spectators reported that "it appeared to give them more to talk about."[101]

At least one constructive countermeasure was put into effect by WAAC Headquarters. A planeload of the most prominent leaders of the Catholic, Jewish, and Protestant faiths had been carried to Fort Oglethorpe and Fort Des Moines for a visit shortly before the newspaper columnist's attack. They were given opportunity to inspect chapels, hospitals, clubs, and barracks, to talk to the staff and trainees, and to hold religious services. As a result, immediately after the attack, these clergymen issued a signed statement to the newspapers, saying, "We feel that parents concerned about the moral and spiritual welfare of their daughters can be reassured. A hopeful harbinger of the new world is evidenced by the sacrificial contribution which American women are making through the WAAC. [It] will strengthen their character." Monsignor Michael J. Ready told the Waacs, "We're proud of you." Dr. Carroll C. Roberts wrote the Director, "We were all deeply conscious of the earnestness of the young women and the quiet dignity with which they carried out their work."[102] These statements were reinforced by articles and

[99] Min cited n. 28(1).

[100] Sec F 22 d, "Frauen," Pol. 5, 2. Bd., Frankreich, FRA; file of German Intelligence, Army High Command, G-2 Sec, Foreign Armies West, H 2/2.

[101] Memo, 2d O D. L. Meyer for PRO, 3d WAAC Tng Cen, 7 Jun 43, sub: Rpt of Speech on Rumors by Rctg Off. SPWA 319.1.

[102] (1) Office of the Chief of Chaplains, Military History of the Second World War: The Corps of Chaplains, p. 95, on ministrations to WAC. OCMH.

CLERGYMEN VISITING FORT DES MOINES. *They are accompanied by (left) Maj. Margaret D. Craighill, Col. Frank U. McCoskrie, and the Director of the WAAC.*

pictures in religious magazines by a group of religious writers and editors who received a similar plane trip.[103]

There was also some evidence that the newspaper blasts concerning Axis sympathizers, even though mistaken in where they placed their blame, had some effect in convincing rumormongers that they might be aiding the enemy.

Congressional committees were so alarmed by the rumors that, two days after the 8 June "Capitol Stuff" column appeared, they summoned Director Hobby to appear and bring statistics on the actual cases of pregnancy and venereal disease. After seeing these, they expressed a desire to be of assistance, and suggested that the Director publish the figures and some of the actual cases of rumor spreading, to prove to the public that WAAC morality was superior to that of civilian women.[104] One Congressional committee stated:

(2) Releases in numerous papers; for example, *Washington Post*, June 11, 1943. The group included Dr. Barnett R. Bricker, National Jewish Welfare Board; Bishop W. W. Peels, Council of Bishops of the Methodist Episcopal Church; Rt. Rev. Msgr. Michael J. Ready, National Catholic Welfare Conference; Dr. Almon R. Pepper, National Council of the Protestant Episcopal Church; Dr. S. Arthur Devan, General Commission of Army-Navy Chaplains; Dr. Joseph C. Hazen, Northern Baptist Conference; Dr. J. Quinter Miller, Federal Council of Churches of Christ; Dr William Barrow Pugh, Presbyterian Church; Dr. Carroll C. Roberts, International Convention of Disciples of Christ.

[103] M/R, 29 Apr 43, sub: Religion. SPWA 000.3.

[104] WAAC Daily Journal, Vol. I, 10 Jun 43.

The committee wishes to express its unqualified endorsement of this organization . . . [and] feels constrained to voice its condemnation in no uncertain terms of those who malign this splendid group of patriotic women. Certainly no self-respecting patriotic American would indulge in such a cowardly, contemptible, despicable course.[105]

Army commanders at many stations also immediately made their stand known to the troops; one stated to the men:

We have gained aid and help from the women of America. If we do anything to make their lot uncomfortable or unhappy, if we fail to give them a maximum amount of respect, if we fail to behave always as gentlemen, we have not only lost a contribution towards winning this war, but we have also lost some of the finer qualities of manhood.[106]

Others in Congress and elsewhere took the attitude that the stories were merely good clean fun and that the Waacs were spoilsports to object to them. According to a member from Alabama, WAAC stories were "like traveling salesman jokes" and he looked upon them as "the way this country keeps its sense of humor." [107] General Marshall was later to suggest that responsible Army officers might be able to distinguish between clean fun and a dirty joke by asking themselves if the stories would appear equally funny if circulated about Army wives and daughters instead of about Waacs. To Director Hobby for the WAAC, General Marshall wrote:

To me, one of the most stimulating aspects of our war effort has been the amazing development of the WAAC organization in quality, discipline, capacity for performing a wide variety of jobs, and the fine attitude of the women themselves. Commanders to whom the Waacs have been assigned have spoken in the highest terms of their efficiency and value. The best evidence in the matter are the demands now being made on the War Department for increased allotments of WAAC organizations. . . .

I wish you would reassure your subordinates of the confidence and high respect in which they are held by the Army.[108]

WAC leaders were later to conclude that the whole slander campaign had been to some degree unavoidable. In this, the British precedent of identical slander in two world wars lent strong support. One WAC leader commented, "Men have for centuries used slander against morals as a weapon to keep women out of public life." Mrs. Hobby, after a lapse of some five years, remarked, "I believe now that it was inevitable; in the history of civilization, no new agency requiring social change has escaped a similar baptism. I feel now that nothing we might have done could have avoided it." [109]

[105] HR Rpt 566, 78th Cong, 1st sess, Rpt of Com on Appropriations to Accompany HR 2996, Mil Establishment Appropriation Bill FY 1944. Hobby files, 18 Jun 43.

[106] Ltr, Base Surg to Mil Pers of Base Hosp, Tinker Field, Okla.; reprinted in Ltr, Base Comdr to All Mil Orgns . . . at Tinker Field, 23 Jun 43; copy in Folder, Slander Campaign, OCMH.

[107] *Washington Post*, June 16, 1943.

[108] Personal ltr, CofS to Oveta Culp Hobby, 15 Jan 43. Gen Marshall's personal files; read by author by courtesy of his secretary, December 1945. The Chief of Staff was in Africa at the time of the columnist's charges, and had just learned of the attack.

[109] Comments to author. British published reports reached the same conclusion. See Appendix B.

CHAPTER XII

July–September 1943: The Conversion to Army Status

By the middle of June 1943, after delay caused by the slander campaign, the WAC bill was at last nearing approval by Congress.[1] The Auxiliary Corps faced its final and most severe test, for every member must choose honorable discharge or immediate enlistment in the new Women's Army Corps, Army of the United States. Neither the modern Army nor any other women's service had ever faced such a trial, and no precedent existed as to how many of the enlisted personnel would depart when given such an opportunity.[2]

It had been known for some time that certain clauses in the pending WAC legislation would probably give every member of the WAAC the right to elect immediate honorable discharge rather than to enlist in the new WAC when it should be established. The Army's Judge Advocate General ruled that it did not lie within the Army's power to transfer women from the WAAC to the WAC without their consent, since only a selective service law could place a citizen under court-martial jurisdiction without his consent.[3] At first, few departures had been anticipated, but in June, following the newspaper attack, Colonel Catron was obliged to inform the War Department: "It looks as though more members of the old WAAC than we had anticipated have it in mind not to join the WAC."[4]

Reporters and editors, sensing a spectacular story in the offing—possibly the end of the Corps itself—demanded confirmation of the fact that "the Army is very much worried about the WAAC situation." When estimates of probable losses were refused them, some intimated that they would publish estimates of their own.[5]

Director Hobby stated, just before the bill passed, "My feeling is that we don't want them if they don't want to stay in. I would hate to lose a great many but I would rather have a smaller Corps of women who are dedicated to this service. This is something that a woman must want to do if she stays in."[6] She therefore refused to take steps to persuade women

[1] Min, Gen Council, 31 May 43.

[2] The Army Nurse Corps a year later secured similar legislation placing it in the Army of the United States, but it was not necessary to discharge those members who refused AUS status, since there was an authorized peacetime component of the ANC which continued to exist and in which members could be required to continue service.

[3] M/R, 8 May 43. SPWA 300.3 sec 1.

[4] Memo, Exec WAAC for G–1, 20 Jul 43. SPWA 314.7 sec 3.

[5] File of Tp Convs, esp Jul–Aug 43.

[6] Min, Stf Dirs Conf, Chicago, 15–17 Jun 43. SPWA 337 (6–1–43).

to stay by emotional appeals or any other means.

The WAC bill itself, as it neared passage, offered women the mixed blessing of Army status, but few other inducements. Little was left of the simple version proposed by the War Department, which would have placed few limitations on the Secretary of War's power to decide administrative matters. The Congressional situation had continued unfavorable all spring. In April the War Department became so alarmed at the rejection of the Navy bill to send Waves to foreign countries that it withdrew the WAC bill entirely and submitted it again in May.[7]

Even so, a number of amendments had to be accepted: the Corps was to last only for the duration plus six months; it was limited to women aged 20–50 years; its commanding officer could never be promoted above the rank of colonel and its other officers above the rank of lieutenant colonel; its officers could never command any men unless specifically ordered to do so by Army superiors; physicians and nurses could not be accepted because of the Medical Department's insistence on autonomy.[8] Many even more hampering amendments had been narrowly averted—the Senate's proviso that the Corps expire on 1 January 1945, and the House's attempts to limit the Corps to 150,000, to set the top age at 45, to forbid WAC officers to be assigned to Army jobs, and to deny all dependency benefits to women—but all these were eventually eliminated in conference. Director Hobby stated that she was willing to accept any amendments if only the bill might pass.[9]

Just as they had been a year previously, the efforts of the Army's highest authorities were required to secure Congressional approval. General Marshall urged passage of the bill in a letter to Chairman Andrew J. May of the House Military Affairs Committee, and also intervened to secure the new age limit of 20–50, which would admit more recruits than the previous 21–45 limit. General Somervell was also summoned to testify when the Military Affairs Committee seemed about to disapprove the bill because of the erroneous idea that Wacs would be used to displace Civil Service workers in Washington. This had actually been done on a large scale by the WAVES without Congressional disfavor, but never by the Army. "They've got the Waacs and the Waves confused," noted the Office of the Secretary of War.[10]

The First Wac

The bill to establish a Women's Army Corps *in* the Army of the United States was passed by the Senate on 28 June and

[7] Min, Gen Council, 28 Apr 43.

[8] Whether to commission women doctors in the WAC or in the Medical Department was for a time a controversial problem. The WAAC had consented to commission women doctors recommended by The Surgeon General, and did commission several—the only exception to its rule that all WAAC officers must attend OCS. (WD Cir 23, Sec II, 18 Jan 43.) The Secretary of War had personally directed Director Hobby to include a clause in the WAC bill permitting commissioning of women doctors in the WAC. (WAAC Daily Journal, Vol. I, Feb 43.) However, Congress refused this and passed a separate law permitting direct commissions in the Medical Corps.

[9] (1) HR Rpt 595, 78th Cong, 1st sess, 24 Jun 43, sub: Establishing a WAC for Serv in Army of U.S., Conf Rpt to accompany S 495. SPWA 314.7 (1-7-43)(1) sec 1. (2) Speech, Min, Tng Cen Comdts Conf, Washington, 31 Mar 43, SPWA 334.8 (3-20-43).

[10] (1) Ltr, CofS to Hon A. J. May, 1 Mar 43. CofS 291.9. (2) Memo, Dir WAAC for Leg&LnDiv, WDSS, 24 Apr 43. CofS 291.9; also in SPWA 341 (4-7-43), with note re Gen Marshall's desire for this plan. (3) Quotation from M/R, Tp Conv, Mr. Mitchell and Dir Hobby, 11 Jun 43. SPWA 314.7 (1-7-43)(1) sec 1. See also pp. 557–59, below.

signed by the President on 1 July 1943.[11] On 5 July, in the presence of General Marshall and other dignitaries, Director Hobby took the oath of office as a colonel, Army of the United States, thus becoming the first woman to be admitted to the new component of the Army.[12]

The passage of the bill was the cause of general pride and pleasure to Waacs throughout the world. The commanding general of the Hampton Roads Port of Embarkation wrote to Colonel Hobby:

> You would have been amused and, I think, pleased at the reaction of our personnel here to the news that the legislation had gone through. The Port Band at mess time, accompanied by a large part of our Headquarters Detachment of enlisted men, proceeded to the WAC barracks and serenaded the girls with "You're in the Army Now." and "This is the Army, Mr. Jones," and a general jollification ensued.[13]

Actually, "You're in the Army Now" was somewhat premature, for the WAAC did not automatically become the WAC. The law gave the Army ninety days to arrange for the dissolution of the Women's Army Auxiliary Corps. By the thirtieth day of September, all Waacs must be enlisted or commissioned in the WAC, or discharged, for the WAAC would cease to exist.

The Conversion

The following ninety days of the summer of 1943, informally called The Conversion, were perhaps the busiest in the history of the Corps. WAAC Headquarters' fourteen-hour day now approached a twenty-four-hour one as WAAC officers worked throughout the night to type, sort, and fold informational material and stuff envelopes. Director Hobby canceled all travel plans for July and August, and the headquarters braced itself for three months of the most intense activity the Corps had yet experienced.[14]

WAAC Headquarters was immediately hit by "a flood of requests for information regarding the changeover."[15] Stations in the field had had no instructions or advice as to the nature of the conversion, and knew only what they read in the newspapers. Some stations immediately began to let women go home if they wished, without proper authority;[16] others began to swear them into the Army, also without authority—in fact, some enlisted women were sworn in by WAAC officers and served for months without ever knowing that they were not in the Army.[17]

Director Hobby had foreseen this confusion and had asked the War Department to permit her to send the complete conversion plan to the field far in advance of the event. This conversion plan had actually been carefully prepared over the past six months by WAAC Headquarters and The Adjutant General's Office, and included a detailed timetable based on T Day or Transfer Day. Women who desired to transfer to the WAC must complete a

[11] PL 110. Approved by WD in Ltr, SW to Bur of Budget, 1 Jul 43. SPGAL 291.9 Gen (6-30-43)-5, in WDCSA 291.9.

[12] (1) WDBPR Press Release, 5 Jul 43. (2) GO 1, Hq WAC, sub: Assumption of Comd. SPWA 300.4 (7-5-43)E (1943). She was not the first woman admitted to the AUS; this honor went to Major Craighill, the first woman doctor to be admitted under Medical Corps legislation of the previous month.

[13] Ltr, Brig Gen John R. Kilpatrick, CG HRPOE, to Col Hobby, 3 Jul 43. SPWA 314.7 (6-29-43).

[14] (1) Memo Routing Slip, Dir's secy to OTI, 25 Jun 43. Vol. I Gen Policy, WAAC Hist File, 1943. (2) Intraoffice Memo to Asst Exec WAAC, 28 Jul 43. SPWA 314.7 (1-7-43)(1) sec 3.

[15] WAAC Daily Journal, Vol. I, 1 Jul 43.

[16] Rectified by Change 2, 14 Jul 43, to AGO Memo W635-10-43, 7 Jul 43.

[17] Rectified by AG Ltr, 21 Aug 43. AG 320.2 WAAC PR-W, in SPWA 314.7 sec 4. Women later demanded release but were refused on grounds that they had been "constructively enlisted."

certain application form, must pass a stricter Army medical examination, and must also be recommended by their commanders. It was intended that this processing be completed by 1 September 1943, on which date Waacs all over the world would be sworn in with mass ceremonies.[18]

G–1 Division of the General Staff had refused the request to communicate this plan to the field in advance of the passage of legislation, stating that such action might be illegal or might adversely influence Congress, and had told the Director, "There will be plenty of time." [19] The Director persisted and finally, only a week before the passage of the bill, succeeded in getting a confidential letter sent to all commands, giving an enlistment plan that could be put into effect under any probable wording of the legislation.[20]

On 1 July 1943, a few moments after the bill was signed, Director Hobby telegraphed all Army commands telling them to begin the processes of this letter, and to abide by WAAC Regulations until new ones could be published. Unfortunately, the letter itself had not yet had time to filter down to most Army stations, since it had to be relayed through several command echelons, and those women who wished to leave, and their parents, became immediately frantic for reassurance. When the Director saw how little information had reached the field, she chose six of the Corps' best officers and sent them on tour to carry correct procedures to every WAAC company; she herself visited as many as possible.[21]

Even after stations received instruction, the apparently simple process soon proved unbelievably complicated: all of the administrative problems of the Auxiliary here converged and demanded a final definitive solution as to the exact status of its members in all legal matters before WAAC records could be closed out. Some of these questions, but not many, could be answered by WAAC Headquarters. A number of red-bordered immediate-action letters were published by The Adjutant General, at WAAC Headquarters' request, to get answers to the field before 1 September, while others were answered by telephone and telegraph.[22]

In addition, there were about twenty major issues which neither the Army nor WAAC Headquarters had the power to decide; these were submitted to the Comptroller General, but decision was not received on some cases until the following October and November. Most stations desired especially to know whether WAAC service might be counted as Army service for purposes of longevity pay. Most Waacs and Army stations felt strongly that it should, since in actual conditions of daily living there was virtually no difference in WAAC and WAC life, which continued without a break in its routine. The Army Service Forces blamed the Director's

[18] (1) M/R, sub: Chronology, Conversion of WAAC to WAC. SPWA 314.7 (1–7–43)(1) sec 3. (2) Memo, Capt Bandel for Maj George G. Wolfe, AGO, 26 Apr 43. SPWA 300.3 (1–7–43) sec 1. (3) Memo, Dir WAC for Dir Pers ASF, 30 Jun 43. SPWA 291.9. (4) AGO Memo cited n. 16. (5) WD Cir 146, 26 Jun 43, sec III, Physical Standards.

[19] Memo, WAAC Hq for TAG, and Ind, 12 Jun 43. SPWA 314.7 sec 1.

[20] AGO Ltr, 24 Jun 43. AG 210.1 WAAC (6–23–43), in SPWA 314.7 sec 1.

[21] A week later, more specific instructions were published by TAG. The Director also telegraphed all WAAC training centers to issue simultaneous press releases. (1) TWX, Dir to all SvCs, Tng Cens, Ports, and Def Comds, 1 Jul 43, and Ltrs to AAF, AGF, and MDW; (2) TWX to all Tng Cens, 1 Jul 43. SPWA 314.7 sec 3. (3) Speech, Dir. WAC, Min, Stf Dirs Conf, New York, 1–3 Dec 43. SPWA 337 (11–10–43).

[22] W635 series, copies in Vol. AGO Memoranda, WAAC Hist File. Also Change 1, 13 Jul 43, and Change 2, 14 Jul 43, to AGO Memo cited n. 16.

Office for the delay in answering such questions, and one of General Somervell's consultants pronounced its organization "inadequate," saying, "Simple questions such as 'Does my service in the WAAC count so far as longevity benefits are concerned?' are unanswered. It is possible that five minutes would be required to settle a question of this type; the only possible answer is yes." [23] As a matter of fact, the only possible legal answer was *no*, and after some years of controversy and attempted corrective legislation, the answer was still *no*.[24]

Some Comptroller General decisions decided such issues as the fact that Waacs were not entitled to re-employment rights even if they enlisted in the WAC, since they had not gone directly from their civilian jobs to the Army; this issue was not finally settled until 1946 when corrective legislation was secured. It was also found that Waacs who enlisted in the WAC forfeited their eventual travel pay home upon discharge, and were entitled only to pay to the place where they joined the Army, their present station; correction of this matter luckily did not require legislation and was made by the Comptroller General in time to prevent wholesale failure to enlist.[25]

Still worse field confusion resulted when the physical requirements were changed in mid-August. For two months before passage of the bill, Director Hobby had argued with The Surgeon General over the physical standards, which she wished to adjust so as to admit every Waac who had been qualified upon her enrollment and who had since given satisfactory service, even though her physical condition might have declined since her enrollment. She felt strongly that these women should be admitted to the WAC and given remedial treatment if necessary. This The Surgeon General refused, stating that routine granting of administrative waivers would negate the desired screening by virtually blanketing the WAAC into the WAC. Director Hobby especially objected to the high visual standards, seldom needed by enlisted women, and directed that her office "go to bat with the Surgeon General." Toward the end of June, G–1 Division decided in favor of The Surgeon General, and directives for physical examinations were published accordingly.

By August the numbers of Waacs rejected for physical reasons became so excessive that the decision was reversed, Colonel Hobby's plan accepted, and all discharges for physical reasons were stopped until the requirements were corrected. This action could not authorize recall of women who had already been discharged under the old standards, and considerable inequality of treatment to individuals resulted.[26]

[23] Memo, "WCM" for Dir Pers ASF, 10 Aug 43. SPAP 341 WAAC (2–26–43), ASF files.

[24] Comptr Gen to SW, Decision B–35441, 4 Aug 43. SPWA 242.2 (6–10–43). Service did count for Yeomanettes, Marinettes, Army field clerks, WAVES, SPARS, Women Marines. Army or Navy Nurse Corps service could be counted by enlisted women but not officers.

[25] (1) Memo, JAG for WAAC Hq, 19 Aug 43, sub: Eligibility for Re-employment Benefits, etc. SPJGA 1943/11282 in WAAC file. (2) Comptr Gen to SW, Decision B–36362, 26 Aug 43. SPWA 245.6 (7–1–43). (3) For a summary of controversy on legal status of WAAC, see Appendix C.

[26] All except (2) and (7) in SPWA 314.7 (1–7–43), secs 1 and 4 (1943): (1) Memo, Dir WAAC for Dir Pers ASF, 3 Jun 43. (2) Memo, Dir Hobby for SG, 26 Apr 43. SPWA 300.3 (1–7–43) sec 1. Director Hobby felt that nurses' standards on weight and eyesight were too high for enlisted Wacs, since Wacs often did more sedentary jobs. (3) Office Memo, WAAC Contl Div for Exec WAAC, 14 Jun 43; (4) M/R, 23 Jun 43; (5) WD Cir 146, sec III, 26 Jun 43; (6) Memo, CofS ASF for Dir WAAC, 21 Aug 43, and Ind, with AG Ltr 220.8 WAAC PR–W, 23 Aug 43. (7) Memo, Gen Styer for TAG, 23 Aug 43. SPEX.

First Companies in the WAC

Meanwhile, station commanders were alarmed to note that, during the two-month period of indecision and confusion, many women who had at first intended to enlist were changing their minds. The delay between passage of the bill on 1 July and the scheduled date for mass swearing-in on 1 September gave parents, friends, and former employers time to urge a return home. Moreover, Army supervisors were found to be offering certain women high civilian salaries to leave the Corps and continue in the same jobs as civilians—the advantage being, from the officers' viewpoint, that the women could then date officers and live where their quarters and hours were not supervised. This practice became so widespread that Colonel Hobby was able to collect evidence of it sufficient to secure an Army Service Forces directive forbidding such offers. Meanwhile, at stations where such transfers had already occurred, remaining Waacs were found less disposed to continue under military restrictions and enlisted wages.[27]

In late July, as the evidence of suasion by relatives and employers mounted, it was directed that each station swear in its company as soon as it was ready, instead of waiting for the scheduled 1 September date for mass enlistment. Only WAAC officers were required to wait until 1 September. Where 95 percent or more of a company intended to enlist, it was suggested that a mass ceremony be held with appropriate publicity.[28]

This move proved successful and field companies raced to be the first to enlist. One Army Service Forces and two Army Air Forces companies tied for first place, each with 100 percent enlistment.[29] Others had only a few losses, chiefly for physical disability, and throughout August these were sworn in as fast as possible.

As the WAC waited to discover what the total loss would be, it became evident that there was considerable variation from station to station—some losing none, some losing 5 or 10 percent, many about 20 percent, and others as much as 50 percent. When the loss figures from their several stations were compared, higher commanders were presented with what amounted to a chart of relative efficiency in personnel management. Air and service command headquarters at once sent inspectors and WAAC staff directors to the stations showing high losses, in the hope that conditions might be corrected while there was still time for the women to change their minds before discharge. Commanding generals of major commands became generally aware of the matter; some actually went in person to speak to the women.[30]

Why Some Went Home

The combined discoveries of these inspectors gave WAAC Headquarters a reasonably clear picture of the reasons for which women were leaving. The most revealing discovery was that the majority was not quitting because of the hardships of Army life. On the contrary, at many

[27] (1) Memo, Dir WAC for Dir Pers ASF, 23 Aug 43, and Memo, Dir Pers ASF for TAG, 27 Aug 43. SPWA 314.7 sec 4. (2) WAAC Daily Journal, Vol. I. 28 Jul 43. Also M/R, 13 Aug 43. SPWA 230.14 (7-22-43).

[28] AGO Memo W635-14-43, 30 Jul 43.

[29] On 2 August 1943: Moody Field, Valdosta, Ga.; Mitchel Field, N. Y.; Edgewood Arsenal, Md. WDBPR Press Release, 3 Aug 43.

[30] Folder, Congratulatory Letters, SPWA 314.7 (6-29-43) 1943.

MASS ENLISTMENT CEREMONIES. *Camp Atterbury, Indiana, 10 August 1943, above, and Fort Oglethorpe, 11 August 1943, below.*

isolated and uncomfortable stations in deserts and swamps fewer left than at large comfortable installations. In one case, Waacs were stationed at two airfields near the same Texas city—one the handsome permanent headquarters of a commanding general, the other a hastily expanded training field; yet thirty-seven women left the first, none the second. Inspectors noted that the more isolated or less comfortable stations could not get civilians and therefore needed and appreciated the Waacs.[31]

One staff director reported, "The happiest girls are the ones doing the hardest work." [32] Inspectors' conclusions could be summarized in one statement: *Waacs remained at stations where they were wanted and needed in their jobs.* Even before the conversion started, Director Hobby had informed commands employing Waacs that only two things would be needed to make a woman wish to remain: "These are, first, that her job shall keep her busy and utilize her abilities to the best advantage, and second, that she shall feel that she is making a real contribution and that the Army is interested in her work." [33]

A frequently identifiable cause of loss was hostility toward Waacs on the part of a post commander, which set the tone for the entire post and was usually associated with poor job assignments for women. When Las Vegas Army Air Base lost 54 of its 151 members, inspectors reported:

> The Unit Commander stated that she had met the post commander only on two occasions. Neither the Inspecting Officers nor the Staff Director was able to see him. . . . Women were assigned in the Officers' Mess and the Post Exchange. The Unit Commander has been allowed little part in the assignment of WAAC personnel.[34]

At Camp Lee, Virginia, 40 of 146 were leaving. Inspectors reported:

> All records were satisfactory . . . [but] morale and company spirit were found to be very low. . . . Women stated that they did not have the proper respect of the enlisted men. They said they had discussed this with the post c.o., but he told inspectors he was unaware of it.[35]

At many other stations the fault did not lie with the post commander, except in failing to recommend removal of the WAAC commander, for inspection reports clearly placed the blame on unsuitable WAAC company commanders. At Richmond, Virginia, inspectors found records, buildings, mess, supply, recreation, and church facilities all adequate, but through the fault of the company commander many Waacs were malassigned, and "punishments have been given that are out of proportion for offenses committed." [36] At Fort Monmouth, New Jersey, inspectors noted: "The c.o. does not appear to have a sympathetic understanding of her troops as individuals, and many complaints were received that the EW could not readily contact her for confer-

[31] WAAC Daily Journal, Vol. I, 28 Jul 43, re Randolph Field and Brooks Field, San Antonio, Texas.

[32] Min cited n. 6.

[33] Personal ltr, Oveta Culp Hobby to each CG, by name, Air Comds, SvCs, AGF Comds, Def Comds, Ports, Chiefs of Servs, and Comdts of Schs, 29 Jun 43. SPWA 314.7 sec 1.

[34] Rpt of Insp, Las Vegas AAB, 31 Aug 43. SPWA 333.1 (8-20-43).

[35] Memo, 1st O Dora F. Petmecky and 2d O Lucia T. Hudgens for Dir Fld Insp Div, WAAC, 28 Jul 43, sub: Insp of Hq Co (WAAC) formerly designated as the 96th. SPWA 314.7 sec 3.

[36] Memo, Petmecky and Hudgens for Dir Fld Insp Serv, WAAC, 28 Jul 43, sub: Insp of 709th Post Hq Co. SPWA 314.7 sec 3.

ences."[37] Another station lost 35 percent of its women because the women believed their WAAC commander to be misconducting herself with the station commander, although inspectors could find no proof of this.[38] On the other hand, it was found that when companies had good WAAC commanders many enlisted in a body, even though every other outward circumstance was unfavorable.[39]

Many unsuitable WAC commanders were now relieved and replaced with women of understanding and integrity. Later, unsuitable WAC commanders could not be removed because the new Office of the Director WAC no longer had command power.[40]

When questioned, women gave various reasons for departure. Some of the more frequent excuses tabulated were:

Changed family and home conditions: "Husband now stationed in U.S. and I wish to be with him." "Mother is ill." "Enrolled expecting husband to be drafted, but he was rejected." Almost half of the excuses given fell into this category. Some women sought leave to try to pacify relatives and men friends before giving up all hope of enlisting.

Dissatisfaction with job assignment: "Feel I can do more good in war industry." "I can do more good teaching." "My qualifications are not of as much value to the Army as they are in civilian life." "Do not feel I am doing what I came in to do." "Resent working under civilians."

Emotional or physical difficulties: "Feel physically unfit." "Cannot accustom myself to military life." "Unable to concentrate since my husband reported missing." "My two sons killed in action. It has made me very restless."[41]

Impatience with Army errors: Two women, removed from motor transport school and sent to cooks and bakers school, said, "We expected the Army to be a well-regulated organization that didn't make mistakes." Others took basic training several times, were put in the wrong specialist school, sat six weeks in staging area, and said, "We see by the papers that requests are on file for 600,000 Wacs and here we sit. What's the payoff?"[42]

Other miscellaneous undesirable conditions: These were analyzed by Colonel Catron as

> Overstatements and unfulfillable promises by recruiting officers; inadequate physical examinations and consequent inclusion of too many physically and psycho-neurotically unfit; faulty classification and assignment; . . . some instances of assigning Waacs to jobs of less importance or dignity while civilians are used in the better positions; some cases of delay in having Waacs replace and take over the duties of soldiers and of appointing them to the grades presumably vacated by the men; . . . the public attitude that the war is all but won; the trait of women which, once they have decided to join a cause, demands that they be kept busy and that their abilities be used to the fullest; . . . recent unfavorable publicity and the indica-

[37] Memo, 1st O Frances M. Holbrook and 2d O Kathryn D. Neely for Dir Fld Insp Serv, 15 Jul 43, sub: Insp of WAAC Det, ESCTC, Ft Monmouth, N. J. SPWA 314.7 sec 3.

[38] Memo, Capt Frances M. Holbrook for Dir Fld Surv Br, 23 Oct 43, sub: Rpt of Investigation at Bergstrom Field, Tex. SPWA 331.1.

[39] See Ch. XXXIII, below.

[40] WAAC Daily Journal, Vol. I, 28 Jul 43.

[41] Reasons in three categories above are from tabulation at Fort Des Moines, copy in Folder, Classification Officers, 1943 WAAC Planning file. See also Memo, Exec WAAC for G–1, 20 Jul 43. SPWA 314.7, (1–7–43)(1) sec 3.

[42] Examples from Daytona Beach and Fort Des Moines, "Survey on Problems Most Frequently brought to Chaplains and Resident Counsellors," 9 Sep 43. SPWA 231.28.

tions that officers, enlisted men, and their families have not accepted the Waacs wholeheartedly, to say the least.[43]

Relatively fewer WAAC officers applied for discharge, although some were eliminated by a final screening board, called the AUS-WAAC Board, set up in Washington.[44] The board was unable to eliminate many of the unsuitable young officers just commissioned, since they were still in the officers' pool and could hardly be rejected until they had been given a chance at one assignment. The board was also hampered by the inconsistent recommendations of some field commands, which asked that certain WAAC officers not be admitted to the WAC, but at the same time gave them excellent and superior ratings.

Why the Majority Stayed

By this time the full extent of losses was fairly well tabulated. WAAC leaders had never ceased to predict that the majority of Waacs would not quit in spite of ample provocation. This prediction was now upheld to a surprising degree. Field stations' records of discharges were so uneven that it was not possible to determine how many of the losses were due to failure to pass the physical examination, how many represented women whose commanders had not recommended them, and how many had actually not applied; nevertheless, the total of all such losses combined was less than 25 percent. More than 75 percent of the WAAC had chosen to enlist and had been found physically and morally acceptable.[45]

Among the three major commands, the lowest average loss—about 20 percent—was found in the Army Air Forces. General Arnold, during the conversion period, directed all air inspectors to make WAC job assignment a special subject, in order that "the job assigned each individual will keep her busy and employ her abilities to best advantage and that she is given an opportunity to make a real contribution." General Arnold also took personal action by appearing in a camp newsreel sequence in which he delivered a message to Air Forces Wacs, emphasizing the AAF's need of them; this was shown to all Waacs at air bases before the final dissolution of the WAAC.[46]

The highest losses among the three major commands—about 34 percent—were suffered by the Army Ground Forces. Here, General McNair had refused to accept an AGF WAAC officer or any staff directors or WAAC specialists; he had repeatedly made known his opposition to employment of Waacs by the AGF and to the admittance of the WAAC to the Army. AGF assignment techniques had also been notably poor. In July, as manpower shortages became more serious, the AGF reconsidered its refusal to accept a WAAC staff director, and on 1 July Maj. Emily E. Davis, one of the first WAAC graduates of the Army's Command and General Staff School, reported to fill the position of AGF WAAC Officer, comparable to

[43] See tabulation and memo cited n. 41; also, Office Memo, Maj Gordon C. Jones for Exec WAC, 4 Aug 43, with Col Catron's comment in pencil note. SPWA 342 (7-30-43).

[44] Memo, Dir WAC for Dir Pers ASF, 7 Jul 43; also Memo, MPD ASF for TAG, 20 Jul 43. SPWA 314.7 sec 2.

[45] Losses totaled 343 officers and 14,607 auxiliaries. There is no available indication of the portion lost for medical reasons, rejection by commanding officer, or at own request. WDBPR Press Release, 1 Oct 43.

[46] (1) AAF Memo 121-1, 3 Aug 43. (2) AAF WAC Hist, p. 47; see Ch. XVI, below.

that of Major Bandel in the AAF. Major Davis at once made a hurried series of staff visits to AGF WAAC units.

The worst situation was at Fort Knox, where she found women of the Armored Replacement Training Center, declared unassignable for lack of typing skill some three months before, still sitting in the barracks without employment. She immediately secured their reassignment to other stations to duties commensurate with their Specification Serial Numbers. Major Davis made efforts to correct varied undesirable conditions in other WAAC units, but she had come too late to prevent extensive departures. Major Davis later noted:

> The negative attitude of the command toward the utilization of women in the Army, which could not help but be transmitted to the field commanders, and through them to their staffs and to the enlisted women, had a part in this loss of personnel.[47]

WAAC observers had no difficulty in identifying the motivating force that had kept 80 percent of Waacs in the Army in the most friendly commands, and 66 percent even where employers were hostile. Most women felt that to leave would be moral desertion under the fire of the slanderers, and a betrayal of the ideals that had caused them to enlist. Enlistment was, like the original enrollment, ordinarily a matter of idealism rather than of logic. Observers at a training center reported that women who had asked discharge for good cause suddenly dropped their packed suitcases and ran weeping across a drill field to join their old platoons and take the oath of enlistment. Some who left soon asked to return.[48] Others who did not enlist later regretted it; one wrote two years later: "If I could only have that chance again. I haven't felt fine every time I think of it."[49]

Director Hobby said:

> We lost a great many fine women because they could not pass the Army physical, but we also lost a great many women who would never have been soldiers. It makes me very happy to know that those who stayed, stayed because they know what true obligation is, and I for one feel that what we now have is a firmer core to build around.[50]

As bulwark against its inherited problems, the new Women's Army Corps had but one asset, but that an invaluable one—some 50,000 loyal members who now, in spite of full knowledge of its problems, followed the Corps into the Army of the United States.

The End of the WAAC

On the first of September, 1943, eligible WAAC officers were sworn into the WAC, in grades equivalent to their WAAC grade and with date of rank the same as in the WAAC. On the same date all WAAC units were officially redesignated as WAC. By this time enlistment of enrolled women was almost complete. Instructions were issued in September for the disposition of remaining WAAC members. Throughout the month, there was considerable scurrying both in headquarters and the field as stray Waacs were rounded up.[51]

On 30 September, the Women's Army

[47] AGF WAC Hist, particularly Vol. I, pp. 12–13 and 22.

[48] TWX, Hobby to 7th SvC, 10 Sep 43. SPWA 314.7 sec 4. AGO Memo W635-23-43, 28 Sep 43.

[49] Ltr, "Ex-WAAC" to CofS, 11 Oct 45. WDWAC 220.8.

[50] WAAC Daily Journal, Vol. I, 12 Jul 43.

[51] By WDGO 42, sec 1, 26 Jul 43, effective 1 Sep 43. AG Ltr 320.2 WAC PR-W, 2 Sep 43. SPWA 314.7 sec 4.

Auxiliary Corps ceased to exist. For the women who chose to remain with the Women's Army Corps, Colonel Hobby suggested, and got, a green and gold service ribbon to show their WAAC service. She said:

It is probable that this period will prove to have been the most difficult in the life of the Corps and the one which, because of its pioneering aspect with all that this implies, will be the source of the greatest pride to, and the reason of the strongest ties among, the members of the WAC.[52]

[52] Memo, Dir WAAC for G-1, 25 Jun 43. SPWA 200.6. Director Hobby intended the ribbon only for those whose service was continuous, but by an error in wording it also could be awarded to those who left the Corps and later returned. She was not able to get the wording changed to prevent this, as some such awards had already been made, and G-1 Division held it illegal to withdraw them. See Memo, Dir WAC for G-1, 19 Feb 44. SPWA 200.6 (1943).

CHAPTER XIII

Attempts To Revive Recruiting

The new Women's Army Corps began in spirit early in the summer, when groups of worried recruiting experts gathered in the Director's office to plan the first campaign. The whole future of the Corps quite evidently hung on the success of this effort. WAAC recruiting closed on 8 August 1943, and WAC recruiting opened without delay on the following day, yet in the entire month only 839 recruits were reported.[1]

At the same time, the Army's need for personnel had never been more urgent. The Adjutant General estimated that 2,000,000 more men must be drafted within the next year, of which 446,000 would be fathers of families. The Adjutant General stated that more than 600,000 of the jobs could more efficiently be done by Wacs, thus preventing the necessity of drafting any fathers whatever. General Marshall, after the ceremonies in which Colonel Hobby was sworn into the WAC, asked her if she could get the Army the 600,000 women it needed. The Director avoided a public reply.[2]

In fact, the real question was whether the WAC could survive at all, after the twin blows of the slander campaign and the conversion. In its first year, the Corps had recruited almost 65,000 members, but at the July recruiting rate it would take almost three years to get another 65,000; at the August rate, five years. At this point normal attrition overtook recruiting.

As the Corps waited for War Department solution to the recruiting problem, excess training centers were closed, surplus instructors were assigned to officers' pools, and recruiters were told to suspend their efforts pending new instructions.

The Search for a Recruiting Theme

In view of the gravity of the situation, both General White of G-1 Division and Brig. Gen. Joseph N. Dalton of ASF's Military Personnel Division personally participated in recruiting conferences during the summer. The ASF donated the full-time services of Maj. Robert S. Brown, rated as one of its best publicity experts, who was to give the problem concentrated attention until his death the following year. The advertising agency of Young & Rubicam sent numerous experts, including its adviser, Dr. George Gallup.

These planners fell into anxious debate over the theme of the next campaign,

---

[1] Figures as available to the War Department at the time. Table 2, Appendix A, gives accessions figures as revised by The Adjutant General's Office, 1 February 1948. (1) Memo, Dir WAC for TAG, 26 Jul 43; (2) ASF Memo S635-6-43, 30 Jul 43; (3) TWX, TAG to SvCs, 5 Aug 43. AG 341 WAAC PR-I; (4) Ltr, 6th SvC to Dir WAC, 13 Aug 43, sub: WD Memo S635-6-43. SPKOG 341-1 WAAC; (5) AGO Form 721, 1 Jul 43, Enlmt Rec, WAC, AUS. All in SPWA 341 (1943).

[2] (1) Min, Conf with Chief, Classif and Repl Br AGO. SPWA 337.2, incl Tables A, B. (2) WDBPR Press Release, 5 Jul 43.

which if not successful might well be the last. Dr. Gallup advocated an appeal to a woman's self-interest, with emphasis on the advantages of military status, the value of WAC training, and free clothes, food, and medical care. Colonel Hobby disliked such an approach, and held out for the straight patriotic appeal. Her favorite advertisement was a "shocker" which showed a wounded soldier dying on the battlefield while women played bridge, with the caption, MEN ARE DYING ON THE BATTLELINE—CAN YOU LIVE WITH YOURSELF ON THE SIDELINE?

This approach was vetoed by Generals White and Dalton, who were obliged to inform Colonel Hobby that the censors forbade showing dead American soldiers to the American public. Colonel Hobby next produced an advertisement showing soldiers' graves, with the caption, THEY CAN'T DO ANY MORE—BUT YOU CAN. She argued, "Something like this is needed to drive home a sense of shame to women not doing anything." Results would be better, she said, "the more difficult this job was painted to a woman, the more of a challenge it was to her . . . telling how hard, how drab, how routine it is. . . . Has she the courage to do the commonplace as against the courage of the spectacular?" Dr. Gallup disagreed with her, saying:

> When you take into account all groups, I am afraid that would not be as good as some other appeal. There is too much competition from other things; there are too many ways by which people can rationalize their own situation. Almost every person has, in his own mind, become important, even though only growing a couple of cabbage heads.

Representatives of Young & Rubicam pointed out also that "what appeals to the manufacturer seldom appeals to the consumer," and finally decided to run only one advertisement of the shocker type to see if it "made the women mad." Colonel Hobby replied that, as far as she was concerned, they could be made as mad as hatters if it only drove them to the recruiting station.[3]

As a matter of record, when the soldiers'-graves advertisement was used in October, it was banned in Boston newspapers because it was "too gruesome" and might have caused "sad week ends" to sensitive citizens. Most later advertisements therefore emphasized the WAC's attractive jobs and its material advantages.

It was decided that advertising emphasis would be put on, first, the hundreds of types of Army work now open to Wacs, and second, WAC benefits, which now offered the greater advantage of full military status. It was noteworthy that the same decision concerning "glamorization" was also deemed essential, at different times, by both British and WAVES recruiters, in both cases over the objections of women directors.[4]

The Manpower Theme

Colonel Hobby took the occasion of her oath of office to launch another theme more congenial to the American public than the ill-fated "Release a Man" slogan; this was the WAC as the preserver of the American home. Pointing out that every

[3] All comments from min of conf, WAAC Daily Journal, Vol. I, 13 Jul 43.

[4] (1) Ltr, C. W. Rogers, Advt Dir, Boston *Post*, to Mr. Jim Mullins, Kelly-Smith Co., New York, 29 Oct 43. SPWA 000.7. (2) Draft, 2 Jul 43, sub: Notes for Publicity re WAAC Becoming Part of Army. Folder, WAAC to WAC, WAAC Planning files. (3) WAVES Hist; also *Conditions in the Three Women's Services*.

woman in uniform enabled a father to stay with his family, Colonel Hobby informed the press, "Women as a group have always been the exponents of family life. They may now preserve and protect this family life, the core of American civilization and culture." [5] Newspaper writers readily took up this idea, stating, "In the absence of a draft, it is just possible that the WAC will be filled by young women who would rather join the Army than see their married brothers taken from their children." War Manpower Commissioner Paul V. McNutt tentatively indorsed the theme, saying, "The number of Wacs and Waves has been increased. As a consequence the month when it would be necessary to take fathers has been pushed back. We do not yet know what the yield will be." [6]

This potentially powerful approach could not be pushed as far as was desired because of the refusal of the nation's highest manpower agencies to lend it any practical support. Attempts were made to get the national Selective Service system to credit WAC enlistments against local draft quotas instead of against national totals, so that women might see the results of their enlistment in their own towns, but this was repeatedly refused by Maj. Gen. Lewis B. Hershey because of the extra effort and the administrative changes involved.

The War Manpower Commission also, in spite of Mr. McNutt's public statement, continued to act on the assumption that WAC recruiting was a threat to the industrial war effort.[7] During these summer months, recruiters in the field forwarded surprising information concerning WMC's part in the recent failure of the Cleveland Plan. Colonel Hobby telegraphed all service commands for confirmation, and discovered that, during the Cleveland Campaign, the War Manpower Commission had effectively stopped all WAAC recruiting publicity in areas including about one half of the population of the United States.

The news appeared the more inexplicable in that the Cleveland Campaign had been cleared with national WMC headquarters before ever being launched. To obtain clearance, considerable concessions had been made by the armed forces in a joint Army-Navy agreement in April, in which they had renounced the right to recruit women engaged in aircraft production, shipbuilding, ordnance or signal work, war research and technical teaching, and wire or radio communication. Nevertheless, service commands indicated that, as soon as the Cleveland Campaign had been launched, the War Manpower Commission had directed its regional agencies not to allow the services to recruit any women, employed or unemployed, in labor-short areas.

While the WMC could not prevent recruiters from setting up offices in a city, it could, through its control of the Office of War Information, prevent them from making their presence known by way of radio, newspapers, or other news media. The OWI disclosed that it had received orders from the War Manpower Commission not to give any publicity to armed forces recruitment of women, and stated, "The War Manpower Commission is the ultimate authority on recruitment. The recruitment of women is limited to such areas as are agreed upon with the Re-

[5] *The New York Times*, July 5, 1943.

[6] Both from *The New York Herald Tribune*, July 11, 1943.

[7] For discussion of Selective Service's role, see Chs. XV and XXXIV, below.

gional War Manpower Commission Office."[8]

These areas were extensive, containing perhaps one half the population of the United States. It was also discovered that almost all large cities had been included: San Francisco, Los Angeles, Atlanta, Indianapolis, Baltimore, and many others—a matter that was serious to understaffed recruiting offices, since cities were the most easily accessible areas. For example, in New York state the OWI directed, "Recruitment of women WAVES, WACS, SPARS, and Marines is not considered advisable in Albany, Buffalo, Rochester, Schenectady, Syracuse, Troy, Utica, and [other cities.]" The stoppage was the more inexplicable in that, while affected areas were chiefly those labeled Class I or II—"labor-tight" areas—others were in Class III or IV, where there was supposedly no labor shortage.[9]

The Army at once took issue with this action, informing the War Manpower Commission that it was highly questionable whether the WMC Enabling Act gave the Commission the authority to limit recruitment of women by the armed forces. The executive order that gave WMC control of the Selective Service system directed only that "no male person," 18–38, should be inducted except through Selective Service. The WAC was not mentioned except that its training programs, with all others of the armed forces, should, if using nonfederal educational institutions, be co-ordinated to ensure full use. It did not appear to the Army that female soldiers could be considered "workers" in the passage which required "the hiring, rehiring, solicitation, and recruitment of workers . . . in any plant, facility, occupation, or area" to be done through the United States Employment Service.

In reply to the Army, the WMC retorted that WAC enlistments "shall be subject to the conditions of employment stabilization plans of the War Manpower Commission." It claimed authority not only over publicity media, but also to direct the Army not to send WAC recruiting teams to any area, or to use drill teams, bands, or local advertising, or to set quotas or take any other means of encouraging women to join the WAC. It also added thirty-five more exempted fields from which the WAC could not accept recruits, some of which were described in such broad terms as "industries" and "education services, public and private."[10]

The Army and Navy joined in protesting this action. They pointed out that it was extremely shortsighted to draft skilled male factory workers and teach them to type, thus forcing industry to replace them with women who had to learn the factory trade; both armed forces and industry would benefit by allowing a woman typist to enlist so that a skilled factory worker need not be drafted. Since competitive recruiting was admittedly to be deplored, the armed forces suggested instead a giant combined campaign, with resources of all agencies directed toward drawing women

[8] (1) Memo, Pers Div ASF for Enl Sec, 29 May 43, with incl copy of Memo, WMC for OWI, picked up by Army officer when a New York radio station refused to broadcast the WAAC news release. SPWA 000.77 (1943). (2) Almost identical Ltr from Georgia, SPWA 311.23. (3) Ltr, R&I Serv, 2d SvC to WAAC Hq, 30 Jun 43, saying that the WMC in Newark asked radio stations to hold off all WAAC publicity "owing to the womanpower drive of the WMC." SPWA 000.77 (1943).

[9] *Ibid.* Also Memo, WAAC Hq for WDBPR, 7 Jul 43, sub: WAAC Rctg, Class I and II Cities [incl telgs and lists of cities]. SPWA 000.77 (1943).

[10] (1) Exec Order 9279. (2) Draft of Ltr, SW to the President USA and to WMC. SPWA 000.77 (1943). (3) Ltr, Dep Chmn WMC to USW, 9 Jul 43. SPWA 341 (1943).

MAJ. JESSIE P. RICE

into war work, with the choice of job left to the individual. Otherwise, the Army pointed out, many unemployed women in labor-short areas would never be reached, since they would not take factory jobs and could not be offered Army jobs.[11]

As the War Manpower Commission refused action, the armed forces in August appealed the issue to the nation's highest manpower authority—Justice James F. Byrnes, the Director of War Mobilization. On 12 August 1943, the Director of War Mobilization informed the contenders that the WMC was forbidden to interfere in national recruiting campaigns even in labor-short areas. Its list of thirty-five essential industries was rejected in favor of the Army's shorter one of eight. The Army was in turn directed not to conduct any "intensive local campaigns" in Class I and II areas without clearance with the WMC.[12] It was only with this reassurance that the Army approved funds for the advertisements already selected, and for a new recruiting campaign to be co-ordinated with them. The campaign, unlike the advertisements, was shaped about the patriotic motif that the Director preferred.

The All-States Plan

Both the idea and the energy for the new campaign were supplied by a newcomer to WAC Headquarters, Capt. Jessie P. Rice of the Third Service Command, who was reassigned to the Director's Office in the summer. Captain Rice—a major in late August—was a woman of 42, described by reporters as "cheery, no-nonsense," and by all who knew her as "dynamic." She brought to the problem what proved an appropriate background—that of an industrial sales manager, history teacher, sports reporter, and native of a small Georgia town; she had also had a year's field experience as Staff Director, Third Service Command. For the next twenty months, first as head of WAC Recruiting and then as Deputy Director, she was to be, next to Colonel Hobby, the chief author of WAC policy. She was later credited by Director Hobby with, quite literally, saving the WAC from public defeat on two different occasions, of which the first was the present critical period. In her approach to the current problem, as to all later ones, staff members noted a rare combination of commonsense practicality with unabashed idealism

[11] Memo, JND for Gen Somervell, 12 Jul 43. SPAP, in SPWA 341 (1943).

[12] (1) Memo, USW to Dir OWM, 21 Jul 43. WDGAP 320.2, in SPWA 341 (1943). (2) Memo, SW for CG ASF, 3 Aug 43. Same file. (3) Ltr, Dir OWM to SW, SN, WMC, OWI, 12 Aug 43. SPWA 000.77 (6-4-43), 1943.

for the nation's goals and the Corps' contributions to them.[13]

Major Rice strongly believed that the WAC would achieve lasting success and overcome the slander campaign only through the support of the American community, which was merely alienated by dazzling supersalesmanship and could be won only through giving community members a personal stake in the outcome. They must not be worked upon by outsiders but must do the work themselves, and its outcome must be made a competitive matter of local pride.

She therefore christened her plan the "All-States Plan" and credited the idea to a local success achieved by Baltimore recruiters. It called, quite simply, for each state to recruit a state company, which would carry its flag and wear its armband in training. The trick was in the mechanics, which involved participation of the states' most reputable citizens: General Marshall was to ask state governors to appoint local committees consisting of mayors and other prominent citizens in each town, who would be responsible for aiding recruiters to meet the quota. Major Rice said:

> You will find in this plan nothing new—it is based on the oldest and most fundamental appeal any nation has ever used, and the device it uses is the oldest recruiting device in America, stemming from Colonial days. The appeal is the straight patriotic appeal. . . . By making this appeal by States we hope to reinforce it by the appeal of State pride. We have a big country with many state and sectional traditions . . . and they can be made to play their part only if the operations are conducted by people who know the States and the people in them.[14]

Another appeal was added by Colonel Hobby concerning the setting of the quota; each town would list the number of its sons who had become battle casualties and would recruit a woman to replace each.[15]

The All-States Plan, Major Rice predicted, would not bring in astronomical numbers of recruits, but it could if properly handled pull the WAC out of its recruiting slump, ensure its continued existence, and help to squelch the slander campaign.

The Recruiters

The WAC recruiting machinery could scarcely have been in worse condition for a new effort. Recruiters had borne the brunt of a year of public hostility, and were so fatigued and discouraged that their relief would have been essential had any trained replacements been available. A typical example was one very junior North Carolina recruiter, who reported bitterly that she had made recruiting trips to all the main towns in nine different counties, secured seventy-three free articles and advertisements, got free radio time, secured lists of prospects in every town and sent them all letters, developed a trailer film and display boards, sponsored a state-wide high school poster contest, and distributed 10,000 copies of the winning poster. In return, she succeeded in giving out only ninety application blanks, received thirteen applications, and enrolled two women. Meanwhile, her four enlisted assistants became discouraged

[13] (1) WDBPR Press Release, 28 Oct 42. (2) "The Nine Old Women," *Time*, 1944. (3) Intervs with Mrs. Hobby, 29 Jun 48, and numerous others.

[14] Entire discussion from Min, WAC Rctg Conf, 7 Sep 43. WAC Daily Journal, Vol. II, SPWA 337 (9-1-43).

[15] Approved by D/F, G-1 to ASF through CofS, 2 Sep 43. WDGAP 341 WAC, in WDCSA 291.9 WAAC.

and applied for discharge from the WAAC to get defense jobs in California. For these tribulations, her only reward was a letter of reprimand from WAAC headquarters concerning the winning poster, which had been in such poor taste as to come to national attention.[16]

The burst of energy demanded by a new recruiting drive therefore presented a particular problem. Colonel Hobby wrote to the discouraged recruiters:

> I know the difficulties, the loneliness, the heartbreaking apathy you have faced. I know how hard it is, after weeks and months of discouragement, to dip deep into inner resources and start again with fresh enthusiasm. But I know, too, that you have such inner resources. . . . If you can give fully of your own strength, your own conviction, your own faith and heartfelt patriotism, we will not fail now.[17]

For the success of the campaign, the WAC leaders depended frankly, as always in the Corps' history, upon these "inner resources," and upon motives which were matters of ideals and spirit. Major Rice took charge of the hundreds of conferences and interviews required by the All-States Campaign, quoting freely from Tom Paine concerning "summer soldiers and sunshine patriots," and using a modern quotation also favored by Colonel Hobby as typifying the Corps' present slim chances of success:

> God of the hidden purpose
> Let our embarking be
> The prayer of proud men asking
> Not to be safe, but free.[18]

As the Corps poised upon the threshold of this crucial effort to restore itself, Major Rice informed a conference of service command recruiters:

> We know that there are millions of women who will join us now, if we can touch that spring of inner compulsion that causes people to forget self and fear of the unknown and strike out along the path that leads to their ideals.[19]

Sources of Assistance

The campaign's organization was nevertheless highly practical, and was to stand as a model for all future efforts. To avoid misunderstandings, Army recruiting and induction officers from each service command and their WAC assistants were thoroughly briefed in a Washington conference on the entire plan. At the same time, the Army's Recruiting Publicity Bureau, in New York City, set its presses rolling to produce, in an unprecedentedly short time, portfolios of instructions for recruiters, with brightly colored and attractive posters, window cards, stickers, manuals, envelope stuffers, clip sheets, mats, and advertisements of all sizes.[20]

Also, the War Advertising Council was approached and agreed to arrange free sponsored national advertising for the WAC as it was doing for other agencies. Recruiters were also instructed in how to get free advertising on a local level; many firms were found to be willing to aid the war effort by contributing advertising space, or by including a Wac in their usual advertisements. Many posters and other aids were also sponsored by various business groups. For example, the Stand-

---

[16] Ltr, WAC Hq to Rctg Off, Ft Bragg, N. C., 2 Aug 43, incl rpt of Lt Mary J. Norton, 22 Jun–20 Jul 43. SPWA 415.

[17] Ltr, Col Hobby to all WAC recruiters, 20 Sep 43. SPWA 341 (8–6–43) sec 1*b*.

[18] Henry Morton Robinson, "Litany for a New AEF," reprinted in *WAC Newsletter*, Vol. I, No. 8, 15 Jan 44.

[19] Speech. Min cited n. 14.

[20] Office Memo, "F" for "R" [Capt Ruth F. Fowler for Maj Rice, both of WAC Rctg Sec], 21 Sep 43. SPWA 341 (8–6–43) sec 1*b*.

ard Oil Company of Indiana sent its dealers 19,000 posters, "The Girl With the Star-Spangled Heart," to replace the previous "Drain Old Oil Now." [21]

National civic and business organizations by the dozen were approached by Colonel Hobby's staff, and agreed to mention the WAC in their trade publications and bulletins: the American Retail Association, the National Dry Goods Association, the National Association of Theater Owners, five press associations, most magazine publishing firms, several film companies, the Mayor's Conference, the National Association of Broadcasters, the U.S. Chamber of Commerce, and others as varied.[22]

Organized clubwomen of thirty-six national organizations announced their support after visits of their representatives to Des Moines.

The Executive Committee of the Governors' Conference also gave a preliminary indorsement which assured the Army that state governors would respond when officially called upon.[23]

Paid advertising was also authorized to supplement all the free publicity that the above agencies could contribute. Colonel Hobby asked that the Army allot $10.00 per needed recruit for this purpose, chiefly for systematic coverage of Sunday newspapers. General Dalton disapproved this on the grounds that "no business which is forced to operate at a profit would consider such an appropriation," but he was overruled by General Somervell, who considered that the situation merited the expenditure. As a matter of fact, the sum was somewhat less than that later used in postwar Army advertising to get comparable numbers of male recruits.[24]

As an added precaution Director Hobby in two different meetings with high War Manpower Commission officials cleared the All-States Plan. She was informed that, from 27 September to 7 December, recruiting was authorized even in Class I and II areas, provided that no woman was taken from essential industry. A representative of WMC likewise stated this fact in a speech to the recruiters conference, and the Army published it to recruiters. In return the Army agreed to urge rejected WAC applicants to accept factory work.[25]

The Air-WAC Plan

Because the WAC's continued existence was virtually staked on this one campaign, the War Department decided to accept assistance from a source that it had previously rejected—the Army Air Forces. The AAF had for months begged to be allowed to send out its own personnel to assist regular recruiters, the only proviso being that recruits so obtained could be promised assignment to the Air Forces. Such special-promise recruiting was not unknown to the Army; the Air Forces' recruiting teams, in a recent campaign to get 57,000 male aviation specialists, had

[21] (1) Memo, Dir WAC for Dir Pers ASF, 28 Aug 43. SPWA 000.7 E, in ASF SPAP 341 WAAC (2-26-43). (2) Ltr, WAC Rctg Off, Chicago, to Dir WAC, 27 Oct 43. SPWA 415.

[22] Memo, Capt Fowler for Maj Rice, 21 Sep 43. SPWA 341 (8-6-43) sec 1*b*.

[23] (1) WDBPR Press Release, 20 Sep 43. (2) Min cited n. 14.

[24] (1) Memo, WAC Hq for CG ASF, 2 Aug 43: (2) Memo, Dir WAC for Gen Dalton, 25 Aug 43. SPWA 000.7 (8-25-43). (3) History: Office of the Director of Personnel, Army Service Forces 20 Jul 42-1 Sep 45. OCMH. Hereafter cited as ASF Pers Hist. (4) For army postwar advertising contract, see *Washington Post*, February 18, 1949.

[25] (1) Memos, Dir WAC for G-1, 16 Oct 43, 24 Aug 43, and 30 Sep 43. SPWA 000.77 (1-4-43). (2) Speech of Mr. Stephen W. Wood, WMC. Min cited n. 14. (3) Ltr, TAG to all SvCs, 12 Sep 43. AG 341 PR-I-E.

recruited 128,000 almost before the campaign could be shut off. Nevertheless, Colonel Hobby had consistently rejected these offers, pointing out that special promises to recruits—their choice of branch or station—were undesirable and that difficulty and hard feeling would later result, particularly since earlier recruits had no such choice.[26]

Army Service Forces recruiting experts now informed Colonel Hobby that the WAC would have to abandon its stand against special promises, saying, "The persistent decline in WAAC recruiting results, along with the increasing need for women in the Armed Forces, make it essential to open the door immediately for special WAC recruiting plans." [27] It was therefore decided to adopt the Air-WAC Plan to supplement the All-States Plan; the Air Forces would not compete, but would bring in women who would be part of the state companies until trained. Branch recruiting thus became an accepted fact, although the other two Army branches—AGF and ASF—did not at this time decide to give any personnel to the project. The AAF at once prepared to throw hundreds of additional recruiters into the struggle.[28]

The Quota

On the eve of the campaign, Major Rice estimated that the take would be 10,000 in the approximately two months between 27 September, when the campaign opened, and 7 December, when it closed. Colonel Hobby optimistically hoped for 35,000, and asked that Fort Sheridan be made available in case the three remaining WAC centers were inadequate. The Adjutant General insisted that the higher goal of 70,000 be set, since 70,000 casualties had been sustained to date. Colonel Hobby deplored this move, saying, "It is negative psychology to have an impossible quota"; she recommended a quota of approximately 17,000 for the entire period, which was disapproved by the Army Service Forces. Major Rice attempted to use merely the number of those killed in action instead of all casualties, but the number 70,000 was chosen by the Secretary of War and used in all publicity releases. Actually, G–3 Division of the War Department had hopefully included 200,000 Wacs in the Troop Basis for 1943, but with the provision that more men could be drafted if women failed to volunteer. The most impossible figure of all came from The Adjutant General's Office, which calculated that 631,000 Wacs were needed at once to prevent the drafting of fathers. The figure of 600,000 was used by General Marshall in his public speeches, to emphasize the extent of the Army's need, but was never considered a real quota.[29]

The Campaign

On the last day of August 1943, the All-States Campaign went into action with the precision and co-ordinated effort of a

[26] Memo, AAF for G–1, 29 Jun 43. Published 4 Sep 43, AGO Memo W635–21–43, Participation of AAF in WAC Rctmt. SPWA 341 (8–6–43) sec 1*b*.

[27] Memo, Lt Col C. C. Curtis for Dir WAC, 11 Aug 43. SPWA 341 (8–6–43) sec 1*b*.

[28] Ltr, TAG to SvCs, 12 Oct 43. AG 341 WAC PR–I, in SPWA 341 (8–6–43) sec 1*a*.

[29] (1) Rice estimate made to author and coworkers. (2) Hobby estimates in 1st Ind to Dir Pers ASF to Memo cited n. 27. (3) Hobby quote from Min cited n. 3. Also Memo, Dir WAC for Dir Pers ASF, 19 Aug 43. SPWA 341 (8–10–43) R, in SPAP 341 WAAC (2–26–43). (4) Ltr, SW to Bur of Budget, 15 Jul 43. WDGAP 320.2 WAAC, in CofS 291.9 WAAC. (5) By Exec Order 9364, top limit to strength of WAC was set at 200,000, as published in WD Bulletin 17, 4 Aug 43.

ALL-STATES RECRUITING CAMPAIGN. *Governor Leverett Saltonstall of Massachusetts cuts the ribbon opening a WAC recruiting booth. Standing next to the Governor is Mayor Maurice J. Tobin of Boston.*

military invasion. Exactly twenty-seven days were allotted to softening up the public before the recruiters landed. General Somervell opened the barrage on 31 August with a personal telegram to the commanding generals of all service commands stating, DIRECTOR HOBBY WILL CONTACT YOU ON WAC RECRUITING PLAN WHICH HAS MY APPROVAL. URGE FULL SUPPORT. On the same day and the next, Colonel Hobby made personal long-distance telephone calls to these commanding generals, explaining the plan. In the following week, full details were placed in the hands of recruiters assembled in Washington.[30]

On 10 September, General Marshall sent a personal letter to each state governor, saying, "I am appealing to you for assistance. . . ."[31] This was followed by immediate visits to the governors by service command representatives. In the

[30] (1) TWX, CG ASF to SvCs, 31 Aug 43. SPWA 341 (8-6-43) sec 1*b*. (2) Min cited n. 14.

[31] The following discussion is based on: (1) Ltrs, CofS to the 48 governors, 10 Sep 43. SPWA 341 (8-6-43) sec 1*a*; also in CofS 291.9 with replies. (2) Ltrs, WAC Stf Dir 7th SvC, 11 and 30 Sep 43. SPWA 341 sec 1*b*. (3) Ltrs, Maj Gen Kenyon A. Joyce, 9th SvC, and others. SPWA 341 sec 1*a* (1943). (4) Proclamation, 18 Sep 43, with other proclamations. SPWA 341 (8-6-43) sec 1*a*. (5) Copy in Ltr, Young & Rubicam to Maj Jess Rice, 19 Oct 43. SPWA 000.7.

ALL-STATES RECRUITING CAMPAIGN. *Texas state flag is held by a WAC officer during a ceremony for recruits in front of the State Capitol, Austin, Texas.*

Seventh Service Command, Maj. Gen. Frederick E. Uhl took a plane and personally visited each state. The governors' responses were in most cases immediate; they appeared eager to be called upon for a further part in the war effort. Many issued proclamations such as

> Whereas the successful prosecution of the war requires the enrollment of adequate manpower and [five other whereas's]. . . . Now therefore I, John W. Bricker, Governor of the State of Ohio, do hereby call for volunteers for service in the WAC. . . . I do further call upon mayors and other governmental authorities . . . upon all citizens, newspapers, and radio stations . . . to the end that the State of Ohio may fully achieve her goal in this great national objective.

The governors appointed committees in every community, consisting ordinarily of local officials and prominent citizens, whose efforts were co-ordinated by WAC recruiters. Many governors and mayors allowed the use of their names in local advertisements, for example:

> WOMEN OF FITCHBURG
>
> A Message from your Governor:
>
> Throughout the history of America, men and women of the Commonwealth have never failed to answer the call of service to their country. . . . I am confident that once again the daughters as well as the sons of Massachusetts will have proved their right to our great heritage.
>
> [Signed] Leverett P. Saltonstall

At the same time, the weight of General Marshall's personal prestige was thrown into the campaign in a national press release which, on 15 September, described the plan to the public. This was supported by paid advertisements on a co-ordinated theme placed in Sunday papers by Young & Rubicam.[32]

On 27 September, the doors were formally opened to All-States recruits. An unprecedented concentration of recruiting activity resulted: it appeared doubtful that any inhabitant of the United States failed to hear of the WAC if he could read, listen to the radio, or be reached by a citizen's committee. The Office of War Information for the next three weeks devoted to the WAC most of the free facilities offered it by radio stations, ordinarily used for such projects as bond sales and food conservation; 17,000 messages were broadcast over 891 radio stations, plus 150 plugs on network programs and special, national, and spot programs. The Army Air Forces on 18 October threw teams totaling 358 officers and 1,136 enlisted personnel into the campaign. The WAC Band toured; governors continued to issue statements, which Young & Rubicam caught up and published; governors whose states led were congratulated, others were chagrined, and some even went on personal speaking tours. In general, recruiters seasoned by earlier campaigns seldom fumbled, and publicity was uniformly good.[33]

State companies were shipped to training centers as fast as enough for a company were recruited. The women were ordinarily assembled at the state capitol building or some other historic spot, where they were sworn in with much ceremony, martial music, parades, crowds, and publicity, and presented by the governor with a state flag to carry at the training center. Some states thus recruited several companies in succession. No one WAC training center had sufficient capacity to hold forty-eight companies, but the spirit of the idea was preserved, and each training center parade had a display of from fifteen to twenty state flags. Competition among state companies was keen, and most recruits were kept in a state of high morale by the favorable attention given them by home-town newspapers and by their introduction to the governor and to other prominent citizens.[34]

Altogether, WAC and Army authorities were highly pleased with the prospects. Colonel Hobby called the plan "the best recruiting device we have ever had." The campaign seemed well on the way to achieving some of the more optimistic predictions. Well-nigh faultless in its organization, and highly effective in achieving the desired effect upon the public, it appeared that it could have been stopped only by one thing—an order from authority higher than the Army. And this was, in effect, what stopped it in many areas.

Opposition of the War Manpower Commission

No sooner was the All-States Campaign well launched than the Army discovered

[32] (1) WDBPR Press Release, 15 Sep 43. (2) Speech. Mr. Lenz, Young & Rubicam. Min cited n. 14.

[33] (1) Speech. Mr. Ferbanks, OWI. Min cited n. 14. (2) AAF WAC Hist, p. 57; also AG Ltr cited n. 28. (3) TWX, Comdt 1st WAC Tng Cen to Dir WAC, 4 Oct 43. SPWA 341 sec 1*b*. (4) Ltrs of thanks, Dir WAC to CGs of SvCs, 11 Sep 43. SPWA 341, sec 1*b*; and Ltrs, Governors to Gen Marshall, Nov–Dec 43. CofS 291.9. (5) Memo, Dir WAC for Secy GS, 12 Nov 43. SPWA 341 (8–6–43) sec 1*a*. (6) Speech, Maj Rice. Min, WAC Stf Dirs Conf. New York, 1–3 Dec 43, SPWA 337 (11–10–43).

[34] Ltr, Atty Gen of Ohio to Dir WAC, 5 Oct 43, and reply. SPWA 341 sec 1*b*.

that the War Manpower Commission was again ignoring not only its two agreements with the Army but the August directive from its superior, Director of War Mobilization Byrnes. In spite of WMC clearance of the All-States Plan, regional WMC offices in five of the nine service commands not only suppressed local WAC recruiting publicity, but, speaking as "The United States," released headlines such as one in Connecticut: U.S. FORBIDS WAC DRIVE IN CITY AS THREAT TO WAR EFFORT. In Macon, Georgia, bewildered recruiters were confronted with an ultimatum from the WMC area director saying, "I have decided it would not be feasible to continue with this [recruiting] program. This is a result of Macon being an extreme labor shortage area, and every woman who would be acceptable to the WAC could be used to advantage in war industries." The Office of War Information withdrew radio support in such localities as directed by local WMC officials saying that it had no other choice.[35]

The most charitable interpretation that could be put upon this action was that the War Manpower Commission had been slow in informing its regional offices of the Byrnes decision and of its own official clearance of the All-States Plan. When questioned by telephone, WMC officials at first gave this explanation, saying, "Frankly, they should have gone out quite some time ago." However, a few days later an executive staff meeting was held, of which Major Rice was informed, "Fact is, it was a complete nullification both in the chairman's office and the executive staff of the things we had discussed as exploratory." Major Rice at once pointed out that the WMC's commitments had not been exploratory, that WMC representatives had stated them at official meetings, that the Army had spent a million dollars and staked the Chief of Staff's prestige upon them, and that "we are in a pretty desperate condition, with time running out on us."[36]

Director Hobby called to the War Department's attention "the fact that we had begun the All-States Campaign believing that we had clearance . . . the fact that with the best recruiting device we have ever had, we have been barred by the WMC area directors from making any effort to reach many thousands of women in areas of essential industry who are not themselves engaged in essential industry."[37]

High-level negotiations with the War Manpower Commission began again. The WMC now refused to inform its area directors of its previous clearance of the All-States Campaign unless the Army agreed to use the WMC list of thirty-five essential industries. which Director Byrnes had already disapproved, instead of the shorter Army list which he had approved. Mr. McNutt's office also demanded that its area directors be given power to add industries to the list at their discretion, to veto any objectionable activity, and to supervise all WAC advertising copy.[38]

Rather than thus renounce the advantages that the Byrnes decision had decreed for it, the Army chose to fight the matter

[35] (1) Sources cited n. 25. (2) Memo, Dir WAC for G–1, 4 Oct 43. SPWA 000.77 (1943). (3) Ltr, Ralph E. Haines, Area Dir WMC, Macon, Ga., to CO Cochran Field, 8 Oct 43. SPWA 341 sec *1b*. (4) Memo, Maj Rice for Dir Hobby, with rpt of 5 Oct conf with TAG, G–1, and OWI; (5) Bridgeport (Conn.) *Post*, in Ltr, Maj Gen Sherman Miles, 1st SvC, to WAC. SPWA 000.77 (1943).

[36] (1) Tp Convs, Maj Rice and Mr. Broughton, 5 Oct 43, and another, no date. G–1 WAC 312.9, Drawer 6, DRB AGO.

[37] Memo cited n. 25(1).

[38] Tp Conv, Col Hobby and Mr. Broughton, 15 Oct 43. G–1 WAC 312.9, Drawer 6, DRB AGO.

through. The negotiations dragged on for some time and were not to reach a workable conclusion until the following year. Meanwhile, the brief All-States Campaign, which had depended for effect on its impetus and concentration, was somewhat blunted in many areas, and reached its final deadline on 7 December without feeling any benefit from the negotiations.

There was no way of computing the exact loss which had been occasioned by the War Manpower Commission's action. Major Rice believed that it was perhaps not as serious as the initial setbacks had indicated. In any case, it was clear that while the Army had at last made plain to the public its united support of the WAC, the government as a whole had not done so. Instead, the nation had approached a grave emergency without any clear definition of the role of womanpower and its apportionment among competing agencies.

Public Apathy

The lack of support by civilian manpower control groups of the United States Government explained to some degree the public conviction that the matter was not important, and the resulting apathy, which recruiters encountered as the All-States Campaign drew to a close in early December of 1943. One annoyed general reported that in his service command there appeared to be three types of women: (1) patriotic, who had already taken essential civilian work or enlisted in the WAC; (2) patriotic but mercenary, who were interested only in the highest-paying war jobs; and (3) "The leisure class, who like to go down to the USO, dance with the boys on Saturday night, and think they are making their contribution to the war effort that way. They make all these talks, appear in theaters, put up placards, appear before women's clubs, talk to everybody in the world, but when it comes to getting them to sign up, they say no soap." [39]

The Army's highest-ranking officers now became personally aware of the factors affecting WAC recruiting as never before in the Corps' career, since the All-States Campaign put the responsibility upon them as well as upon community leaders. General Dalton personally telephoned the commanding generals of service commands to read them a long list of ideas to get recruits. He told each pointedly, "I wonder if you've gotten word down to all ranks that General Marshall and the rest of the top people have an intense interest in this WAC recruiting." [40]

The difficulty in getting a woman recruit at first amazed, then exasperated, then spurred to action the commanding generals concerned. Even the service commander who led the field was dissatisfied and observed to General Dalton: "I don't know what is wrong. It just doesn't seem to take. . . . Well, we will put all we got in it, Joe." Another visited WAC recruiters at work and reported that "these people are up against a hard proposition" and needed all the co-operation that the Army could give. Still another noted how high-paying industry had stripped the area of qualified women, and stated, "If they're not just children, overaged, or a little moronic or something, they've quit, they've gone away." The interest and indignation of these officers worked to the WAC's advantage in that, recruiters reported, they were now able to get from their headquarters better transportation,

[39] Tp Calls, Dir Pers ASF to CGs of SvCs, 1 Dec 43, Reply of 2d SvC, ASF, SPAP 341 WAAC (2-26-43).

[40] *Ibid.*

personnel, and station locations than had previously been made available.[41]

The Army Air Forces recruiters received a similar shock. Employing tactics that had recruited 128,000 male mechanics in a short time, they were able to get, in the entire month of November, only about 1,500 Wacs. One AAF officer reported:

> I never saw anything as tough. When our team hit one Florida city, we closed the place—actually got the mayor to close schools and stores so everyone would come to our parade and rally. We got important speakers, generals, war heroes. Better than 50 planes flew overhead forming the letters *WAC.* That night we had a big dance for prospects, with good-looking pilots as dates. Results? We didn't get one Wac in that town. Some of the local girls put in applications to please their escorts, but withdrew them the next day.[42]

It soon was apparent to the Air Forces that there was one important difference in the mechanics' campaign and the WAC campaign—the men had been faced by eventual draft and so had been happy to have a choice of branch, job, and station near home, whereas the women were not faced by any such compulsion. Nevertheless, the Air Forces persisted, and by December were getting more than half of the total intake.[43]

The Success of the All-States Campaign

In spite of the opposition of the War Manpower Commission and the public apathy, the All-States Campaign proved the most successful the Corps was ever to undertake. The intake of 4,425 in its final month was better than anything that had ever been achieved with comparably high enlistment requirements.[44] The final tally showed that the drive had netted 10,619 recruits, slightly over the 10,000 that Major Rice had predicted. No later campaign, even the highly successful General Hospital Campaign in 1945, was ever to achieve equal numbers in so short a time.

Even better than its immediate results, in Colonel Hobby's opinion, was its continuing effect in overcoming the slander campaign. Major Rice praised the "wonderful civilian cooperation, that will reap results for months to come."[45] A new Gallup survey in the closing months of the year showed that, the Director stated, "Public attitude is now definitely favorable to the WAC—more so than at any time since its organization."[46] Surveys showed that, among eligibles, 83 percent now knew that the WAC was part of the Army, and 85 percent knew that the WAC needed more women. When asked, "Which service needs women the most?" eligibles replied: WAC—58 percent; WAVES—14 percent; Marines—14 percent; SPARS—4 percent; Don't know—5 percent. If drafted, women stated that in making a choice they would give the WAC a slight edge on other enlisted women's service.[47]

General Dalton, while objecting to the cost per recruit, stated that the campaign had led to a better organized and successful recruiting service. Recruiting proved to be permanently restored, even though

[41] *Ibid.*, replies of CGs, 2d, 5th, and 9th SvCs.

[42] Related to author by Capt Dewey Couri, EFTC, at WAC Rctg Conf, Chicago, Feb 44.

[43] Memo, Maj. Rice for G-1, 28 Dec 43. SPWA 341 (1-1-43) sec 2.

[44] Higher totals had been achieved only in the winter of 1942-43 when enlistment requirements were drastically cut and unassignables admitted.

[45] Min cited n. 33(6).

[46] Memo, Dir WAC for CG ASF, 27 Nov 43, sub: Plan for WAC Rctmt, 7 Dec 43-15 Apr 43. SPWA 341 (11-27-43)B, in ASF, SPAP 341 WAAC (2-26-43). Also Min cited n. 33(6).

[47] Min cited n. 33(6).

to a modest level, and was to average more than 3,000 a month until the last days of the war. The campaign had thus answered the question of whether the WAC could survive: it could. On the other hand, the days of expansive planning in terms of millions were obviously ended, although it was always difficult for planners completely to relinquish visions, which had been held since the first World War, of the effortless and inexpensive acquisition of thousands of female volunteers.[48]

Revival of Plans to Draft Women

The ruinous expense of recruiting competition was now so clear that the Army, even before the end of the campaign, revived the question of drafting women, which had been rejected by the General Staff less than a year before as impossible. The Chief of Staff in early autumn personally directed Colonel Hobby to prepare a memorandum on the "selection of women for military duty." [49] This she did, noting, "A year's experience indicates the improbability of attaining, through voluntary enlistments, an adequate supply of womanpower to perform essential military services." [50] The Director was now thoroughly convinced that the American public looked on the draft as the proper means of meeting military needs. To a military conference, she stated, "I believe that it is as fair to ask volunteers for service as to ask for volunteers to pay taxes. . . . When people are asked, 'Why aren't you in the Service?', they remark, 'Oh, come now, if the Government really needs us they will draft us.' " [51]

The plans now made were far more extensive than those that the Army had attempted in its similar legislative study during the manpower crisis of the year before. It was noted that by this time the British Government already had registered women in the 19–40 age group and had power to direct some 17,000,000 of them into the armed services or industry. The British had found it advisable to give local industrial jobs to women with domestic responsibilities, and to emphasize mobility for the armed forces; younger mobile women were withdrawn from retail trades, the clothing industry, and even the government, and replaced by the nonmobile. By this means thousands of women were being successfully called up each week. Some 91 percent of single women and 80 percent of married women of ages 18–40 were already reported to be working directly in the war effort in Britain.[52]

The prospects in the United States looked even more favorable. Statistics from the Bureau of the Census showed that there were an estimated 18,000,000 American women in the 18–49 age group without children under 18, of which over 5,000,000 could probably meet the WAC's high requirements and pass all tests and investigations. Of these, the Adjutant General estimated that the Army would need to draft only 631,000.[53]

[48] ASF Pers Hist.

[49] (1) Memo, Miss B. E. Lies for Dir WAC 9 Sep 43. Folder, Plng Proj 27, WAAC Plng files. (2) Memo, Dir WAC for TAG, 11 Sep 43. SPWA 327.07.

[50] Memo, Dir WAC for CofS, 15 Sep 43. SPWA 327.02E.

[51] Remarks, Stf Dirs Conf, Chicago, 15–17 Jun 43. SPWA 337 (6–1–43).

[52] M/R, Summary of Certain Aspects of British Womanpower Policy, 19 Aug 43. SPWA 327.02 Great Britain (1–19–43).

[53] (1) Memo cited n. 49(1) contains varying figures from Bur of Census, Bur Labor Statistics, 1940 Census, and WMC. Some estimates, with deletions for lack of education, etc., still left over 3,000,000 qualified. (2) Memo, TAG for Dir WAC, 15 Sep 43. SPWA 337.02.

The submitted version of legislation was therefore extremely liberal, and would have exempted all married women living with their husbands, all women with children or other dependents, women in essential industry or agriculture, nurses, physicians, public officials, and others in like category. The proposed law stated, "An obligation rests upon women as well as men . . . to render such personal service in aid of the war effort as they may be deemed best fitted to perform." [54]

The Director's recommendation was that selection be scientific—by Army requirements, not numerical—since 80 percent of the jobs required skill and experience, and the other 20 percent required the ability to acquire skill with short training. Unskilled and low-grade womanpower was not wanted.

The American public seemed to be solidly behind the idea of drafting women. Gallup polls showed that 73 percent in October of 1943 and 78 percent in December believed that single women should be drafted before any more fathers were taken. Single women of draft age themselves endorsed their draft by a 75 percent majority, although stating that they would not voluntarily enlist so long as the Government did not believe the matter important enough to warrant a draft. Every section of the nation had large majorities favoring the drafting of women.[55]

Nevertheless, even its advocates soon realized that the plan would never be adopted unless serious military reverses occurred. Director Hobby stated in conference, " I think that what happens on the question of Selective Service depends on how long the war lasts. I do not think there is any thought of it in Congress at this time." [56]

Her prediction proved correct. The General Staff's G–1 Division, after some consideration, decided that the only possible strategy was to wait until national service legislation was sponsored elsewhere and then to request that the Army be included. Some sponsorship was immediately forthcoming in the Austin-Wadsworth bill.[57]

The Austin-Wadsworth Bill

Scarcely three weeks after the close of the All-States Campaign, there ensued a national manpower crisis which caused an actual "universal service" bill to be considered. In the winter of 1943–44, the Army's major training effort ended and its all-out offensive began. Overseas, the invasions of Normandy and of New Guinea were in the making. In January, the War Department announced that by the end of the year some 6,000,000 men still on training stations must be shipped overseas; station overhead was to be combed for physically fit, well-trained troops.[58]

The nation's manpower economy was in poor condition to support the offensive, being at the moment in possibly the most critical condition of the war. Secret and pessimistic debates in the General Council showed that Selective Service calls were not being met and that there was small prospect that more than 80 percent of future calls would be met; planners con-

[54] Memo cited n. 50.

[55] Gallup Poll results in *Washington Post*, April 2, 1944.

[56] Remarks cited n. 51.

[57] (1) Memo, G–1 for CofS, 23 Sep 43. WDGAP 327 WAC. Handwritten note: "I agree. GCM." (2) HR 1742 and S 666, 78th Cong, 1st sess, 8 Feb 43. *Congressional Record*, Vol. 89, No. 21, A494.

[58] (1) *WAC Newsletter*, Vol. I, No. 8, 15 Jan 44. (2) Washington *Evening Star*, January 27, 1944, "Army Shakeup to Double Forces Abroad."

sidered calling up 400,000 men previously deferred as farmers.[59] The preliminary successes in Italy in the early fall had raised civilian hopes of immediate victory to such an extent that a stampede from war industries began. War Manpower Commission Chairman McNutt denounced the workers who "seek out a safe berth in nonwar work of a permanent character." Ship deliveries declined by 40 percent; war industry noted a severe slump in production goals; and the recruiting of workers by the armed forces and defense industry redoubled in difficulty.

Secretary of War Henry L. Stimson told the Senate Military Affairs Committee that "the home front is on the point of going sour," and that the morale of fighting men would be seriously affected if civilians were left free to snap up the best postwar jobs. Therefore, Secretary of the Navy Frank Knox recommended that workers be frozen in their iobs in order to prevent disastrous absenteeism and turnover of personnel. The Chief of Staff publicly warned the nation that it still failed to grasp the magnitude of the effort needed for victory.[60]

Heeding these pleas, Congress in January turned to active consideration of the pending Austin-Wadsworth bill for "civilian selective service." [61] The heads of the armed forces threw their support behind the measure as a means of equalizing the sacrifices of military and civilian personnel, warning that the war's great effort lay ahead of the nation and not behind it. From the viewpoint of the women's services, the measure appeared a sure means of ending recruiting costs and ensuring thousands of members. While the bill did not provide for the induction of women into the armed forces, it exempted enlistees from being drafted into farms and factories. British reports alleged that, under a similar act, some 40 percent of women had attempted to choose military service.[62]

Unhappily for such hopes, the Austin-Wadsworth bill was opposed by the American Federation of Labor—whose President, William Green, called a labor draft "servitude" and "unconstitutional"—and by the Congress of Industrial Organizations. Finally, when the Senate's Truman Committee announced its opposition, the bill was dropped. Wide-open competition for womanpower continued, under circumstances which made the manpower shortages of 1942 and 1943 appear mere preliminaries.[63]

No accurate estimate was ever possible as to the cost to the nation of the absence of some such law to direct women into industry or the armed forces. The Army's best estimate of the recruiting cost per Wac was $125, including advertising, recruiters' salaries, transportation, station rent, and all other expenses. This sum did not appear prohibitive when compared with the cost to the Army—$107—of recruiting a female civilian worker who might resign at any time. While no average figure was announced by civilian industry for its recruiting, the amount spent on rallies, bands, scouts, and other recruiting activities by the aircraft industry and others was undoubtedly comparable. The

[59] Min, Gen Council, 24 Jan 44.

[60] *Washington Post*, January 10, February 2 and 4, and March 3, 1944.

[61] HR 3944, 78th Cong, 2d sess, 11 Jan 44.

[62] National Service Act of 1941 (no. 2). Memo, MPD SOS for G-1, 4 Nov 42. SPGAM 320.2 WAAC (9-15-42).

[63] *Washington Post*, February 17 and 26, and March 5, 1944.

Army, in return for its expenditure, had at least the assurance that its Wacs must stick to their jobs, whereas Mr. McNutt reported that in industry eight women were quitting for every ten who were recruited.[64]

Army Manpower Crisis

As a result of the manpower situation, an Army directive advised all stations that they could not expect any additional military strength to replace, through the draft, the men shipped overseas. Commanders were told to replace their January losses and keep posts going as best they could by the use, in order of preference, of: (1) civilians, (2) Wacs, (3) men disqualified for overseas service, returned from overseas, over 35 years of age, or with less than twelve months of service.[65]

Misunderstandings at once arose over this order of listing, with Wacs and limited service men complaining that civilians were taunting them with being the less-desired type of personnel. The War Department hastened to explain that the situation was just the opposite: it had been intended that the few remaining men be saved for the jobs which only men could do, the Wacs for other completely military work, and civilians used for all else. When asked by a staff officer, "Why did the War Department, in their list of replacements, list civilians first and Wacs second?" General Dalton replied, "For the simple reason that Wacs are included in the military strength and we need the military for other purposes. We are trying to use as many civilians as possible to release the military, male and female."[66] The War Department's G–1, General White, added, "We never use military personnel on a job that can be done by civilian employees if civilians can be obtained. . . . There is a serious shortage of WAC personnel. Military personnel must be used on assignments where we cannot use civilians."[67]

Sudden WAC Popularity

With unexpected suddenness WAC stock soared to new heights at Army commands and stations. Greatly increased requisitions for Wacs came from sources new and old. The Medical Department, which two years before had refused to employ Waacs, now stated, "With the expansion of hospitals and the withdrawal of enlisted men, there will be an urgent need for at least 50,000 additional Wacs to work in medical installations." The Chief Signal Officer asked that Wacs be increased from 9 percent to 16 percent of the military personnel under his jurisdiction, complained of the "special appeals" with which Air Forces was securing recruits, and called for "aggressive action" to get more for the Signal Corps.

Other administrative and technical services were in a similar predicament. Director Hobby returned from a brief trip to Europe in January of 1944 with further requests from the European and Mediterranean theaters for more Wacs. At the same time a request for 10,000 Wacs was received from the Pacific theater, which had refused to employ them up to this

[64] Memo, TAG for G–1, 28 Feb 44, sub: Estimated Per Capita Cost of WAC Rctg. PR–E, WAAC classif files, under "budget."

[65] WD Cir 293, 11 Nov 43; also ASF Cir 21, Sec II, 24 Jan 44.

[66] Min of Rctg Conf, Feb 44. SPWA 337.

[67] Ltr, G–1 to Hon Francis Case, 9 Mar 44. WDGAP 322 WAC, WDWAC 330.14.

time. The War Department, which had previously banned the use of Wacs in the Washington area, shortly reversed the decision and gave itself first priority on the most highly skilled recruits, to the considerable annoyance of field agencies. The agency habitually the most cautious about employing Wacs, the Army Ground Forces, had only a few months earlier asked that its Wac quota be cut from 5 percent to 3 percent of the intake; in a quick reversal it now asked for a "more equitable distribution" and that its quota be raised from 3 percent to 8 percent.[68]

The situation prompting this sudden popularity was graphically stated to the Air Forces by General Arnold:

> If there is a man left in your organization who does not recognize that the bulk of the men he now has working for him are going to be shipped abroad, and that if he does not train and use the available Wacs, civilians, and soldiers physically unfit for combat, he will have no one to do his job—if there is such a man, he is a definite danger to your organization.[69]

In the Army Service Forces, General Somervell warned his service commanders that "You must . . . give your personal attention to the eventual taking over by the Wacs of many of the jobs in your command."[70]

The Chief of Staff himself went so far as to authorize release to the general public of the precedent-shattering statement that, for American women, "aside from urgent family obligations, *enlistment in the military takes precedence over any other responsibility.*" He added, "It is important that the general public understand the Army's urgent need for women to enable the military effort to go forward according to the schedule of operations in prospect."[71]

The Job-and-Station Promise

In response to the Chief of Staff's plea, the WAC was forced into its final compromise for recruiting success: the adoption of the plan labeled SJA or Station-Job Assignment. At an earlier date, the Army Air Forces had proposed that its recruiters be allowed to promise recruits not only assignment to the Air Forces, but assignment to a particular station for a specified type of work—a system quite comparable to the hiring of civilian employees. Such a plan was generally admitted to be the best possible means of securing parental approval to a woman's enlistment, or of recruiting women with responsibilities that made them unwilling to leave the area of residence. The system also had the advantage of making it possible to fit the stations' needs exactly, and to recruit no skills useless to the Army; it had in fact been developed originally in the days of expansion planning as a means of preventing overrecruiting in any one skill. The job promise was also expected to attract the

[68] (1) Memo, SGO for Dir Pers ASF, 11 Feb 44, with Inds, etc. SPAP 341 WAAC (2-26-43). (2) Memo, SigC for CG ASF, 7 Jun 44. SPSMP 320.2 WAC. SigC 322 WAC 1944-45, DRB AGO. (3) Ltr, SigC to CG ESCTC, Ft Monmouth, 15 Jun 44. SPSMP 320.2 WAC. (4) Min, Gen Council, 7 Feb 44. (5) D/F, G-1 for TAG through CofS, 31 May 44; Incl and M/R. CSA 324.5 WAC. (6) Memo, ASF for TAG, 11 Jul 44. SPGAC 341 Gen (7-11-44)-69. in SPAP 341 WAAC.

[69] Ltr, CG AAF to Air Forces and Comds, 22 Feb 44 (Tab D to Memo, Dir WAC for CofS, 16 Mar 44). WDWAC 341.

[70] (1) Min, ASF Conf of CGs of SvCs, Chicago, 22-24 Jul 43, p. 3. (2) ASF Ltr to all SvCs, 22 Sep 43. SPGAP 201.6 Gen (9-18-43). Folder, Plng Proj 26, WAC Plng files.

[71] Memo, Secy WDGS for WDBPR, 21 Mar 44, CofS Personal file, made available to author, December 1945, by courtesy of General Marshall's secretary. Author's italics.

skilled woman who had previously feared assignment to kitchen duties. A civilian survey revealed that 52 percent of the women questioned, although not planning to join the WAC, said that they might if they could be assured the job of their choice. Of this 52 percent, 50 percent wanted medical, radio, weather, clerical and other technical or mechanical work; only 2 percent wanted to be cooks or orderlies.[72]

At the time, the proposal had been rejected by Director Hobby, on the grounds that such promises were exceedingly difficult to fulfill; the promised job and even the station itself might have vanished before the recruit reported for duty. Also, such promises would plainly cause ill will among previous recruits who desired to choose another job or station.

A reversal of this decision was shortly proposed by the Army Service Forces. By the end of the All-States Campaign it was apparent that the Army Air Forces was, by means of the branch-assignment promise, getting about two thirds of the monthly take as AAF branch recruits. In addition to these branch recruits, the AAF under the War Department ruling got its original 40 percent of the general service recruits who elected no branch. A somewhat insignificant remainder was left to be divided on the customary basis of 57 percent to the ASF and 3 percent to the AGF.[73]

General Dalton, speaking for the ASF and AGF, demanded of Colonel Hobby why the AAF was getting more publicity. He was reminded that neither the ASF or the AGF had chosen to participate in branch recruiting when the choice was offered them the previous October, while the AAF now had some 1,500 of its personnel in the struggle. If anything, the War Department had hindered rather than helped the AAF in its publicity making, for it had forbidden the practice of saying, "Join the Air-WAC," which had occasionally caused the public to believe that a new and superior corps had been set up. The AAF was thenceforth obliged to say, "Join the WAC and Serve with the AAF," although the name Air-Wac was to stick as a title for the recruit herself.[74]

General Dalton therefore secured War Department approval for the publication of a circular giving the AGF and the ASF—but not the AAF—the authority to make the job-and-station promise to WAC recruits. An applicant's choice was limited to stations within the service command in which she applied, since before making the promise it was necessary to determine that housing and a job of the desired type existed at the chosen station. This bit of strategy gave the ASF and the AGF only a momentary advantage, since the AAF within a month secured G-1 Division's authority to make the same promise for air stations within designated air command areas.[75]

Station commanders now had a new and personal incentive to aid in WAC recruiting, since it was largely through their own efforts in the territory around their stations that their personnel needs might be filled. Many stations organized

[72] (1) Office Memo, "R. S." for Exec WAAC, 18 May 43. Proj 4, WAAC Plng files. (2) Folders 4, 19, 21, 23, 1943 WAAC/WAC Plng files.

[73] Weekly Rpt 22, O Dir WAC for ASF, 2-8 Feb 44.

[74] (1) Memo, Maj Rice for Dir Pers ASF, 15 Nov 43. SPWA 341 sec 1*b*, in SPAP 341 WAAC (2-26-43). (2) Memo, Col Catron for TAG, 27 Sep 43. SPWA 341 sec 1*a*.

[75] (1) WD Cirs 286, 296, 298, and 340 (1943). (2) AG Ltr 320.3 (12-28-43) PR-I, 28 Jan 44. SPWA 341 (8-6-43) sec 1*a* (1943).

part-time teams of station personnel to comb the surrounding area for recruits. General Arnold informed air bases, "The sole opportunity for station commanders to fill vacancies created by overseas shipments is to undertake active participation in Air-WAC recruiting." [76]

Considerable commotion and some confusion resulted as competing teams took the field. Airfield teams were especially numerous, but ports, hospitals, Signal Corps stations, and many others also participated in station-job recruiting. Posters for Air-Wacs, Port Wacs, and other varieties of Wacs offered a choice of advantages.

As expected, the administrative failure to fulfill these promises was higher than in the All-States Campaign: papers became lost; women insisted that they had been promised assignments which did not appear on their papers; Colonel Hobby's office was besieged by letters from Congressmen and parents, and was obliged to instigate endless investigations as to whether certain claims had any foundation. Some failures to comply with promises were inevitable, as the changing military situation caused certain jobs or even the stations themselves to be closed out before recruits finished training and reported for duty, or as women proved unqualified for the jobs which they had been promised.

Another disadvantage was the schism created in the previously close ranks of the Corps. Women who were approaching eighteen months' service, and especially those who were vainly applying for transfer to a different type of work, station, or climate, or to be nearer home, were likely to look with less than sisterly affection upon recruits who had not enlisted until the job-and-station promise was made. The newcomers themselves were often ignorant of recruiting history and thus bewildered at their chilly reception by their barracksmates. Deputy Director Rice wrote all staff directors to appeal to the women's soldierliness: "I have the faith in the Corps that makes me believe that the vast majority of its personnel have learned that there is more to being a good soldier than just putting on a uniform and learning to drill and salute." [77] The veteran Wacs were to have need of this soldierliness often during the remaining months of the war, as more and more attractive promises came to be offered to meet the competition for womanpower.

Because of the women's reaction, Colonel Hobby consistently refused repeated recommendations that the WAC make an even more alluring promise: that of overseas assignment. Since it remained impossible to give overseas assignments to many Wacs who had continuously sought them during months of service, she believed it unwarranted to grant these scarce plums to attract latecomers.

Results of the Job-and-Station Plan varied with the energy and the natural advantages of the station. Some Air Forces stations, particularly the more glamorous ones of the Air Transport Command, doubled and tripled their WAC population, to the limits of their available housing. Others were not so fortunate. The commanding officer of the Infantry Replacement Training Center at Camp Roberts sent a letter to each soldier saying, "If each man at this center would assist in recruiting just one woman for the WAC, there would be a sufficient number for a

[76] Ltr, AAF Hq to Air Comds, 15 Feb 44. SPWA 341 sec 1*a* (1943).

[77] Personal ltr, Maj Rice to all SvC Stf Dirs, 10 Jan 44. SPWA 341 sec 1*b* (1943).

STATION-JOB RECRUITING DEMONSTRATIONS *used in Macon, Georgia, to enlist women for duty at Cochran Field, Georgia.*

complete Division." However, in the following months only two recruits were obtained and the proposed unit had to be closed out. Still other stations recruited well but not wisely; the Ground Forces replacement depot at Fort Meade found that it had selected women to do clerical work who were unskilled and untrainable for anything except car-washing. Furthermore, some stations found themselves with a group of women who needed frequent leave and holidays to manage outside responsibilities, or whose anxious parents attempted to take over company administration by telephone or in person.[78]

On a national scale, the returns were in direct proportion to the effort expended and to the attractiveness to the public of the various services. The Air Forces, which employed 1,500 recruiters on regular assignments plus uncounted others on part-time teams, got 5,000 of the first 5,270 branch and station-job recruits; the Service Forces 250, and the Ground Forces only 20. However, the Ground Forces had assigned only one officer, a first lieutenant, to recruiting duty, and had refused his request that returned combat heroes be added to his team; such, said the AGF, would be wasteful of personnel. The Air Forces never regretted the expenditure of recruiting personnel, and stated at the end of the war that the process "secured for the AAF over 27,000 workers, many of them highly skilled and many in the prized clerical field; and it tapped a manpower reservoir that could not otherwise have been tapped, so long as women were not directed into national service by some such system as selective service."[79]

Final judgments of the usefulness of the station-job system as a recruiting tool varied: on the one hand, it resulted in disappointments and administrative annoyances; and on the other, it continued to secure recruits, in the face of increasingly desperate competition, who could have been obtained in no other manner. For this reason, in spite of its obvious drawbacks, the system was to continue up to the closing days of recruiting, and was to be picked up in postwar Army recruiting with even more sweeping promises to male recruits, such as choice of overseas stations.

Women-in-War Campaign

In early January of 1944, the long negotiations with the War Manpower Commission yielded the united recruiting drive that the Army had long been proposing. Labeled "Women in War," its aim was to urge women to accept some kind of full-time war work, whether in industry or in one of the military services. The OWI estimated that there were still five and one-half million idle women eligible for such work. However, questionnaires showed that the basic restraining factor was no longer a lack of knowledge of the need but a lack of interest.

The OWI leveled all of its very considerable forces against this apathy. Newspapers, motion pictures, magazines, radio, and all other civilian news media were fully at its command, and for several weeks all devoted their plugs to the theme of Women in War—urging women impartially to join one of the armed services or to take a job in industry. Fifty national radio shows a week and countless local shows used the OWI's Fact Sheet. Stars and producers of the motion picture industry co-operated by making two-and-

[78] See basic AAF, AGF, ASF refs in Chs. XVI, XVII, and XVIII.

[79] AAF WAC Hist, p. 72.

a-half-minute bulletins to be attached to newsreels, and 16-millimeter films for showing in churches, schools, and war plants.

In the magazine field, 548 editors simultaneously used the womanpower theme. Through its news bureau, the OWI placed free stories with every type of newspaper: news items were given to the wire services, human-interest stories to the syndicates, and special articles to the rural press, business press, labor, Negro. women's. and foreign-language press. Cartoonists and editorial artists at OWI's call featured an attractive Wac, Wave, nurse, or factory worker. As for posters, the nation's best artists contributed designs, while the OWI's Boy Scout Distribution Service handed out 750,000 posters bi-monthly, as well as 50,000 car cards a month. The Outdoor Advertisement Association contributed twenty-four different billboard spreads.[80]

All of this united activity had little perceptible effect on public apathy or on the armed forces recruitment of women: the WAC returns merely held steady at slightly below the level set by the All-States Campaign. Recruiters were inclined to believe that, had armed forces and industry earlier achieved a united front in harmonious publicity, greater results would have been obtained.

Campaign Conclusions

The All-States Campaign, the Station-Job Plan, and the Women-in-War drive had to some extent counteracted the slander campaign and restored recruiting to a level that permitted the Corps' continued operation. Neither they, nor any plan ever used by the naval women's services, could do more. The final lesson for the armed forces, after all the furor of effort and speculation had died, appeared to be merely the same that it had already learned in previous wars: the American public would not, without conscription, furnish the required number of volunteers for military success. In retrospect it became hard to understand why—in a nation that had been obliged to conscript men since the Civil War and in which women's service had never been popularly accepted—it had ever been imagined that women would make a better response to recruiting appeals than had men.

[80] Memo, Asst Dir Rice for Dir WAC, 12 Jan 44. ASF SPAP 341 WAAC (2-26-43).

CHAPTER XIV

Integration Into the Army

The business of changing the WAAC to the WAC had been virtually suspended while the All-States Plan was rushed into the line to make sure that there would be a WAC to continue. In October and November of 1943, as some measure of recruiting success seemed assured, planners turned back to the business of converting the obsolete WAAC organization, with its own operating headquarters, into the new WAC.

Upon the moment of conversion, an unidentified enthusiast in The Adjutant General's Office issued to the field the blanket directive that all Army Regulations would apply to the WAC until new WAC Regulations could be published. This superseded Director Hobby's plan to use WAAC Regulations in the interval. Outcry from the field soon made embarrassingly clear that the matter was more complex than expected. The Adjutant General hastily rescinded the directive and replaced it with one reinstating WAAC Regulations on clothing, housing, training, medical care, and all other matters with the exception of enlistment, discharge, and military justice, where Army rules would apply. The embarrassment continued: even this limited application of unqualified Army Regulations proved impracticable. In the matter of enlistment, there was a different legally established age and citizenship requirement and several practical differences concerning dependency and disqualifying defects; in the matter of discharge, there were differences in at least eight matters, including minority. As for pregnancy discharge, admission of infants to Army status was narrowly averted by resourceful field stations which gave discharges for disability to expectant mothers.[1]

Having thus perceived that some amendments to men's rules would be necessary, various Army agencies began to fight out the necessary differences. Disputes over the uniform regulations became so heated that it was necessary to detach them from the remainder of the regulations for later publication in the new year. Other matters were gradually agreed upon during the last months of 1943.[2]

Abolition of Separate Grades

In the months before complete agreement could be reached, a makeshift circular in late September took the first step toward integration by abolishing separate WAC grades, a step that appeared more startling to Army personnel officers than its nature merited. Civilian personnel allotments were of course not divided to specify which positions must be filled by

[1] (1) AGO Memo W635-14-43; changes 1 (31 Aug 43) and 2 (17 Sep 43). (2) Ltr, 2d SvC to O Dir WAC, 22 Sep 43, and reply. SPWA 314.7 (1-7-43)(1) sec 4.

[2] Memo, G-1 for G-1 Enl Br, 31 Oct 43. SPWA 300.3 (1-7-43) sec 2.

males and which by females, but for military personnel the merger of allotments aroused alarming visions of a merger of unit command and housing.

The Air Forces took the lead in pointing out that this need not necessarily follow: that women might well be assigned to the organizations for which they worked, while residing in their own barracks with a WAC commander possessed of full command powers. Major Bandel, Air WAC Officer, stated:

> It had become apparent that the [accounting] troubles which arose came from thinking of WAC personnel as a special type of personnel which had to be assigned, counted, and administered apart from all other types of military personnel. Insofar as their housekeeping went, Wacs obviously had to be administered separately; insofar as their jobs went, if they were to be utilized with maximum efficiency they had to be administered as a part of the unit performing the job in question. Abolition of WAC grades would result in careful utilization, since Wacs would clearly count against a command's overall allotment of grades and personnel; maximum freedom of assignment, since Wacs could be assigned without delay to any job for which a military grade vacancy had been established; and elimination of duplication in manning tables and reports.[3]

Major Bandel therefore proposed to the AAF that each station, instead of having two types of Tables of Allotment, male and female, have only one, and that either a man or a woman be placed in any authorized job at the station's discretion, according to the qualifications and availability of personnel.

Major Bandel was obliged to do considerable sales-work before her headquarters assimilated the plan, but the idea appealed to the Air Forces once it was understood and was proposed to the War Department in August. A few months later the Air Forces also adopted a system for its men under which grades were not broken up among living units, but were retained at station level and distributed among various offices, while the squadron commanders retained all command and disciplinary powers. Major Bandel later received the Legion of Merit for her numerous contributions to the Air Forces' program, with particular reference to originating this idea for Wacs.

In late September the War Department approved the AAF proposal for both the AAF and itself. Thereafter, all War Department grade allotments to the major commands were without reference to the sex of personnel. Any command might place a Wac in any suitable job for which it had an authorized military vacancy, without further formality or bookkeeping. The only limitation was upon the number of Wacs that could be furnished any command; gradeless numerical WAC allotments continued to be made to Air, Ground, and Service Forces based on their usual shares.[4]

The Air Forces system of bookkeeping now became so simple that scarcely any extra load was incurred by the presence of women. Also ended was the whole cumbersome system of WAC requisitions, which had so frequently been outdated before Wacs could be recruited, trained, and shipped. Instead, each station merely reported its current personnel shortages, by specialty number, and its housing capacity, by sex, and was likely to receive either a man or a woman, according to availability. In filling such shortages, Air

[3] AAF WAC Hist, p. 33ff.

[4] (1) WD Cir 226, 22 Sep 43. Directed by D/F, G-1 to TAG through CofS, 13 Sep 43. WDGAP 220.3, in CofS 291.9 WAAC. Rescinds AG Memo W635-3-43, 31 Mar 43. (2) Memo, MPD ASF for TAG, 14 Oct 43. SPWA 320.2 sec 6.

Forces headquarters was no longer obliged to compute WAC requirements and place a requisition. Instead, The Adjutant General merely reported each week the numbers and skills of WAC graduates to which the AAF's percentage entitled it, and these were ordered at once to the station having the most immediate need and the necessary housing. There was almost no WAC skill so unusual that the woman could not be utilized somewhere among the thousands of military jobs thus opened to Wacs. There was also no longer any tendency for a station to think of Wacs as an additional allotment, since it clearly had only one military quota.

One major discrepancy in this integration system was the failure to allot grades for WAC overhead. While the size of a men's unit might be somewhat reduced by the arrival of Wacs, it still required a first sergeant, a supply sergeant, a company clerk, and other key cadre members. Part of the WAC cadre thus did not replace men and constituted an extra administrative load on the station.

Major Bandel's plan had included a provision for increasing each station's allotment by an allotment of WAC grades for cadre only; instead, the War Department went further and abolished all WAC grades. Stations were thus forced to make up an allotment for the cadre of the new unit by somehow withdrawing grades from other allotted activities. This was usually possible in the lower grades, but there was seldom a spare first sergeant's rating or a captaincy at any station, while other cadre members were more or less at the mercy of chance for their grades. Such a system not only lowered morale, but made it impossible to keep competent women in cadre jobs or as company officers.

Major Bandel repeatedly tried to secure the recognition of WAC administration as a duty that required an allotment, but without success. The Air Forces was able to solve a part of the difficulty by authorizing the promotion of WAC first sergeants to the grade warranted by their job regardless of station strength allotments, and their retention thereafter as authorized overstrength. No such solution for company officers was ever achieved.[5]

There was only one other flaw in this system: the fact that men's unit commanders were accustomed to command all assigned personnel, and frequently attempted to usurp the normal powers of the WAC company commander in such matters as discipline, uniform regulations, and hours of bed check. Such usurpation, where it occurred, invariably reduced unit discipline to a farce, divided the WAC company into quarreling factions under different rules and regulations, and resulted in insubordination to the WAC commander. The AAF solved this problem without difficulty by spelling out in its regulations the fact that all normal command powers were reserved to the gradeless WAC unit. Actions in which both office chief and company commander were interested could be originated by either and co-ordinated with the other. Discipline was reserved to the WAC commander, either on her own initiative or upon receipt of a report from the section chief.

One other difficulty, a psychological one, at times prevented full application of the system. The intention of the directive was that military personnel, like civilian, receive the grade allotted to the job each

[5] AAF WAC Hist. Also D/F, G-1 for Dir WAC, 31 Aug 43. WDGAP 320.2 WAC, in SPWA 320.2 (7-23-43).

was filling. Actually, at almost all Army stations, there appeared to exist a deep-rooted idea that there should be a "women's share," which should be equal to the "men's share"; that women should in fairness have the same percent of each grade that men held. In practice this caused widespread inequities, since all women sent to a station might be so unskilled that they merited no ratings, or they might all be so skilled that all merited ratings.

In checking on the application of the directive, the Air Forces found male office helpers promoted above the Wac who ran the office, and WAC cooks refused the rating that male cooks received, all on the grounds that the "women's share" was held by stenographers in headquarters. Another Air Forces directive promptly made it clear that the grade went with the job, regardless of the sex of the occupant. Elsewhere the idea of a share appeared too well fixed to suppress; even such a WAC authority as General White, on a visit to North Africa, directed that the theater provide "the same ratio of grades for WAC personnel as is provided for male." [6]

The Army Service Forces and the Army Ground Forces, for this reason, never chose to follow the Air Forces' system, but divided military grades into male and female, and maintained separate allotments for the WAC unit. The motive was frankly stated as a fear that women, if not limited, would eventually collect more than their "share" of grades, since they usually filled skilled jobs and were more permanent employees. As a matter of fact, in application the result was exactly the opposite: Wacs were fairly latecomers in the Army, and many commands were already so overstrength on rated men that promotions were frozen; thus the Air Forces Wacs fared no better than male latecomers, while the Service and Ground Forces Wacs had a reserved quota and were promoted even though men inducted at the same time could not be. The well-integrated Air Wacs were shortly outstripped in ratings by the segregated ASF and AGF Wacs, and complained frequently at the loss of "our grades." [7]

The Office of the Director WAC

Second of the integration problems to be resolved was that of the reduction of WAAC Headquarters to form the new and smaller Office of the Director WAC, which had advisory duties but no operating functions. Only about 25 officers and 20 civilians were allotted the new Office of the Director; the remaining 75 officers and 46 civilians were scattered among the various Army Service Forces offices which now handled all operations.[8] With them went the files of their sections. All individual 201 files were sent to The Adjutant General, who was thereafter responsible for cases concerning individuals: initial assignment, transfers, promotions, filling of requisitions, classification, discipline, and discharge, as well as for answering inquiries from the public concerning the status of individuals. With this responsibility The Adjutant General received Capt. Clara G. Han and her assistants, who had been managing classification and assignment in the WAAC.

As for policy decisions concerning any

[6] D/F, G-1 for TAG, 18 Oct 43. WDGAP 320.2 WAC (9-10-43), in WDCSA 291.9.

[7] See details in chapters on services concerned.

[8] M/R, 15 Feb 44, Justification for Civilian Positions for FY 1945. WAC files.

of these personnel actions, the responsibility went to General Dalton's office, Military Personnel Division, ASF, with Maj. Florence Jepson to assist. Full authority over training policy as well as operation went to Military Training Division, Army Service Forces, accompanied by Maj. Elizabeth Smith. General Faith's Training Command was disbanded, and training centers were given to the service commands for administration, becoming Class I installations.[9] Similarly, recruiting, housing, supply, overseas selection, and other functions were all transferred.[10]

Such removal of operating functions was in general enthusiastically received by the WAC staff, since it was the same action that they had recommended a year before and that had then been impossible in an auxiliary status. In the first burst of enthusiasm for decentralization, the staff ruthlessly swept out of the office every discernible operating duty, and forwarded public inquiries and field requests unanswered to other offices.

In only two cases did the Director attempt to retain some supervision. She first asked that training centers be made Class IV installations, not under the service command for training doctrine and assignments, but this was refused by the Army Service Forces.[11] She also asked that she retain the power of final approval over the moral character of women officers selected by Army commanders for overseas shipment; this was also objectionable to the Director of Personnel, ASF, and was rejected by the War Department.[12]

The vague relationship of the Office of the Director WAC to the operating agencies was a source of confusion which Director Hobby in November requested the Army Service Forces to remove by publishing a circular defining responsibility. While she was permitted to recommend policy on all WAC matters, the other offices had always been finally responsible for policy, and there was no requirement that any co-ordination take place. The Director therefore asked that ASF divisions be charged with the responsibility of keeping her office fully informed of WAC matters and of WAC policy originated by them. This request was rejected by Brig. Gen. Clinton F. Robinson of the Army Service Forces' Control Division, on the

[9] (1) Routing Slip, Asst Exec WAAC to Chief, Operating Div, 26 Jul 43. SPWA 311.23 (1943). (2) Speech, Maj Jepson, Stf Dirs Conf, New York, 1-3 Dec 43. SPWA 337 (11-10-43). (3) AGO Ltr 320.2 WAC (10-7-43) PR-W-SPMOU, 31 Oct 43, incl WAC Tng Comd, effective 31 Oct 43. (4) Memo, O Dir WAC for MTD ASF, 19 Nov 43, SPWA 210.63 (11-19-43). (5) ASF Cir 110, 4 Nov 43.

[10] WAAC Supply, Housing, and Mobilization Sections to Mobilization Division, ASF; Control Division, Historical Section to Control Division, ASF; WAAC Office of Technical Information to Office of Technical Information ASF (later WD Bureau of Public Relations); WAAC Special Services to Special Services Division, ASF; WAAC Training Division to Military Training Division, ASF; Policy Functions of WAAC Personal Branch to Military Training Division, ASF; Classification, Assignment, Testing, Reassignment to AGO Classification and Assignment Branch; Recruiting to AGO Appointment and Induction Branch; Discharge, etc., of enlisted women to AGO Enlisted Branch; Discharge, etc., of officers to AGO Officers Branch. (1) ASF Adm Memo S-64, 14 Oct 43; (2) Drafts of same, Aug 43; (3) Memo, TAG for Dir Pers ASF, 5 Nov 43. SPWA 314.7 (1-7-43)(1) sec 4.

[11] (1) Memo, WAAC Hq for DCofS ASF, 17 May 43. SPWA 353 E; (2) Memo, Dir Tng ASF for DCofS for SvCs ASF, 21 May 43, 1st Ind, 22 May, DCofS ASF, 2d Ind, 27 May, MPD, 3d Ind, 1 Jun, Dir WAAC. WAAC Training Centers were made Class II installations by ASF Memo S210-14-43, 10 Jun 43, under the control of CG ASF and the technical supervision of the Director WAAC. ASF file, SPAP WAC (5-17-43).

[12] (1) AGO Memo W635-11-43, 8 Jul 43; (2) Memo, WAC Hq for G-1, 17 Sep 43; (3) Memo, TAG for Dir Pers ASF, 7 Sep 43. SPWA 320.2 (7-5-43) sec 1.

grounds that ASF divisions would ordinarily give the Director information about WAC matters of special importance without being directed to do so.[13]

The success of the several shifts in responsibility was therefore variable depending upon the degree of voluntary co-ordination between the offices concerned. The Adjutant General's Office was highly successful in removing a tremendous load of routine operations from the Director's Office, such as the receiving of availability reports and the issuing of assignment orders.

On the other hand, Military Training Division habitually took unilateral action in training matters, publishing training courses and circulars that had never been seen by the Director. Training center commandants complained of a perpetual conflict between training directives and other portions of the WAC program, such as recruiting. In a few cases the transfer resulted in total obliteration of a phase of the WAC program; work on the WAC history was stopped by General Robinson of ASF Control Division, who relegated the previous excellent WAAC collection to the files, disapproved collection of further material, and asigned the former WAAC Historian, Lt. Virginia Smithson, to a clerk's job.[14]

Several divisions, although assuming responsibility for part of the WAC program, did not accept assignment of a WAC officer or other specialist to help perform it. The Morale Services Division, for example, took the all-important responsibility for orienting soldier opinion, and its director, Maj. Gen. Frederick H. Osborn, after personal conference with Colonel Hobby, agreed to accept assignment of a WAC officer; but in spite of efforts of Colonel Hobby's staff to remind him, no such assignment was made during the six months in which the WAC remained in the ASF. Similarly Brig. Gen. Joseph W. Byron of Special Services Division informed the Director that he intended to accept assignment of a WAC officer, but no assignment was accepted during this period.

Such reluctance appeared generally to be due to doubts concerning the propriety of employing a specialist for such a small group as the WAC, and to the belief that the regular staff could care for the additional responsibilities. However, within a few months it was apparent that little action was being taken on WAC problems in offices that did not have a WAC specialist. Her staff informed the Director, "There are certain operating divisions, ASF, which seem not to be fully aware that responsibility has fallen to them for matters concerning WAC personnel."[15]

The Director attempted to overcome such omissions by including, in her new office, a WAC liaison officer for each major ASF division. This system also did not work very well. Liaison officers noted that, even when a division was extremely co-operative in taking any action suggested by the Director's Office, it was inappropriate for their responsibilities to en-

[13] Memo, Dir WAC for CG ASF, 5 Nov 43, sub: ASF Cir, and 1st Ind, 13 Nov 43. SPWA 314.7 (1-7-43)(1) sec 4.

[14] (1) Speech, Maj George Martin, Chief, Contl Div WAAC. Min, Stf Dirs Conf, Chicago, 15-17 Jun 43, SPWA 337 (6-1-43). (2) Memo, Maj Anne Alinder for Exec WAC, 10 Sep 43. SPWA 314.7 (1-7-43)(1) sec 4. (3) Interv with Capt Virginia Smithson, Hist Div, AAF Hq, Dec 45. (See Bibliographical Note.)

[15] Memo, Well-Being Officer for Dir WAC, 20 Dec 43, with atchd drafts and buck slips. SPWA-W.

listed women to be constantly called to the divisions' attention by the Office of the Director, and also inefficient in view of the "complexity of the organization. . . . and the impracticability of one liaison officer from this office being aware of all the functions or activities of the many branches and sections of each division." The WAC liaison officer noted that, in order to get Wacs included in these divisions' activities, she was forced to undertake "action more in the nature of operations than advisory." [16]

In addition, such outside interference was not always popular. The Navy noted, concerning an identical situation, that so long as the WAVES liaison officer's assignment, or even her desk, was in the Director's office, other divisions tended to regard her as "a sort of polite espionage service," and that the only workable system was the location of a specialist in the division office itself.[17]

Especial deterioration was noted in the handling of public relations, never well co-ordinated. WAAC Headquarters' small Office of Technical Information was transferred to the ASF Office of Technical Information, and subsequently to the War Department Bureau of Public Relations, where material concerning Wacs was handled, with even less co-ordination than had prevailed before, by the respective assistant directors of the bureau for ASF, AGF, and AAF. In the Army Service Forces' opinion, the "overall WAC public relations job" would now have to be performed by the service commands.[18] Although General Somervell took steps to call this problem to their attention, there resulted during this six months a series of unfortunate releases which did little to assist the current recruiting drive.[19]

Even at best, considerable confusion prevailed during the period of transfer of operating functions, and there were unavoidable lapses in continuity of operation. Files of the various transferred offices were shipped off bodily to offices in which they could not later be located, and other documents were destroyed for lack of space to house them. Continual office moves, an old Pentagon custom, added to the commotion. After two moves and one consolidation in the preceding twelve months, the office moved again in November of 1943 from the second floor of the Pentagon to the third; two months after this it was moved back to the second floor.[20]

General Impression of Director's Status

There was general failure on the part of Army and WAC personnel in the field to understand the Director's new status, and total failure on the part of the public. Few could grasp the fact that she no longer had any command power over individual women; fewer still realized that she was no longer consulted about every WAC directive nor informed of the policy in every matter. As a result, the Office of the Director constantly received many inappropriate requests, inquiries, and demands for corrective action. Service commands still telephoned the Director's Office with complaints, such as the failure of Army depots to distribute WAC Regulations on time. Others wrote her about matters under their own command, such as at what hour Wacs should go to bed and whether dating

[16] *Ibid.*

[17] WAVES Hist.

[18] Memo, CG ASF for DCofS for SvCs ASF, 18 Dec 43. SPWA 014.13 E.

[19] See Ch. XXXIV, below.

[20] (1) Hq Memo, 23 Nov 43; (2) Memo, MDW to CG ASF, 14 Jan 44. SPWA 310.2 (10-5-42).

should be permitted in the women's unit dayrooms.[21]

Following the conversion to Army status, the Army Service Forces refused to let the Director's Office have any further reports on WAC clothing status, training, station lists, or rank; it approved reports only on WAC strength and recruiting returns. The Director's Office continued negotiations in an attempt to get more.[22]

After this time, the Director had no means of discovering the problems of field units except by such visits as her staff was able to make, or by Congressional or other complaints; hence her later knowledge of field conditions tended to be based on a sampling rather than on a complete compilation. She succeeded in having written into the forthcoming regulations a provision, resembling that of certain British services, that any WAC commander might write directly to the Director on matters of women's well-being, provided that she routed her letter through the post commander. This provision proved virtually inoperative; the Director's Office noted at the end of the war that it had almost never been used, since a company commander obviously jeopardized her own local advancement by using the privilege, and therefore ordinarily did so only in the gravest emergencies.[23]

First WAC Regulations

Among all integration problems, the hottest debate concerned the matter of the special Army Regulations, if any, to be issued for the WAC. For almost a year, since the introduction of the WAC legislation, General Somervell's office and Director Hobby's office had been at odds on this matter.

The basic issue was the matter of minimum safeguards for women. Director Hobby maintained that this would require a directive stating that Army enlisted women would be assigned only to units commanded by a woman officer. The ASF disagreed, pronouncing any special regulations favoritism for women. Maj. Gen. Lorenzo D. Gasser, of the War Department Manpower Board, added that Army personnel officers would be hampered in exercising their prerogatives of transfer and assignment if they could assign enlisted women only to stations where a WAC officer was located.[24]

An ASF committee, appointed before the conversion to consider the matter, informed the Director that no changes in Army Regulations would be made, since "Experience has dictated that the effect of a regulation is often resented, especially if it seems to curtail power." [25] General Dalton, ASF's Chief of Military Personnel, also informed the Director: "It is highly undesirable that policies and regulations governing the control of WAAC personnel be more restrictive than those governing officers and enlisted men." [26] Upon reconsideration, the Army Service Forces decided that one basic Army Regulation for

[21] (1) Speech, Maj Mary-Agnes Brown, Min, Stf Dirs Conf, New York, 1-3 Dec 43. SPWA 337 (11-10-43). (2) Ltr, 9th SvC to WAC Hq, 17 Nov 43. SPWA 353.8.

[22] (1) Memo, Dir WAC for Col Conrad G. Follansbee, MPD ASF, 26 Nov 43; 1st Ind, MPD to AGO, 29 Nov 43. SPWA 201.7. (2) Memo, Miss Lies to Exec WAC, 15 Dec 43. Copy given author by Dr. Jessie P. Rice.

[23] Remarks, Dir WAC, Conf of Overseas Theater G-1's, 15 Oct 46. WDWAC 337 A-S.

[24] Memo, CofS ASF for Chiefs of Servs, 1 Mar 43, and 3 Inds. SPWA 333.1, now in Dir Pers ASF files.

[25] Memo, 2d O Mary Vann Racey for Dir WAAC, 1 Mar 43. SPWA 300.3 (1-7-43) sec 1.

[26] Memo, MPD ASF for Dir WAAC, 26 Apr 43. SPGAE 322.5 WAAC (4-26-43) G-1, in SPWA 300.3 (1-7-43) sec 1.

the WAC would be required: that under no circumstances would women command men.[27]

Director Hobby replied that, if less than a company with its own WAC commander was used at any station, it was not economical to provide women's clothing stocks, hospital equipment, post exchange items, recreational programs, and other needs that differed from those of men. Where these were not provided, the women did not receive care equal to that normally provided by the Army for its men. The Director also asked for a written definition of her own new noncommand duties in the WAC, which should, she stated, be those of advising on women's well-being, "defined to mean conditions of employment suitable for women, which shall remain the primary concern of the Director WAC." [28]

The Director's view was several times upheld by the Chief of Staff, General Marshall. In the spring of 1943, over ASF nonconcurrence, he authorized publication of the revised WAAC Regulations (1943), which contained the desired safeguards.[29] These governed the WAAC and WAC until late October 1943, when the General Staff was able to reconcile conflicting viewpoints and produce a final version of the WAC Regulations. The circular as published was drafted by G–1 Division, over the Army Service Forces' objection, personally discussed with Colonel Hobby by General White, and revised to incorporate her suggestions. On 9 November 1943 it was published as War Department Circular 289, the WAC's first important publication, which was to last, with amendments, for over a year, and to serve as a model for all later regulations.

Covering only six pages, it pointed out which Army Regulations were applicable unchanged, and authorized certain exceptions and special provisions. The required differences were:[30]

1. WAC units would contain only women and be commanded by WAC officers, exactly as men's units were composed of and commanded by men.

2. Wacs would not be confined in the same building with men, except a hospital.

3. WAC messes would not be combined with men's messes except with War Department approval. [General Dalton almost immediately repealed this restriction without informing the Director.]

4. Wacs would not be used in "restaurants or cafeterias in service clubs, guest houses, officers' clubs or messes." [Such assignments were ordinarily not authorized for any military personnel.]

5. WAC officers would not be promoted to the grade of colonel. [By act of Congress.]

6. Wacs would not command men unless specifically ordered to do so. [By act of Congress.]

7. Wacs would not be employed as physicians or nurses. [By act of Congress, to avoid infringing on existing organizations.]

8. WAC officers would be appointed only from officer candidate school graduates, and officer candidates would be selected only from women already in the Corps.

9. Enlistment standards would differ from men's in the age and citizenship requirements set by Congress, and in a different physical examination; venereal disease was also disqualifying, and women with dependent children were ineligible.

10. Discharge was mandatory for minors [by act of Congress]; authority was included for discharge for pregnancy.

---

[27] Memo and Inds cited n. 24.

[28] Min cited n. 14(1).

[29] *Ibid.*, incl ltrs, SvCs to Dir WAAC, May–Jun 43.

[30] (1) Contl Div Rpt. Min cited n. 14(1). (2) Memo, Dir WAAC for Sp Serv Div, Chief of Chaplains, SG, CofFin, JAG, and Contl Div ASF, 11 May 43, SPWA 300.3 (1–7–43) sec I. (3) Memo, "MGW" (Maj Gen Miller G. White, G–1) for G–1 Enl Br, 31 Oct 43. SPWA 300.3 (1–7–43) sec II. (4) Memo, G–1 for TAG, 3 Nov 43. WDGAP 300.3 WAC, in CofS 291.9. (Rescinds Cir 226.)

In addition, Director Hobby succeeded in getting spelled out certain other requirements which were identical with requirements for men, but which were still frequently violated because of the Corps' early ambiguous status: Wacs were to fill only military jobs, could not replace civilians, would not be assigned as permanent kitchen police, and were eligible for membership on courts and boards. It was required that courts and boards hearing WAC enlisted women include at least one WAC officer, in line with the general practice of including members of an enlisted man's arm or service. It was also required that commanders using WAC troops employ WAC staff directors, part of whose duties was "continuous inspection."

Rights and Benefits of the WAC

None of the circulars published made clear the differences between the rights and benefits of enlisted women and those of enlisted men. The Comptroller General decisions, requested in July, began to arrive in late September and defined some of these. On the question "May Wacs get allowances for dependents?" the Comptroller General ruled that it was clearly the intent of Congress to grant such allowances, since an amendment to forbid dependency allowances had been defeated. This, however, resulted in "material discrimination" against Waves and nurses, for whom the Comptroller had rendered an opposite opinion. The Comptroller decided that Wacs could not get allowances for husbands but could for parents and children in certain cases.[31]

The Comptroller General also reversed a decision of the Veterans' Administration that husbands were not eligible to receive the death gratuity, and made other related decisions. In addition, other decisions were made by the Army's Judge Advocate General and the General Staff. Many of these concerned the benefits toward which WAAC service might be counted, which were in constant question by every Army station. Wherever possible the General Staff attempted to give credit for WAAC service: it was decided to count it for time-in-grade for promotion purposes, for relative rank within grades, toward accrued leave, and toward overseas theater ribbons. As for the Good Conduct Medal, The Adjutant General first ruled that WAAC service did not count toward the time required to earn it, but was overruled by the General Staff. For other questions the answer had legally to be negative, since WAAC service had not been actually military service, and could not be counted for longevity pay or other such financial benefits.[32]

Colonel Hobby requested that these decisions be compiled in a new circular entitled Rights and Benefits of the WAC, for ready reference by puzzled field stations. The War Department decided against publishing such a compilation on the grounds that this would be a special publication for women and that it would be better to insert each reference in appropriate Army Regulations as they were revised from time to time. The Army Air Forces nevertheless obtained a copy of the rejected circular, mimeographed it, and

[31] Memo, Gen Styer (CofS, ASF) to Dir Fiscal Div, 21 Sep 43; Decision B-36497, Comptr Gen to SW, 24 Sep 43, SPEX, in SPWA 314.7 (1-7-43)(1) sec 2.

[32] See Ch. XII. Also (1) Decision B-35441, Comptr Gen to SW, 4 Aug 43. SPWA 242.2 (6-10-43). (2) Memo, JAG for Dir WAC, 3 Jan 44. SPJGA 1943/19509. Published in WD Cir 24, 1944. SPWA 314.7 (1-7-43)(1) sec 2. (3) Ltr, 3589th SU, Richmond, Ky., to TAG, 4 Oct 43, and 1st Ind from TAG, SPWA 200.6. (4) Memo, Col Hobby for G-1, 29 Dec 43; atchd to WD Cir 36, 28 Jan 44. SPWA 200.6.

sent it informally to AAF staff directors for their assistance in locating answers quickly. Many of these decisions never reached other field authorities; as late as 1946 War Department separation centers were discovered refusing to credit WAAC service for various benefits toward which it was creditable.[33]

Army Advisers Depart

The job of advising on Corps well-being under its new regulations now fell to an Office of the Director consisting of twenty officers, all women. The preliminary reduction, in November, had left the office with twenty-five officers, including Colonel Catron as executive and General Faith as head of the Field Survey Branch, but in December Colonel Hobby recommended that the office be still further reduced.[34] For service with the WAAC, General Somervell secured for General Faith the award of the Distinguished Service Medal, saying "To him, in large measure, belonged the responsibility of success in the whole program of women's participation in service with the Armed Forces." [35] To replace the departing officers Colonel Hobby gave key office positions to women brought in from staff directorships in the field. Major Rice left the recruiting office to become deputy director; Maj. Mary-Agnes Brown, formerly of the Eighth Service Command, was made executive officer. In civil life Major Brown had been an attorney for the Veterans' Administration and past president of the Women's Bar Association of the District of Columbia. Maj. Katherine R. Goodwin of the First Service Command became the personnel officer. Major Goodwin's civilian experience had been as head of business administration in the Hartford, Connecticut, public schools.

Field Needs After Integration

Army commanders simultaneously took over the task of operating the Corps under Army Regulations. One WAC authority noted:

> Generals, lieutenants, and sergeants who had long been privately certain that this business of utilizing women as soldiers would be perfectly simple, if the War Department would just let them handle it, now had a chance to prove their point. They went into action with all shades and degrees of concepts, from that which held that all women, in or out of uniform, were ladies who should be shielded from the rough ways of the world, to that which held that all soldiers were soldiers and should be treated alike, whether or not they were women. . . . Even for the many men who did not go to either of these extremes . . . women as soldiers were something new and different and required considerable getting used to.[36]

In spite of the inestimable advantages of Army status, field commanders faced certain disadvantages in attempting to begin operations under men's regulations, which were still largely unmodified. At Camp Gordon, Georgia, inspectors found that a WAC unit was getting all of men's T/BA equipment, including fifteen water tanks, spare gas tanks, a litter for carrying the wounded from the battlefield, and other property which it was unable to store in the space provided. Elsewhere just

[33] Memo, Dir WAC for TAG, 7 Oct 43; Memo, G-1 for CGASF, 4 Dec 43, WDGAP 243; SPWA 314.7 (1-7-43(1) sec 2. The informal AAF publication is in Policy Book, property of author.

[34] Memo, Dir WAC for CG ASF, 29 Dec 43. SPWA 320.2 E.

[35] Memo. Gen Somervell for WD Decorations Bd, 15 Jan 44. Personnel Abstracts, WAC files, OCMH.

[36] AAF WAC Hist, pp. 17-20.

the opposite idea prevailed, and because WAC units were gradeless, stations refused to issue them guidons, unit vehicles, or other standard equipment.

Army records and reports were now used for women and had to be modified in many cases. Officer candidates already processed and at the head of the waiting list had to be reprocessed under Army Regulations and take the test used for men, as well as appear before Army boards which sometimes differed considerably in judgment from the previous all-Waac boards. Commands were embarrassed when they tried women by court-martial and sentenced them to confinement without first providing a place of confinement. Stations and training centers that had previously weeded out prostitutes and other undesirables found that Army Regulations gave them no authority to discharge military personnel for prostitution.[37]

A major, if temporary, complaint was that no uniformity of personnel action was thenceforth possible for the "Corps"—now a corps in name only. Promotion was in the Army channel, and therefore could not be consistent from one command to another; junior officers were promoted before senior, women in less important jobs before those in more responsible, according to varying command and station policies and opportunities. Discipline was also a command matter, and major offenders went unpunished in some commands while minor matters in others were dealt with severely. Policies on dress, leaves, restrictions, and such matters varied. Training of various sorts, including physical training, was required by some stations and not by others.

Similarly, Corps-wide personnel programs were made impossible, including rotation of training center officers to the field, relief of recruiters, or systematic advancement of company officers. The difficulty was intensified in that, while most Army commands had thousands of male personnel and could effect normal rotation among them, the average command had only a handful of WAC units and perhaps one staff job of field grade.

Another new difficulty was the lack of information. At Fort Moultrie inspectors noted, "Complaint was made that few notices of any nature were reaching the detachment from Headquarters Fourth Service Command"; at Camp Gordon, "The WAC Commanding Officer felt a need for more information about her job." It would obviously hereafter be more difficult to get WAC matters down to company level through the more complicated Army echelons than through the shorter WAAC channels.

To solve such new problems, there was noted a general tendency for the field to request directives from Washington to solve field problems and set Corps-wide policies. At a staff directors' meeting in December 1943, representatives asked the Director for a standard operating procedure to secure uniform punishments for women in various commands, adequate physical training, proper officers' quarters, and other items. Violations of uniform regulations by female personnel were in particular looked upon as a matter that some WAC authority should correct. Many asked to be allowed to assign worthless personnel to some sort of central WAC

[37] (1) Memo, WAC Hq for MPD ASF, 15 Sep 43, with atchd file. SPWA 315 (1943,) (2) WD Cir 240, 4 Oct 43, sec IV. (3) Memo, Dir WAC for G-1, 1 Jan 44. SPWA 424.2, 1942. (4) Folder Rpts of Stf Visits, esp rpts 24-29 Nov 1943. WAC Plng Serv files. (5) Ltr, WAC Stf Dir AAF WFTC to Exec WAC, 20 Nov 43. SPWA 320.2. (6) Min cited n. 21(1).

pool for disposition, as it proved extremely difficult to get male personnel to discipline and reclassify women. The difficulty of lack of uniformity was especially acute at places like Fort Riley, which had three sets of regulations for the three WAC units assigned respectively to the cavalry school, the replacement training center, and the post headquarters.[38]

Such difficulties, while very real, presented a peculiar problem in view of Colonel Hobby's determination to integrate the WAC in the Army. She repeatedly rejected any proposal which, she said, "tends to give the impression that the WAC is something apart from the Army." [39] Thus, while she marked for correction certain legitimate needs for new policies or amplified regulations, in general she replied to such requests that they were matters of command. The lack of uniformity of action among Army commands, although especially conspicuous among the WAC minority, was not a WAC problem but an Army one, concerning men as well as women, and inherent in the system of command responsibility.

Close of 1943

At the end of 1943, Wacs at a staff directors' conference pointed with pride to the WAC's progress since the previous December. Then, ten staff directors had attended the meeting; in 1943, fifty-three were present. Then the WAAC had boasted only two post headquarters companies plus the Aircraft Warning Service in the field; now it had more than 260 companies. More Wacs were presently in each of two foreign theaters than had a year before been in the whole field. Then the headquarters and training centers had been run chiefly by men; now, they were run chiefly by women. Then, the training center had been the WAC world; now, it had receded to relative insignificance. Then, only service commands had employed Wacs; now all commands did. Wacs had been taken into the Army and had generally done away with separate grades, T/O's, and channels.[40]

The first of January 1944 found the WAC with a reported membership of 62,859 as against 20,943 a year before. Losses of the conversion period had almost been regained, several months before this had been believed possible. "We are very much encouraged," Colonel Hobby said. The Director added:

> It is quite a consolation and encouragement to compare the difference in the questions and problems that arose at this Staff Conference with the ones that arose about a year ago and the ones that arose six months ago—partially a tribute to your indoctrination and understanding, and partially a tribute to the Army for the very fine and splendid assistance it has given the WAC, now one of its components.[41]

[38] Min cited n. 21(1).

[39] Memo, Dir WAC for CG ASF, 27 Oct 44, sub: Dr. Durfee's Tng MS on Leadership. WDWAC 353.

[40] Opening speech by Air WAC Off, Maj Bandel. Min cited n. 21(1).

[41] (1) WDBPR Press Release, 6 Jan 44. WAC Planning files. Strength reported at time. (2) Rpt, Machine Recs Div, AGO, 4 Mar 44. WDWAC 320.2 sec 4*b*. For adjusted total of 61,355 for WAC strength in January 1944, see Table 1, Appendix A.

CHAPTER XV

Removal of Director's Office to G–1 Division

There remained one step to be taken in the evolution of the WAC organization. This was the removal of the Office of the Director from the Army Service Forces to the General Staff level. During January and February of 1944, a number of contributing causes led directly to this move, the last in the series by which the Corps was fully integrated into the normal command channels of the Army.

There had for some time been general agreement within the War Department that the past two years' method of handling Corps' problems was unsatisfactory, and that some sort of remedial action was called for, with more high-level attention to its problems than the Corps had yet received. A prominent civilian special assistant, who was called upon by the Army Service Forces to diagnose the situation during the conversion crisis, bluntly informed it that its headquarters had in fact killed the WAAC with its left hand while publicly sponsoring it with its right. He wrote:

> The situation with respect to the WAC is such that I believe immediate action is necessary, and that such action must be of rather far-reaching nature. Does the Army really want the WAC to succeed? I am not at all convinced that the answer to this question is "yes." . . . I have seen no real evidence of such a desire. . . . Are civilians really convinced that the WAC is an important organization? Civilians will be convinced of this only when the Army is convinced of it. . . . It should not be necessary to review arguments why the WAC must succeed. The Army started this program, and it is obviously a good idea to be associated with successful programs.[1]

A similar view had been voiced by Colonel Catron, after ten months of assignment to WAAC Headquarters. Later, "acting on my own initiative, but with the knowledge of Director Hobby," he had bypassed his former associates in the Army Service Forces to inform the General Staff directly, "In my opinion, the situation calls for . . . the military authorities to recognize an obligation to the WAAC, to the Army, and to the women of the country, to help make the WAC a go."[2]

G–3 Recommendation

At the time of the conversion, the Deputy Chief of Staff, General McNarney, therefore directed the General Staff to recommend a new location for the Director's Office. After analysis, G–3 Division

[1] Memo, "W.C.M." (Wilbur C. Munnecke, Vice Pres. of Univ. of Chicago, and wartime sp asst, ASF) for Gen Dalton, Dir Mil Pers ASF, 10 Aug 43. SPAP 341 WAAC (2–26–43).

[2] Memo, Exec WAAC for G–1, 20 Jul 43. SPWA 314.7 sec 3.

reported that the location in the Army Service Forces was not suitable. Although the Director WAC was a special adviser to the Chief of Staff, and was responsible for advice concerning Air, Ground, and Service Forces, she was currently obliged to submit all papers to or through the Army Service Forces. The G–3 planners noted: "This form of organization is not entirely satisfactory, as the scope of WAC matters is Army-wide and determination of all basic policies should properly be made at the War Department level." [3]

G–3 Division therefore recommended that the office be placed under G–1 Division of the General Staff, comparable in place to the Military Intelligence Division under G–2, or else, and preferably, be made an independent Special Staff division, like Civil Affairs Division. The Special Staff location, although it was never to be adopted, was always the one recommended by organizational experts. G–3 Division pointed out that the Director's responsibilities were not limited to G–1 matters and that the Special Staff location "will expedite direct consultation with appropriate divisions of the General Staff." [4]

Director Hobby concurred in either move, but General Somervell nonconcurred and presented a counterplan that would keep the Office of the Director under his jurisdiction. He proposed that it be attached to his own office, and not be placed under any subordinate office such as that of General Grunert or of General Dalton. He agreed also that the Director would be allowed to consult directly with Air and Ground Forces on matters that did not concern the Service Forces.[5]

General Somervell's plan prevailed; at the time of the conversion, General McNarney verbally advised Colonel Hobby to try the plan for several months and see if it would work. As a matter of fact, it was to be exactly six months before all concerned were convinced that the plan would not work.

Handicaps on ASF Level

During the early weeks of 1944, General Marshall called to his office General White and other officers to express his dissatisfaction with the situation, particularly as it affected recruiting.

Among issues leading toward a move from the ASF level, the major one was that of supply. In the midst of debate over the Austin-Wadsworth bill, members of Congress began publicly to criticize the WAC uniform. Criticism from Army stations, particularly air bases, was equally severe. Very little honest defense against the criticisms was possible. In spite of the easing of the supply problem, which had presumably resulted from the collapse of expansion plans eight months before, the situation remained approximately where it had been at that time.

Severe shortages still existed on field stations; recruits in training centers still did not receive complete clothing issue, and still suffered high sick rates; and the appearance of the uniform had not been improved in any respect. Since the Army Service Forces had now been responsible for the WAC uniform for almost two years, it did not appear that either inexperience or rapid build-up could any longer be offered in explanation. However, eight separate proposals for improving the situation—made by the Army Air Forces, by

[3] Memo, G–3 for DCofS, 17 Aug 43. WDGCT 322 WAC (8–17–43), in CofS 291.9.

[4] *Ibid.*

[5] See pencil notation, *Ibid.*

The Quartermaster General, and by the Director—were all disapproved by Requirements Division, ASF, by the end of February of 1944, without any workable counterproposals. The last two proposals were addressed to the General Staff by authority of Circular 289, as a matter of the well-being of women in all major commands, but were disapproved by the Army Service Forces and turned back without reaching the General Staff.[6]

In the matter of public relations, a rapid deterioration had been noted after the Army Service Forces took over operating duties at the conversion. None of the Director's suggestions made at the time of the slander campaign were accepted, and in the first weeks of 1944, a series of publicity releases were so poor as to come to the attention of the Chief of Staff himself. General Marshall in January sent a stiff note to the Bureau of Public Relations: "It seems to me that very poor use is made of the best publicity possibilities in the WAC organization. . . . Who is handling this business?"[7]

As for soldier opinion, the Director had been unable, as late as February 1944, to secure the assignment of specialists to either ASF Morale Services or Special Services Divisions, to work on the improvement of soldier opinion or WAC morale.

Similarly, in November of 1943 General Marshall had directed that, in order to improve the supply and health situation at Fort Oglethorpe, a ranking WAC officer be made commandant. Although the Director twice reminded General Somervell's office of this directive, no action had been taken by late February.[8]

Also in November, General Marshall directed that some WAC officers now be made lieutenant colonels, to fill position vacancies that had existed for eighteen months. However, Air Forces proposals to this effect were subsequently refused by the Army Service Forces upon General Somervell's personal written advice. In a seperate study, General Somervell instead recommended giving direct commissions in this rank to socially prominent civilian women. It was this action that led the Assistant Chief of Air Staff to appear personally in the War Department seeking the removal of the Director's Office from General Somervell's jurisdiction.[9]

A particular cause of interservice friction was the handling of papers on WAC matters, which received Army Service Forces comment in cases that, had they involved men, would not have been handled by the ASF at all. For a time it had been hoped that the objections of the Air and Ground Forces, and of overseas theaters, could be met by the new status of the Director's Office as a part of General Somervell's own office, with authorization for direct communication with the Air and Ground Forces and the General Staff on matters that did not concern the Service Forces. Unfortunately, it soon developed that in actual operating practice there was little difference in the office's former position, since it became General Somervell's custom to route all outgoing WAC papers through the office's former superior, Gen-

[6] See Ch. XXVI, for details.

[7] Memo, "G.C.M." for WAC Recruiting Section, WDBPR (nonexistent section), 26 Jan 44. OCS 324.5 WAC.

[8] Memo, Dir WAC for DCofS ASF for SvCs, 26 Feb 44, with pencil notations. SPWA 210.31 (2-2-44).

[9] Entire discussion from: (1) Memo, Dir WAC for G-1, 27 Nov 43. SPWA 210.2, in SPAP 210.2 WAC. (2) Memo cited n. 8. (3) Entire file on controversy is also in CofS 324.5 WAC, and in SPWA 210.2 WAC Promotions, 1943. (4) Ltr, TAG to AAF, 18 Nov 43. 210.2 WAC (11-16-43) PR-W-WDGAP, in SPWA 210.2.

eral Dalton, as well as through other ASF divisions. Disapprovals of policy recommendations by General Dalton or by Requirements Division, ASF, excited increasingly unfavorable comment from other commands in matters which concerned them solely, or would not normally have received Army Service Forces' comment.[10]

Move to General Staff Authorized

In February of 1944, as a result of these complaints, Director Hobby went personally to General Somervell to ask clarification of her office's status. He asked her to present the problem in writing, which she did, attaching as corpus delicti some three defunct policy papers which she had addressed, supposedly directly, to the General Staff, all of which had been disapproved upon General Dalton's recommendation. All three concerned policies that were Army-wide, and one was solely an Air Forces proposal.[11]

Upon receipt of the Director's protest, General Somervell immediately recommended in writing to General Marshall that the Office of the Director be moved from his jurisdiction to that of the General Staff. General Somervell stated, "The office of the Director WAC is concerned primarily with matters affecting policy for the WAC which are applicable on an Army-wide basis in three major commands." [12] When later acquainted with his action, Colonel Hobby stated that, if allowed to comment, she would have expressed ready concurrence.[13]

As a result, General Marshall called Director Hobby in for a conference and directed that the Office of the Director be moved to G–1 Division, War Department General Staff, effective 1 March 1944.[14] Because of the small size of G–1 Division, the Director's Office again had to be reduced in size, this time to the Director and four—later only two—other officers, including Major Rice as Deputy.

The Meek Report

The move was still in process, with office furniture not yet installed, when General Somervell's office received a lengthy written report, which it forwarded to General Marshall. The report was in the nature of a sweeping criticism of the conduct of the WAC program to date, alleging that Director Hobby and her advertising advisers had overlooked many obvious means of improving recruitment, had failed to improve the WAC uniform, and had allowed the WAC to fall far below the WAVES in public esteem. This study came to be called the Meek Report after its author, Mr. Samuel W. Meek, a member of the advertising agency of J. Walter Thompson, a competitor of the firm that handled official WAC advertising, Young & Rubicam.[15]

The Meek Report alleged, among other things, that the public was not aware of the urgency of the Army's need for Wacs;

[10] Specific examples in following text.

[11] Memo, Dir WAC for CG ASF, 5 Feb 44, sub: Functions of Dir WAC, with atchd papers. SPWA 314.7, Hobby files.

[12] (1) Memo, CG ASF for CofS, 9 Feb 44. Styer Files, Drawer 2197, DRB AGO. (2) Incl to Memo, Dir WAC for CofS, 21 Mar 44, sub: Comment on Public Opinion Concerning WAC, Tabs A to J atchd. WDGAP 341 (1944). (3) Min, Gen Council, 21 and 28 Feb 44.

[13] According to statement of Deputy Director to author.

[14] Memo, G–1 for CofS, 14 Feb 44. Approved by CofS, 19 Feb 44. WDGAP 320 WAC, in CofS 324.5.

[15] Manila Folder with 45 pages of surveys and statistics with atchd note, "HJK" to Gen Somervell, 21 Mar 44, and pencil notation: "Not to be use [sic] without my personal OK. BS." Somervell Files, Hq ASF, A46-257, DRB AGO.

that the WAC did not have the public standing of the WAVES; that three out of four women said they would choose the WAVES in preference to the WAC; that the public thought Waves got better treatment, more suitable jobs, more attractive uniforms; and that the WAVES had commissioned more outstanding women educators, whereas the WAC had not offered direct commissions to prominent women. Mr. Meek recommended that publicity tie-ins with high-ranking officers be used, that more prominent women be directly commissioned, and that society news pages then publicize the backgrounds of these new WAC leaders.

General Marshall called Director Hobby to his office on 15 March 1944 and, over Mr. Meek's objections, handed her the surveys and directed her to study these charges and prepare recommendations. This she did, first in a preliminary report and then in a confidential staff study which, because of its outcome, was to rank as perhaps the most important WAC policy paper of the Corps' career.[16]

The charges in the Meek Report proved, in themselves, unfounded and easy to counter. The only new survey of "public" opinion for which it contained documented statistics proved to consist chiefly of interviews with 111 college girls, 94 of them from Vassar, and with several prominent women, including women's magazine editors.[17] By contrast, the Young & Rubicam campaign had been based on a confidential Gallup survey of a scientifically chosen national sample of 1431 eligible women and 1415 parents of eligibles.

The two surveys often differed. Meek stated that the public was unaware of the urgency of the Army's need for Wacs; Gallup found that a record 85 percent said they were aware of it, and when asked "Which women's service is in the greatest need of more women?" 58 percent said the WAC, only 14 percent the WAVES. The WAC advertising under Young & Rubicam thus appeared to have been more effective than was alleged. Meek stated that the public liked the WAVES better than the WAC, and that 66 percent of eligible women said they would choose the WAVES as against 20 percent the WAC; Gallup's non-Vassar eligibles answered: War work—51 percent, WAC—12 percent, WAVES—10 percent. The only statement which no one disputed was that about the uniform, for Gallup had always reported that only a few women preferred the WAC uniform.

Director Hobby was also able to prove that the WAC had, some months before, taken every action recommended by Mr. Meek except direct commissions and short-term enlistments: it had used publicity from every prominent person named, offered station assignment, used society pages, publicized overseas assignment as far as surveys showed desirable, obtained two Gallup and two scientific Army opinion surveys, mailed literature to parents, and so on. None of these steps had ended the Corps' problems or come near its basic difficulties.

[16] (1) Preliminary Rpt, Dir WAC for CofS, 16 Mar 44, sub: WAC Rctg. WDWAC 341, in CofS 324.5 WAC. (2) Memo cited n. 12(2).

[17] Following sources were listed in Meek Report: (1) 20 interviews with prominent persons, including ranking officers in the WAVES and Women Marines. (2) Unidentified surveys made by American Institute of Public Opinion and *Fortune* Magazine, Feb–Oct 43; statistics, dates, and scope not given. (3) Some 1,208 persons (473 young women, 377 young men, 358 parents) "interviewed in a national survey" in November 1943; surveying agency or method of selection not given. (4) Questionnaire to 240 staff members of *McCalls*, *Time*, and *Reader's Digest* and "advertising personnel." (5) "College Survey" of 94 students of Vassar and 17 from Sarah Lawrence.

In countering the various charges, it proved easily possible to discuss those of the Director's previous recommendations that the Army Service Forces had disapproved. The Director, perceiving this opening, at once drove straight through with two memoranda and ten appendixes.[18] These consisted of important parts of her own program, previously suppressed by the Army Service Forces, for improving the WAC's situation. This was the first opportunity to bring most of these proposals to the attention of the Chief of Staff.

When the Chief of Staff's final verdict was rendered, more of the Director's proposals concerning major issues had been approved in two weeks than had previously been done in the past two years in the Army Service Forces. The proposals were concerned chiefly with soldier attitude, public relations, and the uniform.

Corrective Action by the Chief of Staff

The first two issues, soldier opinion and public attitude, Colonel Hobby unhesitatingly named as the basic difficulty of the WAC program. Dismissing all lesser theories, which attributed recruiting difficulties to ignorance, WAVES prestige, the uniform, the failure to grant direct commissions, poor jobs, the desire for shorter enlistments, and other petty causes, she said flatly, "The two greatest deterrents to WAC recruiting are the attitude of soldiers toward women in the military service and the apathy of unmarried non-working women." [19]

General Marshall saw for the first time some of the slander campaign material, including literature which soldiers had mimeographed and circulated to almost every Army installation at home and abroad.[20] He saw a sample of the many letters, with which the WAC files were now bulging, from Wacs and eligible women.[21] He also saw cartoons from soldier publications all over the nation.[22] The most unexpected exhibits were certain anti-WAC statements made by high-ranking Army officers, all except one of whom were combat officers without experience in WAC employment. These Colonel Hobby revealed by name only after being twice expressly commanded to do so.[23] One general remarked, in a national magazine, "Fortunately I've no experience with that particular species [Wacs] and what's more I don't want any of them around here." [24] The Director commented, "The attitude of the officers and enlisted men in the field will never change to the degree desired as long as key personnel, whose expressions can be assumed to reflect the War Department attitude, make statements such as these." [25]

To remedy the Army attitude, Colonel Hobby proposed a campaign of re-education similar to that which General Bowley had proposed twenty years before. She asked that the Army attempt to change the soldiers' attitude by including material on the Corps' usefulness in the Army orientation course, by which other soldier attitudes were successfully influenced. She asked that available films on the WAC be shown in this course, and a new film

[18] Memo cited n. 12(2). Each Tab was a separate signed memorandum so that recommendations could be acted upon separately.

[19] Rpt cited n. 16(1).

[20] Memo cited n. 12(2), Exhibit I: "Characteristics, functioning, care, and preservation of the WAC, M-1 (model 1942-43)."

[21] *Ibid.*, Tab D.

[22] *Ibid.*

[23] *Ibid.*, Tab C.

[24] Maj. Gen. Lewis H. Brereton in *Liberty*, October 9, 1943.

[25] *Ibid.*

made, specifically designed to reach the soldier. She asked that a Wac officer be assigned to Morale Services, ASF, to direct the distribution and clearance of WAC news and cartoons to camp newspapers; also, that an enlisted woman be assigned to the staff of *Yank* magazine, previously unfavorable editorially. She asked that courses on the organization of the Army mention the WAC. And particularly, she asked that general officers clear with the Bureau of Public Relations before making public statements concerning the WAC.

General Marshall very soon approved much of this program. General Somervell now concurred in the orientation course material, and in the assignment to Morale Services and Special Services Divisions of a WAC officer and an enlisted woman. He protested that to require general officers to clear their remarks on the WAC was too drastic. He suggested instead that General Marshall write a letter to general officers discussing the problem, similar to the Marine Corps Commandant's letter sent out some seven months earlier concerning the treatment of Women Marines.[26] This alternative was acceptable to General Marshall, who immediately wrote not only to all generals but to all Army commanding officers, pointing out that the Army was about to launch another drive for Wacs to meet "imperative" needs, and that he expected to have the drive supported. He added:

> The Women's Army Corps is now an integral part of the Army and a highly essential part of our war effort. . . . However, reports indicate that there are local commanders who have failed to provide the necessary leadership and have in fact in some instance made evident their disapproval of the Women's Army Corps. The attitude of the men has quickly reflected the leadership of their commanders, as always.
>
> All commanders in the military establishment are charged with the duty of seeing that the dignity and importance of the work which women are performing are recognized and that the policy of the War Department is supported by strong affirmative action.[27]

In the matter of the uniform, Colonel Hobby presented General Marshall with a list of her previous recommendations to the Army Service Forces, from August of 1942 to date. It appeared that General Somervell had not been aware until after the Meek Report was forwarded to General Marshall that the condition of the WAC uniform resulted from disapproval of the Director's proposals by various echelons of his own command. Colonel Hobby informed General Marshall that the Service Forces had already begun restudying some of her requests by direction of its commanding general.[28] General Marshall within a few weeks approved eight separate policy recommendations made by the Director and The Quartermaster General, all previously rejected by the Army Service Forces. These were shortly to work a considerable change in

[26] (1) Memo, CG ASF for CofS. SPAP 324.5 (3-29-44) ASF. (2) Ltr, Comdt USMC to all COs of posts and stations, 14 Aug 43, sub: Responsibility for Behavior of Men Toward Women of MCWR. "Information reaching this Headquarters indicates that in some posts and stations officers and men of the Marine Corps treat members of the Women's Reserve with disrespect. . . . In some cases, coarse or even obscene remarks are being made without restraint by male Marines in post exchanges, moving picture houses, and other places in the hearing of members of the Women's Reserve. . . . This conduct . . . indicates a laxity in discipline which will not be tolerated. Commanding officers will be held responsible to this Headquarters. . . ." [Signed] T. Holcomb.

[27] Printed Memo, CofS to all Hq divisions, overseas commanders, AGF down to tactical units, AAF, ASF, and Defense Commands down to posts, camps, and stations, 6 Apr 44. WDCSA 324.5 WAC.

[28] Memo cited n. 12(2), Tab I.

the appearance of the uniform, although many were unfortunately too late to reach the field in the one remaining year before the end of the war in Europe.

In the matter of public relations, the Director asked General Marshall to direct the formation of a co-ordinating board, which would not only initiate and direct an Army-wide constructive informational campaign, but would comment on the possible public relations effect of decisions concerning the WAC before the decisions were released. General Somervell stated, when asked to comment, that such a board was not necessary.[29] Colonel Hobby replied that General Somervell's attitude was "not in accord with fact or experience," and that "The present situation has been unfortunate and it can reasonably be expected that a continuation of uncontrolled and uncoordinated activity on the part of literally hundreds of issuing agencies will continue to be unsatisfactory. . . ."[30]

General Marshall, who had supposed that such a board already existed, overrode objections and directed General Surles of the Bureau of Public Relations to set up the co-ordinating group, saying, "I have personally taken every opportunity to support the WAC. . . . It is necessary that a central agency of the War Department be charged with the problem of keeping the public informed of matters concerning the WAC and of controlling the issuance of such information."[31]

Colonel Hobby also took this opportunity to secure the Chief of Staff's approval of several lesser projects which had previously been rejected at lower levels. One concerned equal provision for the WAC in the Army's demobilization plans, which were already being drawn up, and in which planners had made no provision for women. General Marshall approved the greater part of what she asked, which was to be of considerable importance at a later period.[32] Another proposal was for the centralization of recruiting control in the hands of specialists in The Adjutant General's Office, similar to that just approved in Public Relations. These and other recommended reforms in the Recruiting Service were all approved by General Marshall, with far-reaching later effects.[33]

The Director also concurred in one of the suggestions made in the Meek Report, the establishment of a civilian advisory committee composed of prominent women in each service command, modeled on the group that she had established in the Women's Interests Section of the Bureau of Public Relations, when she had organized that office in prewar days. This recommendation, providing for both local and national committees, was also approved. General Somervell concurred, provided that he be allowed to choose the prominent woman to head the national committee.[34]

Within a few weeks, by General Marshall's action, there were six WAC lieutenant colonels. In April, a WAC officer was also assigned as commandant of the Third WAC Training Center.[35]

In a final comment on the Meek Report, General Somervell sent the Chief of Staff another study recommending that key WAC officers be replaced by promi-

[29] Memo cited n. 26(1).

[30] Memo, Dir WAC for CofS, 4 Apr 44. WDGAP 341 (4-4-44), and in WAC Gp, WDBPR, under "Organization of the WAC Group."

[31] Memo, CofS for Gen Surles, 11 Apr 44. WAC Gp, WDBPR, and in CofS 324.5 WAC.

[32] Memo cited n. 12(2), Tab J.

[33] Memo, Dir WAC to CofS, 10 Apr 44, with CofS approval, 11 Apr 44. CofS 324.5 WAC.

[34] (1) Memo cited n. 12(2), Tab H. (2) Memo cited n. 26(1). (3) Memo, CofS for CG ASF, 11 Apr 44. CofS 324.5 WAC.

[35] Memo, ASF for TAG, 13 Mar 44. SPGAG 210.3 WAC (3-13-44)-129.

nent women to be selected by him and directly commissioned. He also stated that WAC leaders had too much control of WAC affairs, and that Colonel Hobby's current powers should be reduced by eliminating those of direct dealing with the major commands, of handling staff co-ordination on WAC matters, of taking any action, and of maintaining separate records.[36]

None of these recommendations was adopted. The Army's new policy toward the WAC, as forcibly inaugurated here by General Marshall, was keynoted by the introduction of specialist groups or specialist officers in almost every major office that handled WAC matters. The new location also ensured that Army-wide recommendations from the Office of the Director would not be rejected without consideration by the General Staff. From this time throughout General Marshall's tenure of office, specialist attention to women's welfare was to be the rule rather than the exception—with, however, the specialists concerned being fully integrated into their respective Army sections.

With the completion of the move to G-1 Division, and the accomplishment of the changes as approved by General Marshall, the Women's Army Corps completed its stage of organization and development, and entered a new phase of its career. Freed of the burden now assumed by the specialist groups, the small Office of the Director was to devote most of its time toward planning for the solution of the more intangible social and welfare problems which, although pushed into the background until now, had from the beginning attended the integration of women into military service.

The two years just ended had seen the discovery of the Corps' problems in public relations, clothing supply, administration, and many others. The next sixteen months in G-1 Division, before the end of the war and Director Hobby's resignation, were to see workable solutions devised for all but five of the problems now on the Director's calendar.

The Corps was, by its second anniversary in May 1944, already world-wide in distribution and approaching peak strength in numbers. Corps strength in June was almost 77,000, with the peak of 100,000 not too far distant. In the United States, Wacs were present at 193 Air Forces installations, and 176 Ground Forces and Service Forces stations. Almost 10,000 women were already overseas in every major theater of operations—North Africa, Italy, England, Australia, New Guinea, India—as well as in Hawaii, Alaska, and other smaller bases.[37]

In May of 1944, the authorized strength was reallotted among the major commands:[38]

Army Service Forces	[a] 60,000
Army Air Forces	60,000
Army Ground Forces	7,500
War Department	1,050
Overseas theaters	21,450
	150,000

[a] Including training centers

In these widely scattered Army commands, with their varying needs and policies, lay the further history of WAC employment, and success or failure on Army jobs.

[36] Memo, CG ASF, for CofS. SPAP 324.5 (3-29-44) in Dir Pers ASF files.

[37] (1) Memo, Dep Dir WAC for CofS, 21 Jul 44. WDCSA 324.5 WAC. (2) WDBPR Press Release, 17 Jun 44.

[38] D/F, G-1 for ASF, 8 May 44, WDGAP 320.2 Misc, WAC Cl files, based on Memo, Dir WAC for G-1, 17 Apr 44. For previous quotas, see: Ltr, TAG to CG ASF, 22 Oct 43. AG 320.2 WAC (11-19-43) PR-W-SPGAS.

PART TWO

WORLD-WIDE EMPLOYMENT

CHAPTER XVI

The Army Air Forces

Among the various Army commands, the Army Air Forces had been the first to employ Waacs and was now the first in field strength of WAC units. The history of the Air Forces Wacs was the story of almost one half of the Women's Army Corps—some 40,000 of the eventual 100,000 women.[1]

In spite of its enthusiasm for the use of womanpower, the Air Forces faced greater initial handicaps in employing Wacs than did other domestic commands. Subordinate air commands did not cover definite geographical areas as did service commands, but were strictly functional in nature, each having control of air bases scattered throughout the United States. Staff directors thus found the problem of supervision and supply more complex, and were obliged to take to the air to cover the distances between units assigned to their commands. The AAF's great Training Command had flying schools and technical schools from Florida to California; its Troop Carrier Command stations were less numerous but as widely separated; Air Service Command depots were scattered from coast to coast. The First, Second, Third, and Fourth Air Forces together covered the United States in their mission of forming men into combat groups for the numbered Air Forces overseas. The AAF's Materiel Command had one large WAC unit at Wright Field; the School of Applied Tactics had an even larger unit at Orlando, Florida; and the Proving Ground Command had one isolated group at Eglin Field, Florida.

Although air bases were generally prepared to receive the women, they were caught without advance warning when the War Department ended the T/O system for the WAC some months before the Air Forces did the same for its men—a move that suited the Service Forces but left the AAF without means of accounting for the women. The new Air WAC Officer, Major Bandel, upon her arrival in 1943, found headquarters agencies shuttling the War Department directive back and forth from A–4 to A–1 and again to A–4. She noted later:

[1] (1) First Wacs to the field were in AWS units. While AAF WAC strength was greater in the field, ASF was greater if training centers were included. (2) The chief source of reference for this chapter is the AAF WAC Hist, by Lt. Col. Betty Bandel. (See Bibliographical Note). On the basis of 1945 reports, Colonel Bandel estimates 42,000 women as the AAF WAC top, including those in pipeline from training centers or to overseas theaters. Later AAF statistics set the actual top at about 40,000. Statistics used here do not always agree with later TAG re-evaluation of reports, but were those available to the AAF in its wartime planning. See Table 9: "WAC Asgd to AAF," and Table 13 (for 1946), in AAF Statistical Digest, World War II, prepared by Office of Statistical Control, AAF Hq. Hereafter cited as AAF Stat Digest. On file at U.S. Air Forces Historical Division, Air University, Maxwell Air Force Base, Alabama. Hereafter referred to as USAF Hist Div. See also Table 4, Appendix A.

To a headquarters wrestling with the problem of throwing thousands of fighting men and aircraft around the world, the entire WAAC program must have seemed at that moment a troublesome gnat buzzing around the head of a giant.

Promptly snaring the gnat on its return trip through A-4, the Air WAC Officer worked out a series of directives and accounting procedures which allowed the program to continue without interruption and which eventually led to full integration.

Acceptance of Wacs on Airfields

Wacs arriving on air bases ordinarily reported an enthusiastic reception. One company commander noted:

> The friendly atmosphere and co-operative spirit prevalent throughout this post have made the Wacs feel that they are a definite part of the military life here. Everything is being done to further the comfort of the group.[2]

By mid-1944, Wacs became even more popular when, the supply of infantry replacements having proved insufficient, the War Department required the transfer of thousands of combat-fit men from the Air Forces to the Ground Forces. Airfields thus lost many of the specialized medical and technical personnel upon whom they had depended for normal operations, and Wacs became increasingly sought after as replacements.

Receptivity was also greatly increased by the official action of General Arnold and his staff, whose policy from the beginning was to neglect no measure that might impress upon the public the AAF's real need for Wacs and its cordiality toward them. When the Air-WAC recruiting campaign was launched in October of 1943, General Arnold issued a public statement that "members of the WAC have made an enviable record through their work at Air Force installations."[3] In November the Deputy Chief of Air Staff, Brig. Gen. William E. Hall, journeyed to Philadelphia to see the first all-Air-WAC company sworn in. In December Maj. Gen. Barney McK. Giles, the Deputy Chief of Air Staff, wrote to all Air Wacs that "it is the record you have made which has convinced the AAF of the great value of Wacs."[4] In the same month the AAF declared a national "Air WAC Week" in which airfields everywhere honored their Wacs at retreat parades and other ceremonies. Upon receipt of evidence that military personnel were involved in the slander campaign, General Arnold denounced such undercutting of his policy in an angry letter to the field.[5] With such constant and public declaration of policy, the Army Air Forces came to have the reputation in the eyes of the public and of the Wacs themselves of being friendly toward the employment of Wacs, and appreciative of their services.

The Air WAC Division

This attitude showed itself in AAF headquarters by a willingness to consider favorably the recommendations of the Air WAC Officer. A working WAC staff organization was set up which, in Lt. Col. Betty Bandel's opinion, was near-perfect under existing conditions. The Air WAC Division was granted a strength of six officers plus clerical help, and was eventually placed in the office of the Assistant

[2] WAC CO, Mather Field, Calif. AAF WAC Hist, p. 20.

[3] WDBPR Press Release, 20 Oct 43.

[4] AAF WAC Hist, pp. 62-64.

[5] Ltr, CG AAF to all Air Forces and Comds, 22 Feb 44. WDWAC 341 (3-16-44).

Chief of Air Staff for Personnel (A–1). Although in Colonel Bandel's opinion a location in the Special Staff would have insured easier co-ordination with other offices, in actual practice the policies and personal assistance of Maj. Gen. James M. Bevans and his staff in A–1 made for almost equal freedom and efficiency of operation.[6]

The generally co-operative spirit emboldened the Air WAC Division to divest itself at once of all operating powers, confining its activities to study of field conditions, formulation of policy recommendations, and co-ordination of the WAC program. It was able to take this action chiefly because each major operating division of AAF headquarters early accepted the assignment of a WAC specialist to perform the operating duties, either full-time or in addition to other duties. These offices included AAF's Military Personnel Division, both Officer Branch and Enlisted Branch; the Assistant Chief of Air Staff for Training; headquarters and regional offices of the Air Inspector; the Air Surgeon; and the Air Provost Marshal.

Together, representatives in these offices formed a good example of what was known to Wacs as the Tel-a-Wac system, reputedly faster than telephone or telegraph. The Air WAC Division was seldom left in ignorance of developments in the larger AAF program, and in return the various division chiefs, to whom the other WAC officers concerned owed primary allegiance, were seldom caught short by higher authorities concerning knowledge of WAC matters for which they were responsible. Policy papers were originated either by the Air WAC Division or by the division that discovered the need for them, and were mutually co-ordinated before publication. A headquarters regulation required all publications mentioning the WAC to be co-ordinated with the Air WAC Division. Offices were also reminded in writing that, in drafting all publications, they should remember that AAF Regulations were equally applicable to Wacs, unless specifically excepted, and that if any explanatory passages were needed they should be added.[7]

AAF "Firsts"

Under this system, a notable series of AAF "firsts" resulted, to which that organization was not backward about calling attention. The Air Forces was first in full integration and abolition of WAC grades; first to attempt branch recruiting; first to require WAC inspectors at all command levels employing Wacs. It was the first to make a woman a lieutenant colonel, and was stopped from going further only by Congressional limitations; first to request the employment of WAC officers in non-WAC or "operational" jobs; first and only domestic command to propose extensive improvements in the WAC uniform, and its recommendations in this field were the ammunition with which Colonel Hobby finally won many arguments. It was the first major command to allow its own insignia to be worn by Wacs and to recommend that they be detailed in the Air Corps or the appropriate arm or service, a move that did much to make the women feel an accepted part of the Air Forces.

The AAF was the first to admit enlisted women to all men's noncombat schools, including some previously deemed unusual for women; first to propose, even if unsuccessfully, that women officers be

[6] Before its assignment to A–1, the Air WAC Division had been briefly in A–4 (ACofAS for Matériel, Maintenance, and Distribution).

[7] AAF Hq Office Instruction 5–13, 29 Nov 44.

allowed to go to men's noncombat officer candidate schools if they were not intended as WAC administrators; the first major command to support a school for the advanced training of WAC troop officers. It unsuccessfully recommended that Wacs be sent to the Army's School of Military Government as a preparation for duty in occupied areas. It was the first to bring to the War Department's attention the need for interpretation of certain discharge regulations, and for establishment of a system of maternity care for discharged women. It was the first major command to supervise the overseas placement of its own women by studying the needs of its components overseas and guaranteeing to fill them at its own expense. As a result, the Eighth and Ninth Air Forces in England, and to a lesser degree the air forces in Italy, the Pacific, and India, received well-selected, Air Forces-trained Wacs in numbers greater than the War Department had proposed to supply them.[8]

The AAF did not share other commands' distaste for specialized explanatory publications in any matter in which Colonel Bandel reported widespread misunderstanding in the field. The Air Forces was first to publish a circular defining the WAC company commander's job, and thus to end the friction and misunderstanding prevalent over that point in other commands where section chiefs claimed the commander's normal powers.[9] Another directive cautioned local commanders against placing enlisted women on commutation of quarters and rations or ordering them to stations where no housing existed. In six separate publications the Air Forces discussed all phases of successful selection of women for overseas shipment.

After a series of poor selections of WAC officer candidates, AAF headquarters amplified the War Department regulation and spelled out a precise procedure, including the often-neglected provision that selection boards would contain one WAC officer—the AAF made it two—and that rating lists based on merit alone would be maintained at Air Command level, not lower. Likewise, the AAF published a WAC inspection manual, listing standards which should be expected in such matters as women's medical care, uniform and appearance, leadership, and the commander's handling of marital and social problems. Similarly, the AAF included in its personnel officers' handbook a chapter on WAC administration.[10]

AAF headquarters also permitted the

[8] In order: (1) Speech, Maj Bandel, Min, Stf Dirs Conf, New York, 1-3 Dec 43. SPWA 337 (11-10-43). (2) AAF WAC Hist, pp. 12, 49, n. 1, 67, 84; also AAF Ltr 35-83, 29 Apr 44, and 21 Aug 44, pp. 65, 85, 86-87. (3) Memo, Exec WAC for CG AAF, 22 Feb 44, returned, 9 Mar 44, by CG AAF for further study. (4) See overseas chapters below, particularly Ch. XXI.

[9] AAF Reg 35-44, 10 Nov 43, revised and amended 14 Sep 44, 18 Oct 44, and 19 Jan 45: "WAC squadrons will be commanded by WAC Officers, who will have administrative authority and duties analogous and equivalent to those of a company commander. . . . The WAC squadron commander is charged with housing, messing, supply, discipline, training, recreation, morale, well-being, maintaining WAC squadron records, and rendering necessary reports. . . . Recommendations for changes of status, including promotion, reduction, transfer, discharge, etc., may originate with the Officer in whose section or unit the Wac is working, or with the WAC squadron commander. In all such cases the recommendations must bear the indorsements both of the Officer in whose section or unit the Wac is working, and of the WAC squadron commander."

[10] (1) AAF Reg 35-49, 13 Jan 44. (2) AAF Ltrs: 35-119, 3 Aug 44; 35-152, 6 Nov 44; 35-175, 27 Dec 44 and 14 Feb 45; 35-175*a*, 6 Mar 45; 35-47, 6 Apr 45; 35-65, 12 Sep 44; 35-89, 9 Apr and 28 May 45. (3) AAF Manual 120-2, WAC Insp Manual, Hq AAF, 10 Apr 44. (4) AAF WAC Hist, pp. 45, 46, 67, and 84; AAF Reg 120-12, 25 Dec 43.

Air WAC Division to send an informal monthly mimeographed letter to its staff directors, explaining the reasons behind current actions by headquarters. The recipients of these letters were responsible for passing on this information to their WAC company commanders. This service was so popular with Wacs in the field that non-Air Forces staff directors overseas asked to be included on the mailing list, stating that they otherwise did not see even War Department regulations for months, if ever; and non-AAF company commanders set up regular correspondence with AAF WAC squadron commanders to learn what they could, stating that War Department circulars were seldom otherwise available to company commanders.

Flying Jobs for Women

In the field, the Air Forces' progressive attitude was demonstrated chiefly by an equal lack of inhibitions in the assignment of women to new and unconventional jobs. No AAF schools were barred to women except combat schools, and no AAF jobs for which they could qualify, however unusual for women. It was not even the Air Forces' intention to exclude women from the most extreme masculine province: its flying schools and assignment as pilot.

Women quite early had been hired as ferry pilots, on a Civil Service status. Air Forces headquarters had planned that this group, first called the WAAF and later the WASP (Women Air Service Pilots), would be placed in the Army as part of the WAC as soon as Congress had made the WAAC a part of the Army. Such an addition was agreeable to WAC authorities, including both Colonel Hobby and the Air WAC Officer, since the number of women pilots was relatively tiny—some 800 as against Colonel Bandel's 40,000 Air Wacs—and could have been administered without additional expense. However, the proposed merger was blocked by the Director WASP, Mrs. Jacqueline Cochran Odlum, who recommended that women pilots not be grouped with other Air Corps officers, but given a separate military corps on equal status with the WAC, with a director equal in rank to Colonel Hobby. Legislation to this effect was attempted, but failed, and Congressional sentiment against the use of women as pilots was so strongly stirred up that the Air Forces was forced to disband the WASP without ever giving its members military status and benefits. Some WAC officers and enlisted women were also found to be qualified pilots, but it was deemed unwise ever to assign them as such in view of the emphatic nature of the Congressional decision.[11]

Enlisted flying jobs for Wacs, other than that of pilot, were never forbidden, although no such jobs had been considered by early planners because of the greater need for women elsewhere. As soon as assignment was turned over to local authorities, it proved impossible to keep a few women out of certain flying duties for which they had peculiar qualifications. Within two months Mitchel Field reported that it possessed the first two "flying Wacs," radio operators who were participating in B–17 training flights. This

[11] (1) Memo, CofAS for G–1, with Ind, WAAC Hq, 17 Apr 43. SPWA 314.7 (1–7–43) sec 1. (2) Memo, Arnold for Marshall [Jun 43], with reply from G–1, 20 Jun 43. SPWA 324.5 CAP (3–25–43). (3) Hist Recs Rpt 319, History of ATC: Women Pilots in ATC. USAF Hist Div. (4) For further discussion, see Summary of WASP Controversy, Appendix D.

precipitated scores of requests from other Wacs for flying duty. A few more WAC radio operators, mechanics, and photographers were soon assigned to regular runs. Several such Wacs actually received Air Medals, including one in India for her work in mapping the Hump, and one posthumously after the crash of an aerial broadcasting plane.

Both the women and the airfields concerned were ordinarily so pleased with themselves for accomplishing flying duty assignments that Air Forces headquarters, when it discovered them, published a directive authorizing such duty provided that the flights concerned were not for purposes of combat training, and that Wacs did not replace any man who might be receiving combat training on these flights.

The Air WAC Officer, while not discouraging such duty for women well suited for it, did not encourage any extension of it, or widespread training of women for it. Her opposition was due partly to the difficulty of housing a female crew member when nonscheduled landings were necessary, and partly to the danger of public outcry if any Wac was assigned to a position that even remotely involved combat training. Later in the war the Air Transport Command repeatedly proposed to use Wacs to replace male flight clerks on scheduled passenger runs. This was at first discouraged because of the irregular housing arrangements involved, but was eventually approved on a small scale and proved successful in flights such as those from Paris to London.[12]

Even including these flight clerks, the number of Wacs on flying duty, wearing wings and drawing flying pay, was never great. Only twenty Wacs held the Specification Serial Number of Air Crew Member, and flying radio operators and other technicians were equally scarce.

This comparative absence of a relatively exciting and popular type of Air Forces duty was one factor with which commanders of female troops had to reckon in maintaining morale. Many air-base commanders, recognizing the difficulty, directed that Wacs be allowed to go on local flights as passengers where space permitted, or for at least one orientation flight upon arrival at a station. Nevertheless, most AAF Wacs seldom saw the inside of a plane, and many were in the unfortunate position of Air Transport Command Wacs in Scotland and Wales, who were shipped overseas by boat and complained that they knew there was an outside world only by remarks dropped by transient men. Because of the value of occasional flights to women's morale, the Air WAC Officer secured a directive making clear that women were to have the same privileges as other military personnel in "hopping rides" or flying on military orders.[13]

Specialist Training

AAF headquarters from the beginning showed enthusiasm for training Waacs in other technical and mechanical jobs which were not ordinarily considered a woman's field. An elaborate training system was originally planned, with 8,000 women a month to be fed into AAF schools for training as everything from

[12] (1) AAF WAC Hist, pp. 19 and 41. (2) AAF Reg 35-45, 12 Nov 43. (3) WDBPR Press Release, 14 May 46. (4) Memo, AAF for Dir WAC, 30 Aug 45. WDWAC 319.1 AAF.

[13] (1) For list of Specification Serial Numbers see AAF Stat Contl Rpt SC-PS-123, sub: Strength of AAF Pers in WAC Within Continental U.S. by Comd, SSN, and Race. (2) Hist Recs Rpt 315*c*, Wacs in European Division, ATC, June 1944-August 1945, by European Division, ATC. USAF Hist Div.

TRAINING AT LOWRY FIELD, COLORADO, *in February 1943. Waacs learning camera operation, above, and making photographic mosaics, below.*

armorers to weather forecasters, in addition to 5,000 a month to be received already trained in WAAC schools as clerks. This plan was very similar to the system actually used on a smaller scale by the Naval Air Service, which determined in advance to employ set numbers of Waves as Link trainer instructors, control tower operators, and other specialists, and trained them accordingly. The Army Air Forces carried the plan so far as to set up Wac classes at the AAF Photolaboratory School at Lowry Field, which trained several all-WAC photolaboratory companies. The AAF also earlier encouraged the WAAC to set up its radio training school at Kansas City, and contracted to employ all graduates.[14]

However, as women began to arrive, the AAF was surprised to discover that the majority of women, unlike the 17- and 18-year-old boys who were being inducted, had a usable job skill before they entered the Army. To employ the average Wac successfully, it proved necessary only to place her, fresh from basic training, on almost any airfield, where, if not immediately snatched into too many pieces by competing section chiefs, she ordinarily soon found useful employment. The number of recruits arriving was never great enough to make placement a problem.

The Air Forces therefore soon dropped all but a few of its formal technical training plans for women, and adopted a system which was believed to be the most natural, speedy, and economical under the circumstances. Most AAF Wacs finishing basic training were shipped immediately to air bases having need of personnel in their civilian skills and were assigned to immediate duty or to on-the-job training, which was believed less wasteful of time than the average technical course. If a station needed a specialist it was unable to train, or if a Wac showed such aptitude that higher specialized training was clearly indicated, the station was authorized to send her to any noncombat AAF school that could arrange to house her. In such cases the station ordinarily had to take the trainee back upon graduation and employ her in the new skill; training was thus seldom employed merely to gratify individual desires for unnecessary training or to get rid of troublesome individuals.

Unfortunately for the compilation of statistics, the attendance of Wacs at AAF technical schools was so well integrated and came to excite so little comment that it was later impossible to determine exactly how many women had attended each course and how their record compared with that of male students. It was known only that approximately 2,000 women successfully completed courses in AAF technical schools, including weather observers, weather forecasters, electrical specialists of several kinds, sheet metal workers, Link trainer instructors, cryptographers, teletype operators, radio mechanics, control tower specialists, parachute riggers, bombsight maintenance specialists, clerks, airplane mechanics, photolaboratory technicians and photo-interpreters, chaplains' assistants, and physical and military training specialists. Lt. Gen. Barton K. Yount, commanding general of the AAF Training Command, wrote, "Their record of accomplishment reflects great credit upon the Women's Army Corps and the Army. . . . Their contribution to the Training Command's mission has been invaluable."[15]

[14] AAF WAC Hist, pp. 41-42.

[15] Ltr, Gen Yount to Dir WAC, 5 May 45. WDWAC 201.22.

Conventional Clerical Jobs

At the peak of enrollment, in January of 1945, it was estimated that about 50 percent of AAF Wacs in the United States held administrative or office jobs, being assigned as typists, stenographers, and various sorts of clerks.[16] This was a percentage of women in conventional jobs considerably smaller than that in most other commands or overseas theaters, where the numbers in "feminine" work ranged as high as 90 percent in commands such as the Eighth Air Force in England. The work of such Air Wacs did not ordinarily differ from clerical work in other commands, except that it required a knowledge of Air Forces nomenclature, organization, and administrative procedures.

The different air commands differed slightly from each other in certain uses for these clerical workers. The largest, the AAF Training Command, with its hundreds of flying training schools and technical training schools, found Wacs particularly useful in keeping the complicated records of trainees, training courses, and flight hours. The domestic air forces—First, Second, Third, and Fourth—likewise used Wacs to keep their personnel and flight records, and also to staff the assembly lines by which crews and equipment were processed for overseas shipment. The Air Technical Service Command used WAC clerks at its large depots to keep stock records of the thousands of items of Air Forces technical equipment. The Air Transport Command, running the Army equivalent of a commercial airline system, used numbers of Wacs at its information desks, in its dispatching offices, and to process passengers and supplies through aerial ports of embarkation. Toward the end of the war, the new AAF Personnel Distribution Command used Wacs to process personnel in replacement depots, redistribution centers, and convalescent centers.

Army Air Forces headquarters itself lagged behind its commands only because of the War Department ban on bringing Wacs into Washington. After repeated efforts to circumvent this policy, an opening was found in December of 1943 by the authorization to bring in twenty WAC messengers to handle secret and confidential mail, which had been repeatedly going astray under civilian operation. These were soon followed by weather observers, statistical machine operators, and others to a total of several hundred, housed at Bolling Field and working in the Pentagon.[17]

Technical Assignments

Technical jobs tended to increase in number toward the end of the war, as it became increasingly difficult for the AAF to get enlisted men in the top AGCT brackets necessary for some of the work. An especial competition for women of high ability developed between Air Forces agencies, which were frequently hard put to find men of suitable qualifications for technical work who were not immediately lost to officer candidate school or needed for overseas stations.

Weather Observers

AAF's Weather Wing, which sought Wacs long before it ever saw one, finally obtained a quota of 500. A few were sent

[16] Rpt cited n. 13(1). Administrative jobs totaled 46 percent, plus undoubtedly some of the group marked "Duty."

[17] AAF WAC Hist, p. 66.

to the regular Weather School, but most were trained on the job, five or six Wacs taking over the weather station at air bases where they could be housed with a WAC company. The majority of such Wacs were weather observers; only a few were ever trained for the more complicated job of weather forecaster, for which there was no shortage of men. The quota of 500 was never entirely filled in spite of all efforts.[18]

Radio and Cryptographic Work

The Army Airways Communications System (AACS), which had a quota of 1,450 and which competed for high-type personnel, chiefly sought radio operators. At the height of a shortage of this personnel, AACS discovered that several hundred WAAC radio operators and mechanics had been shipped under the old T/O system to airfields that did not need them. An individual search by name was instituted to unearth them. Some women were found suitably assigned to air-base radio stations, or as code instructors in schools such as the AAF Navigation School at Hondo, Texas, but others who were doing clerical work were carried off by AACS. These were assigned in groups of six or eight to operate AACS radio stations at airfields where a WAC unit was located. Most female radio operators recruited by the AAF were also given to AACS; when its needs still remained unfilled, AACS was also assigned Wacs without radio skills but of high intelligence and promising background. At a special AACS screening center, these women were tested and those with any radio abilities were sent to intensive training courses. Those with college or research backgrounds were trained as cryptographers, for which women proved especially suitable because of the exacting and monotonous nature of the work.[19]

Control Tower Operators

Those with good diction and good nerves, such as former teachers and telephone operators, were trained by AACS as control tower operators, after a brief dispute concerning whether females should be given a position of such responsibility. Even the advanced British services did not at this time use women in this work.[20] Some objectors feared that a woman's voice would not be audible, that women would become hysterical in emergencies and be unable to give the necessary landing directions, or that women's training as control tower operators would interfere with men's training. One AACS region, in fact, put out a directive that "under no circumstances will WAAC personnel be permitted to operate the microphone in an airways station or airdrome control tower, or to operate on the net in an airways station." AACS headquarters quickly overruled the regional headquarters, stating that its need for personnel would require all qualified women as well as men.

From the end of 1943 onward, Wacs served successfully in these positions. Pilots seemed to experience no difficulty in following women's voices, and while no major emergencies occurred to test WAC operators' self-control, in several minor

[18] *Ibid.*, pp. 12, 42, 66–67. Appendix shows 242 WAC weather specialists.

[19] *Ibid.*, pp. 13, 31–32, 66.

[20] Memo, Sgt Goodwin for Col Hobby, 17 Apr 44, re inquiry from Air Chief Marshal Frederick W. Bowhill. WDWAC 211 Control.

crashes they rerouted aircraft and directed rescue apparatus in a manner that received praise. Colonel Bandel later reported, "Wacs love the work and have been highly commended. There seems to be no difference between men and women in adaptability to this work." [21]

The only difficulty experienced by AACS and Weather Wing in any of these technical assignments was a purely administrative one: they considered their Wacs, like their men, to be high-grade specialists who should perform no kitchen police or other duties for the unit with which they lived, whereas the WAC units, being smaller than men's units, seldom appreciated such exalted boarders, and demanded that these women share company rules and company duties. When differences threatened to result in eviction, both AACS and Weather Wing acquired WAC staff directors, whose chief duties were to travel continuously, persuading AACS supervisors that specialists could without damage be housekeepers, and convincing WAC commanders that some shift workers on exacting technical work might properly be given a few concessions.[22]

Business Machine Operators

In 1944 another shortage developed, that of skilled operators of business machines for AAF's statistical control units throughout the United States. Available trained men were required for mobile overseas units, and the need eventually became such that priority of assignment was given for 1,000 Wacs for statistical control work. Since few women skilled in this work could be recruited, women with suitable intelligence and aptitude were chosen and sent to a six-week intensive training course at the AAF School at Orlando, Florida. Women proved especially apt at the work, although many complaints were received concerning its monotony and the refusal of statistical control units to vary the duties or permit transfer or rotation between stations.[23]

Link Trainer Instructors

Another popular technical job for Wacs was that of Link trainer instructor. As early as November of 1942, AAF headquarters had proposed the use of Waacs in this work, since "Men suitable to act as such instructors are rapidly disappearing into officer candidate schools." Men qualified to do the work often were pilots or potential pilots themselves, and were not content to remain as enlisted instructors, once trained. A number of Wacs selected by local authorities were therefore trained in the men's training school for Link trainer instructors, and others learned the work on the job; the duties proved both successful and satisfying to the women.[24]

Mechanics

Had hundreds of thousands of Wacs been available, the AAF had planned to train many women as mechanics, but since all recruits obtainable were needed for other jobs, and since the work was less suitable because of women's lesser physical strength, no particular attempt was made to train large numbers of them. Nevertheless, several hundred women

[21] Memo, Air WAC Off for Asst Chief WAC Br G-1, 19 Apr 44. WDWAC 211 Control.

[22] Ltr, forwarded to Hq AAF by 1st Ind, Hq AACS, 4 Apr 45. AACS 220/1.

[23] AAF WAC Hist, p. 67.

[24] *Ibid.*, p. 13.

with civilian skills, requiring no further training, were assigned to the work. One station, by way of experiment, successfully set up an entire flight line staffed by some sixty WAC mechanics.

Some WAC staff directors were extremely doubtful about such work because of what they termed the coarsening effect on women of the average conversation encountered from men on most flight lines. However, the assignment was the only economical one for skilled women from civilian employment in the aircraft industry, and such women usually displayed a keen interest in assignment in their specialty and in the advanced mechanical training open to them, regardless of its effect on their vocabulary.

Medics

Some 6 percent of Air Wacs were medical specialists. Medical companies for larger AAF hospitals were requested as early as April and May of 1943, eighteen months before the ASF adopted WAC general hospital companies, but these had to be refused at the time for lack of trained WAAC personnel. As a result, no formally organized WAC hospital companies were ever employed by the AAF, which instead used whatever workers were available—clerks, receptionists, orderlies and medical technicians—as part of regular squadrons. These women were trained on the job instead of receiving special courses—a system that proved inferior to school instruction only in that it was necessary for inspectors constantly to prod hospital authorities to continue upgrading Wacs instead of leaving them as untrained orderlies. Nevertheless, twelve different medical Specification Serial Numbers came to be held by AAF Wacs.[25]

Aircraft Warning Service

Another group of specialists, veterans in length of service, were the approximately 1,000 women, remnants of the Aircraft Warning Service, who continued to perform duties with Fighter Command stations along the eastern seaboard. Theoretically these duties were clerical; actually, until near the end of hostilities the Wacs also continued to help in operations and filter centers when civilian volunteers defaulted. At the AAF School of Applied Tactics at Orlando, Florida, 500 more AWS veterans assisted in the aircraft warning portion of the school's training function. Women proved particularly suitable for this work, which was both monotonous and exacting, but as the danger of aerial invasion became more and more remote, women began to complain of fatigue after two or more years at this work, and to request transfer to other duties, which was seldom forthcoming.

Miscellaneous

In addition, some 5 percent of Air Wacs were drivers; 4 percent, supply experts; 3 percent, photographic technicians; 2 percent, telephone and teletype operators. Lesser numbers were radar operators, armament specialists, carpenters, dietitians, and interpreters. College women and others with responsible civilian experience were frequently assigned as librarians, classification specialists, instructors, personnel clerks, vocational advisers, occupational therapists, reporters and editors, and historians. The AAF's flexible system of assigning Wacs made it possible to fill special needs or to place un-

[25] *Ibid.*, pp. 13, 36.

usual skills found in only a handful of women: these included topographers, cartographers, sanitary inspectors, geodetic computers, chemists, and many others even more unusual.[26] A former watch repairwoman was trained in secret bombsight maintenance; six photographic airbrush artists illustrated technical manuals at Patterson Field; Russian interpreters found employment at ATC's Montana terminal. In California, the WAC's only dog trainer found that the AAF could also provide dogs to train.[27]

Officers

WAC officers assigned to the Army Air Forces were admitted to more than sixty different types of jobs in addition to that of company officer.[28] These were all full-fledged officer jobs; as soon as WAC officers arrived in AAF headquarters, a written directive made it clear that they were to replace commissioned officers only, and specifically not warrant officers or civilians.[29]

The majority were given administrative jobs in the fields of personnel, office management, mess, supply, and finance. Numbers were also found useful in jobs that involved dealing with people—special services, information and education, and public relations. A few WAC officers by the end of the war had worked up to positions of some responsibility; there was one air-base judge advocate, one inspector general, one base executive, and thirteen who held the coveted Specification Serial Number 2260, personnel staff officer.

Grades and Ratings

The wide scope of Air Forces jobs open to Wacs, both officer and enlisted, could be attributed chiefly to the AAF policy of full integration and the resulting abolition of all separate WAC grades, allotments, and Tables of Organization. An AAF Regulation stated, "Practices which tend to apportion grades and promotions by sex rather than by actual duty assignment are undesirable, as they serve to impair efficiency and morale." [30] Actually, when the total was reckoned, men were found to be far more highly rated than women, for most AAF Wacs had come on the scene in 1943 and 1944 when most high grades had already been distributed. Air Forces men in the United States still had four times the percentage of master sergeants' ratings that women did; twenty times the percentage of technical sergeants; ten times that of staff sergeants, and twice that of sergeants. Among officers, men had five times as great a percentage of field grade officers as did women.[31]

It was Colonel Bandel's opinion that there was no reason why late-arriving Wacs should have had higher ratings than the men who arrived in the same years, and that Air Forces procedure "resulted in greater freedom in utilizing Wacs and men doing similar jobs," even though in fewer guaranteed ratings for women.[32]

Toward the end of the war, officers of AAF's Military Personnel Division became concerned about the fact that WAC

---

[26] Memo, Chief, Classif and Rcpl Br AGO, for Dir WAC, 24 Apr 44. SPX 220.3 WAC (11-30-43) OC-II, in WDWAC 201.6.

[27] See Table 5, Appendix A, for division by occupational groups of total enlisted AAF Wacs in zone of the interior, 31 January 1945. The complete breakdown for both enlisted women and officers by SSN, race, and air command is given in the report cited n. 13(1).

[28] For a condensed summary of jobs held by AAF WAC officers in the United States, 31 January 45, see Table 6, Appendix A.

[29] AAF Hq Office Instr 35-25, 25 Sep 43.

[30] AAF Ltr 121-48, 29 Mar 45.

[31] AAF Stat Digest, Tables 9, 10, and 15.

[32] AAF WAC Hist, p. 34, n. 1.

officer grades included more first lieutenants and fewer second lieutenants than the men, and wished to forbid promotion of any more female second lieutenants. Air WAC Division killed this idea by pointing out that about 1,700 of the AAF's worldwide total of 1,900 WAC officers had come to it in 1943. After this, the WAC officer candidate school was virtually stopped, and while the supply of newly commissioned male second lieutenants continued, few more female second lieutenants were received. The WAC officers had received no more promotions than men who came to the AAF in 1943. Air WAC Division again pointed out that "separate" grades for women would have led to injustice.

Overseas Assignment

In 1944 several air commands received the hitherto unprecedented authority to move Wacs to overseas stations without reference either to the War Department or to AAF headquarters.[33] The AAF's Air Transport Command was the first to win permission to move its Wacs overseas as it moved its men, without reference to usual staging methods.[34] Air Transport Command had soon picked up Wacs from domestic stations, staged them, and set them down in Hawaii, Alaska, Bermuda, Labrador, England, Scotland, Wales, France, Africa, and India. Army Airways Communications System and AAF Weather Wing followed suit, attaching their Wacs to any convenient ATC unit. The several thousand women in these units, which remained under the command of the Army Air Forces, were in addition to some seven thousand Wacs furnished higher headquarters of combat air forces overseas.[35]

Such freedom of movement did not of course apply to combat units; AAF headquarters was kept busy refusing the requests of fighter and bomber groups trained in the United States who desired to ship out complete with two or three borrowed Wacs to whom they had become accustomed. After the defeat of Germany the U.S. Strategic Air Force was given permission to ship out its Wacs as it pleased, in its program to swing the bulk of its air power to the Pacific. This was the only combat organization to receive such authority and only a few women were ever shipped out in this manner.

Conclusion of Program

The high point of the AAF WAC program was reached in January of 1945, with a peak strength of almost 40,000 women. In Colonel Bandel's opinion there were few remaining difficulties in the employment of womanpower which the AAF had not solved.

Her final report listed only four remaining problems. One of these was the housing situation. As a matter of economy, initial mistakes in design could never be corrected once barracks were built, although both AAF inspectors and surgeons repeatedly objected. It was found that permanent employees like the Wacs suffered greatly in health and efficiency from being housed for two or three years in crowded

[33] Granted 10 Jun 44. AAF WAC Hist, pp. 74-81.

[34] Granted November 1943, ahead of blanket authority cited in preceding note. First ATC unit overseas was one to Hawaii, March 1944.

[35] WAC Pers with Air Forces Overseas, Jan 45, Table 28, AAF Stat Digest, 1946:

European Theater	2,835
Mediterranean	457
Pacific Ocean Areas	2
Far East Air Force	694
China-Burma-India	287
Air Transport Comd	2,755
Other (AACS, Etc)	285
Total	7,315

and flimsy temporary barracks with their hand-fired coal stoves, designed primarily for transient combat trainees. Another unsolved problem was that of the selection of WAC officers. The AAF felt that the War Department system of sending all candidates to a school for WAC troop officers was outdated, and that women should have been selected and trained as men were, in men's schools, except for those who would actually administer Wacs. Other unsolved problems were that of friction with civilian women workers, and that of disputes over fraternization of officers and enlisted personnel of opposite sexes; these could not be resolved within the limits of Army organization in World War II.[36]

None of these remaining problems was, in Colonel Bandel's opinion, a serious threat to the efficiency of the AAF WAC program. She concluded:

> There can be little doubt that women did contribute many critically needed administrative and clerical skills—not generally possessed by men—to the AAF at a time when there was a very great need for such skills. . . . The WAC program in the Air Forces during World War II was a part of the natural evolution toward the full employment of a nation's manpower during a modern war.[37]

[36] See Chs. XXV and XXVII, below.

[37] AAF WAC Hist, pp. 94, 96, 97.

CHAPTER XVII

The Army Ground Forces

Although equal in echelon to the Army Air Forces, the Army Ground Forces offered a less favorable field for the employment of womanpower.[1] Ground Forces units were chiefly tactical in nature, depending upon the Army Service Forces for supply and other services. Except for its schools and replacement depots, the AGF had few permanent installations under its jurisdiction in the United States. There was some early debate within AGF headquarters as to whether Waacs could not be used to staff the rear echelons of separate corps, thus giving "a continuity of administration in the event a Corps is ordered overseas," but this idea was not approved. In general the WAAC was regarded as part of the Services of Supply, even when a few units were attached for duty at AGF installations.

It therefore came as an unexpected problem when, on 1 May 1943, the War Department formally assigned to AGF and other commands the units previously on loan from the ASF. The standard ASF Table of Organization units were not particularly suitable for the needs of the AGF, since most routine services were performed for it by the ASF, and only highly skilled typists and stenographers were wanted.

Upon her assignment as AGF WAC Officer during the period of conversion to Army status, Maj. Emily E. Davis visited all AGF WAC units and found that the only one with good morale was that at Fort Riley, which had been sent out after the end of the T/O system and thus fitted the needs of the station. Others suffered from malassignment and underutilization, and also had not received full uniform issue or been able to get salvage or repair of clothing. Some units were found to lack adequate housing and dayroom space, and had no recreational or special services program. Two units were under the command of WAC officers whom Major Davis pronounced "very weak" and immediately replaced; in three months these commanders had not made the most elementary arrangements for the welfare of the units.

For the remainder of the war the employment of Wacs in the AGF increased slowly. Replacements for the 34 percent losses of the conversion period were not forthcoming until early 1944, and additional requisitions were at first few. AGF headquarters voluntarily limited the number of Wacs by refusing all requests from the District of Columbia or installations "which are near large cities where they can procure civilian personnel." Requests from installations such as the Desert Training Center were also disapproved as unsuitable. Negro Wacs were rejected, since it was believed that they could not

[1] This chapter draws heavily upon the AGF WAC Hist, by the AGF WAC Officer, Colonel Davis. All direct quotations in this chapter, unless otherwise indicated, are from this history.

FIRST OFFICER EMILY E. DAVIS, *right foreground, with Colonel Hobby and Colonel McCoskrie on a tour of Fort Des Moines in June 1943.*

be employed in sufficient numbers to justify a detachment unless the War Department reconsidered their use to replace civilians in the operation of kitchens, dining halls, and cafeterias. For these reasons, the total number of AGF Wacs was never to be very great, at its peak strength amounting to less than 2,000, one twentieth of that employed by the Air Forces or the Service Forces.[2]

Lt. Col. Emily E. Davis remained the sole WAC representative in Headquarters, AGF. Although her responsibility was parallel to that of the Air WAC Officer, smaller numbers of women were assigned to AGF, which never approved her request for an assistant in its own headquarters. At a later date a WAC staff director was assigned to the AGF's largest field command, the Replacement and School Command, and another to the Antiaircraft Command.

[2] (1) Ltr, Hq AGF to CofS, 1 Apr 43. GNGCT 320.2 WAAC. (2) For complete statistics on receipt of personnel by station by month, see AGF WAC Hist, Vol. II, App. XX. Peak strength in this chart appears as 1,950 women. See also Table 4, Appendix A.

WOMEN ASSIGNED TO THE ARMY GROUND FORCES *work alongside soldiers at Camp Davis, North Carolina, above; WAAC ex-school teachers outline courses for the enlisted personnel elementary school at Fort Sill, Oklahoma, below.*

AGF WAC Jobs

Wacs with the Army Ground Forces worked gradually into positions more diversified than the original assignments had promised. AGF requisitions continued to ask for women with a high degree of skill in some administrative or technical field, with the result that only 14 percent of its total enlisted women were in the two lowest AGCT classifications; the other 86 percent were in the top three groups, with 47 percent of enlisted women having the scores required of officers. These considerably surpassed the scores of Wacs in any other domestic command. A special arrangement was made whereby, if the AGF was accidentally assigned a Wac lacking in skill or aptitude, she was immediately reassigned to another command.

These women gravitated naturally toward a station's more responsible positions, especially since they were relatively more permanent employees than the enlisted men, not being subject to withdrawal for combat service. Wacs were finally assigned to every major training installation of the Army Ground Forces, both schools and replacement training centers: infantry, cavalry, field artillery, coast artillery, antiaircraft, tank destroyer, airborne, and armored force. They were also assigned to two overseas-replacement depots. Over 70 percent of the Wacs' jobs in these installations were clerical and administrative; others were technical, professional, mess, supply, or automotive.[3]

Type of Duty	*Percent*
Total	100.0
Administrative and Clerical	72.0
Automotive	8.9
Technical and Professional	8.2
Mess	5.4
Supply	4.8
Radio and Communications	0.7

Replacement Depot

One of the largest AGF WAC units was at the Ground Forces Replacement Depot, Fort Meade, Maryland, where over 400 women were assigned. To handle the tremendous volume of overseas replacements, a "belt line" process was established, by which each man's records were passed down a line where each checker took care of a single item. This process resulted in extreme accuracy but extreme monotony; it was performed on a twenty-four-hour basis, in shifts of from eight to twelve hours, depending on the load. It was reported that "WAC personnel became proficient in this type of operation, and as general-service men were released for overseas duty, they became the section heads of many of these processing departments."

In addition, many Wacs were used in training regiments in clerical capacities. WAC tabulating machine operators were in great demand by the Machine Records Unit, so much so that some showing aptitude were especially trained in the work, but the total number desired was never obtainable. Instead, many of the women recruited under the station-and-job plan with a promise of assignment to Fort Meade were so lacking in clerical aptitude as to be untrainable for the processing of papers, and 115 of them were used in the motor pool as dispatchers, drivers, mechanics, and attendants.

Antiaircraft Artillery School

In the Antiaircraft Command Wacs performed a variety of duties on the firing ranges where trainees were taught the

[3] Complete list of installations is given in AGF WAC Hist, Vol. I, pp. 47–48.

principles of antiaircraft defense. At some antiaircraft firing points the entire training operation, with the exception of technical supervision by male officers, was handled by enlisted women: WAC control tower operators kept the tow-target plane on the course, while other Wacs gave the fire signal over the field telephone, and WAC mathematicians computed the correct angle of fire and the accuracy of fire.

Other AGF Schools

At the Field Artillery School, a few Wacs were used as control tower operators to guide liaison planes. At the Armored, Field Artillery, and Cavalry Schools, Wacs were employed as radio mechanics, taking care of records and requisitions involving radio equipment, and repairing and installing radios in tanks, bantams, and other vehicles, both in camps and in bivouac areas. Other Wacs were radio operators and instructors in the Cavalry and Armored Schools, and in the communications school of the Field Artillery School, training men in code sending and receiving.

At the Parachute School, Fort Benning, Georgia, over one hundred women were utilized in the parachute rigging and maintenance sections. Women originally recruited for this work, under the station-and-job plan, were former textile mill workers and did not prove very apt, some 40 percent being unable to complete the training, with the average female rigger being able to rig only six chutes daily, as compared to ten rigged by the average man. Women were not allowed to jump with the chutes they had packed, as men were required to do, on the grounds that this was not an essential part of the job for a woman, but were allowed to fly with paratrooper trainees and watch them jump with chutes the women had packed. After station-assignment recruiting was discontinued and the requirements raised, women developed greater proficiency in this work and ultimately were considered "equally qualified with male personnel in rigging and better qualified in maintenance."

Women served also at various stations in such varied jobs as proofreaders on technical manuals, as photographic technicians, as sewing machine operators in a book stitching shop, as translators for the preparation of language courses, and as motion picture projectionists in training classes. One woman, as chief clerk of a department, was responsible for keeping track of its officer and enlisted personnel and 300 trainees, for checking the contents of all incoming and outgoing papers, for the computation of rations needed, and for maintaining records of completion of qualification courses. Another, a classification clerk, assigned all male cadre returned from overseas. One was personal secretary to two generals and a colonel. Another secretary prepared reports and took all testimony concerning cases brought before a board of officers.

Staff directors on field visits made a point of checking such job assignments against actual skills. While typists were usually well assigned, stenographers were frequently found to be dissatisfied, and staff directors reported, "The failure of staff officers to use stenographic ability persisted through the entire period of the war." Many complained of not being busy, saying, "I feel that the Wacs could handle much more work than is given them," or, "It is said there are five persons to each Army job and that statement seems believable." With these and a few other exceptions, the classification of enlisted women was, by 1945, pronounced "excellent" by the staff director.

Experiment With Mixed Tactical Units

The only official experiment with mixed tactical units in the United States was the early one made with antiaircraft artillery units under the actual command of the Military District of Washington. This experiment was made in spite of the objection of the Judge Advocate General that it would involve "assignment to combat units" and was thus probably illegal. However, the British experience in such mixed units had proved satisfactory, and the Antiaircraft Artillery (AAA) desired the action.[4] Civilian volunteers would otherwise have been required, and this involved difficulties of administration, discipline, control, and efficiency. Also, as the Army Service Forces noted, "it takes about ten volunteers to take the place of one soldier."[5]

No publicity was permitted on this project, partly because the unit was itself highly secret in nature, equipped with radar, height finders, fire directors, and other such apparatus, and partly lest WAAC recruiting should be harmed by mistaken notions that women were to fire guns or sleep in the men's barracks. The experiment actually offered little grounds for alarm, for the "tactical" unit in question appeared firmly rooted in the vicinity of Washington, D.C., and was "mixed" only during the working day, the Waacs enjoying their customary separate housekeeping at night.

In accordance with General Marshall's verbal instructions, two WAAC companies, totaling 10 officers and over 200 enrolled women, reported to the area in December 1942. With these, the AAA was directed to organize two composite batteries using women in at least 55 percent of the jobs, and to work out Tables of Organization showing exactly which jobs in AAA batteries, battalions, and regiments could be held by women, and whether the whole idea was workable.

The women concerned were carefully selected for their high AGCT test scores, although actually the AAA discovered that women of less ability could have performed certain routine jobs as well and felt less bored. They were never assigned a tactical mission, but did participate in the Eastern Defense Command's tactical problem. The women were not assigned to fire guns, nor were they given small-arms training. They were not assigned to outlying searchlight positions because of the isolation and the fact that the public would have observed the experiment. They were assigned to operate various instruments and machines, as well as to perform clerical duties, and the entire range section was completely manned by Waacs. The AAA reported, "WAAC personnel exhibited an outstanding devotion to duty, willingness and ability to absorb and grasp technical information concerning the problems of maintenance and tactical disposition of all types of equipment." Little administrative difficulty was reported, although separate housing was of course required, and there were separate dayrooms for men and women plus one for mixed groups.

---

[4] Cable 1101, 12 Aug 42, Eisenhower for Marshall, replies to Cable 2890, Marshall for Eisenhower, and gives data on use of ATS in AAA mixed batteries. Handwritten Ind, CofS, directs information be sent to G-1, G-3, Mrs. Hobby, Brig Gen Thomas H. Green, Maj Gen John T. Lewis, CG MDW. CofS 291.9.

[5] Memo, DCofAdm Serv for CG SOS, 18 Nov 42. SPAAS 320, Sp Coll DRB AGO.

At the conclusion of the experiment some months later, a full report was rendered to the War Department as to the possibilities of using women in such duties. The report concluded:

> WAAC personnel can be used in performing many of the tasks of the Antiaircraft Artillery. They are superior to men in all functions involving delicacy of manual dexterity, such as operation at the director, height finder, radar, and searchlight control systems. They perform routine repetitious tasks in a manner superior to men. . . . The morale of women used in the AAA was generally high due to the fact that they felt that they were making a direct contribution to the successful prosecution of the war.[6]

As a result, the AAA requested that it be allowed to retain WAAC personnel and be given ten times more, to the extent of 103 officers and 2,315 enrolled women.

Since the need for antiaircraft protection in the area had subsided, G-3 Division decided instead that the most profitable immediate use for Waacs was in overhead installations to replace combat-fit men. With Colonel Hobby's concurrence, G-3 therefore directed that the AAA units be dissolved, their personnel reassigned, and records preserved to permit reactivation of such units should future need arise. This action was taken, over the protest of the AAA. Records were filed for future use, and G-3 Division concluded that "The experiment which has been conducted of employing WAAC personnel in antiaircraft artillery units has demonstrated conclusively the practicability of using members of the Corps in that role."[7]

Problems of Full Integration

Except in the one tactical unit, considerably more difficulty was encountered in achieving full integration of women into the Army Ground Forces than was the case in the AAF. From the enlisted women's viewpoint, the most visible distinction was the matter of insignia, in which it was the Ground Forces policy for women to wear WAC insignia instead of that which would have been worn by a man in the same assignment.

A more important handicap to integration, in the AGF staff director's opinion, was her inability to secure a merger of personnel allotments. Although allotments from the War Department no longer were divided into male and female, AGF headquarters continued to divide them, setting aside for each WAC detachment the exact number of grades the old WAAC Table of Organization had specified. These were designed for an ordinary ASF unit with many unskilled workers, and were somewhat modest for the average highly skilled AGF unit, allowing only 5 or 6 in each unit of 150 to hold one of the first three grades, and requiring about 80 to remain privates. These grades, however modest, were the sole property of the Wacs and were not interchangeable. Stations soon protested this lack of flexibility. The Armored Replacement Training Center wrote:

> When we first received this company it was felt that our women were not qualified for the ratings in competition with the men, and our staff officers desired to give the ratings to men, which of course, we could not do. Now, with our expansion, the reverse is true . . . there are 41 privates in the WAC detachment, and we feel that most of these would be T/5's if all our ratings were pooled and they were in direct competition with the

[6] MDW Narrative Rpt, 10 Jul 43. 324.5 (151st WAAC Tech Co). Cited in mimeographed history, The Military District of Washington in the War Years (1942-1945), by William H. Cartwright, Jr. Copy in OCMH.

[7] Memo, G-3 for CofS, 7 Jul 43. WDGCT 291.9 WAC.

> men. . . . Eventually our most competent administrative personnel will undoubtedly be women; some thought should be given to this and the War Department should initiate some policy.

The AGF WAC Officer therefore recommended that no separate grades be allotted for women, except cadre, but that grades be open to men and women alike. AGF's G–3 did not concur, and its system was continued throughout the war, with the modification that a revision of the WAC allotment could be obtained by any station upon formal request to Headquarters, AGF. Concerning this, AGF's Replacement and School Command protested that "such a recommendation requires extra paper work and loss of time. Normally, changes will be made only when a number of desired changes have accumulated, and individuals deserving of higher grades suffer from the lapse of time." AGF Replacement Depot 1 sent a similar protest. These protests from the field convinced AGF's G–1 Division, but not G–3, which maintained:

> Since WAC personnel are in a more permanent status . . . and since they are used in large proportions in administrative job assignments where higher ratings tend to be concentrated, the proposed directive will tend to filter WAC personnel toward the higher grades and will exclude opportunity for male personnel returning from combat who may be qualified for such jobs. Although opportunity should be given all WAC personnel to qualify for promotion, it is believed that the present system should be continued as a means of control.

The separate allotments were therefore continued. This matter of ratings was a common cause of complaint among women. One wrote:

> Many Wacs have taken over positions vacated by enlisted men and are performing the duties well, but due to current directives cannot be promoted to the grade they are occupying. . . . I believe that a Wac should be given the grade which belongs to the job.

The actual result of this policy was the opposite of that expected. As large numbers of highly rated men were returned from overseas, all male grade allotments were soon exceeded and male promotions were frozen, whereas Wacs, with a separate allotment and few WAC returnees, still received promotions. AGF Wacs soon came to have higher ratings than those in any other domestic command. Even so, the AGF WAC Officer continued to recommend that allotments be merged, saying that to promote AGF Wacs when men working with them in identical jobs could not be promoted "obviously had unfair implications from a practical point of view in the field."

Well after the end of the war, in October of 1945, Colonel Davis finally obtained publication of her 1943 recommendations: "Grades for enlisted men and enlisted women, other than detachment overhead, will be interchangeable without reference to this headquarters."

In the opinion of AGF staff directors, the failure to integrate womanpower, either in insignia or in grades, caused some loss of *esprit* among AGF Wacs. One enlisted woman wrote, "If the Army Ground Forces wants us to like it, why not give us an incentive?" Another said, "I am sorry to say that the girls in the other branches of the Service appear to have more to do, more responsible work, and are given more credit for the jobs they do." Like men serving under the same conditions, many objected to the more severe AGF discipline. One wrote, "I believe the Wacs in the AGF are disciplined far more than the Air Wacs and Service Force Wacs," and another added, "I cannot understand why we have less privileges and more severe discipline than the Air and Service Force Wacs. Restrictions concerning bedcheck, uniforms, dating officers, are more rigidly enforced, bordering in some cases on downright pettiness and hair-splitting."

One of the Ground Forces' characteristics that women found peculiarly objectionable was its policy of permitting very little transfer or rotation for women, and almost no overseas assignment. When the War Department directed the release of women for shipment overseas, the newest arrivals on a station were usually sent, partly because their low ratings fitted the requisitions, and partly because they were not as yet highly trained and indispensable.

One woman wrote, "Wacs do not rotate every year as the enlisted men do; the coveted overseas assignments go to the lower-ranking women, so here we sit year in and year out, same post, same job, which might account for our sinking morale." Dozens of women supported this view; one added, "as a matter of fact, this detachment has begun to be known as the Lost Battalion." Many who had no desire to go overseas believed that rotation of some sort was desirable. One enlisted woman summarized the general lack of *esprit* by concluding that "the lack of imagination on the part of some of the Regular Army men and their failure to understand that morale is based, among other things, on recognition of ability and the proper reward for such ability, has undoubtedly played a big part in the failure of the AGF WAC recruiting drive."

Effect of Manpower Shortages

Toward the end of the war, there began to be evident in the AGF a noticeable improvement in the integration of female military personnel. In part this change originated with the commanders of AGF installations in the United States at the time of severe personnel shortages. The manpower shortages of late 1943 and 1944 found the Ground Forces in the worst personnel situation of any major command, eventually requiring the emergency transfer of thousands of men from the Air Forces and Service Forces. At this time, and especially when the War Department required the shipment overseas of all general service men, many AGF stations began to show increased interest in obtaining WAC units. There was by this date little prospect of obtaining them, because of recruiting difficulties and increased competition from other commands.

The War Department's decision was that incoming WAC recruits would be divided among the three major commands according to the number of requisitions on file from each; the AGF's requisitions being less than 3 percent of the total, only 3 of every 100 women recruited were sent to AGF. This trickle was reduced still further when the War Department authorized branch recruiting, since the Army Ground Forces did not at first elect to recruit. In the first four months of 1944, the Ground Forces got only 20 branch recruits, as against the Air Forces 5,000; in addition, some 3,000 women enlisted without promise of a branch, and of these the AGF got only 3 percent, or about 90 women.

The AGF considered the assignment of recruiters as wasteful, and turned to efforts to get more Wacs without expenditure. First, the War Department was persuaded to give the AGF some 100 Wacs left surplus when the Eastern Defense Command was inactivated. Next, in April of 1944, the AGF requested the War Department to give it 8 percent instead of 3 percent of all recruits. This the War Department refused to do, since the increase would have meant a reduction in the shipments to other commands which had earlier demonstrated faith in Wacs and had made their housing plans accordingly.[8]

In July, the Army Ground Forces appealed again to the General Staff for a "more equitable distribution," asking that it be given almost all of the general assignment recruits, which were then being divided among the three major commands. This was opposed by other commands and by The Adjutant General, who pointed out that it would penalize the AAF and ASF, which had sent out recruiting teams when the AGF had not. However, it was agreed to grant the AGF a special favor, in recognition of its manpower shortages; it might recruit for job and branch in all service commands, whether it had a station there or not.[9] AGF now at last assigned small numbers of its own personnel to obtain recruits, and sent out station teams under the station-and-job recruiting system. The teams did not prove particularly adept at selection, and the desired type of recruit was not always obtained until after the station-assignment plan was ended.

As AGF personnel shortages continued, the War Department was finally obliged to increase its quota from 3,690 to 7,500, thus giving it a larger share of incoming recruits.[10] In early 1945 it was necessary to give AGF a further concession: that of dropping the previous system of apportionment of recruits, and acceding to the Ground Forces' request not only for top priority on WAC recruits but for the first choice on the scarce clerical skills. By this means, stenographers, typists, and clerks were diverted from other commands and sent to fill the AGF quota.

At the time of the Ground Forces' greatest crisis, the Battle of the Bulge, AGF WAC units were ordered withdrawn from its less active stations, such as those of the Antiaircraft Command, and rushed to stations where a heavy load was expected, particularly to combat infantry replacement training installations.[11]

Wacs also were relied upon in the great overseas replacement depot at Fort Meade, where it was stated that the women now "formed the backbone of the administrative organization of the installation, particularly since enlisted men in the headquarters were comparatively new because of the enforced rotation policies for men established by the War Department." Immediately after V-E Day, the women at Fort Meade were hastily transferred across the country to Replacement Depot 4 at Camp Adair, Oregon, to meet the expected heavy shipments to the Pacific. When V-J Day halted the expected rush of replacements westward and the west coast depot was inactivated, the Wacs were promptly shipped back to Replacement Depot 1, now at Camp Pickett, upon the insistence of the commanding officer of that installation.

At about this time the last major headquarters refusing to use Wacs—Army Ground Forces headquarters in Wash-

[8] D/F, G-1 for TAG through CofS, 31 May 44. See Incl and M/R. CofS 324.5 WAC.

[9] Memo, ASF for TAG, 11 Jul 44. SPGAC 341 Gen (7-11-44)-69. History of proposal summarized here, and disapproved by SW, in SPAP 341 WAAC (2-26-43).

[10] AGF WAC Hist, Vol. I, pp. 42-46, gives history of allotments.

[11] By May of 1945, 42 percent of AGF Wacs were in Infantry installations, 21 percent at Overseas Replacement Depots, and 30 percent divided among Armored, Field Artillery, Cavalry, and Tank Destroyer Training installations.

ington, D. C.—also yielded to the problem of the shortage of expert clerical help, and brought enlisted women into Washington. These women quickly became an integral part of a number of offices, so that the headquarters decided to retain them permanently and move them with it to Fort Monroe.

The close of the war was to see a complete reversal of original policy, with the Army Ground Forces taking the lead in urging a postwar WAC component, and spearheading the demand for a Regular Army WAC. "The services of these women proved of direct assistance in winning the war," observed one staff study, "A far greater number could have been effectively employed." Another study alleged: "Economical, efficient, and spirited results are achieved in military installations where both male and female personnel are on duty."

CHAPTER XVIII

The Army Service Forces

The Army's third major command, the Army Service Forces, had by mid-1944 been assigned some 40 percent of the WAC's enlisted women, including training center cadre. Upon the departure of Colonel Hobby's office for the General Staff level, the position of ASF WAC Officer was established, parallel to that of Colonel Bandel in the Army Air Forces and Colonel Davis in the Army Ground Forces. The first and only incumbent was Lt. Col. Katherine R. Goodwin, formerly staff director of the First Service Command and personnel officer in the Office of the Director WAC. Her office was located in that of the ASF Director of Personnel, General Dalton, but she had advisory powers direct to General Somervell on nonpersonnel matters. Because of its late beginning and because the ASF did not desire separate statistical studies for women, Colonel Goodwin's office had only one assistant.[1]

In the field, each of the nine service commands and the Military District of Washington continued to have its own WAC staff director. For a time the staff director's office was under the service command's director of personnel, until later raised by an ASF circular to the immediate office of the commanding general of the service command. In those commands which Colonel Goodwin considered the most successful in recruiting and public relations, notably the Third and Seventh, the staff director also conducted periodic conferences of WAC company officers and recruiters in the area in order to improve efficiency and spread the latest information on personnel practices.[2]

The numbers of Wacs assigned to each service command varied greatly, from scarcely more than 1,000 in the First to 2,000 in the Eighth and more than 4,000 in the Fourth, which operated a training center. In general these were organized in regular post headquarters companies which, although no longer required to be identical, contained similar clerical and administrative personnel to help staff the post.[3]

Service Command	*Officers*	*Enlisted Women*
First	110	1,150
Second	120	1,371
Third	180	1,287
Fourth	600	[a] 4,600
Fifth	65	1,500
Sixth	170	1,140
Seventh	480	[a] 2,800
Eighth	240	2,190
Ninth	300	1,500
Military District of Washington	50	1,050

[a] Including training center.

[1] Distribution, Versatility, and Excellence of Wacs Serving with ASF, a 10-page summary prepared by Colonel Goodwin a year before the end of the war. Atchd to Memo, Dir WAC for CG ASF, 23 Sep 44, WDWAC 341, in SPAP 341 WAC (11-6-43). Unless otherwise stated, references are to this, by brief title, ASF WAC Summary. See Table 4, Appendix A.

[2] Interv with Lt Col Katherine Goodwin, 2 May 51.

[3] ASF WAC Summary.

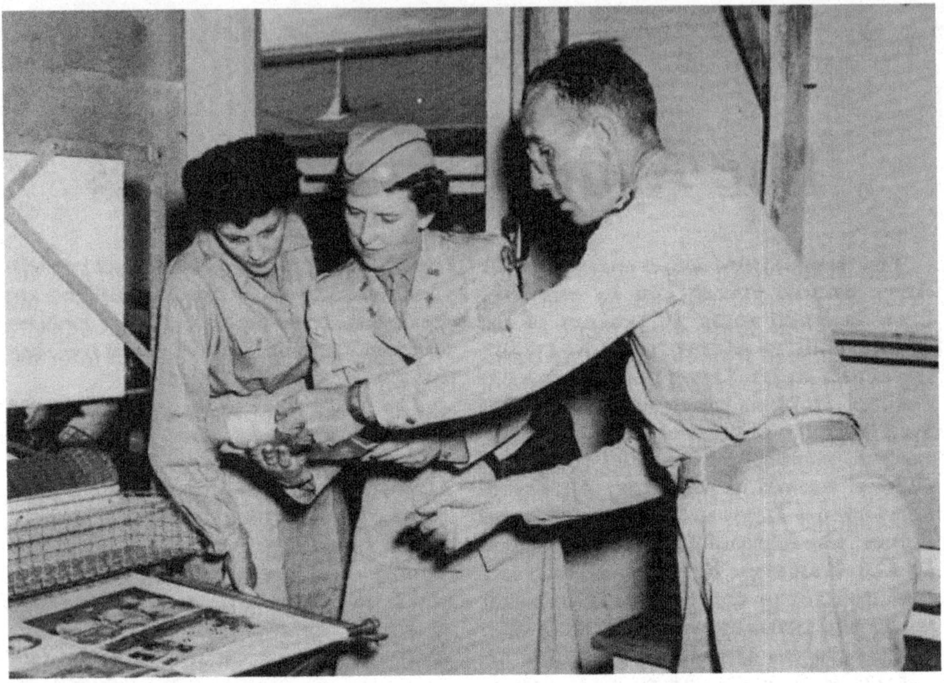

LT. COL. KATHERINE R. GOODWIN *(center) visits the Signal Corps photographic laboratory at Camp Hood, Texas, August 1944.*

Other Wacs were allotted directly to the various ASF technical services such as the Signal Corps, for duty on Class II installations under the jurisdiction of the technical service, although located within the service command. Each of these services ordinarily had a WAC staff director, and sometimes several, as in the case of the Transportation Corps and its various ports. The ASF administrative services, such as The Adjutant General's Department, seldom had staff directors or assigned WAC personnel, since they had few if any Class II installations; personnel in their specialty was ordinarily assigned as part of post headquarters companies.

In this far-flung organization, existing records gave little clue to the exact jobs filled by Wacs and the exact numbers of Wacs in each job. Wacs on Class II installations, such as those of the Signal Corps, might accurately be assumed to be working for that service, but not all were necessarily in communications duties. Similarly, Wacs in post headquarters companies might be working for the post headquarters, but they might also be in suboffices of the administrative services, such as the post judge advocate, or in Quartermaster Corps, Signal Corps, Medical Corps, or other technical work not under the technical services allotment.

Even though exact statistics thus were never gathered, it was known that the Army Service Forces, as its name implied, offered large numbers of noncombat jobs intrinsically suitable for women, and that Wacs at the height of the war had worked their way into a wide variety of them. Especially from the individual administrative and technical services, some conclusions could be obtained as to the relative success and suitability of the duties.

The Signal Corps

The first of the Army Service Forces' agencies to request Wacs was, according to its own undisputed claims, the Signal Corps.[4] Even in World War I the Chief Signal Officer reported:

> Our experience in Paris with the untrained and undisciplined English-speaking French women operators and experience elsewhere with the willing but untrained men operators was almost disastrous. The remarkable change in the character of the service at General Headquarters and other points when the American women operators took over was one of the features of the Signal Corps work of the times.[5]

In spite of this early enthusiasm, the Signal Corps experienced numerous delays and difficulties in launching a successful program to employ Waacs in World War II. In part this early delay resulted from a strong distrust of the Auxiliary status. In the earliest conferences the chief exponent of this view was Col. Henry L. P. King, Military Personnel Branch, Office of the Chief Signal Officer, who feared loss of trained personnel if the WAAC Director had command powers. In addition to seeking assignment and transfer control, the Signal Corps asked that the women wear Signal Corps insignia, that their units be designated Signal Corps and not WAAC, and that communications training be given by the Signal Corps and not by WAAC training centers.

Such proposals for integration were deemed by the War Department illegal under the Auxiliary status, with the exception of the matter of technical training. WAAC Headquarters proved agreeable to abandoning its plans for a communications specialists school at WAAC training centers in order that the Signal Corps might send qualified personnel to whatever Signal Corps schools were indicated. The Director also gave reassurance that—as was later to be demonstrated—she would never move any such unit, once assigned. At this, the Chief Signal Officer expressed in writing his appreciation of the concession and added:

> It is felt that as a result of the co-operation given . . . by the Women's Army Auxiliary Corps, the way has been opened for the release of needed enlisted men to field units and for a more extensive use of WAAC personnel.[6]

[4] Unless otherwise specified, statements through 1943 are from (1) Narrative OCSigO, Activities in Connection with WAAC or WAC Pers for Signal Corps Duty, October 1943, DRB AGO. Hereafter cited as SigC WAC Narrative. (See Bibliographical Note.) (2) Bound folder, WAAC Pers on Signal Corps Duty, Tab 1, Policy of CSigO Regarding WAAC Pers; loaned historian by Capt Jane Reeves, SigC WAC Stf Dir, 3 May 46, present location unknown. Hereafter cited as SigC Stf Dir file. (3) Interviews with: Maj Walter F. McDonald, SigC, OCSigO; Col Donald H. Nelson, OCSigO, Maj James L. Clark, OCSigO; Mr. Bruce W. Quisenberry, PRO, OCSigO; Lt Norma J. Fisher, OCSigO; Maj Charlotte E. Rhodes, former CO 334 1st Sig Serv Co, Germany, and 17th Sig Co, Ft Myer, Va.; all on 19 Sep 49; Capt Barbara Rodes, WAC Stf Dir, G–2, 22 Mar 46.

[5] Rpt, CSigO to SW, 1919, extract in Memo, Chief MPD SigC for Adm Div SigC, 29 Sep 42. SigC 320.2 WAC.

[6] Memo, Brig Gen Charles M. Milliken, OCSigO, for Dir WAAC, 26 Oct 42. SPWA 320.2 (10-26-42), unnumbered folder, Signal Corps, G–1 WD WAC files.

Proceeding with enthusiasm, the Signal Corps not only requisitioned Waacs for its Class II installations, but requested that all service commands be directed to requisition Waacs in every one of certain listed communications duties under their jurisdiction. Additional sweeping estimates were submitted to the Army Service Forces, as they had been periodically since the WAAC was organized, proposing to use women in literally every Signal Corps activity, from mixed tactical units to pigeon companies.[7]

The most important immediate proposal was that the WAAC aid in what came to be called the WIRES plan (Women in Radio and Electrical Service). Currently enrolled in Signal Corps schools were some 30,000 civilians, receiving free training and allowances during three- to six-month courses in radio and switchboard operation and related skills. Men in such courses were in the enlisted reserve and could be called to active duty upon successful graduation, but women graduates were necessarily hired as civilians, with the possibility that once trained they would seek better-paid jobs elsewhere. Use of the WAAC inactive reserve thus appeared a logical solution, and was generally agreeable to the women; a group at one school organized themselves as the first WIRES and appealed to Director Hobby to admit them. Upon the request of the Signal Corps, WAAC officers were dispatched to make necessary arrangements.[8]

Unsuitable Organizations and Locations

These extensive plans were shortly checked by the discovery that, while most Signal Corps jobs were perfectly suitable in themselves, they could not be given to Waacs without violation of numerous War Department policies on assignments for women. Since most Signal Corps jobs on Army posts were performed by civilians, the Signal Corps informed the Director that it would be necessary to rescind the ruling that Waacs would not replace civilians. There were also about 1,500 Signal Corp enlisted men who might have been replaced, but these were in small scattered detachments, the majority at stations where no WAAC unit existed. Therefore, the Signal Corps also asked the War Department to rescind the ruling that enlisted women would not be sent to any station without a WAAC barracks and commander for their adequate housing, supply, and protection.[9]

Both of these efforts were unsuccessful, since neither of these policies was ever to be relaxed by the War Department. Another point of dispute was the fact that Waacs, as civilians, could be assigned only to a WAAC unit and not to the Signal Corps; thus, the assignment of Signal-trained Waacs was a prerogative of the post commander and not the Signal Corps, and the WAAC Director could not guarantee their correct assignment.

On the other hand, the Signal Corps insisted that the WAAC must "guarantee them assignment on a communications basis and none other," and Colonel King wrote, "Tell the WAAC that since Signal

[7] (1) Stf Study, Maj J. K. Cunningham for G-1, 4 Nov 42. SPGAM 320.2 WAAC (9-15-42), in WD G-1 321 WAC. (2) Memo, CG ASF for CofS USA, 17 Jun 42. WD G-1 320.2 WAC (6-12-42), Tab A.

[8] (1) WDBPR Press Release, 9 Jan 43. (2) Ltr, Miss Stella Gunning, Pres WIRES, Trinidad Junior College, Colo., to Dir WAAC, 20 Nov 42, with Ind, Dir WAC to SigC. SigC 320.2 WAAC.

[9] (1) Memo, Maj. W. K. St. Clair, MPD OCSigO, for Chief MPB, 14 Oct 44. SPSMP 320.2 WAAC, in unnumbered folder, Signal Corps, WAC files. (2) Memo, Chief MPB OCSigO for Actg CSigO, 19 Oct 42, SigC 320.2 WAAC. (3) R&W Slip, Action 1, SigC Aux Br for MPD SigC, 9 Jul 42. SigC 320.2 WAAC.

Corps money is being used to train Wires we mean to keep same either as civil service employees or Wacks." [10]

Colonel King therefore returned to efforts to secure transfer of command from the Director to the Signal Corps, in the belief that the Signal Corps could then assign Waacs to replace civilians and to stations without WAAC companies. He wrote memoranda and made personal visits to most of the other technical and administrative services, urging their concerted action against the War Department policies involved. Securing the agreement of some, and noncommittal replies from others, he then attempted to call an ASF conference to force the issue.[11]

The Army Service Forces did not comply with the request, since it was at the time awaiting General Staff decisions as to the Corps' future status and size, and was unable to give the Signal Corps much more satisfaction than had WAAC Headquarters. At this time the Signal Corps noted, "The situation became so complicated that it was not known how requisitions were to be submitted, or how to procure Waacs for any type of installation under the Chief Signal Officer." [12]

Such squabbles over maximum replacement proved doubly futile, in that the numbers of WAAC recruits were never to be such as to warrant extension of employment to questionable locations or to replace civilians. The immediate result was that the Signal Corps delayed in opening its schools to Waacs, the WIRES plan lagged for some six months without action, and the Signal Corps established only the schools already agreed upon—three radio schools at civilian institutions, to provide radio operators for the AAF.

Thus, the first WAAC communications specialists to reach the field went in post headquarters units and—other than radio operators—without Army training in Signal skills, relying wholly upon their civilian background. Recruiters continued to accept volunteers in Signal skills, and training centers to send them out, without much knowledge of actual needs.[13]

Signal Duties in the Field

Among the first communications specialists to reach the field were the bilingual telephone operators in the first overseas unit, which reached North Africa in January of 1943. Two WAAC officers also went to the Eighth Air Force in England, even before arrival of the European theater staff director, to plan for Signal Corps Waacs in that command. A considerable percentage of the first WAAC unit to reach England consisted of telephone operators and other communications personnel, sent with the approval of the Chief Signal Officer but without Army technical training.

At about this time, the first units began to reach the Signal Corps' own installations in the United States. First to arrive,

[10] (1) Comments to R&W Sheet, MPD OCSigO to Chief, 17 Nov 42; (2) R&W Sheet, Entry 1, Capt Kenneth E. BeLieu to Col King, 8 Dec 42. SigC 320.2 WAAC.

[11] Ltrs, Chief MPB OCSigO to Chiefs of Mil Br, CE, Ord Dept, 10 Nov 42, with incl; replies by 1st Inds. SigC 320.2 WAAC.

[12] SigC WAC Narrative; also R&W Slip, Capt BeLieu, MPB OCSigO for Chief, 9 Dec 42. SigC 320.2 WAAC.

[13] (1) Story of further WIRES planning: Stf Study, Capt BeLieu, MPB OCSigO, 15 Dec 42, with covering memos. SigC 322 WAC 1943. Memo, 2d O M. McCull for Dir WAAC, 5 Jan 43. Unnumbered folder, Signal Corps, WAC files. R&W Slip, Entry 1, Maj John W. Ramsey, SigC, for Exec OCSigO, 6 Jan 43. SigC 322 WAC 1943. Memo, 2d Os Brainerd and McCull for Dir WAAC, 11 Feb 43, with Incl and Inds. Unnumbered folder, Signal Corps, WAC files. (2) Two WDBPR Press Releases, 10 Dec 42, and 14 Apr 43.

in March of 1943, were units assigned to Camp Crowder, Missouri, which by the end of the WAAC totaled some 489 enrolled women and 100 officers. Another unit under Signal Corps control went to Fort Monmouth, New Jersey, in June, and still others were on requisition. However, most women in these units were not technical specialists, but clerical and overhead workers to aid in running the schools.[14]

From the first there was no doubt of these units' success in the assigned duties. In Allied Headquarters in North Africa, the assistant chief signal officer stated:

> The manner in which they picked up the work was outstanding. . . . Had we had enough of them, we could have used them to operate all our fixed communications installations—telephones and telegraph—throughout the rear areas. Every one . . . released some man for Signal Corps duty up in the combat zone.

The former chief signal officer of the AAF in North Africa added:

> Don't tell me a woman can't keep a secret. Why, their own company commander doesn't even know where the board is located—we've tested the girls again and again.[15]

The Chief Signal Officer, Maj. Gen. Dawson Olmstead, on a visit to the area, was informed by the theater commander that approximately 1,300 more Signal Corps Waacs in ten different specialties would be required if obtainable. The same good reports were received from the European theater, where, after a one-week indoctrination course, 100 WAAC telephone operators were assigned to operate multiple switchboards at five command headquarters. The officer in charge reported, "Very soon I received praise of the improved service, as the good results were noticed almost immediately."[16]

Another early success was the participation of twenty-nine WAC telephone operators in the Quebec Conference in August of 1943. Rushed from train to switchboard, the women were on continuous duty for eighteen hours and doubled as secretaries and stenographers when a shortage of these occurred. After seventeen days of secret duty, each member received an individual commendation. Wacs also served successfully on the switchboards at the Cairo Conference and the Potsdam Conference.[17]

Requests from Signal officers in the United States continued to indicate an urgent need for more Waacs, since current directives required release of their general service men. Civilian replacements were scarce, and available limited service men often were unsuitable.[18] In general such requests also indicated a desire for Signal Corps technical training for the women. In England the using agencies, desiring teletypists, had been obliged to train WAAC typists in the theater, which they protested was a task more properly assumed by the Signal Corps in the United States. Studies showed that women could, with training, fill seventy-eight different Signal Corps jobs. Therefore, in May of 1943, Brig. Gen. Henry L. P. King's successor, Col. Duncan Hodges, put the long-delayed WIRES plan into effect. The women were to be trained first in Signal Corps duties in an inactive WAAC status, and then sent to basic training. Objectives were set at some 2,000

---

[14] SigC Stf Dir file.

[15] The WAC in the SigC, Speech by Col Oveta Culp Hobby. SigC file, OCMH. See Ch. XX, below, for North African Theater of Operations details.

[16] Appendix to SigC WAC Narrative.

[17] (1) *Ibid.* (2) Speech cited n. 15. (3) Interv with Capt Jane Reeves, SigC WAC Stf Dir, 3 May 46.

[18] Ltr, Post SigO, Ft Bliss, Texas, to Chief MPB OCSigO. SigC 322 WAC 1943.

QUEBEC CONFERENCE, AUGUST 1943

women, the first of whom reported in late May to a training course at the State Teachers' College at Livingston, Alabama, with eleven other schools scheduled for later opening.[19]

Effect of Army Status

The Signal Corps was highly pleased by the WAC's new Army status and the system of assignment, instead of attachment, to the using command. Signal Corps Wacs could now be assigned to the signal service company and not to the post, while the simultaneous multiplication of WAC units in the field made it possible to attach these women to existing companies in adequate numbers without violating War Department housing policy. On the other hand, Army status wrecked plans for WIRES training, since the new WAC legislation provided for no inactive reserve component. A mere 276 women were already in schools; these were given a choice of discharge or enlistment, and all enlisted.

The Signal Corps immediately began to seek some other system of earmarking acceptable recruits in order to retain Signal Corps control of qualified women during their basic training. For this purpose, the newly devised plan for branch recruiting appeared made to order, and it was with some disappointment that the Signal Corps received the Army Service Forces' decision not to join the AAF in branch recruiting at this time. The disappointment was alleviated within a few months by the authorization for station-and-job recruiting, which achieved virtually the same result. Under this plan, women received basic training first and then reported to the Signal Corps for either immediate assignment or specialist training.[20]

Some difficulty in obtaining recruits, even with the station-and-job promise, was caused by the high requirements of the Signal Corps. Women for attendance at the three radio operators schools were required to have near-officer qualifications: AGCT scores of 100 for radio operators and 110 for repairmen, as well as good scores on the radio operator's aptitude or mechanical aptitude tests.

By September of 1943 some 1,750 women had been trained, chiefly for the Air Forces, and the schools were discontinued. Thereafter the Signal Corps generally accepted only women with high AGCT scores as well as civilian skill "sufficient to qualify them for assignment with only nominal on-the-job indoctrination in Army procedure." Thus, women to be trained as tabulating machine operators were required to have clerical skill and an AGCT score in Grade III or better; women to be key punch operators required the same plus average typing speed; cryptographic clerks needed officer qualifications—an AGCT score in Grade I or II plus good education and mathematics or language background. Full civilian training in their fields was required of radio operators, radio and electrical repairmen, telephone operators, teletypewriter operators, draftsmen, and various photographic experts.

Such highly qualified personnel was not only hard to find, but the subject of competition from other agencies, particularly the WAVES, which offered officer status

[19] (1) Appendix cited n. 16. (2) Stf study cited n. 13(1). (3) Memo, Col James S. Willis, SigC for Brig Gen Frank C. Meade, OCSigO, 17 May 43. SigC 322 WAC. (4) WDBPR Press Releases, 11 and 21 May 43.

[20] (1) Ltr, Hq AAF to AACS, 31 Jul 44, and incls. WD G-1 220.3 WAC (8-7-44). (2) SigC WAC Narrative. (3) Memo, Dir MPD ASF for Dir Pers ASF, 26 Oct 43. SPAP 341 WAAC (2-26-43).

to cryptographers and other specialists. By June of 1944 the Office of the Chief Signal Officer was still highly discontented with the returns from station-and-job recruiting, which had produced Wacs only to the extent of some 1,100, or about 9 percent of the strength of the Signal Corps instead of the 16 percent that it had been authorized.[21]

Signal Corps Training

With a few exceptions, the Signal Corps did not find it necessary to give specialist training to the women who reported to it. It was stated that "substantial" Signal Corps classes, chiefly for overseas shipment, were conducted at Fort Monmouth and at Camp Crowder, but these were chiefly refresher courses in Army procedure for already-qualified women. For skills scarce or nonexistent in civilian life, such as cryptographer, classes were eventually held at Fort Monmouth. In general it was reported that women could master Signal training without alteration in course content, and presented no problems except in the matter of housing.[22]

Only one training difficulty was reported, that concerning the first cryptographic school. In the early months of this course, the officer in charge reported:

> The WAC personnel we have received so far are almost without exception woefully uninformed about the personalities, the geography, and the matériel of the war. It is also evident that their intelligence, on the average, is decidedly inferior to the men they are replacing.

He asked that personnel be better selected according to prescribed standards: officer AGCT score (110 or better), wide vocabularies, technical skill, and knowledge of world events, and that they be chosen from among teachers in higher grades, librarians, editors, authors, and similar occupations. "They are not likely to be found among sales clerks, packers, restaurant employees, machine tool operators, beauty operators, and music teachers." [23]

With higher WAC enrollment standards, higher-grade personnel became more readily available. With the virtual cessation of the WAC Officer Candidate School at a time when the Signal Corps Officer Candidate School still received large quotas, it became still easier to get women of officer qualifications but enlisted status. Final reports indicated good results in training female cryptographers, and the instructor at Fort Monmouth reported that the women "far exceeded expectations" and could "hold their own with the best students." [24] Some WAC supervisors were of the opinion that cryptography would prove the most important future use for Signal Corps Wacs in occupied areas, since women did not fraternize with enemy nationals to the extent that men did, and were therefore less apt to betray secrets inadvertently.

Signal Corps Jobs

One of the first and most important assignments of Signal Corps Wacs was that in the War Department Signal Center in Washington, called "the world's most im-

---

[21] (1) Speech cited n. 15. (2) Quotation from Ltr, TAG for six SvCs, 25 Mar 44. SPX 341 WAC (3-22-44) PR-I. (3) WD Cir 340, 29 Dec 43. (4) Memo, Chief Pers and Tng Serv, OCSigO, for CG ASF, 7 Jun 44. SPMSP 320.2 WAC, in SigC 322 WAC.

[22] (1) WDBPR Press Release, Enlistment. (2) R&W Slip, Maj John W. Ramsey, SigC, for Exec OCSigO, 6 Jan 43. SigC 322 WAC.

[23] Memo, Lt Col Matthew G. Jones, SigC, OIC Code Cen, for Capt Louis F. Fuchs, 20 Jan 44. Unnumbered files of Career Management Br, OCSigO.

[24] The SigC *Message*, Ft Monmouth, N. J., July 21, 1944, p. 2.

portant communications center,"[25] and hub of the Army's entire wire, radio, and cable system. With its outlying transmitting, receiving, and operating installations in the vicinity, it made up station WAR, in immediate contact by radio with all theaters and by land-line circuits with every major installation in the United States. Wacs for this service met the highest of intelligence and aptitude tests, and their loyalty records were minutely investigated. With the first assignment of a few Wacs to this work, the Chief Signal Officer, Maj. Gen. Harry C. Ingles, reported "an immediate improvement. . . . The loyalty and application to duty of these Waacs created an example which resulted in a noticeable increase in production."[26] By the end of 1944, some 235 Wacs were employed, chiefly in teletype operation, including radioteletype, and transmission and relay of messages translated into perforated tape on highspeed perforators. The chief of the Army Communications Service, Maj. Gen. Frank Stoner, stated, "We have found the Wacs conscientious, efficient, and dependable . . . thoroughly capable of filling many positions which formerly required trained men."[27]

The 2d Signal Service Battalion, which was under G-2 Division of the War Department for operations, employed some 1,000 Wacs at its stations at Arlington Hall, Vint Hill, and Two Rock Ranch. In general the secret work was routine, noisy, but of a nature requiring perfection in detail. The WAC staff director reported:

> It was proven over and over again that women were far better equipped than men for routine but detailed work. They were not qualified for highly technical jobs in most cases, but were qualified for routine research and for specialists' jobs such as teletype and radio operator.

About fifty women were also assigned to the battalion's Miami detachment, which distinguished itself chiefly by attempting to operate with a male commander for the WAC unit, in violation of War Department regulations, to be found out only when improper administration resulted and was investigated.[28]

Other Signal Corps stations also employed appreciable numbers of Wacs. The Eastern Signal Corps Training Center at Fort Monmouth employed an eventual total of about 300, chiefly clerical and administrative personnel to run the center.[29] The Holabird Signal Depot near Baltimore employed some 100 women for similar duties, and a quota of 75 was allotted to the Army Pictorial Service and Signal Corps Photographic Center on Long Island, for work in film libraries and laboratories.[30]

The average Army post in service or air commands also employed a few Signal Corps Wacs, although in lesser numbers than enlisted men and civilians. For example, the staff of the Signal Corps office at Fort McClellan, Alabama, included 3 officers, 8 enlisted men, 5 enlisted women, and 55 civilians. All positions appeared to

[25] WDBPR Press Release, Wacs in the Signal Center; in *Washington Post,* November 12, 1944, and *Army Times,* December 16, 1944.

[26] Ltr, Gen Ingles to Col Hobby, 12 July 43. WAAC file and SigC 322 WAC.

[27] Release cited n. 25.

[28] (1) Interv with Capt Barbara Rodes, WAC Stf Dir G-2 WD, 22 Mar 46. (2) WDBPR Press Release, Enlistment. (3) Memo, Dir Pers ASF for CSigO, 6 Dec 44. SPAP 320.2 WAC.

[29] M/R, sub: Fld Trip to Ft Monmouth; also R&W Slip, MPB OCSigO for Pers and Tng Serv, 15 Jul 43. Folder, WAC Surveys, unnumbered files of SigC Career Management Br.

[30] WDBPR Press Release, Enlistment. (2) Interv with Capt Jane Reeves, SigC WAC Stf Dir, 3 May 46. (3) Memo, MPB OCSigO for CG SigC Photographic Cen, Long Island, 26 July 43. SigC 320.2 WAAC.

be essentially civilian in nature. The experience of WAC training centers, where Wacs necessarily composed most of the staff of the post signal office, indicated that Wacs could if necessary perform most of the Signal duties on an Army post. The question as to whether all or any of such jobs required military status was a technical one for higher authorities, but from the WAC viewpoint either all-WAC or all-civilian offices were preferable to mixed ones, which spotlighted differences in pay and privileges.[31]

Officer Employment

Employment of officers was in general seldom as efficient as that of enlisted women. Both in the United States and overseas the Signal Corps showed a propensity for assigning women officers to secretarial and similar duties scarcely requiring an officer's training. Eventually three WAC officers were allowed to attend the G–2 school, with the promise that if all graduated in the top 10 percent more Wacs would be accepted; all did. Several WAC officers also attended the three-month Advanced Radio Communications School at Arlington Hall and the Signal Corps Supply School at Camp Holabird.[32]

Signal Corps Jobs Overseas

Extensive employment of Signal Corps Wacs, both officer and enlisted, was made in all overseas theaters, although not to the extent that the Signal Corps desired, Brig. Gen. Jerry V. Matejka, the new head of Signal Corps personnel, complained:

> Substantial requirements for Wacs on Signal Corps duty have existed in the various theaters, yet the furnishing of WAC personnel has been either retarded or abandoned because of apparent lack of the same co-ordination of interested agencies in the War Department as is applied to furnishing enlisted men as fillers and replacements.[33]

Some measure of the difficulty appeared due to the Signal Corps' own reluctance to part with Wacs on its stations in the United States. Thus, when requests from the European theater for a WAC signal service battalion of approximately 550 Wacs were approved by Operations Division, War Department General Staff, the Army Service Forces protested that male personnel should be sent in order to avoid disrupting the work of field stations using trained Wacs on key jobs. On Operations Division's insistence the unit was authorized for activation in the theater, but with the requirement that 200 of the Wacs be supplied by the theater.[34] Even when trained radio operators were found malassigned as clerks on Signal Corps installations in the United States, the Chief Signal Officer refused to release them for radio work overseas.[35] General Matejka further remarked that "use in overhead installations has been disrupted by the withdrawal of WAC personnel for overseas."[36]

In spite of the alleged difficulty of getting Signal-trained women for overseas duty, most overseas theaters by the end of the war had acquired relatively large

[31] (1) *The McClellan Cycle*, April 13, 1945, "SigC Job Big One on Post." (2) Press Release, 2d WAC Tng Cen, 1 Jul 43.

[32] (1) See overseas chapters. (2) Intervs cited n. 4. (3) Memo, Chief Pers and Tng Serv OCSigO for CSigO, 23 Sep 44. SPMSP 320.2 WAC, in SigC 322 WAC.

[33] Memo cited n. 21(4).

[34] Memo, Exec, G–1 Div for G–1, 12 Oct 44; Memo, "H.E.K." for Maj Gen Stephen C. Henry, G–1, 18 Oct 44. WD G–1 322 WAC.

[35] Memo, CofT for CG ASF, 30 Sep 44, SPTPM 220.3 WAC; with 2d Ind from CSigO and 3d from MPD ASF, SPGAC 210.3 WAC.

[36] Memo cited n. 21(4).

numbers, many trained in the theater. Some one third of all Wacs in the Mediterranean theater, or about 700 women, were estimated to be employed on communications duties—a percentage notably higher than that achieved in the United States.[37] In Europe, the 3341st Signal Service Battalion eventually reached a peak strength of about 28 officers and 738 enlisted women, and in Paris its women were employed in every department of what claimed to be "the largest message center in the world . . . with the exception of Washington, D. C." [38] Other units also contained some Signal Corps personnel, which brought the number in Europe to an estimated one fourth of the theater's WAC strength, or some 1,700 women. The Pacific theater reported that only about 3 percent of its Wacs, or 150 women, were in communications work, including secret duties with the Signal Intelligence Section. Lesser numbers were also used in smaller theaters.[39]

Peak of Employment

In September of 1944, some months before the WAC reached its peak strength, the Chief Signal Officer estimated that there were about 1,500 enlisted women at six installations under his own jurisdiction, or some 16 percent of the total enlisted strength. This, he believed, exceeded all other technical services in percentage of replacement with the possible exception of the Chemical Warfare Service. In addition, there were 236 enlisted women in the War Department Signal Center, and more than 700 performing Signal Corps duties in the service commands. To this could be added at least 1,700 in Europe, 700 in Italy, and 150 in the Pacific. It thus appeared likely that Wacs on Signal Corps duties totaled at least 5,000, some 5 percent of the WAC strength.[40]

The general job groupings for Wacs, at least in the United States, were estimated by the Chief Signal Officer to include:

	Percent
Signal Corps Technical Specialists	63
Administrative and Clerical	25
Mess and Motor Pool Personnel	9
Miscellaneous	3

Special Problems of Signal Corps Wacs

In spite of such wide employment and consistent success on Signal Corps duties, it was evident by the end of the war that in employing women the Signal Corps faced several problems. The women described most Signal Corps technical jobs as "nerve-wracking"; after a few months they tended to become a nightmare of beeping earphones, clicking keys, or clattering machinery. A WAC inspector reported:

> Severe nervous strain was imposed by the necessity for constant attention . . . no thought was required for much of it, yet the worker did not dare to let the work become automatic, as one slip wrecked everything, and the fear of costing men's lives was always with them.[41]

[37] For details, see Ch. XX, below.

[38] (1) Press Release, Hq ComZ ETO, Receipt No. 92782, 10 Jul 45. (2) Interv with Capt Jane Reeves, SigC WAC Stf Dir, 3 May 46. (3) M/R, 9 Oct 44, sub: Summary of Radiograms on WAC Sig Serv Bn for ETO. MTD OCSigO 322 WAC Sig Serv Bn. (4) Ltr, CSigO to Dir WAC, 5 May 49. PRO file, OCSigO.

[39] See Chs. XXI, XXII, below.

[40] (1) Memo cited n. 21(4). (2) Memo cited n. 32(3).

[41] Discussions of WAC staff officers' opinions are based on: (1) Interv with Capt Barbara Rodes, WAC Stf Dir G-2 WD, 22 Mar 46. (2) Intervs with Lt. Col Dorothea A. Coleman, WAC Stf Dir, NATO and MTO, 21 Jan and 8 Feb 46. (3) SigC refs in Chs. XX–XXIII, below. (4) Intervs cited n. 4(3).

Added to this was the ever-present strain of keeping security, and sometimes the knowledge that the operator, alone with a few of the Army's top authorities, shared secrets in which a slip of the tongue in public could cost the lives of hundreds of soldiers.

Most Signal duties also involved shiftwork, which constantly rotated so that regular sleeping and eating habits were impossible. British inspectors had pointed out long before the WAC was formed:

> A policy is present in the service of working watches or shifts with irregular hours of duty . . . it should have been known by now that the irregular rhythm of working hours is injurious to health.[42]

WAC inspectors repeatedly recommended that shift changes be less frequent than the customary two weeks, but this was nowhere possible under established Signal Corps policies. On ordinary posts the handful of Signal Corps Wacs was obliged to try to sleep at odd hours while barracksmates engaged in normal activities. On all-Signal Corps stations, company commanders attempted to reserve one building for each shift, a move which seldom proved practicable since women worked in different offices on different schedules. At only one station, Arlington Hall, was some measure of relief achieved by the use of cubicle housing, four women to a cubicle. One WAC adviser noted:

> One of our biggest problems was that of shift workers . . . we had to watch their health carefully. There was always the danger of emotional upsets caused by difficult sleeping arrangements.

The company commander's usual efforts to provide feminine-type sports, recreation, orientation, or even hot food and sympathetic counsel were often thwarted by the women's peculiar working hours and by the impossibility of keeping cadre on 24-hour duty.

In addition, at most stations the shortage of Signal Corps personnel was such that no leaves, furloughs, or passes were possible for months. WAC advisers recommended, without effect, that "it is most important that officers and enlisted personnel, both men and women, be required to take regular leaves twice a year."[43] There was on many stations also a certain reluctance to release women for any cause, even for officer candidate school, and Wacs were not eligible for the direct commissions that men of equal qualifications received in certain duties, or for admission to Signal Corps Officer Candidate School. Even appropriate enlisted grades were chronically scarce; remarkably consistent complaints were received from stations all over the world that enlisted Wacs had replaced men in top enlisted grades, or even officers and warrant officers, without receiving any ratings at all.

On restricted installations, particularly the several detachments of the 2d Signal Service Battalion, the situation was especially difficult. Here a woman, once assigned, was a virtual prisoner for the duration; men might look for enforced rotation to overseas service, but even this was not permitted for a woman until after the end of the war. If a woman failed in or was unequal to technical duties, she was retained on the station in menial or service duties on the grounds that she "knew too much" to permit her to be released for proper assignment elsewhere.

WAC inspectors reported that women for such installations should have been originally selected from volunteers who had been clearly informed of their pros-

[42] *Conditions in The Three Women's Services.*

[43] Sources cited n. 41.

pects, a method that was successful in other technical services. Instead, Signal Corps personnel went to Forts Des Moines and Oglethorpe and selected women who afterwards alleged that they had received promises of grades, overseas service, officer training, or "the glamor of working for intelligence." Women recruited directly for the station were not informed of its restricted nature. Neither had some of the company officers been well chosen, in the opinion of WAC inspectors. One observer recommended that in the future all company officers be chosen from among operational officers who had themselves worked shifts and were "aware of aching feet and backs and being unable to sleep in the daytime."[44]

Even if capable, the company officers on secret installations were often not "cleared" to visit the men or women on the job, and thus understood neither the working conditions nor the measures needed to provide outside rest and quiet. Inspection privileges were likewise denied to Army classification survey teams, staff visitors, and medical officers. Because of the highly secret nature of the installations, neither the Signal Corps nor G–2 Division believed it possible to permit interference by medical officers or inspectors on the grounds of health and well-being for either men or women. The Signal Corps employed a rapid succession of WAC advisers and staff directors—among them Capt. Susanna Turner and Capt. Zelma Hanson—but did not clear these individuals to visit secret work.[45]

Eventually the health and morale situation of the 2d Signal Service Battalion became so bad that G–2 Division, in the closing months of the war, was obliged to accept a WAC staff director and to clear her for visits to secret units for which the Signal Corps staff director was not cleared. This officer, Capt. Barbara Rodes, expressed her belief that many of the problems had been preventable, and that "If they had had a WAC Staff Director in the beginning to pick personnel after a study of G–2's needs, they would never have had low morale in the WAC unit." Captain Rodes succeeded in getting a few women released for overseas duty and officer candidate school, and in getting ratings for others. She also gave considerable effort to helping women "to visualize the importance of a job and to get the whole war picture so that they could keep their perspective."

Independent reports from virtually every overseas theater showed an almost identical pattern of difficulty in Signal Corps work, mitigated only to some extent by the more stimulating nature of life overseas. In the Mediterranean theater, by the end of the war, some 8 percent of Signal Wacs, or an estimated fifty women, had suffered mental breakdowns, and others were believed to be on the verge of physical or psychiatric illness. In the Southeast Asia Command, the majority of all illnesses occurred in the minority of women who were assigned to Signal Corps work.

Since the end of the war intervened before women in most theaters and domestic stations had been on Signal jobs for more than eighteen months, there was no exact knowledge as to how long their efficiency could have been sustained under existing conditions. There were some grounds for belief that, under unchanged Signal Corps personnel and housing policies, two or two and a half years might be the maximum serviceability of the

[44] *Ibid.*

[45] Memo cited n. 21(4).

average Wac, after which disability discharge would be required and attrition might outweigh efficiency. Civilian experience indicated that the duties themselves were not at fault, since civilian women made careers of such work, with the aid of vacations, normal recreation, and private housing. One WAC commander reported, "When demobilization came, my women swore they would never come back to the War Department or to shift work, but about 25 percent eventually did—as civilians." [46]

Evaluation of Success in Signal Jobs

Apart from the question of how long it could be sustained, there was no doubt that WAC performance on Signal Corps jobs had been successful. A board convened in late 1948 at Fort Monmouth to consider the employment of the WAC in communications activities stated that it was "an accepted conclusion" that women were "more adaptable and dexterous than men in the performance of certain technical specialties." [47] A survey made of Signal officers in Army areas and overseas commands produced the opinion that women could fill approximately 50 percent of all operations, communications, and photographic positions within a communications zone. The consensus was that many maintenance functions could also be performed by women, with the exception of those involving heavy physical labor, and that "Wacs should be used as far forward as Army Headquarters."

It was concluded that women should be used in as many such activities as possible, with immediate emphasis on training in Army schools as teletype and telephone operator, message center clerk, radio and radio telephone operator, cryptographic technician, and facsimile operator. Similar studies within the office of the Chief Signal Officer indicated likewise that Wacs could be used in communications centers, on switchboards, in photographic detachments, in film libraries, and—provided that a WAC detachment was already at the installation—in Signal Corps administrative offices at Army posts, camps, and stations. The proper division of employment between civilian and military women was left for future resolution. [48]

Chemical Warfare Service

Second among the Army Service Forces agencies to obtain a WAAC staff director and the promise of WAAC personnel was the Chemical Warfare Service (CWS). [49] While the numbers involved were relatively small, never exceeding about 700 Wacs, the percentage of replacement achieved was higher than that in most larger agencies, eventually reaching, at different times, from 13 to 22 percent of the Chemical Warfare Service's domestic

[46] Intervs cited n. 4(3).

[47] SigC Bd Study 17, Employment of WAC in Communications Activities. FMSCB 324.5, 5 Oct 48, Hq Ft Monmouth, in SigC 200.3 WAC No 1, 1948.

[48] D/F, SigC Career Management Br to Maj. James L. Clark, 23 Mar 49, sub: WAC Problem, SIGMP-4 Comment 1, with reply, Maj Clark to Col Nelson, 24 Mar 49. Corresp folder, Career Management Br files, OCSigO.

[49] Unless otherwise specified, all references are from the following sources: (1) ASF WAC Summary, Sec With Chemical Warfare Serv. (2) Memo, WAC Stf Dir CWS for Asst ASF WAC Off, 10 Aug 44, copy furnished author by Stf Dir. (3) Interv with Maj Helen E. Hart, CWS WAC Stf Dir, 26 Mar 46. (4) CW Bull, Mar–May 45, Vol. 31, No. 2, "Here Are the WAC," pp. 4–11. (5) Comments, undated, by Lt Mary B. Warner, OTI CWS, on first draft of this section submitted to her for CWS comment. WAC files OCMH.

INSTRUCTION IN RADIO THEORY *for women attending the Midland Radio and Television Schools, Inc., Kansas City, Missouri.*

enlisted strength.[50] In early April 1943 the first CWS WAAC unit reached Pine Bluff Arsenal in Arkansas. A second WAAC unit reported in late April to Dugway Proving Ground, Utah, and a third and fourth in June to Camp Detrick, Maryland, and to the Chemical Warfare Center, Edgewood Arsenal, Maryland.[51]

Commanding officers of field installations did not entirely share the enthusiasm of the chief of the Chemical Warfare Service for requisitioning Waacs, and admitted later that "men busy on research projects where the slightest error might ruin months of experiments at first took a dismal view of female assistants in laboratories and proving grounds." [52] However, tentative inquiries a few weeks after the units' arrival brought the reply that "the Waacs at these installations are rendering excellent service." An early thousand-hour experiment with a detail of sixty at

[50] (1) Strength of the Army, STM 30, 5 Jul 45, p. 19. (2) Memo, Chief MPB OCSigO for Chief Pers and Tng Serv OCSigO, 16 Jun 44. sub: WAC Enl Str of Various Tech Servs. SigC Career Management Br files, OCSigO.

[51] Memo, Col Harold E. Brooks, CWS, for Maj Gen William N. Porter, 2 Jun 43. CWS 324.5 WAC (Apr–Jun 43).

[52] Bull cited n. 49(4).

WOMEN DECONTAMINATION EXPERTS *in the Chemical Warfare Service at Gowen Field, Idaho.*

Edgewood Arsenal, to determine to what extent Waacs might be used in the work of impregnating protective clothing with chemicals, met with "such surprisingly good results that we are about to recommend that some all-WAAC chemical impregnating companies be organized."[53] The WAAC staff director, Lt. Helen E. Hart, later stated that her chief contribution to the CWS program at this time was her success, after much pleading, in persuading her commanding general to change the name of this unit, which was finally given the name of WAAC Processing Company, instead of that which the general had first directed—WAAC Impregnation Company.[54]

Very shortly, CWS inspectors were surprised to find the entire WAAC detachment at Pine Bluff Arsenal plunged into a serious state of low morale by what appeared to male inspectors to be a minor circumstance—the fact that Waacs had not replaced soldiers. As a matter of fact, the shortage of civilian personnel had made it possible to secure a bona fide military allotment for the installation, which

[53] Ltr, Col Brooks, CWS, to CO, Huntsville Arsenal, Ala. CWS 324.5 WAC (Apr–Jun 43).

[54] Interv cited n. 49(3).

eventually would have required enlisted men, but to the Waacs nothing was apparent except that they had filled former civilian vacancies. CWS inspectors noted in some surprise that the "insidious idea" that a woman should replace a soldier appeared to be "prevalent among all WAAC companies in the field." The situation was somewhat alleviated upon conversion to Army status, when "malcontents" departed Pine Bluff Arsenal en masse. Thereafter, the fact of Army status reassured the women that military jobs were being filled.[55]

While the diminishing supply of recruits made it impossible to fill all further CWS requisitions, three more detachments later reported respectively to Rocky Mountain Arsenal, Huntsville Arsenal, and the Office of the Chief of Chemical Warfare Service.[56] The Chief of the CWS, Maj. Gen. William N. Porter, noted that as the Corps ended its second year, the quality and efficiency of WAC personnel was holding up well and that there was "a high degree of enthusiasm in their respective tasks."[57]

CWS Jobs

The majority of these CWS Wacs were employed in routine post headquarters duties which differed from those of other commands only in subject and terminology. For example, at Pine Bluff Arsenal Wacs drove, serviced, and cleaned the large Army buses that connected the widely separated munitions buildings, transporting civilian shift workers twenty-four hours a day. Other Wacs ran the arsenal's military personnel section, handling records of both officers and enlisted men. Wacs in the quartermaster branch handled stock records, requisitions, inventories, shipping tickets, and reports. One WAC sergeant ran the sales commissary; a corporal computed field rations for the post; others sorted the post's mail; the publications section was composed entirely of Wacs. One WAC officer was in charge of the motor pool; another was arsenal librarian; another the assistant to the chief of the payroll section. Two WAC officers inspected the seven post cafeterias and ran the officers' mess. Another WAC lieutenant was chief of the chemical production control section, co-ordinating manufacture and shipment schedules.

A minority of WAC jobs were the technical ones which concerned the real chemical warfare mission. "We got most of the scientifically trained Wacs there were," claimed the CWS staff director. For reasons which could not then be publicized, the CWS wanted not merely chemists but also pharmacologists, neurologists, and toxicologists. Other technical services frequently surrendered to the CWS any WAC scientist whose qualifications were too high to be utilized appropriately elsewhere.

Before admittance to the installations, a four-way loyalty check was required for each woman, which often caused a wait in staging areas of two months or more. Upon reaching CWS installations, women were surprised to find assembled many prominent scientists engaged in closely guarded research projects of great interest and variety. "Many of the women already had master's degrees," the staff director stated, "and I would estimate that they got about the same opportunities that a Ph.D.

[55] (1) Memo, Chief MPB CWS for CO WAAC Det, Pine Bluff, Ark, 17 Aug 43; (2) Ltr, Gen Porter, Chief CWS, to Dir WAC, 5 Jul 43. CWS 324.5 WAC (Jul–Sep 43).

[56] Memo, WAC Stf Dir CWS for CG ASF, Att: MTD, 29 Mar 44. CWS 324.5 WAC (Jan–Mar 44).

[57] Ltr, Chief CWS to Dir WAC, 6 May 44. CWS 324.5 WAC (Apr–Jul 44).

doing research would get." Many quickly became devoted to their projects; one woman, in rejecting a bid to officer candidate school, explained that she had previously shared the instructions of a certain famous scientist with a college class of seventy, but now had his exclusive supervision.

A few such enlisted women were eventually given work of real responsibility. A WAC pfc, with twelve years in public health work and an M.A. in bacteriology, sought new therapeutical methods of handling infections of gas wounds. Another WAC pfc, the former director of the Neuropathology Laboratory at Yale Medical School, investigated the effect of war gases upon the blood. At Johns Hopkins University, Wacs were directed by the staff in research concerning the use of penicillin spray in control of pneumonia which might be caused by chemical burns. At Dugway Proving Ground, a WAC T/4's research in heat radiation received a commendation from the U.S. Weather Bureau. Some of the Wacs were lured away by the Sanitary Corps, which was able to offer direct commissions, but others refused to leave their projects even for officer status. The extent to which such WAC scientists were employed was limited only by the fact that recruiters were unable to furnish them in the numbers desired by CWS.

Where college-trained scientists were not available, CWS requested and made good use of Wacs with lesser professional skills. WAC laboratory technicians were numerous at most installations, and WAC draftsmen were employed to design protective equipment. Dugway Proving Ground boasted the only WAC glass blower extant, who devised special sampling apparatus for experiments. Edgewood Arsenal used WAC photographers to photograph material for technical publications, and a former designer of fur coats to design and make canvas cases and protective coverings.

Women without skills could also be used by CWS with only on-the-job training. "It would be a mistake, in the future, to assume that CWS can use only scientists or technical experts," said the WAC staff director later. "For certain work all that is needed is a liking for outdoor work and for messing around with chemicals, smoke pots, and field instruments. We found that former factory girls were quite good at this, and interested in it." Perhaps the most colorful CWS job at Edgewood Arsenal was that of a WAC pfc who prepared colored-smoke grenades for signaling purposes and detonated them to test for brilliancy and duration of color; when the wind proved capricious this individual was habitually either green, yellow, or red from the—fortunately soluble—smoke clouds. At Dugway Proving Ground, Wacs were trained to participate in field observation during the mortar or rocket shoots, noting wind direction, air temperature, air pressure and humidity, and mastering the principle of the balloon run. They set up and oriented an artillery aiming circle and noted ballistic characteristics during rocket or artillery tests. Most women enjoyed outdoor work of any sort, and when shortage of personnel and periods of twenty-four-hour duty brought WAC office workers near the breaking point, it was found possible to revive them by brief rotation to field tests.

Problems Common Among CWS Wacs

The chief difficulties in CWS assignments for women were quite similar to those noted on secret installations of the Signal Corps. Because of the restrictions surrounding the more technical work, inspectors, staff directors, and company commanders were forbidden to check on any job assignments, even routine. As a result, some installations took advantage of this immunity from supervision to assign women to unsuitable duties contrary to War Department regulations. At various stations the requirement for WAC company officers was ignored. At another, enlisted women were assigned to a post commander's home to cook and act as personal servants to his wife. In several cases, civilian jobs in post exchanges and commissaries were staffed with Wacs. If detected, none of these violations could be reported, since it was alleged that security would have been endangered. The ASF WAC Officer, upon repeatedly receiving complaints from irate relatives, several times attempted to visit such installations, but was always informed that they were officially exempt from ASF inspection. Even the CWS WAC staff director was not cleared for such inspection.

Another factor which, in the staff director's opinion, caused lowered morale was the fact that women could not go into combat-supporting CWS units and that the CWS refused to release women for any other overseas assignments, on the grounds that such action would violate security and deprive installations of irreplaceable personnel. CWS was therefore exempted by the War Department from the WAC shipment quotas allotted to most other commands. The staff director estimated that only about 20 percent of the Wacs were employed on work that was too secret to permit them to go overseas or even on leave. She said, "The women could never understand it. I fought to get just two of them over, to give the others hope of reward for good service, but I failed."

Another problem faced by CWS units was that of obtaining good company officers. "We had a very high type of women on the average," the staff director stated later. "We found that WAC commanders had to treat them like intelligent adults. It was fatal to give unnecessary restrictions or to be too energetic in enforcing petty regulations." In general such restricted units, like those of other technical services, were found to require special attention to suitable recreation, rest, and living conditions.

In spite of various problems, as the war drew to a close ranking officers of the Chemical Warfare Service indicated satisfaction with the women's contribution. Army commanders in the field noted, "The men often consider themselves frustrated combat officers and thus lose interest in research and production work, but the Wacs eagerly plunge into their jobs with all their energy." General Porter, chief of the CWS, stated that Wacs had exhibited "all of the soldierly qualities of obedience, initiative, and devotion to duty. . . . The WAC has permeated our entire organization, and we owe them a great debt."[58]

The Corps of Engineers

The fact that Wacs had been assigned to the Corps of Engineers since early 1943 was almost as unknown, until August of 1945, as was the project on which they were employed, the atomic bomb. Not until then was it revealed that 422 Wacs, approximately eight and a half percent of the enlisted personnel under the command of the Chief of Engineers, had been employed on the project. Maj. Gen. Leslie R. Groves, commander of the project, wrote to these women:

[58] Bull cited n. 49(4).

> Little is known of the significance of the contribution to the Manhattan Project by hundreds of members of the Women's Army Corps. . . . Since you received no headline acclaim, no one outside the project will ever know how much depended upon you.[59]

Manhattan District

The term *Manhattan District* was deliberately deceptive, for Engineer offices employing Wacs were located not only in New York, Washington, and Chicago, but also at Oak Ridge, Tennessee; Los Alamos, New Mexico; and Pasco, Washington. Here, the volume of classified material had recently become such that maintenance of security was a problem, and it was held essential that certain key secretaries and file clerks be under military control. A first cautious request was made for only 74 Waacs, who reported in June of 1943; others followed shortly.

The WAC staff director, who commanded the Oak Ridge company for most of its history, reported that she was sent to WAC training centers to select many members personally. The women interviewed were told frankly that, while the work was secret and vital, they would do a hard job, would never be allowed to go overseas or to officer candidate school, could never receive any publicity, and would live at isolated stations with few recreational facilities. A surprising number of highly qualified women responded. Women were rejected if they did not actively desire the work, or did not appear to possess "experience, judgment, earnestness, and sincerity of purpose." The staff director highly recommended this system of selection, to which she attributed her detachment's good fortune in avoiding many of the problems experienced by Wacs in secret installations of other technical services. As a matter of fact, it later proved possible to send a few women to OCS and overseas without violation of security.

There was likewise no pretense that women would do glamorous scientific jobs; the Corps of Engineers did not desire to use Wacs as engineers or technicians, but rather in general administration and paper work. Those assigned were therefore chiefly skilled stenographers, typists, telephone operators, clerks, and teletype and cryptographic technicians.

Exact job descriptions were never recorded, for reasons of security, but it could be stated that the women handled detailed records and technical reports. These included the classified files, communications, the production section, and all offices that handled scientific reports. The District Engineer, Brig. Gen. Kenneth D. Nichols, stated: "Members of the WAC were charged with a major portion of the responsibility for preserving project security."[60]

A minority of the women were assigned as chemists and other scientific technicians. A few, by the end of the war, had worked into various technical duties for which they proved apt and in which they received on-the-job training. WAC metallurgy technicians became qualified through work in ceramics, plastics, and powdered metals. Electronics technicians worked chiefly on the construction of needed electronics equipment. WAC photographers became specialists in photo-

[59] All references, unless otherwise stated, are to the following sources: (1) Ltr, Asst to Dist Engr, U.S. Engr Office, Manhattan Dist, Oak Ridge, Tenn, to Hist Div WDSS, 19 Apr 46. EIDMP-44, now in WAC files, OCMH. (2) Unit History, WAC Det, 9812th TSU-CE, Manhattan Dist. Copy in WAC files, OCMH. (3) Interv with WAC Stf Dir CE, Capt Arlene G. Scheidenhelm, 14 Feb 46. (4) STM 30, Strength of the Army, 5 Jul 45; WAC strength is from ASF WAC Summary.

[60] Commendations cited n. 59(1).

graphing metals and metallurgical processes. Spectroscopic technicians were eventually qualified for advanced work with the spectroscope.

Most of the Wacs' positions carried considerable responsibility, and as the project had a liberal allotment of grades, many of the women had received good ratings by the end of the war. Both the responsibility and the ratings were a definite factor in aiding morale, according to the WAC commander, but even greater was the fact that the work was hard and most women were always kept busy.

Problems of Engineer Corps Wacs

Only one major problem was encountered by Wacs serving with the Corps of Engineers, quite similar to that already reported independently by other technical services. This was the fact that the secrecy surrounding all installations prevented inspectors and WAC staff officers from checking on the accuracy of job assignment or the compliance of post commanders with War Department directives.

Serious morale situations soon developed in the Pasco and Los Alamos units, where Wacs were habitually used as babysitters, servants for civilians, and other nonmilitary jobs. In one, a number of highly qualified communications experts were assigned as barmaids to serve beer to civilian laborers. The women stated that they were subjected to rough language and insulted by the civilians, until one fled to Washington to appeal to the Air Forces to obtain their transfer to bona fide communications work.[61] The case was reported to the Chief of Engineers, who refused to transfer the women to the Air Forces; the ASF WAC Officer and the Ninth Service Command WAC staff director were both denied permission to visit the installation. At the second WAC detachment, an even more widespread morale problem developed when a young WAC officer without troop experience, sent to the station to do an office job, suddenly found herself in command of a newly acquired WAC company in addition to doing full-time office work, and without the assistant required by War Department regulations.

The problems presented at both of these stations eventually became so serious that, late in the war, the Corps of Engineers was obliged to appoint a WAC staff director. To avoid bringing in an outsider, Capt. Arlene G. Scheidenhelm was named staff director and was cleared to visit the other stations. At one station the detachment was reorganized; at the other, the staff director stated:

> I merely told the post commander that he shouldn't use Wacs as barmaids or we should all get in trouble; he asked which women to reassign, I named them, and he reassigned them at once. All that was needed was a WAC Staff Director who was a part of the Corps of Engineers.

Commendations

The day when the A-bomb news could be released was dubbed by the Wacs "our day," and they were, according to the staff director, "the proudest Wacs in the Army," feeling that they had done more than most other Wacs to shorten the war and save the lives of American soldiers. At this time, the WAC units were awarded the Meritorious Unit Service Award; twenty women were presented with the Army Commendation Ribbon; and one received the Legion of Merit. Congratulatory messages were also received from the Secretary of War, from General Groves,

[61] Specifically, to the Exec, Air WAC Div, Maj Treadwell.

from the District Engineer, and from the WAC Director. General Groves wrote:

> As one who observed your performances month after month, I wish to express my personal appreciation. . . . You can well be proud of your service with the Manhattan District and the part you played in saving countless American lives.

To this the District Engineer added:

> No WAC assignment anywhere exceeded in importance their mission with the Atomic Bomb project.[62]

Ordnance Department

Some 700 enlisted women—over 5 percent of the enlisted strength—and 80 WAC officers were eventually employed in ten Ordnance installations in the United States, including five depots, two arsenals, two Ordnance districts, and the Office of the Chief of Ordnance. Other requisitions met with less success in that they sought Wacs to replace civilian employees, who "require fixed and steady hours of work" and would not accept undesirable shifts.[63]

The first official reference to Corps jobs at these installations concerned a controversy over the size of trucks which women might properly drive. WAAC Headquarters wished to set this limit at one and a half tons, but, upon the objection of the Army Service Forces and the reassurance of The Surgeon General, the limit was raised to two and a half tons. This provoked further objections from Aberdeen Proving Ground, which stated that the odd types of vehicles being driven and tested by its Waacs could not best be judged by tonnage. The Chief of Ordnance therefore asked an exception for this installation, which was concurred in by WAAC Headquarters.[64]

In the ballistics testing laboratories where ordnance equipment was examined, Wacs were employed in computing the velocity of bullets, measuring the weight of bomb fragments to determine the degree of fragmentation of bombs, mixing gunpowder, and loading shells. They also worked as draftsmen, mechanics, and electricians, and received practical training in ordnance engineering. A good percentage of the women were employed in ordinary clerical duties. Reports from overseas theaters revealed that Ordnance Wacs were found especially valuable in handling the complicated stock records of ordnance equipment, its procurement, storage, and shipment.[65]

Ordnance Department also expressed itself eager to employ WAC officers in highly technical skills—engineers, mathematicians, analysts, designers, artists, statisticians, lawyers, and physicists. Unfortunately, the WAC Officer Candidate School was designed for troop officers and turned out very few such specialists, most of them considerably below the desired qualifications.[66]

The Quartermaster Corps

Contrary to the general impression, those units and installations under the

---

[62] Commendations cited n. 59(1).

[63] (1) ASF WAC Summary. (2) Memo, CG Erie Proving Ground for CofOrd, 30 Jun 44. Ord 322/731. (3) Places of employment: Ordnance Depots: Atlanta, Frankfort, Mt. Ranier, Red River, and Savannah, Ordnance Districts: Philadelphia, St. Louis. Arsenals: Raritan, San Antonio.

[64] (1) Ltr, 151st WAAC Tech Co to Dir WAAC, 10 Feb 43; (2) Memo, WAAC Hq for TAG, 19 May 43; (3) Memo, WAAC Hq for SGO, 2 Jun 43; (4) AGO Memo W635-3-43; (5) Ltr, Aberdeen Proving Ground to CofOrd, 27 Aug 43. WA 451 (7-25-42) 1942, Vehicles for WAAC.

[65] (1) ASF WAC Summary. (2) Account by Col Merle H. Davis, Ch. XXII, below.

[66] (1) Memo, Tech Div OCofOrd for MPB Ord, 16 Nov 43. Ord 322.436. (2) Memo, Dir Pers ASF for CofOrd, 5 Jun 44. SPAP 210.3 (8-30-43).

control of The Quartermaster General did not offer particularly suitable conditions for the employment of Wacs.[67] An early survey by the Office of The Quartermaster General pointed out that it was not advisable to use Waacs in graves registration companies; they could not replace men in salvage collecting companies because personnel had to be armed and had to perform heavy work in battlefield cleanup; they could not be employed in laundry battalions, companies, or platoons because these were mobile units operating in combat areas. It was suggested that Waacs could be employed in fixed laundries in the continental United States, which normally employed 89 percent women, but these were civilian job vacancies. As for bakery units, it was decided that bread-baking was definitely a man's assignment because of the strength required. It was not felt that Waacs should take over operation of light motor vehicles, which were used to train men for combat areas.[68]

Furthermore, as in the Signal Corps, employment was restricted by the requirement that enlisted women be assigned only in units. Most quartermaster depots employed only a small detachment of enlisted men, and in many depots only civilians were employed. The Office of The Quartermaster General stated that, if civilian replacement and use of small numbers were authorized, Waacs might be used in places where civilians were scarce, or to take night shifts and other work that civilians would not accept. The Quartermaster General did not particularly recommend such assignment, pointing out that provision of suitable housing and messing would be a problem for small numbers, and that it would be undesirable to mix Waacs and civilians because of the difference in pay and privileges.[69]

Therefore, by the time of the conversion to Army status, the Quartermaster Corps employed only one WAAC unit, located at Camp Lee, Virginia, where the women worked in the Quartermaster School and the ASF Training Center. Although The Quartermaster General assured Director Hobby at this time that "On the whole the girls are pleased with their assignments and are performing them in a satisfactory manner," the unit at Camp Lee suffered more than usual losses at the conversion. One possible explanation was The Quartermaster General's admission that some 20 percent of the women had at that time not yet replaced men. The women at Camp Lee eventually came to be employed also by the Quartermaster Board and the Quartermaster Replacement Training Center at that installation. Only one additional unit was recorded—that of the Fort Robinson Quartermaster Depot.[70]

The employment of WAC officers, who could be scattered about in small numbers, was more suitable to the Quartermaster Corps than that of enlisted women. In early 1944 the Office of The Quartermaster General stated that "within the next few weeks a shortage of QMC officers will develop. A number of units are scheduled for activation and demands for over-

[67] Text is based on (1) Seven documents found in OQMG files. (2) ASF WAC Summary.

[68] (1) Memo, Serv Installations Div OQMG for Mil Plans Div, 26 Jan 43, and 4 Incls. QM 321 WAC 1943. (2) Memo, by Brig Gen Henry D. F. Munnikhuysen, OQMG. QM 320.2 WAC 1942. (3) 1st Ind, "K.H." to SigC, 16 Nov 42, to Bsc Ltr, Chief MPD CSigO to OQMG, 10 Nov 42. QM 320.2 WAC.

[69] *Ibid.*

[70] (1) Ltr, Maj Gen Edmund B. Gregory to Dir WAC, 16 Jul 43, QM 321 WAC. (2) Memo, Capt Laura Asbury, WAC Asst, for MTD ASF, 29 Mar 44. QM 320.2 WAC.

seas requisitions are heavy." [71] Accordingly, it was suggested to depot commanders that they requisition Wacs, and a combined requisition for 340 WAC officers was placed with the Army Service Forces, which by mid-1944 had netted some 100 WAC officers. These officers ordinarily required no training other than a brief orientation course, although some were admitted to the depot course at the Quartermaster School and at the Columbus ASF Depot. A WAC staff director, Maj. Laura Asbury, was eventually employed.

The only available strength figures indicated that the Quartermaster Corps eventually employed at least 250 enlisted women and 125 WAC officers on its own installations. An unknown number, probably much larger, was also employed on posts, camps, and stations by local quartermasters in offices, warehouses, and sales commissaries—these being under the jurisdiction of an air or service command or overseas theater, and not counted apart from the rest of the overhead. In 1944, The Quartermaster General stated:

> Personnel of the Women's Army Corps have become an integral part of this [QM] organization, and we now have unfilled requisitions from QM installations requesting more WAC personnel.[72]

The Transportation Corps

Early pessimists feared that ports of embarkation were no place for Wacs, who would undoubtedly be invalided by the rough physical work or seduced by the transient combat troops. Actually, the Transportation Corps proved second only to the Medical Department among Army Service Forces agencies in the number of Wacs employed, and first in the number employed on its own Class II installations. With a peak strength of over 5,000 women, some 6 percent of its total enlisted strength, the Transportation Corps was able to maintain large WAC detachments in the eight major ports of embarkation—New York, Boston, Hampton Roads, Charleston, Los Angeles, San Francisco, Seattle, and New Orleans. Each port had one or more WAC detachments and its own WAC staff director, whose work was co-ordinated by a senior staff director, Lt. Col. Mera Galloway, located in the Office of the Chief of Transportation.[73]

These 5,000 women were obtained by the Transportation Corps by means of enthusiastic pursuit along every possible avenue of recruiting. Since most ports were in labor-short areas, they were hard hit by the ASF's directives that general service men be replaced by Wacs and civilians. Civilians were scarce, and Wacs were at once requisitioned for every manner of duty. Some of these uses were approved, while some, involving shipboard service, were at the time considered too unusual.[74]

The number thus acquired proving insufficient, other women were obtained by transfer from commands which found

[71] (1) Ltr, Gen Munnikhuysen, OQMG, to CG Boston Depot, 12 Feb 44. QM 210.31 WAC. (2) Ltr, OQMG to 20 QM depots, etc, 21 Mar 44. QM 210.31 WAC. (3) Memo, OQMG for TAG, 2 Mar 44. QM 210.31 WAC.

[72] ASF WAC Summary.

[73] Unless otherwise documented, this section is based upon: (1) History of Hampton Roads POE, *The Road to Victory*, Vol. II, pp. 174 ff. Copies in OCMH. (2) Intervs with Capt Jane Engleman, WAC Stf Dir TC, 1945–46; Capt Fay Foushee, WAC Pers Off TC, 1946–Oct 1947; Lt Key, WAC Pers Off TC, Oct 1947–48. (3) ASF WAC Summary. (4) STM 30, Strength of the Army, 5 Jul 45.

[74] Ltr, TC to Dir WAC, 16 Dec 43; also 1st Ind and Ltr, Dir WAC to CofT, 28 Feb 44. SPTPM 220.3 WAC, in SPWA 320.2 (8-18-42)(1) sec 6.

them unassignable because of low aptitude or lack of skill; the Transportation Corps was confident that it could assign and use any sort of Wac. Puerto Rican Wacs, whose possible language difficulties were feared by some commands, were requested and successfully employed. When WAC training centers were forced to relinquish their bands in the interests of personnel economy, the Transportation Corps acquired the musicians. When at last station-and-job recruiting was authorized, the Transportation Corps used its own recruiters to conduct such a lively campaign for Port Wacs that it, like the Air Forces, had to be restrained from implying that a new corps had been set up.

Transportation Corps Job Assignments

The women assembled by these means were assigned to a variety of duties. Some 700 of them, in New York and San Francisco, were assigned to postal battalions where all V-mail letters were processed. Wacs opened the letters, checked for correct form and for rips and tears, and sent them to the commercial company that photographed them; incoming letters were likewise processed, with Wacs operating the machines that put each letter in an envelope. This work proved extremely monotonous, calling for 24-hour duty on several shifts, and was successfully performed for over two years.

Large numbers of Wacs, both enlisted and officer, also worked on the piers at every port, assisting in the staging of troops for overseas and the reception of returning troops. This type of assignment began in a small way when Wacs were tentatively assigned to assist in staging female personnel—civilians, nurses, and Wacs. More than a hundred Red Cross women thus staged wrote to the commanding general:

> It is our unanimous opinion that we have never been in an area where everything functioned more smoothly. There has been none of the confusion or discomfort which we had come to consider a normal part of the embarkation process.[75]

The use of Wacs in staging was soon expanded to include all types of personnel embarking or landing. WAC checkers on piers and ships processed not only outgoing personnel, but an estimated 800,000 items of supply for overseas forces. For example, in the San Francisco Unit Supply Office, Wacs not only kept records of the types of guns, the proper ammunition for each, and the type to be issued to each man according to his destination, but they also actually handed the gun to the man. One Wac said, "The soldiers come in unarmed and go out with a gun, and it gives me a pretty good feeling." As the men began to return from overseas, all-WAC bands in New York and San Francisco went out on small craft to meet the ships. As men came down the gangplanks, all-WAC teams met them for roll call and routine checking, after which other Wacs led them onto the proper train. Incoming officers also reported to a Wac's desk to receive any changes in orders. In this reception process, male military personnel were ordinarily employed only to check heavy baggage. The women generally reported no particular difficulty from the troops. One port noted, "Directions given [by Wacs] to troops debarking were cheerfully obeyed for the men 'just wanted to hear them talk.' "

In addition to postal and pier work, Wacs were employed in many other duties

[75] Ltr, Ruth Frederick, Gp Leader, American Red Cross, to CG NYPOE, 12 Nov 46. WAC Stf Dir TC file.

which differed little from those of other commands. At a typical large port, the Hampton Roads Port of Embarkation, it was reported:

> Though pier duty had priority over all other WAC assignments . . . [there were] dozens of other jobs in water, rail, motor vehicle, and mechanical maintenance, machine records, salvage, communications, photo-laboratory, mess, and supply. . . . The Wacs key-punched tons of manifests, cargo loading reports, convoy recaps, and what not . . . proofread passenger lists one inch thick, which were checked and rechecked. . . . There were 45 MOS numbers: butcher, sales clerk, auto-parts clerk, multigraph operator, quartermaster supply technician, tailor, photographer, movie projectionist. . . .[76]

At some ports, WAC medical companies were also assigned to administer shots, assist in processing, and keep records. During debarkations from hospital ships, these women were withdrawn from regular duties to act as guides, type admittance cards, and help in the Finance Department. On their own time, the women were discovered to be running errands and shopping for their patients, and carrying home the men's newly issued uniforms to iron. Most ports were deliberately striving to make wounded men feel that their country gave them a grateful and personal reception, and it was found that Wacs excelled in this effort. The effect on the women's morale was likewise good, as they felt less disposed to make known their own gripes when observing the greater discomforts of the wounded.

The chief difficulty encountered in the administration of Transportation Corps women concerned the fact that they often worked long hours and on night shifts as well as on monotonous duties, and were thus extremely susceptible to fatigue, nervousness, improper sleeping and eating habits, and impaired health. To combat this situation effectively, it was necessary for the cadre to maintain long and irregular hours. One commander of a postal battalion stated that for months she habitually slept for only two or three hours at a time, since the only time to discover the women's needs was when they were relaxing, talking, and eating after completion of their work shifts. As a result, where health and morale were maintained it was at the expense of the company officers, who wore out quickly.[77]

Later Experimental Assignments

As the Transportation Corps gained in experience in WAC administration, it began to show propensities, comparable only to those of the Air Forces, for experimentation with new and unusual duties. Soon after the invasion of Normandy in 1944, the all-WAC 29th Traffic Control Group was hand-picked from experienced Port Wacs and sent overseas. This group omitted the usual WAC overseas training at Fort Oglethorpe, and was entirely trained and equipped at the port, with such success that the European theater WAC staff director pronounced it the best-trained WAC group ever sent to Europe.[78]

Shortly afterward, to cope with the huge backlog of supplies for Europe which developed after the invasion of France, the Transportation Corps in September of 1944 organized a flying squad of approximately 100 Wacs, who were rushed to help clear out the jam at the Elmira Holding

[76] History cited n. 73(1), Ch. 14, written by Sgt Freda Zegman, HRPOE.

[77] Interv with Capt Theresa Davis, wartime CO WAC Det NYPOE.

[78] See Ch. XXI, below.

and Reconsignment Point. Here the Wacs found warehouses and a large open storage area crowded with an apparently hopeless assortment of boxcars, tanks, and other equipment. Much top-secret paper work was found to be involved in sorting manufacturers' lists and routing supplies to proper destinations. WAC clerks, checkers, and drivers worked outdoors for most of the winter, in three around-the-clock shifts, even operating the heavy lifts which moved and sorted supplies, and sleeping in the gymnasium of a local college. WAC officers later were of the opinion that the work involved had been too heavy for women, but no men were available to the Transportation Corps for the assignment and women performed it successfully.

With the end of the war and the resulting load of returning shipping, the Transportation Corps took a step not shared by any other zone of the interior agency, and set up rest camps similar to those used in combat areas. WAC camps were established shortly after men's camps were set up, one located near Camp Stoneman, California, and one at Vassar College. The idea proved extremely successful in combating fatigue, particularly for postal battalions and others in tedious and confining work.

Toward the end of the war, the Transportation Corps also achieved its previously thwarted objective of employing Wacs on shipboard. Earlier WAC objections to placing five or six enlisted women and a WAC officer on shipboard were based on fears that the women would either require expensive special provisions or would have none, since they could not use the men's recreational and living areas. Late in 1944 the Transportation Corps again requested Wacs for such duty, stating that "an urgent need exists for radio technicians for troop and hospital ships." [79] WAC radio operators were by this time as scarce as male ones, and could not be provided by the Signal Corps, but the nature of the idea appealed to the Air Forces, which voluntarily furnished Wacs who merited reward for long and superior service in radio instruction in AAF schools. Three or four enlisted women and one WAC officer were assigned to each of several hospital ships, where they proved successful; one reported later, "It has been weeks since we have been referred to as experiments." [80] With the success of the WAC radio operators, other Wacs came to be employed on shipboard as clerical workers and as medical technicians to assist the nurses.

Quarters for such women did not prove a major problem, since they were usually housed in a cabin with its own toilet facilities, or with the nurses. As feared, the life proved confining and lacking in provisions for recreation for women. Enlisted women were not permitted to fraternize with passengers of either sex or with crew members of the opposite sex. On hospital ships, women not on duty were allowed to visit with patients, which proved of some morale value to both. One Wac reported: "Most of them are anxious for a game of bridge or just talking . . . it makes us feel that our job is worth while if we have a small part in helping to bring them home safely."

Nevertheless, "hospital ship fatigue" often developed; commanding officers believed that two months' unbroken duty without shore leave should be the maximum, although some individuals were able to continue for a year or more with-

[79] Memo, CofT for CG ASF, 30 Sep 44. SPRPM 220.3 WAC.

[80] Ltr, Sgts Brown and Niles and Cpl Kunmick, undated. Folder, Hospital Ships, WAC Stf Dir TC file.

out loss of efficiency. On other ships, where members were at sea only about ten days on each trip, with stopovers in foreign ports, the restrictions of shipboard life proved less important. The only reported casualties were two WAC officers who had to be reassigned for habitual seasickness.

The employment of WAC officers, originally requested only to secure approval of requisitions for enlisted women, soon proved so successful that WAC officers were assigned to other ships. Many such officers were graduates of the Army's School for Personnel Services, and were assigned as reconditioning officer, special services officer, and related duties. They were responsible for conducting discussions in the wards concerning veterans' benefits, hospital policy, and discharge procedure, for preparing a daily news sheet, for conducting the orchestra, and for arranging variety shows in the wards.

War Bride Ships and Dependents' Ships

With the end of the war, the employment of Wacs on shipboard expanded into a new and sometimes controversial duty: that on "War Bride Ships" and "Dependents' Ships." On Army ships bearing Italian, French, German, English, and Australian brides to the United States, one WAC officer was ordinarily assigned as assistant transport commander. This officer was responsible for stocking the ship's exchange with supplies needed by women; for the women's welfare in such matters as housing and feeding; for supervision of the nursery and diet kitchen. She was appointed liaison between the transport commander and the women, to explain his policies; she maintained regular office hours when passengers could present their personal problems, which were usually numerous; she organized discussions on the monetary system, dress and customs, prices, products, and geography. Transport commanders confronted with voluble and excitable groups of passengers soon stated that WAC assistant transport commanders were invaluable, and eventually delegated to them the additional duties of post exchange officer and billeting officer. While Red Cross women, stewardesses, and Army nurses were also on board, the essential need was for a liaison officer in the chain of command and familiar with Transportation Corps policies.[81]

Good results were also obtained from the employment of enlisted women on the same ships, although there were ordinarily only a handful on any one ship. These were chiefly clerks, or medical technicians employed in the formula room under the direction of Army nurses, with little contact with passengers.

A different and somewhat greater problem concerned enlisted women passengers on the same ships. Such women, like enlisted men, were already resentful of the War Department policy which caused them to be allotted worse accommodations than the foreign civilian women who had married American soldiers. Their indignation frequently overflowed when, in addition to their ordinary work details such as kitchen police and latrine duty, transport commanders repeatedly assigned them to details involving day-and-night care of babies whose mothers were seasick, unsanitary in methods of infant care, or enjoying shipboard entertainment. The problem was not alleviated when war brides were supplanted by soldiers' wives and children en route to join men stationed overseas. These families, by Army Regulation, received better accommoda-

[81] For general discussion of War Bride Ships: Ltr, WAC Stf Dir NYPOE to CofT, 19 Jan 46, with incl. SPTAA 324.5 WAC, copy in WAC Pers Off, TC.

tions than male and female replacements on the same ships, and at the same time showed a certain tendency to regard the enlisted personnel as servants rather than as fellow passengers. Again, the Wacs' special objections concerned baby-sitting and other unmilitary work details often deemed especially suitable for enlisted women. In both cases, there appeared to be no solution as long as the Transportation Corps was required to carry civilian dependents while troops were still being transported, and to give civilian women better accommodations than those authorized for military women and men.

Evaluation

In discussions of a permanent women's corps, the Transportation Corps requested that from two to three and a half percent of its Regular Army strength be made up of Wacs in a variety of skills, principally clerks, postal workers, stenographers, typists, and communications experts.[82] Maj. Gen. Charles P. Gross, Chief of Transportation, stated:

> The Wacs have demonstrated thoroughly their value to the Transportation Corps. They have become an integral part of our ports, not only because of the quality of their work, but also because of the enthusiasm they have displayed and their loyalty to the Transportation Corps.[83]

However, with the placing of ocean transport for all services under the jurisdiction of the Navy, the Wacs on shipboard were scheduled for peacetime replacement by Waves, who had not previously been permitted to serve on ships.

The Adjutant General's Department

Although duties of The Adjutant General's Department were almost without exception suitable for women, that Department—like the other administrative services—did not have a number of Class II installations under its command as did the technical services, but merely provided personnel for station overhead. The AGD was co-operative in the matter of training, first admitting officers and then enlisted women to its schools at the request of the WAC. Such training was sometimes desired to qualify women for positions in WAC administration, and at other times to enable them successfully to replace male AGD personnel in various commands. WAC officers attended both the administrative and the classification courses at the AGD school and later the personnel consultants' course.[84]

Possibly the largest single group of AGD WAC employees was that assigned to the Classified Reproduction Section in the Pentagon, with the particular assignment of reproducing the weekly minutes of the General Council as well as other material for the office of the Deputy Chief of Staff—ordinarily top secret in nature. Of this work, the Deputy Chief of Staff, General McNarney, stated in 1944:

> WAC personnel were assigned to the Classified Reproduction Section in order to solve a most difficult problem. I feel that their performance has been a fine tribute to the Women's Army Corps, and I am extremely grateful for their fine work and loyal service.[85]

[82] Memo, CofT for TAG, 14 Jan 47, (Rpts Contl Symbol WDGPA-(OT)-78) file TCPER-MPB-320; in reply to Ltr, TAG for CofT, 26 Dec 46, A6AM-PM-320 (11-27-46).

[83] ASF WAC Summary, TC sec.

[84] (1) Memo, Dir WAAC for TAG, 17 May 43; (2) Memo, Exec WAAC for TAG, 4 Jun 43. SPWA 210.63, in AG file. (3) Memo, AGO Tng Br for Dir WAAC, 25 May 43. AG 210.63 (AG Sch) OT-S. (4) Memo, Exec WAAC for TAG, 23 Jul 43. SPAP 352 (5-19-43).

[85] ASF WAC Summary.

Finance Department

No statistics were preserved as to the numbers, if any, of Wacs on duty with the Finance Department in the service and air commands. Upon request, the Finance Department obligingly set up a quota of six for the WAC for its officers' course at Duke University, beginning in July of 1943. A little later, enlisted women were admitted to eight successive classes of the two-month enlisted finance course, to an eventual total of 305 women, of whom 266—about 87 percent—were graduated.

The only possibly unsuitable Finance duties for women were those in which it was necessary for the paying officer to wear or keep available a revolver or other weapon. Neither the Finance Department nor WAC Headquarters was able to see any reason why an otherwise qualified woman should not be allowed this privilege, inasmuch as the paying officer seldom was required to shoot anyone. However, the War Department refused to make an exception in this case to the general rule about "combat" assignments.[86]

Corps of Chaplains

When the WAAC was established, the Corps of Chaplains noted that there were two possible modes of utilization: that women ministers be used as chaplains for WAAC training centers, and that enlisted women be used as chaplains' assistants.[87] The first of these was never seriously considered. The Corps of Chaplains reported that, when the First WAAC Training Center was established:

> Quick to take advantage of the new situation were the organizations of women preachers, and a considerable amount of agitation followed. . . . The statutes omitted any reference to the sex of chaplains, and at that time War Department directives did not make any specific statement on the subject. However, by tradition, as well as by general Army policy, chaplains were assumed to be of the male sex.

All applications from women ministers for commissions in the Corps of Chaplains were therefore refused. Ministering to the spiritual needs of the Wacs did not prove a problem to male chaplains, who stated, "Actually, the presence of Army women seemed to make very little difference in the work of most chaplains."

The use of enlisted women as chaplains' assistants was likewise originally disapproved by the Chief of Chaplains, who felt that it might prove compromising for a chaplain to be shut up alone in his office with a female assistant, and that even if it did not, the enlisted men who ordinarily revealed intimate problems to a chaplain would not do so if a woman was present. The assignment was not forbidden, but was left to the discretion of stations in the field. Within a few months, such assignments began to take place on the initiative of local chaplains and personnel officers.

Some of the women assigned were clerk-typists and general office workers, while others were musicians, such as organists and choir directors, or organizers of church activities. The Corps of Chaplains, after some experience with such employ-

[86] (1) Historical monograph, Personnel, General. Job A46-156, Drawer 4, ASF files of OCofFin, Jul 1940-Nov 1945. (2) Memo, Fiscal Dir for TAG, 21 Jun 43. AG 210.63 Army Fin Schs (2-9-43)(1). (3) Historical Monograph, Mil Tng, OCofFin, Tab G, Basic WAC Enl Fin Course. Job A46-156, Drawer 4, OCofFin.

[87] This section, including quotations, is based on: (1) Military History of the Second World War: Corps of Chaplains, Part III, pp. 95 ff. OCMH. (2) Memo, Lois Hill, Secy, for Dir WAAC, 10 Mar 43. SPWA 230.3.

ment, reported, "It was soon found that they could be used as Chaplains' Assistants to the same degree as men in the Army."

It was observed that women in the Army played much the same part in a post's religious life that civilian women did in the average community: they attended services more readily, formed more than a proportionate share of the choir, and in general were willing to lend more support to church activities. Under the circumstances, the employment of women as full-time chaplains' assistants ceased to excite comment. Estimates indicated that Wacs formally assigned to work with the Corps of Chaplains did not number more than a few hundred.

Only the Army Air Forces formalized the employment program to the extent of sending women to a school. Women already assigned to this duty were permitted to attend the AAF Chaplains' Assistants School for training in choir work, maintenance of records, and other necessary activities. These women represented all of the three major religious groups in the United States and both the white and the Negro races.

The Provost Marshal's Department

There was no official indication that Wacs were ever employed, in World War II, to replace male military police except at WAC training centers, where such employment was more or less necessary to avoid embarrassing male police. Early in the WAAC's existence, G–3 Division of the War Department had suggested that it was undesirable to assign men in the younger age groups "to units or activities not to be used for combat service" and that the necessary high caliber of military police be secured from older age groups, limited service men, and Waacs—the Waacs to be employed "to the fullest extent where appropriate." [88] Such assignment was never considered by WAAC Headquarters to be especially appropriate except in clerical positions in Provost Marshal offices or to assume custody of women.

Even this latter use was not realized to the extent deemed permissible by WAAC Headquarters. A request was made by Director Hobby in the summer of 1943 for WAC MP detachments at large centers of population where female absentees were often apprehended, but this was refused by the Army Service Forces because male MP's would not have been replaced.[89]

Throughout the war there were frequent rumors and reports that Wacs in scattered stations and overseas were being employed as military police; at times publicity photographs showed women wearing the armbands of MP's or parking guides. Upon investigation these were usually found to be office workers in a posed photograph. No employment involving the carrying of weapons was officially authorized for Wacs.

[88] Memo, G–3 for CG SOS, 26 Sep 42. WDGCT 322.999 MP (9–16–42), in SPAP 320.2 General.

[89] Memo, MPD ASF for Dir Pers ASF, 10 Aug 43. SPAP 320.2 WAAC (5–17–43).

CHAPTER XIX

The Medical Department (ASF)

Slowest of all ASF services to employ Wacs, the Medical Department by the end of the war had become the greatest user, employing some 20,000 Wacs, one fifth of the entire Corps, almost one half of Army Service Forces WAC personnel. These were to be found in general, regional, and station hospitals, ports, air bases, and convalescent centers, as well as in the few installations directly under The Surgeon General's command, such as the Army Medical Center in Washington.

The slow start resulted from the fact that well-trained civilian personnel and enlisted men were plentiful when the WAAC was organized. "It is the opinion of this office," stated The Surgeon General at that time, "that it is not desirable to utilize the WAAC in Army hospitals." [1] WAAC Preplanners had not pressed the issue, since requisitions from other agencies were already more than could be filled.[2]

In 1943, when the Service Forces surveyed the possibilities of expanding the Corps to a million members, The Surgeon General appointed a board of medical officers to reconsider the employment of Waacs. Hospital commanders, queried by the board, requested Waacs to a total of some 10,000 women for the ASF and almost as many for the AAF, to replace from 30 to 50 percent of their enlisted men. Nothing came of the study; Director Hobby informed the board that she was "in accord with the idea that members of the WAAC could be utilized in such a capacity," but the difficulties of recruiting prevented her from supplying the personnel.[3]

During the summer of the conversion, the WAAC was able to make available to The Surgeon General less than two hundred individuals a month, for technical training at the Army-Navy General Hospital.[4] Reports indicated that hospitals in the field were also obtaining certain numbers of women from post headquarters companies, although the women had been neither trained nor equipped for the work.

[1] 1st Ind, SGO to CG SOS, 14 Apr 42. SPMCP, SGO 322.5-1, in SPMCM 322.5-1 WAC, SGO Hist Div.

[2] Later comments from medical officers indicated a universal idea that the Director had attempted to force Waacs on Army hospitals. Actually, only two routine inquiries had ever been made, as a part of surveys—the first before Director Hobby's arrival in 1942, and the second by the ASF in 1943. No follow-up was made for agencies replying in the negative. See Chs. II and V.

[3] (1) Rpt of Proceedings of Bd of Offs, 11 Mar 43, appointed by Office Order 41, SGO, 26 Jan 43; (2) Ltr, SGO to all SvCs, AAF and CofT, 29 Jan 43. SPMCT, SPMCM 322.5-1 WAC, SGO Hist Div.

[4] Memo, SGO for CG ASF, 2 Jun 43. SPMCM 322.5-1 WAC, SGO Hist Div.

In early 1944, the nationwide personnel shortage caused the Medical Department an unprecedented loss: the forced transfer to the Infantry of some 5,000 combat-fit men, trained technicians upon whom hospitals had depended for normal operation. This loss brought about really urgent requisitions for Wacs. "The Medical Department is critically short," The Surgeon General's Office protested. "Efforts to hire civilians in sufficient numbers have as a whole failed. . . . With the withdrawal of enlisted men, there will be an urgent need for at least 50,000 additional Wacs to work in medical installations." The Surgeon General therefore asked an "extensive recruiting program" to recruit Wacs specifically for the Medical Department.[5]

The request for 50,000 Wacs was impossible of fulfillment, nor could the Medical Department be given first priority on even the actual intake without causing withdrawal of Air Forces, port, and other recruiting teams. Instead, Director Hobby was able to offer the Medical Department merely the privilege of recruiting for hospitals under the station-and-job plan, which was a simple matter of adding them to the existing list of participating installations. In the spring of 1944, the Corps' recruiting and publicity media therefore lent their support to what was called the Female Medical Technicians Campaign.

Female Medical Technicians Campaign

The Medical Department's demands were unusually difficult to meet, in that only highly qualified workers were acceptable.[6] Particularly sought were women with college training who could serve as bacteriologists, instructors in lip reading and Braille, pharmacists, optometrists, psychiatric social workers, orthopedic mechanics, and numerous other technicians.

Specifications were set for education, training, and experience which were so exceptionally high that the ordinary recruiting station was deemed unable to evaluate them accurately. The Medical Department therefore followed the WAVES' example and secured the assistance of the Officer Procurement Service, which assisted regular recruiters by filling out a qualification certificate guaranteeing assignment to, and suitability for, a technical job.

For example, a WAC psychiatric social worker (Specification Serial Number 263) was required to have "at least two years of supervised experience in social case-work . . . or a graduate degree in social work," and it was stated that those accepted "will be assigned directly upon completion of basic training and are eligible on the basis of such assignment to attain the grade of staff sergeant, dependent on individual ability and the existence of vacancies." A WAC laboratory technician was required

[5] (1) Memo, Brig Gen Raymond W. Bliss, Chief, Opns Serv SGO, for Dir Pers ASF, 11 Feb 44, SPMCP 322.5-1; Ind, Dir WAC to Dir Pers ASF, 12 Feb 44, SPWA 341. (2) Memo, Col Arthur B. Welsh for Col John R. Wood, re Off Procurement Serv. (3) Memo, MPD ASF for TAG, 11 Apr 44. SPGAP 341 Gen-69. (4) Memo, Chief, Pers Serv SGO, for Dir MPD ASF, 2 Jun 44. SPMCM 322.5-1 WAC, SGO Hist Div.

[6] This section is based on the following: (1) Memo, Lt Col Edward R. Whitehurst, Chief, Enlmt Br SGO, to Chief of Pers SGO, 13 Mar 44; (2) SGO Ltr, Chief, Pers Servs SGO to each SvC Surg, 15 Mar 44; (3) Memo, Chief, Enlmt Br SGO, for Dir Officer Procurement Service, 13 Apr 44; (4) Ltr, ASF Procurement Div for CGs of all SvCs, 24 Mar 44, FR-198; with incl Ltr, SGO to all SvCs, 15 Mar 44; superseded by Ltr, 15 Jun 44, FR-202. SPMCM 322.5-1 WAC, SGO Hist Div. (5) WD Cir 340, 29 Dec 43. (6) Interv with SGO WAC Stf Dir, Capt Maribeth Turnbull, 13 Feb 46. (7) SGO Tng Div, Rpt on Schooling of Enl Pers, MD, 1 Jul 39-30 Jun 44. OCMH. (8) Memo, G-1 for Asst DCofS, 11 Jan 45. CofS 324.5 WAC. June-December 1944 intake: 4.099.

to have a B.S. degree or three years' experience, and was similarly promised eligibility for a staff sergeant's rating. Other specialties were comparable. The Surgeon General agreed in return to accept the Officer Procurement Service's recommendation without change and to provide the guaranteed assignment.

In addition, regular recruiters were authorized to accept other women for three months' training as medical and surgical technicians, provided that the women had AGCT scores of 100 or better and had graduated from high school. Thus, only women in the upper half of the intelligence range of the population, and of near-officer caliber, were channeled to the Medical Department.

In spite of the high requirements, the Female Medical Technicians Campaign proved reasonably successful, the Medical Department's glamor obviously being second only to the Air Forces'. Women were recruited so surprisingly fast that the entire Medical Technicians' School reserved for WAC training proved inadequate. Within a few months some 4,000 well-qualified women had been recruited for Army hospitals—a number which, though short of 50,000, proved to be all that was required at the time.

Success of Training

It was quickly demonstrated that Medical Department training courses offered women no unusual difficulties. In early months, Medical Department schools were already carrying a peak load, and housing and sanitary facilities for women were so inadequate as for a time to cause a serious welfare problem. This condition was remedied gradually as male enlisted personnel for training became scarcer, until by the end of the war over half of the students at the enlisted technicians schools were Wacs. No alteration was required in the usual training courses for men, except for omission of training in catheterization, which was considered unsuitable. Women in most cases qualified in two months for specialties that ordinarily required three months' training, and in three months for other specialties normally requiring four. Morale and discipline were generally high.[7]

Relative Success in Medical Jobs

The actual success of women in various types of hospital work varied with the duties. The most successful women were generally those who performed the more technical duties, especially ones related to patient care and reconditioning. Hospital authorities at first had felt that only male combat returnees should help rehabilitate the wounded, "due to personality changes of combat-wounded patients," which would presumably make them allergic to noncombat soldiers, especially female ones. When the first women began to be assigned to hospital work, authorities discovered, instead, that "the psychological reaction to feminine association has been most beneficial in combating certain prevalent attitudes of overseas returnees." Even untrained Wacs, according to commanders, showed natural ability for the work and obtained satisfaction from it. Authorities concluded that, "due to the personal services rendered and to the enthusiasm and understanding exhibited by

[7] (1) SGO Hist Div, Med Dept Tng Activities, 1 Jul 45–2 Sept 45, A Supplement. OCMH. During this period: *Enrolled:* 1,838 EM, 1,506 EW. *Graduated:* 588 EM, 1,329 EW. (2) Supplement to Rpt cited n. 6(7). (3) SGO Hist Div, Hist of Tng of MD Female Pers. OCMH.

such personnel, there has been a marked increase in the morale of the patients which is definitely considered to be an important part of the therapeutic regimen." In hospitals that pioneered in using Wacs in such work, complete satisfaction was expressed by all concerned.[8]

Psychiatric Social Worker and Psychiatric Assistant

Although earlier surveys had pronounced women unsuitable for any sort of psychological work with men, the Medical Department successfully employed more than 300 women as psychiatric social workers and psychiatric assistants. Women performed case work, interviewed patients and wrote case histories, carried out psychiatrists' mental hygiene prescriptions, and maintained necessary files and records on mental patients.[9]

Educational and Physical Reconditioning Personnel

More than 150 women with college degrees and teaching experience were enlisted and given one month's training in the educational reconditioning course at the Army's Special Services school. They were then employed to teach various school subjects to patients during convalescent periods in Army hospitals. Women were pronounced by medical authorities to be particularly successful in such teaching, since "men are conditioned to female instructors . . . and there is a psychological advantage to their employment."

The same school's course in physical reconditioning was barred to Wacs as inappropriate. One Wac accidentally attended, graduated, and was successfully assigned; others did the work without such training and were pronounced "capable of fully participating in physical reconditioning, without specific limitations." When Wacs conducted games, recreational athletics, and exercises for male patients, there was noted "a marked increase of patient participation and enthusiasm."[10]

Therapy Assistants

Although the Medical Department had no authority to commission the highly trained civilian occupational therapists, as did the Navy, it succeeded in securing several on an enlisted status. In addition, more than two hundred women, experts in some handicraft, were recruited and assigned to convalescent hospitals to lighten the load on civilian occupational therapists by teaching leatherwork, weaving, and other crafts to the mentally or physically handicapped.

To assist commissioned physical therapists, the Medical Department gave a short course which successfully trained some 500 Wacs as physiotherapy aides, for duty in general and convalescent hospitals.[11]

---

[8] Memo, Fld Dir, Reconditioning Consultants Div SGO, for SG, 16 Feb 45. WDWAC 341.

[9] This and following subsections based on: (1) Rpts cited ns. 6(7), 7(1), 7(2). (2) Interv cited n. 6(6). (3) Final Rpt for 29 Jun–4 Aug 45, Lt Col E. R. Whitehurst, WAC in Procurement of Female Med Technicians, SGO Tng Div, 1945. It is not apparent whether this list includes both the Officer Procurement Service recruits in the 1944 Female Medical Technicians Campaign, *and* the General Hospital Campaign recruits of 1945, or only one of the two. In any case it has been used as a safe minimum.

[10] (1) Memo cited n. 8. (2) Rpt cited n. 7(2). (3) Interv cited n. 6(6).

[11] (1) Rpts cited ns. 6(7), 7(1), 7(2), and Interv cited n. 6(6). (2) Memo, Exec SGO for Dir WAC, 14 Aug 44. SPMC/DD-DW 322.5, in WDWAC 720. (3) Ltr, SGO to TAG, 2 Apr 45. SPMCM, in SGO Hist Div 220.1 WAC.

Laboratory Technicians

More than 700 women were recruited and assigned as laboratory technicians. Those with a B.S. degree or three years' experience were assigned immediately, while others were acceptable for the Army school if they had graduated from high school in the upper half of the class and had some laboratory experience. Women seemed entirely suitable for such duties, which consisted chiefly of analyzing specimens, making bloodcounts, preparing reagents, performing basal metabolism tests, and typing blood.

Similarly, some 500 women were recruited for X-ray work—those with experience for immediate assignment, and those with high school education, including physics and photography, for three months' training in Army schools. Technicians not only operated X-ray equipment, but prepared patients for treatment, managed protective measures, and developed film.

Medical Stenographers and Clerks

Recruiters had considerable success in locating recruits with both a knowledge of medical terms and the necessary clerical or stenographic skills. Such women were used in hospital offices to type records, prepare reports, take dictation at conferences, and construct technical charts.

Miscellaneous Technicians

With promises of a master sergeant's rating, recruiters succeeded in enlisting approximately a dozen dental hygienists, able to scale and polish teeth and perform other oral prophylactic work. Larger numbers of women with high school education, and of the proper size and weight to fit into a dental cubicle, were also enlisted for three months' Army training as dental technicians and assistants.

The Officer Procurement Service also secured small numbers of recruits for various other technical duties: some forty pharmacists and pharmacists' aides, a half dozen lip reading technicians, and lesser numbers of Braille teachers, optometrists, hearing aids technicians, and others in rare specialties.

Only one flaw marred the assignment of such specialists during most of 1944. This was a certain lapse in liaison between procurement and employment authorities on the matter of salaries. Thus, the Officer Procurement Service's advertisements in professional journals had headlined the Medical Department's statement that Army specialist pay compared favorably with civilian technician pay, and that promotion to the specified rank for each duty depended upon "the proficiencies of the individual and existing vacancies." It had not been made clear that there were no existing vacancies and would not be for the current war; most hospitals' grade allotments were already exceeded by assigned men and by the daily return of highly rated men from overseas. Women recruited under the Female Medical Technicians Campaign therefore could rarely be promoted at all, and almost never to the specified rank. Hospital commanders naturally refused the requests for discharge of specialists who had enlisted under the wrong impression, and tended to blame the Officer Procurement Service for overglamorization of opportunities. On the other hand, that agency was able to show that it had merely operated on the written specifications furnished it by The Surgeon General's Office, and was not clairvoyant as to the internal conditions in Army hospitals.[12]

Commissioned Duties

Somewhat more fortunate were those few skilled technicians who were recruited on a commissioned instead of enlisted status.[13] Since The Surgeon General lacked Congressional authority to commission females in the Sanitary Corps, a special arrangement was made whereby WAC bacteriologists, serologists, and biochemists were commissioned in the WAC and detailed in the Sanitary Corps. Similarly, a number of WAC officers holding administrative jobs in hospitals were detailed in the Medical Administrative Corps to replace combat-fit male officers.

The Surgeon General had for some time had Congressional approval of commissioning women physiotherapists in the Medical Corps, without reference to the WAC. For a time the training course leading to a commission was open only to civilian women. Finally, at Colonel Hobby's insistence and because of difficulties in recruiting students, qualified enlisted women were offered the same opportunity, and several hundred were commissioned. Eventually the Medical Department's interests in this work were found better safeguarded by WAC recruitment of college graduates for training on an enlisted status and subsequent commissioning, and the greater part of the procurement of commissioned physical therapists was thereafter managed through the WAC.

Medical and Surgical Technicians

Greater in numbers than all of the skilled specialists obtained through the Officer Procurement Service or by commissions were those women obtained direct through recruiting stations—the medical and surgical technicians.[14] These were to number from 6,000 to 10,000 by the end of the war. Such women ordinarily had no medical experience, but upon proof of the required education and AGCT score were promised training in the Army's enlisted technicians schools and subsequent assignment as medical or surgical technicians.

The duties of such ward personnel proved in general only slightly less suitable for women than those of the Officer Procurement Service technicians. Medical technicians made beds, gave baths, took temperature, pulse, and respiration; prepared patients for meals, carried trays, and fed patients if necessary; gave enemas and bedpans; filled icebags and hot water bottles; kept ward records, prepared dressings, and kept linen closets neat. Surgical technicians performed much the same duties, as well as sterilizing gowns and equipment and giving pre- and postoperative care. In general both were intended to assist nurses by relieving them of simple details of patient care.

Wacs ordinarily proved highly adept at such simple nursing duties, and approached the work with an enthusiasm and a sense of dedication not always found among male technicians. On the other hand, it was an accepted principle from the beginning that women could not replace all or even half of the male tech-

[12] Ltr, Dir Tng, MD Enl Techns Sch, Camp Atterbury, Ind., to Dir WAC, 28 Aug 44, and reply. WDWAC 210.3 (1945).

[13] Sources for this section cited ns. 9, 11.

[14] WD Cir 121, 19 Apr 45, defining duties of all SSN's. Ltr, Dir Pers ASF to Maj Gen Russell B. Reynolds, 15 Feb 45, with incl. SPAP 220.3 WAC (7-28-43). Statistics from Rpt cited n. 9(3).

nicians in any installation, since certain duties required great physical strength and others involved services for male patients that were considered inappropriate for women to perform. The accepted ratio was about 40 percent enlisted women to 60 percent men. Where these ratios were preserved, the jobs of medical and surgical technicians appeared extremely well suited for women.

The WAC medical and surgical technicians nevertheless suffered from a particularly obvious comparison with civilian women employed as paid nurses' aides. Trained by the Red Cross, these were almost identical in qualifications and training with the WAC medical technician, and were intended for similar duties. Civilian nurses' aides had figured in Medical Department history since World War I, and were frequently women with home responsibilities who had accepted the work for patriotic reasons. Nevertheless, it was a rare enlisted woman who did not remark the fact that civilian nurses' aides could quit if not given desirable duties, got three times the Wac's salary for half the hours per week, and in addition had officers' privileges on the post, all of which the Wac herself might have had but for the wiles of an Army recruiter.

Possibly no more unhappy example had yet occurred of the unwisdom of recruiting both military and civilian personnel for identical duties and of using them on the same installation. The final indignity to many enlisted women came when all hospital civilian employees were allowed to wear the Wacs' uniform, the surplus blue cotton dress formerly worn by nurses, thus depriving enlisted women of the only visible glory of military status.

The Hospital Orderly

The only hospital job on which women were admittedly unsuccessful was the least skilled: that of hospital orderly, or corpsman. Women had never been recruited for this purpose, but in 1943, when certain low-grade workers accumulated at training centers, the WAC requested the Medical Department to train them as ward orderlies.[15] The Medical Department accordingly set up an experimental training course, but within a few months was obliged to abandon it, pronouncing women unsuitable for hospital orderlies because of "the limited tasks to which this type of personnel might be assigned, and the impossibility of using them to replace enlisted men as ward orderlies." [16] These women, unlike enlisted men of the same category, could not successfully work long hours lifting heavy objects, pushing loaded food and linen carts, and scrubbing wards and corridors.

Two more attempts to train Wacs as orderlies, at Mayo and Nichols General Hospitals, likewise failed, and The Surgeon General informed the Director that such personnel required so much supervision that the nurses' load was increased instead of lightened. In any case women orderlies could not replace men on a one-for-one basis because of lesser physical strength. Therefore, the position of orderly was not represented in the Female Medical Technicians Campaign, and the standards set for recruits were far above that appropriate for such personnel.[17]

[15] Draft of Memo, WAAC Hq for SvCs, 11 Aug 43. SPWA 220.3.

[16] Rpt cited n. 6(7).

[17] Memo, SGO for ASF, 21 Oct 44. SPAP 319.1 WAC (10-9-43).

Under the circumstances, the War Department was at first unable to account for a growing trend that began to make employment of WAC orderlies the rule rather than the exception. Widespread reports indicated that in many hospitals ward orderlies were beginning to outnumber technicians among WAC personnel. Hospital commanders, for reasons not at once apparent, showed an increasing tendency to assign Wacs of high intelligence and technical training to duties that were exclusively those of charwomen and kitchen police, even when such women had in their possession written Recruiting Service guarantees of assignment as medical technicians, which by specification included no work heavier than "keeps bedside tables neat."

Repeated investigations shortly made clear that, although the Medical Department had acted on the assumption that technicians were needed, hospitals were if anything overstaffed on personnel for patient care and short chiefly in low-grade positions that civilians would no longer accept. A civilian observer reported, "Cadet Nurses, civilian nurses' aides, civilian nurses, Army nurses, and enlisted men are all above them [Wacs] in the hospital caste system and do all the actual medical care of the patients." [18] It was clear—at too late a date—that the Medical Department should instead have asked a campaign to recruit women in the lowest aptitude and educational group and without commitment as to duties.

All reports confirmed the situation. The commanding officer of Gardiner General Hospital told inspectors that he "had no use for Wacs except as ward orderlies whom he might use on cleaning and scrubbing duties." [19] At Fort Jackson, South Carolina, the commanding officer published a memo stating, "All [WAC] personnel . . . will be used entirely for cleaning of wards." [20] From Halloran General Hospital, a Wac wrote:

> If I had only known before I joined what I know now they could have shot me before I would have ever joined. Your people at the recruiting office show a beautiful film that shows the girls at work really doing things for the boys. . . . We scrub walls, floors, make beds, dust, sweep, and such only. . . . They should stop this farce of recruiting 'medical technicians' and ask for 'mop commandos.' [21]

Even an indignant civilian worker wrote, "Most of them [Wacs] had medical technician training [but] cleaning is all they do officially." [22] Civilian visitors to hospitals—including Mrs. Robert P. Patterson, wife of the Under Secretary of War—likewise protested to the War Department concerning conditions they saw. Mrs. Patterson commented on the WAC orderlies' makeshift uniforms, bedraggled appearance, overstrenuous duties in kitchens, long hours, and lack of the privileges and training given enlisted men. Military visitors carried the same report. Conditions discovered on one visit by the ASF WAC Officer, Colonel Goodwin, were so alarming that they were personally communicated to The Surgeon General's Office by General Dalton, with the advice that he would follow up on action taken. This he directed to include better attention to the women's appearance, better

[18] Ltr cited n. 14.

[19] Memo, Capt Chance, GSC, for Maj Gen Stephen G. Henry, G-1, 13 Mar 45. WDWAC 331.1.

[20] Memo, Asst CO for all Ward Offs, Ft Jackson Regional Hosp, 18 Feb 44, atchd to Ltr, Dir Pers ASF to CG 4th SvC, 26 Mar 45. SPAP 220.3 WAC (7-28-43).

[21] Atchd to Ltr, Dir Pers ASF to Dir Pers 2d SvC, 16 Feb 45. SPAP 220.3 WAC (7-28-43).

[22] Incl to Ltr cited n. 14.

salvage and supply, provision of one day off a week for kitchen police and orderlies, and an investigation of the chances of upgrading orderlies.[23]

By the summer of 1944, even before the end of the Female Medical Technicians Campaign, the problem had become one of the Corps' most serious. For the few women who had come into the Medical Department before the campaign, the question was merely one of malassignment to a duty possibly above their strength. For those who had been promised assignment as medical and surgical technicians, the matter was the more urgent one of the Army's good faith. A more unfortunate situation, or one more provocative of resentment on all sides, had never before occurred in the Corps' history. To the hard-pressed hospital commander, charged with maintaining patient care, the insistence upon fulfillment of the written recruiting promises seemed an unwarranted infringement upon his command prerogatives in assigning duties to keep the hospital functioning. For the WAC Recruiting Service, it was pure disaster. No other Army command, however much it deplored the necessity of recruiting promises, had ever deliberately violated them, and the Recruiting Service saw the source of supply of Wacs for the entire Army threatened by the personnel practices of one agency.

Request for Survey

By early summer of 1944, with the Female Medical Technicians Campaign still continuing, the number of complaints reaching the Director's Office "from civilian and Congressional sources" was so great that Deputy Director Rice asked the major commands for an investigation to determine the actual situation:

> The few investigations conducted in response to some of these criticisms have substantiated the fact that conditions in these assignments are in many instances contrary to the well-being of women. Realizing the load on the hospitals . . . the Director WAC does not wish to recommend to the War Department that women not be utilized in this capacity, but because of conditions now alleged to exist in both time-length of daily scheduled duties (without commensurate time off) and in actual scrubbing work and other menial tasks, she feels that her responsibility for the well-being of women calls for study of such utilization.[24]

A simultaneous move came from The Inspector General, who informed the Chief of Staff, ASF, that "possibly there is misassignment of WAC personnel," and asked a survey.[25]

Reports of surveys, when received some months later, disclosed that the situation was in some respects worse than expected. In the Army Service Forces alone, over 1,200 enlisted women—about 5 percent of its total WAC strength—were found assigned as orderlies. Almost every station or general hospital was guilty in an appreciable number of cases. At one, ten technicians were used as orderlies, at another seven, at another six—in every case a percentage of the company large enough to cause comment. It was found that nonmedical specialists were also being mal-

[23] (1) Memo, "JND" for Gen Somervell, 13 Jul 44. SPAP 341 WAAC (2-26-43). (2) Routing Slip, Dep Dir Pers ASF to Gen Somervell, 13 Jul 44. Folder, WAC Rctg, ASF Sp Coll DRB AGO.

[24] Memo, Lt Col Rice for CG AAF, AGF, and ASF, 29 Jun 44. WDWAC 720.

[25] Memo, Gen Styer, CofS ASF, for Dir Pers ASF, 6 Jul 44, with Memo, ASF WAC Off for Dir MPD ASF, 31 Jul 44. SPEX, in SPAP 220.3 WAC (7-28-43).

assigned as orderlies, one station alone having converted thirty graduates of the motor transport school to this use.[26]

Hours of Work

Surveys also confirmed another fact which began to suggest that hospital work on an enlisted status was possibly not proper for women. This was the difference in hours of work for military and for civilian personnel. The customary hours in Army hospitals were 40 hours a week for civilian employees or 48 hours with overtime pay, and 8 hours a day for Army nurses. Enlisted men were believed to have more endurance, and traditionally worked a 12-hour day in Army hospitals.

The difficulty for Wacs was that they were both female and enlisted. As females, a 12-hour day was no more appropriate for Wacs than for nurses, and even less appropriate when Wacs were used for strenuous physical labor. As one female medical officer observed, "Wacs are replacing enlisted men but you are not turning women into men by an Act of Congress. . . . Wacs are enlisted women but they are still women."[27] On the other hand, Wacs were enlisted personnel, and if they worked only an 8-hour day the damage to the morale of enlisted men on the same wards was reported to be considerable because of the discrimination involved, and one-for-one replacement was made even more difficult.

The surveys revealed that enlisted women were almost without exception required by hospital commanders to work 12 hours a day and often more. Also, many were working 7 days a week and were not allowed compensatory time off during the work day or any leaves or passes. In addition to the regular 12-hour day, all were obliged to take night calls in emergencies, since civilian nurses' aides were not quartered on the post. Wacs also had to manage on their own time the four-hour-weekly continuation training courses required by the ASF, and other duties such as company fatigue, inspections, personal grooming, and sometimes laundry. As a result, enlisted women in Army hospitals were found to be working a minimum of 72 hours weekly and a maximum of at least 100, in addition to company duties. For example, Camp Wheeler, Georgia, employed trained WAC medical technicians as ward orderlies on a 12-hour, 7-day week, a total of 84 hours a week exclusive of emergency duty, training, and inspection. Camp Croft used a 13½-hour day, 6-day week, or 81 hours a week plus extras. Other variants at different hospitals produced the same general totals.[28]

These discoveries posed one of the most difficult questions to be encountered by the Army in its employment of womanpower: to what extent the Army could be bound by labor laws and civilian rules in the hours of work for women. Eighteen of the states had laws to the effect that 48 hours per week, or 8 hours per day for 6 days, were the maximum for women workers. In England, the Factory Act of 1937 placed the maximum for women at from 48 to 54 hours, and even in the darkest days after Dunkerque, the Ministry of Labour recommended that these hours not be exceeded. The U.S. Army Industrial Hygiene Laboratory, after a study of women in industry, reported:

[26] Memo, MPD ASF for G-1, 12 Sep 44. SPGAM 322.5 WAC (9-6-44)-97, in SPAP 220.3 WAC (7-28-43).

[27] Min, Med Tng Conf, Ft Oglethorpe, 1-3 May 45, pp. 29-30, comments of Maj Margaret Janeway. WAC folder, SGO Tng Div.

[28] Memo cited n. 26.

One of the lessons learned in World War I and again after the collapse of France in 1940 was that excessively long hours of work do not ultimately pay, even when considered solely on the basis of output and apart from the effect on health. . . . All of the available evidence from England and this country indicates that working hours in operations involving a fair amount of physical effort should not exceed 60–65 per week for men, and 55–60 per week for women.

A committee representing the Army, Navy, Public Health Service, War Manpower Commission, Labor Department and other agencies had agreed in 1942 concerning either male or female civilians that "one scheduled day of rest for the individual approximately every 7 days should be a universal and invariable rule." [29]

The Army Medical Department was therefore obviously exceeding all accepted safety limits in requiring a routine 72 hours a week and frequently 84 or 100 hours, and a 7-day week. On the other hand, the Army was in a somewhat different position from civilian agencies. Since it was entitled to ask a man's life in combat, it appeared equally entitled to ask that a woman sacrifice her health if required. There was a strong feeling in the Medical Department that any special regulations or labor laws concerning hours or type of work for women constituted discrimination; also, many Army doctors did not believe that any real damage to health would result. A representative of The Surgeon General stated, in refusing to accept the assignment of a WAC staff director for the Medical Department:

Some think that there has been entirely too much said on the subject of the physical capabilities of males vs. females and I can only say that I don't think any of them work as hard as they will when they become housewives and mothers.[30]

Director Hobby's recommendation, after she had considered the survey, was that any sacrifice was justified, either of health or life, if a true emergency required it; but that, since commanders did not deliberately kill men unnecessarily, the WAC should not unnecessarily sacrifice women's health. As a means of determining whether a true emergency existed, she proposed that the hours worked by Army nurses be considered the criterion. If nurses were not busy enough to require extended hours, and if the situation continued over months and years, there obviously was no emergency but rather a failure of hospital commanders to take aggressive action to improve personnel management. She therefore proposed that WAC and Nurse Corps hours be identical. This theory met with the general disapproval of hospital commanders, and was not to be acceptable for many months.

There was thus by the end of the summer of 1944 considerable doubt in WAC authorities' minds as to whether women should properly be employed at all in hospitals on an enlisted status, but should not instead be hired as civilians on a 48-hour week and with the privilege of quitting if overtaxed, a move that would also avoid the one-for-one replacement problem. To this was added the ethical question of whether recruiters were any longer justified in making commitments as to technical assignments which quite possibly would not be fulfilled.

Cessation of Medical Department Recruiting

Any such decision was postponed when, in late September of 1944, WAC recruiters

[29] (1) Baetjer, *Women in Industry*. (2) Helen Baker, *Women in War Industries* (Princeton University Press, 1942).

[30] Memo, Col Durward G. Hall, SGO, for Col Whitehurst, 15 Nov 44. SGO Hist Div 220.1 WAC.

received a request from The Surgeon General to discontinue the Female Medical Technicians Campaign. At this time, some months before the Battle of the Bulge, war news seemed good, and on the basis of low casualty estimates from overseas The Surgeon General believed that adequate hospital personnel had been obtained. Recruiting of paid nurses' aides was also scheduled for suspension; recruiting of nurses lacked only a few hundred of meeting the original ceiling. On the basis of similar reports from other agencies concerning the approaching end of the war, the entire WAC Recruiting Service made plans to go on a maintenance basis at the end of the year.[31]

It was thus prematurely hoped that a clash between procurement and medical authorities had been avoided for the duration. Surveyors, reporting finally in October, believed that they had corrected half of the cases of broken recruiting promises, and others were scheduled for correction if and when discovered. The ASF was not willing to take further action at the moment, since this "would harass commanders . . . corrective action is not now advisable"; it was hoped that time and the end of the war would bring the necessary readjustment.[32]

Resumption of the offensive in Europe brought greatly increased casualties, plus heavy hospitalization from combat fatigue, exposure, and trench foot. In one month the Army received more casualties from overseas than it had in all other months since Pearl Harbor, and soon the sick and wounded were being returned at the rate of 20,000 monthly. Almost overnight the situation regarding future medical care changed from comfortable to desperate.

A re-evaluation of the Nurse Corps ceiling indicated that installations in the United States would be short more than 8,000 nurses; only 41,839 nurses were in the Army, 75 percent overseas, of the 50,000 now believed needed at once. Also, although the Cadet Nurse Corps had absorbed some 177,000 young women and $192,285,518 in appropriations, it was chiefly designed to meet civilian nursing needs and had at this time furnished the Army only a few hundred cadets. In a final blow, more than 5,000 general service enlisted men were again ordered transferred from the Medical Department to make up Army Ground Forces casualties. This was the situation when, on 16 December 1944, the Battle of the Bulge began.[33]

At about this time a number of drastic measures were taken simultaneously by the Medical Department, any one of which would have been adequate to meet the situation that actually developed. In spite of certain objections from the General Staff, the Nurse Corps ceiling was raised to the total of 60,000 recommended by The Surgeon General. The Surgeon General on 19 November assured the Secretary of War that the nursing situation was "nearly hopeless." In an effort to arouse both the public and the General

[31] (1) M/R, Conf of Representatives of Dir WAC and TAG, 19 Oct 44. SPWA 341. (2) Memo, G-1 for ASF, 26 Oct 44, with Inds and M/R through 8 Dec 44. SPAP 341 WAC (11-6-44).

[32] (1) Memo cited n. 26 and Min cited n. 27. (2) Memo, MPD ASF for Dir Pers ASF, 7 Aug 44, SPGAA 210.3 WAC (7-6-44)-157; (3) Memo, Dir Pers ASF for MPD, 9 Aug 44; (4) Memo, MPD for Dir Pers ASF, 3 Oct 44, SPGAC 210.3 WAC (10-3-44)-157. All in SPAP 220.3 WAC (7-28-43).

[33] (1) Organized Nursing and the Army in Three Wars: A Political and Administrative History of the Army Nurse Corps, by Mary W. Standlee, Walter Reed Gen Hosp, in collaboration with Col Florence A. Blanchfield, R.N., ANC, pp. 461, 479, 606. MS copy furnished author for review and comment. Hereafter cited as ANC Hist. (2) Rpt cited n. 6(7).

Staff to his needs, he went so far as to bypass the Bureau of Public Relations to give alarming statistics, based on the new Nurse Corps ceiling, to a well-known columnist. Great national concern resulted. Near the beginning of the new year, a proposal to draft nurses went from the Secretary of War to the President, and from him to Congress. Meanwhile, to fill the immediate gap, plans were made for the Red Cross to recruit 5,000 more nurses' aides. And even earlier, on 15 November, The Surgeon General's Office discussed plans to reverse the decision on WAC recruiting and to request some 8,000 more Wacs.[34]

Exact plans for these 8,000 WAC recruits developed slowly during the last six weeks of 1944. At first The Surgeon General, Maj. Gen. Norman T. Kirk, after receiving the pledge of higher authorities to do everything possible "to strengthen Kirk's hand," [35] asked for 8,500 enlisted medical and surgical technicians, either men or women. It was stated that each could fill a nurse vacancy by assuming enough routine nursing duties to permit one nurse to handle the professional duties of two. For each of these men or women, The Surgeon General asked a rating of T/4, or sergeant, as befitted the skills involved. After discussion, both The Surgeon General and the Army Service Forces recommended that all of the technicians be Wacs, since the limited service men available for retraining seemed chiefly interested in getting out of the Army, and it was felt that women would have more enthusiasm for caring for the wounded in the long hospitalization and rehabilitation period ahead. Again the Medical Department asked for women in the upper AGCT and educational brackets.[36]

The Service Forces immediately authorized their hospitals to carry one WAC medical technician overstrength for each unfilled vacancy for a nurse. No announcement was made to the field as to how these extra technicians would be obtained, or whether they would get the T/4 rating previously recommended by The Surgeon General.[37]

Over this point, the General Staff disputed during the last days of 1944. Directives to curtail all WAC recruiting went to the field on 20 December, and an immediate decision was necessary if they were to be recalled. Resumption of the Female Medical Technicians Campaign would have been simple, but Director Hobby was not willing to sponsor a drive to induce highly qualified women to enlist unless some better assurance could be made that recruiting promises would be fulfilled. She sent The Surgeon General a list of questions concerning the means of providing the guaranteed job and grade, and in reply was assured only that the hospital training would be very valuable and would give the women an excellent chance, after the war, of being hired by the Veterans' Administration.[38]

The temper of the moment was one that made refusal to recruit for the purpose unthinkable. Director Hobby's telephone

[34] (1) ANC Hist, pp. 526 ff. Also Memo, SW for President of U.S., 30 Dec 44, Mr. Goldthwaite H. Dorr's file, Office of Records and Co-ordination, Office of SW. (2) Memo, Col Whitehurst, SGO, for Maj Gen George F. Lull, Deputy SG, 15 Nov 44. SGO Hist Div 220.1 WAC.

[35] Pencil note on Memo, SG for G-1 through MPD ASF, 23 Nov 44. SPMCM, in SPAP 220.3 WAC (7-28-43).

[36] (1) Memo Routing Slip, Exec SGO for Dir WAC, 22 Nov 44. WDWAC 341. (2) Memo cited n. 35.

[37] Ltr, Dir Pers ASF to SG and all SvCs, 9 Dec 44. SPGAS 320.2 Med (12-9-44)-285, SGO Hist Div 220.1 WAC.

[38] (1) Ltr, TAG to all SvCs, 20 Dec 44. SPXPR-I WAC (12-20-44), in SPAP 341 WAC, sec II. (2) Memo Routing Slip cited n. 36(1).

conversations revealed that she shared the general belief that "those boys are being wounded without people there to do things for them." Nevertheless, she cast about for any alternative to sending more women on enlisted status to Army hospitals. She went so far as to suggest in high places that nursing standards were too high, and that rather than use Wacs with a few weeks' training, the Nurse Corps should employ, as warrant officers or on a similar status, the practical nurses and graduates of smaller nursing schools who were not acceptable for commissions. This, she was informed, was impossible because of the pressure of civilian nursing groups against recognition of such women. Plans to recruit more Wacs therefore continued.[39]

Director Hobby's concern over the situation was such that she went in person to General Marshall to record her protests. General Marshall shortly thereafter interjected into the War Department conferences an opinion that sufficient women of the high caliber desired could not be recruited by the WAC for these jobs unless the Army could back up its guarantee of work involving patient care and a rating. Unless these could be assured the recruits, General Marshall refused to sanction any further recruiting of women as medical or surgical technicians.

Since such assurance was impossible under the current system, he proposed that the new Wacs be sent to hospitals in Table of Organization companies. The rigid T/O unit came equipped with its own extra allotment of grades, and with specifications as to the exact job of each member; neither the job nor the grade could be changed by field commanders. Such units for women had long ago proved impracticable for general duty Waacs, since few stations could use identical inflexible units, but general hospitals were a different matter, since all had almost identical missions and organization, and all used technicians.[40]

Wacs vs. Nurses' Aides

Before such units could be authorized, Director Hobby became aware of another fact which again caused her to consult General Marshall: The Surgeon General was asking the Red Cross for 5,000 civilian nurses' aides to be recruited at the same time that the WAC would be recruiting its 8,000 enlisted technicians. WAC recruiters feared that no WAC campaign, even if it guaranteed a good job and a rating, could succeed in competition with Red Cross offers of higher pay and officer privileges for identical jobs in identical hospitals on a civilian status. Instead, recruiting funds would be wasted and the government placed in an indefensible position.

On 5 January 1945, General Marshall wrote to his deputy, Lt. Gen. Thomas T. Handy:

> I have just talked to the Director of the WAC and she tells me the ASF is going after 5,000 nurses aides.
>
> My guess would be that this will ruin what I want to do in creating General Hospital companies in the WAC organization, because of the pay status and general competition with the Red Cross. . . .
>
> Please look into this business and see that we are not working against ourselves in this enterprise. I want action.[41]

The decision between Wacs and nurses' aides at once involved the War Depart-

---

[39] Tp Convs, Col Hobby. Drawer 6, G-1 WAC files 312-312.9, DRB AGO.

[40] Memo, "JPR" for Dir WAC, 29 Dec 44. WDWAC 341. Describes 27 Dec 44 conf.

[41] Memo, CofS for Handy, 5 Jan 45; copy is Tab A to reply, G-1 for CofS, 5 Jan 45. WDWAC 341 WAC Hosp Units (4-22-45); also in CofS 324.5 WAC.

ment in much bitterness and misunderstanding. In training, the two positions were virtually identical, although graduates of the Red Cross course were considered by The Surgeon General to be "not as well trained" as WAC graduates of the Army's six-week course, and often needed retraining.[42] In relative cost, also, the WAC appeared to have a slight edge. A study made by the Chief of Staff's office indicated that a Wac cost from $1,032 to $1,368 yearly, including pay, housing, food, and clothing, while a civilian nurses' aide received $1,752 to $2,190 for a 48-hour week.[43]

However, civilian advocates of nurses' aides noted that there were always many hidden costs to military personnel and that the Red Cross bore the cost of recruiting and training aides. Army Nurse Corps leaders preferred civilian aides, stating that nurses would be needed to train Wacs, that Wacs were subject to redeployment, that WAC officers had been given Army rank a year before nurses, and that women under WAC command would be subject to call for drill without regard to patient care.[44]

In making its decision the War Department considered all of these arguments of less consequence than the decisive one: Wacs could not quit when the shooting ended. G–1 Division stated:

> Since the personnel necessary for the care of the sick and wounded in Army hospitals must be available not only during the next . . . months but throughout the period when the Army hospitals must care for and rehabilitate all men wounded during the war, it is considered essential that a sufficient complement of military personnel, enlisted for the duration plus six months, be available.[45]

Breaking off with the Red Cross aide program proved a delicate business; the Red Cross for many years had been accustomed to recruit Army nurses as well as aides, and to set the standards of acceptance. Since Red Cross officials objected on the ground that commitments had been made to women already in training, it was agreed to hire all of these before discontinuing the program.[46]

Red Cross leaders co-operated in agreements by which their graduates might be guaranteed a medical technician rating without further training if they enlisted in the WAC. Resentment within some of the Medical Department offices was longer-lived. After the end of the war, Training Division, Surgeon General's Office, stated in its history that the Medical Department had never requested the 8,000 Wacs at all, or planned to use them, until Director Hobby had exerted influence upon General Marshall to cancel civilian nurses' aide recruiting, thereby forcing the Medical Department to take Wacs "to meet the shortage thus created." [47]

T/O Units for General Hospitals

Upon receipt of General Marshall's demand on 5 January, G–1 Division called from their evening meals WAC Deputy

[42] (1) Memo, Pers Div SGO for Tng Div SGO, 10 Feb 45. Tng Div SGO WAC file. (2) Memo, Lt Col Westray Battle Boyce for Col Charles F. Collier (G–1 Div) and Col Hobby, 24 May 45. WDWAC 220.3.

[43] Memo, G–1 for Col Frank McCarthy, Secy GS, 13 Jan 45. CofS 324.5 WAC.

[44] (1) ANC Hist, p. 492 ff. (2) Memo, Maj Edna B. Groppe, O Dir ANC, for Col James R. Hudnall, SGO, 10 Oct 44. File 23, Nurses' Aides, SGO Mail and Recs Br.

[45] Memo cited n. 6(8).

[46] (1) Memo, G–1 for CofS, 5 Jan 45. WDWAC 341 WAC Hosp Units (4–22–45). (2) Ltr, SW to Mr. Basil O'Connor, American Red Cross, 16 Jan 45. CofS 324.5 WAC. (3) Tp Conv, Col Rice and Mrs. Walter Lippmann, 1 Feb 45. Drawer 6, G–1 WAC 312–312.9.

[47] Rpt cited n. 6(7).

Director Rice and representatives of The Surgeon General, in various stages of off-duty attire, and the group worked all night to deliver to General Marshall the completed plan for WAC hospital companies.[48]

The Table of Organization, details to be prepared later by The Surgeon General's Office, was set at 100 women, in an assortment of technical and clerical jobs believed to be adaptable to any general hospital. Since no member was to be less than a skilled technician or clerk, no rating less than T/5 or corporal was included—a slight decrease from The Surgeon General's estimate that a T/4 or sergeant would be required for each, and one which gave the new units no higher spread of grades than that of men already in Army hospitals. Any hospital desiring one or more of such companies was to report the fact, and the women would then be recruited with complete assurance of assignment to that hospital and to a technical job, which would if successful automatically carry a T/5 rating.[49]

G–1 representatives pronounced the T/O an excellent idea, which would not only permit recruiters to make foolproof guarantees, but would also prevent hospitals from frittering away, in luxurious station overhead, personnel given them to care for the sick. The chief of G–1's Personnel Policy Branch stated later:

> These units were developed to furnish medical and surgical technicians and other specialists in the hospitals, to assist the nurses. They were not developed for the overhead personnel of the hospital. Their purpose was to assist and care for the wounded and sick.[50]

To prevent injustice to women recruited earlier, the plan required that every Wac already working at such a hospital be incorporated into the T/O unit, or be transferred from the hospital if she was deemed incapable of performing a technician's duties. Pursuant to this scheme, The Surgeon General's Office was required on the night of 5 January to report the number of Wacs already at each hospital, which was subtracted from the number to be recruited to fill the unit. It was therefore confidently anticipated by planners that the establishment of the T/O units would remedy two ills at once: it would remove women from work as hospital orderly, which was officially considered beyond women's strength, and it would redeem all recruiting promises, past or future. The question of hours of work was also resolved by War Department approval of a circular requiring the same hours for all military women, commissioned or enlisted. Last, but far from least in the minds of enlisted women, the Director eventually secured authorization for a hospital uniform, a becoming rose-beige chambray, which could not be worn by civilians.[51]

General Hospital Campaign

Because the women had to reach the hospitals simultaneously with the peak of returning casualties, the General Hospital Campaign was launched almost at once. Women were required to have a score in the three upper AGCT brackets. White House co-operation was extended when Mrs. Eleanor Roosevelt invited Colonel Goodwin to share a press conference in

[48] (1) Completed plan: Memo cited n. 46(1). (2) Description of planning session by Col Collier, G–1 Div, p. 8, Min cited n. 27.

[49] Tab B of Memo cited n. 46(1). The T/O provided 67 T/5's and 33 higher grades.

[50] Remarks by Col Collier, n. 48(2).

[51] (1) Tab C of Memo cited n. 46(1). (2) Remarks of Col Whitehurst, SGO, pp. 26, 29. Min cited n. 27. (3) WD Cir 121, 19 Apr 45.

MEDICAL RECRUITING DISPLAY

January, and General Marshall again wrote to all state governors, seeking their aid.[52]

The appeal of nursing jobs to women was reported as "terrific," even in what the public now knew were the last days of the war. The General Hospital Campaign was soundly backed by public relations authorities, with numbers of local stories such as "WOUNDED APPRECIATE HELP," and effective local stunts. For example, Fitzsimons General Hospital received a giant bell, which toured Denver, tolling every eighty seconds to mark the return of one more American battle casualty who needed care.[53] A quota of about 6,000 by 1 May was set; in mid-career it was raised to 7,000; yet recruiters met and passed the number by the end of March, a month ahead of schedule.[54]

Training also went off well. The Third WAC Training Center at Fort Oglethorpe was turned over to medical recruits, and a shortened basic course was followed by a six-week medical course on the same post,

[52] (1) Memo cited n. 46(1). (2) Ltr, TAG to all SvCs, 13 Jan 45. AG PR-I 341 WAC (1-13-43). (3) Ltr, AGO to all SvCs, 23 Jan 45. SPXPR-I-341 WAC, copy in WDWAC 341 (1945). (4) D/F, G-1 for CofS, 5Feb 45. WDWAC 000.7 (1945-46). (5) Ltrs, CofS to all State Governors, 7 Jan 45. WDCSA 324.5 WAC.

[53] *Omaha World Herald*, February 19, 1945; *Denver Post*, February 22, 1945.

[54] Memo, Dir WAC for CofS, 5 Apr 45. WDWAC 341.

staffed by Army doctors, nurses, and enlisted technicians from nearby hospitals. Both students and training authorities had only praise for the excellent and practical course and for the instructors.[55]

The recruits themselves were a promising group, extremely enthusiastic. Many were motivated by the fact that relatives had been casualties; they wished to give to some wounded soldier the care they had not been able to give personally to their own wounded. The majority were young single women of good character and intelligence; they successfully completed the three-month course in six weeks, with few failures even in cases where women with lower-than-qualifying AGCT scores had been erroneously recruited. About 5,000 were graduated as medical and surgical technicians, and 1,300 as medical clerks.[56]

However, the difficulties inherent in the whole General Hospital Campaign became obvious even before the women reported for duty. The Surgeon General's representative stated:

> Utter confusion and conflicting policies and directives re ideas as to the utilization of this personnel have been flagrant since the original idea of WAC Table of Organization companies was conceived.[57]

Because of the confusion, repeated conferences were held with Director Hobby and with medical officers from every service command, without much success. Although the hospital T/O had been drawn up by The Surgeon General's Office itself, that office believed that changed conditions merited abrogation of recruiting commitments.

Even before the new recruits reached the hospitals, the Battle of the Bulge was won and January's fright concerning medical care was in the past. Just as V-E Day was in prospect, all of the drastic attempts to get medical personnel were embarrassingly successful and bore fruit simultaneously. The nursing profession, under threat of the draft, had responded with a rush; in the three weeks between 8 and 29 January 1945, over 10,000 applications for commissions were filed. Draft rumors also stimulated 60 percent of senior cadets to choose the Army instead of civilian employment. Hospitals proved resourceful in hiring civilian nurses. In May, the director of the Army Nurse Corps, Col. Florence A. Blanchfield, returned from Europe to find that there were actually too many nurses. She recommended that 2,000 civilian nurses be released, and secured a recall of the proposed draft legislation. As a result, hospital commanders everywhere objected to the WAC T/O units, and pointed out that they no longer needed Wacs to care for the sick, but were in urgent need of more orderlies, kitchen police, and charwomen.[58]

The matter of grades was also a disappointment. It had been guaranteed by G-1 that these would be an extra allotment, but when the ASF appealed to Lt. Col. Westray Battle Boyce, the G-1 WAC officer handling the grades, it was in-

[55] (1) Memo, SGO Tng Div for SGO Resources Analysis Div, 26 Jan 46; (2) Memo, ASF Plans and Opns Div for SG, 5 Feb 45, SPMOC 532; (3) Memo, Capt Turnbull for Dir Tng Div SGO, 23 Feb 45; (4) Min cited n. 27. WAC file, SGO Tng Div.

[56] (1) Rpt cited n. 7(2). (2) Remarks by Lt Col Elizabeth H. Strayhorn, Comdt, 3d WAC Tng Cen. Min cited n. 27.

[57] Memo, Col Hall, Actg Chief, Pers Serv SGO, for Surg Gen, 14 Mar 45. WAC file, SGO Tng Div.

[58] (1) M/R, Extract from SGO Diary, 28 Feb 45, initialed ERW, (2) Min, Mtg in O Dir WAC, 9 Mar 45; (3) Memo, Col Whitehurst to Col Hall. WAC file, SGO Tng Div. (4) Min cited n. 27. (5) Memo, SGO Pers Servs for CG ASF, 9 Jan 45. SGO Hist Div 322.5-1 WAC. (6) Memo, G-3 for CG ASF, 10 Feb 45. WDGCT 320.3 (1-5-45), in WDWAC 341. (7) D/F, G-3 to CofS, 30 Jan 45. CofS 324.5 WAC. (8) ANC Hist, pp. 553, and 584-5.

formed that "you must find them," and, after further appeals, it was six months before these were forthcoming from G–1 Division.[59]

Meanwhile, men and women already in hospitals felt that the new recruits were receiving their grades. They also failed to realize that the T/O grades were no higher than those already allotted hospitals, the only difference being that in the new units every member received a low rating at the expense of fewer high ratings than men in hospitals possessed. In any case the new arrivals could be promoted at once under the T/O while older hands had to wait for a vacancy in the existing overstrength.

The Air Forces solved the problem by getting War Department approval to the promotion of every Air Forces medical technician, man or woman, to the merited grade of T/5 regardless of other overstrength, but The Surgeon General asked instead that the situation be equalized by not promoting the new recruits. The Surgeon General's representatives privately admitted that they had concurred in setting up the T/O with rated technicians, and that "some of the things that you and I object to most strenuously were recommended by The Surgeon General's own office." Nevertheless, SGO Training Division now proposed that hospitals be freed to utilize the new recruits in any duties in which they were most needed, and to promote them on the same basis as other personnel. It was also proposed that some 1,800 recruits still in basic training not be given the technical training they had been promised, but be sent at once to unskilled work.[60]

Director Hobby opposed this plan on the grounds that recruiting promises made to women and affirmed by their state governors at the request of General Marshall could not be broken. The Surgeon General's representative objected:

> Colonel Hobby has the idea that every company should be exactly as the Table of Distribution prescribes. . . . This is an unrealistic and impossible point of view. . . . In one breath, the theme of G–1 and the Director WAC appears to be to give the Surgeon General what he wants and needs and the next is to say that the Surgeon General will take what they give.[61]

Nevertheless, on this point the War Department's final decision was that recruiting pledges could not be broken.[62]

Even greater opposition from The Surgeon General's Office was offered toward the plan to assign all hospital Wacs to these companies. The directive produced after the all-night planning session of 5 January had stated specifically, "The WAC personnel now on duty at General Hospitals will be absorbed into the companies within the T/O strength." The Army Service Forces reported to the General Council as late as 13 March that the personnel on duty at general hospitals would be included in these units and be given at least 90 days to qualify for grades. On the next day it developed that the Office of The Surgeon General had not been aware of this clause; representatives also denied seeing The Adjutant General's

[59] (1) Comment 2, MPD ASF to G–1, 18 Oct 45, to basic D/F, WD Manpower Bd to G–1, SG, ASF. WDSMB 323.3 (Hospital), in WAC file, SGO Tng Div. (2) Interv with Col Goodwin, 2 May 51.

[60] (1) Memo cited n. 57. (2) AAF Reg 35–54 A (Amendment), 28 Jun 45, with notes. WDWAC 220.01. (3) Draft of proposed Ltr, SGO Tng Div to CG ASF, Jul 45. WAC file, SGO Tng Div. (4) Quotation from Comments from Pers Servs, 18 Jul 45. SGO Tng Div. (5) Memo, SG for MPD ASF, 23 Jun 45. SPMOM, WAC file, SGO Tng Div.

[61] Memo cited n. 57.

[62] AG Ltr to SvCs, 1 Jun 45. AG 322 (5–29–45) OB-I-SPMOU-M.

letters to the field that set up the companies on this basis. The Surgeon General had even issued contrary directives saying that WAC orderlies would be retained. On this point, since there had been no public commitment, Director Hobby was overruled and the directive rescinded.[63]

The general hospital T/O units lasted less than a year. Shortly after the defeat of Japan The Surgeon General obtained their dissolution on the grounds that they were too large for the decreasing size of hospitals, and would hinder demobilization. Wacs in these companies remained at hospitals as part of the regular bulk allotment, but the General Staff required, before permitting inactivation of units, that every woman receive the rating promised her. Thus the general hospital recruits could finally be used on any duties, as desired by The Surgeon General, but at least they had corporal's stripes.[64]

Conclusions

Final reports from inspectors in the last months of the Corps did little to dispel doubts concerning the suitability of hospital employment for women on an enlisted status. There was no doubt of its suitability for women on some status. The care of patients held an attraction for them, and even when Wacs were allowed officially to do nothing but scrubbing and cleaning, they devoted off-duty hours to the patients, writing letters, reading, shopping for them, and running various errands. Hospital commanders with few exceptions praised the work of Wacs to inspectors even while objecting to restrictions imposed by recruiting commitments. One said, "I'd rather have one Wac than three civilians."[65]

Final developments gave little indication that Medical Department policy could change to the extent necessary to protect the health of women in an enlisted status. In spite of the War Department circular requiring the same working hours for enlisted women as for nurses, the 12-hour day continued. When inspectors found women at Walter Reed Hospital working an 80-hour week long after the end of hostilities, the commanding general stated frankly that he had "circumvented the provisions of the circular by putting a small number of Army nurses on 12 hours duty."

The prominent women of the WAC's National Civilian Advisory Committee visited general hospital companies at work, after the Chief of Staff had overruled objections to the visit from both ASF and The Surgeon General; they reported themselves shocked at conditions. Barracks were frequently crowded, ill ventilated, and sprinkled with coal dust; workers on all shifts used the same bar-

[63] (1) Min, Gen Council, 13 Mar 45. (2) Min cited n. 58(2). (3) Ltr, AGO to SvCs, 12 Mar 45. AG 322 (3-8-45) OB-I-SPMOU-M. (4) Ltr, SGO to all SvCs, 15 Jan 45. SPMCM, SGO Hist Div 322.5-1 WAC. (5) Memo, Col Boyce for Cols Peter Schmick and Hobby, 21 Mar 45. (6) ASF Ltr SPGAC-210.3 WAC, 26 Apr 45. SPAP 220.3 WAC. (7) AG Ltr cited n. 62, and amendment, 13 Jul 43.

[64] (1) D/F, WD Manpower Bd to G-1 WD, SGO, ASF, 10 Oct 45. WDSMB 323.3 (Hospital), WAC file, SGO Tng Div. (2) Ltrs, AGO to all Comdrs, 3 Dec 45, and later dates. AG 322 (11-28-45) OB-I-SPMOU-M. (3) D/F, G-1 to CofS, 8 Oct 45. WDGAP 320.2 WAC in 320.2 (1945). Incl in Ltr, CofS to State Governors, 17 Oct 45. CofS 324.5 WAC.

[65] (1) Transmittal Sheet, MTD ASF to SGO, 7 Jun 45. Wac file, SGO Tng Div. (2) Memo cited n. 19. (3) Memo, Capt Margaret Stone for Dir WAC, 7 Jul 45. WDWAC 333.1, in WDWAC 319.1 Hosps. (4) Memo, Lt Col Mary-Agnes Brown for Dir WAC, 27 Jul 45. WDWAC 333.1. (5) D/F, G-1 to SG, 3 Oct 45. WDGAP 333 WAC, in WDWAC 333.1. (6) D/F, G-1 for CofS, 4 May 45, and D/F's, G-1 to ASF, 22 May, 29 May, 9 Jun 45. WDGAP 341 WAC.

racks, disturbing those trying to sleep either day or night; and hours had not greatly improved, with women still required to train and frequently to launder their clothing on their own time. There was also no evidence that the practice of malassigning technicians had abated. General Dalton eventually lost patience with repeated violations, and Colonel Goodwin ascribed to him personally most of the credit for redeeming such broken commitments as were redeemable.[66]

By some coincidence, the Navy reported the same experience, and the director of the Bureau of Naval Personnel stated later:

> Experience indicated that in some jobs too much was asked of the women's physical strength. Oddly enough, the outstanding offender was BuMed. . . . As evidence mounted of women breaking under the strain of long hours, the office of the Women's Reserve put increasing pressure on BuMed to remedy the situation.

A 51-hour work week was regarded by WAVES authorities as "the desirable maximum," but the Bureau of Naval Personnel reported, "Efforts to have this cast in the form of a directive failed to overcome the opposition of Naval tradition against 'union hours'."[67]

Final recommendations by the Army Medical Department confirmed the fact that continued utilization of servicewomen was contemplated by Army hospitals. For commissioned positions, The Surgeon General asked and obtained legislation to allow women to be commissioned directly in a women's medical specialists corps, headed by its own colonel, which would include dietitions, physical therapists, and occupational therapists.

The Surgeon General also advocated that enlisted women who worked in hospitals be enlisted in the Medical Department alone and not in the WAC. This move would have obviated the requirement for a WAC company commander and for compliance with War Department restrictions as to the hours and types of duty for members of the WAC. The War Department did not look with approval upon any such proposal for the setting up of independent women's corps in the different administrative and technical services, which tended toward compartmentalization as opposed to integration; it also did not agree to abdicate its Army-wide responsibility for safeguarding the well-being of women. Medical Department utilization therefore continued upon the wartime terms and without much greater clarification of the situation than had hitherto prevailed.[68]

---

[66] (1) *Ibid;* also series of ltrs, SPAP 220.3 WAC (7-28-43) beginning 1 Feb 45. (2) Interv with Col Goodwin, 2 May 51.

[67] WAVES Hist.

[68] (1) HR 1943, 80th Cong. (2) HR 3054, 80th Cong. (3) Final strength report: Memo, Dir WAC for Gen Henry, 18 Aug 45. WDWAC 319.1 Hosp.

ASF:	General Hospitals	12,087
	Convalescent Hospitals	423
	Regional Hospitals	2,203
	Station Hospitals	1,379
	Port of Embarkation	890
		16,982
AAF:	(as of 30 Jun 45)	3,055
		20,037

CHAPTER XX

The North African and Mediterranean Theaters

The North African Theater of Operations was the pioneer in employment of Waacs overseas.[1] The experiment owed its impetus to the first theater commander, Lt. Gen. Dwight D. Eisenhower, who had been impressed with the efficiency of British servicewomen, and to his deputy, Maj. Gen. Everett S. Hughes, who fifteen years before had prepared the General Staff study, Participation of Women in War. On most lower echelons there was a less favorable general attitude, ranging from mild skepticism in headquarters offices to bitter opposition in the great majority of soldiers' letters.[2]

The theater received its first five WAAC officers on 22 December 1942, some six weeks after the invasion of North Africa. The women, qualified as secretaries, were assigned by the War Department to England, but upon General Eisenhower's request were immediately placed by the London headquarters on a ship bound for North Africa. A day out of port, the ship was torpedoed, and the women came into port aboard a British destroyer, which had taken two from the burning ship and three from a lifeboat. The day in the lifeboat was remembered by the Waac participants as a hectic interval in which they lost all their equipment, the only crewman aboard became violently seasick, and the women fished five or six men, one badly injured, from the water. When the Waacs arrived in port, dirty and bedraggled, they were greeted by anxious dignitaries who contributed oranges, toothbrushes, and other emergency items. General Matejka offered half a jar of hand cream, while Maj. Gen. Walter Bedell Smith procured small-sized men's trousers, which the women found they could not get into. General Smith immediately chose as his assistant one particularly filthy member who had lost hairpins and hat and been vomited on; he confided to her later that he "picked the worst-looking so her feelings wouldn't be hurt." The greatest con-

[1] No official WAC history was discovered in theater files. This chapter draws upon: (1) The sections concerning NATO in SPWA 320.2, WAAC classified files, DRB AGO. (2) WAC policy file, in possession of Lt Col Dorothea A. Coleman. Hereafter cited as Coleman file. (3) Intervs with, and written comments from, Lt Col Dorothea A. Coleman, 21 Jan and 8 Feb 46; Lt Col Cora M. Foster, 8 Mar 50; Maj Hortense M. Boutell, 8 Mar 50 and 6 Mar 51; Maj Ruth M. Briggs, 6 Mar 51; Maj Aline Drezmal, 6 Mar 51. (4) Memo, WAC PRO, Hq MTOUSA, for WAC Gp, WDBPR, 12 Sep 45, sub: WAC Redeployment. WDWAC 000.7. Hereafter cited as Redeployment Summary. (5) Clippings and publicity material, cited by name and envelope. DRB AGO. (6) Interv and written comments from Dr. Sidney T. Mathews. OCMH. (7) Review of this chapter by Col Westray Battle Boyce Leslie.

[2] For all excerpts from mail, see NATO AG files 319.1 Morale, Vols. II–V.

tribution came from General Marshall, who met the women at the Casablanca Conference and took home a list of lost equipment. Finding that there was no legal means of free replacement, he personally paid for and forwarded new clothing, refusing to accept repayment.[3]

Arrival of First Unit

The first enlisted women arrived in the theater a month later. The 149th WAAC Post Headquarters Company, called by newspapers "the first American women's expeditionary force in history," was one of the most highly qualified WAAC groups ever to reach the field. Hand-picked and all-volunteer, almost all members were linguists as well as qualified specialists, and almost all eligible for officer candidate school. The company was shipped from the United States on a regular military transport, which encountered no enemy action. The only difficulty was the loss of unit equipment, which never arrived in Algiers. The unit's vehicles were later found to have been issued to a male unit at the port, while cooking equipment, medical supplies, folding cots, recreational equipment, typewriters, and clothing maintenance supplies vanished, a serious loss in view of shipping conditions and the absence of women's supplies in theater warehouses. Such occurrences in later shipments were to be prevented by the prior appointment of a theater WAAC staff director to make advance preparations.[4]

The unit reported on 27 January 1943 to General Eisenhower's headquarters in Algiers, a location now considered safe, except for air attack, from the conflict still raging to the east. The women were housed in the dormitory of a convent school some distance from the headquarters. The unfamiliar climate and the unheated quarters caused almost every member to succumb immediately to colds and other respiratory disorders—a tendency noted also in newly arrived men. Waacs washed in an outdoor trough, carried water in helmets, and worried in their usual manner concerning the impossibility of careful laundry or neat appearance. Working hours were long; women were carried in trucks to the headquarters at an early hour, and home again for an early curfew. The nightly bombings, with brilliant displays of antiaircraft fire, made sleep difficult for the first weeks.

Nevertheless, most women managed a satisfactory adjustment, and most of the sick required no hospitalization.[5] Morale was high, and women called themselves the luckiest in the Corps, commenting variously, "Life has been one thrill after another"; "All my life I wanted to travel and see strange sights and now I am doing just that"; and, of the ack-ack fire and burning barrage balloons, "No Fourth of July celebration could be more spectacular."[6] Ecclesiastical authorities protested when Wacs were housed in church quarters, but soon admitted that they had "never imagined that a unit of American women could be so well-disciplined and so considerate as well."[7]

---

[3] Intervs cited n. 1(3); also two WDBPR Press Releases. (1) "Torpedoed," in Env, Africa—Interesting Backgrounds. (2) 27 Jan 43, in Env, Stories—1943.

[4] (1) Memo, CO 149th WAAC Post Hq Co, APO 514, for Dir WAAC, 8 Feb 43. WDBPR file of NATO correspondence, DRB AGO. (2) For theater strength figures, see Table 7, Appendix A.

[5] *Ibid.;* also Memo, CO 149th WAAC Post Hq Co for Dir WAAC, 23 Feb 43. Same file.

[6] WDBPR Env, Africa—Ltrs Written by Waacs.

[7] Ltr, Col Donald B. Adams to Dir WAC, 23 Mar 44. WDWAC 000.7.

As there were not enough Waacs in this first unit to supply every office's needs, the theater adjutant general reported an "inevitable scramble by chiefs of general and special staffs to obtain additional WAAC personnel." [8] The largest part of the company went to the Signal Corps and to the newly organized Central Postal Directory. Others were assigned, by twos and threes, to various headquarters offices: three to the Office of Psychological Warfare; three to the adjutant general's office; one as General Eisenhower's secretary and one as his driver; more than a half dozen to drive other officers. Ten more were assigned as cooks and bakers to keep food ready for workers on three shifts.[9]

Reports on initial job success were consistently good. Among the first to receive commendation were telephone operators and other communications workers in theater headquarters. The Voice of Freedom—the entire telephone switchboard system of theater headquarters, large enough to service a good-sized city—was eventually manned entirely by WAAC supervisors and operators. Many telephone operators had from ten to fifteen years of civilian experience as operators and supervisors, and some were bilingual. A record for calls handled was set, and the theater chief signal officer, Brig. Gen. Terence J. Tully, stated, "It is highly desirable that the use of Wacs be extended," particularly to message center and cryptographic duties.[10]

The first postal directory workers were likewise successful. Officers in charge reported that "since the Waacs have taken over with an entirely different attitude toward the job than the men who had previously handled it, the percentage of errors has decreased materially." [11] Although the job was considered by some to be a deadly routine, Waacs were somewhat romantically apt to visualize the pleasure of a soldier in getting long-delayed mail, or the anguish of a mother erroneously returned her letters, stamped "missing in action." The postal directory's Army supervisor stated:

> The office was piled ceiling high. These girls came in and took over and worked from 8 a.m. until 9:30 p.m. seven days a week until it was cleaned up. . . . They never thought of asking for time off and I had to order them home nights.[12]

Administrative Difficulties

This early and continued job success was deemed remarkable in view of the administrative difficulties that the unit encountered during its first six months. Theater authorities attributed the confusion that developed to the absence of a WAAC adviser on the staff level, for the women had been sent without a WAAC staff director for the theater—the first and last time that such a mistake was made.[13] In the meantime, certain policy questions had arisen which even General Hughes, with his long study of the problem, felt unable to answer. He sought Director Hobby's advice by mail, but she felt it undesirable to attempt long-distance pronouncements in ignorance of local conditions.[14]

---

[8] Ltr, Brig Gen Thomas J. Davis, AG AFHQ, to Col Hobby. 29 Sep 43. SPWA 320.2 sec 7.

[9] Inez Robb dispatch, WDBPR Env, Africa—Opinion, Gen Offs.

[10] Memo, Gen Tully for CG NATOUSA, 27 Sep 44. Coleman file.

[11] Ltr, Gen Hughes, Dep Theater Comdr, to Dir WAAC, 29 Apr 43. SPWA 320.2 sec 5.

[12] AFHQ Press Release, "Postal," WDBPR, Env, Africa—Interesting Jobs.

[13] Cbl 3093, 20 Jun 43. SPWA 320.2 sec 5.

[14] Ltrs, Gen Hughes to Col Hobby, 17 Apr 43, and to Col Catron, 13 May 43, and reply 3 Jun 43. SPWA 320.2 sec 5.

MAJ. WESTRAY BATTLE BOYCE, *WAAC Staff Director, North African Theater, reads orders replacing enlisted men of the adjutant general's office with enlisted women, Algiers, North Africa.*

Finally, responsible commanders in the theater were stirred to action by the report that a large number of Waacs did not intend to enlist in the WAC. At this, General Eisenhower cabled a request for "a highly competent senior WAAC officer to be sent to this theater without delay." [15] Director Hobby promptly complied with the request and sent a theater WAAC staff director, Maj. Westray Boyce, previously staff director of the Fourth Service Command. Major Boyce, according to General Hughes' account, "arrived to find the girls in a state of mind which required immediate and drastic action. This she took." [16] The previous company officers and key cadre were returned to the United States in a body, since it appeared impossible to place individual responsibility or to restore the women's lost confidence; all later proved successful in a variety of duties in the United States. A request was cabled for "competent, experienced" company officers to fly to the theater.[17]

[15] Cbl cited n. 13.

[16] Ltr, Gen Hughes to Col Hobby, 1 Jan 44. SPWA 320.2 sec 9.

[17] Cbls W7186, 12 Aug 43, and W8669, 30 Aug 43. SPWA 320.2 sec 5.

Meanwhile, the company had been successfully pulled together by one of the five officer secretaries, Capt. Martha Rogers, who had not been previously considered for company duty for lack of "voice and command" presence. Major Boyce secured some reassignment of women who were unsuitably assigned or not kept busy. She also succeeded in slowing up the promotion of Waacs, which she felt that a few section chiefs had been making too indiscriminately. After the reorganization, morale soon improved, and the number of Waacs lost at the conversion to Army status was less than had been anticipated. Gen. Dwight D. Eisenhower personally addressed the women in mid-August and informed them that they were necessary to the fighting ahead, and that "if a single one of you goes home, it's too many." [18]

Because of the general alarm at the results of lack of staff supervision, General Eisenhower and General Hughes personally drafted and put into effect a system that gave Major Boyce greater power than any ever held at any other place or time by any other WAC staff director. Under the title of Theater WAC Executive Officer, she was given virtually complete command jurisdiction over Wacs in the theater, bypassing normal military channels in all matters except supply and routine post administration. She was supreme on all matters of promotion, job suitability, discipline, discharge, and well-being. In fact, the system was a complete reversion to the old WAAC regulations, although few WAAC officers in the United States had ever exercised all of these powers as fully as Major Boyce now did.[19]

These powers did not last long, for a new commander, Lt. Gen. Jacob L. Devers, replaced General Eisenhower a few months later, and soon restored the command prerogative of Army officers. He also placed the WAC staff director's office in G-1 Division; as a result, she could no longer consult directly with the deputy chief of staff. However, in the interim, the theater staff had become accustomed to consulting the WAC staff director on policy matters, and the habit persisted even after WAC command powers were revoked and after Lt. Col. Westray Boyce was succeeded, in less than a year, by Lt. Col. Dorothea A. Coleman. The headquarters eventually directed in writing that the WAC staff director be allowed to comment on all matters affecting Wacs before action was taken by any headquarters division.[20] With this precaution, it appeared to be a matter of little moment what command system was used, and the early administrative problems seldom if ever recurred in later units.

Further WAC Shipments

From the time of arrival of the first unit, theater requisitions for more Waacs had been repeatedly forwarded to the War Department, with the highest shipment priorities. As soon as approaching military status warranted, these began to be shipped, in accordance with General Marshall's policy of priority to combat theaters. The second shipment arrived in May, some four months after the first, and contained chiefly workers for the Central Postal Directory, although fewer than the chief of that agency had tried to obtain. In August of 1943, a platoon of about

[18] (1) Cir 205, Hq NATO, 23 Oct 43. Coleman file. (2) Ltr cited n. 16. (3) Quotations from AFHQ Press Release, 17 Aug 43. WDBPR Env, Africa—Opinion, Gen Offs.

[19] Cir cited n. 18(1).

[20] MTO Ltr, 7 Feb 45, sub: Responsibilities of WAC Stf Dirs. AG 322.01/234 A-O.

sixty members arrived in Casablanca and proceeded to Mostaganem for service with the Fifth Army. In September the 60th WAC Headquarters Company arrived to augment the 149th, and the 61st for SOS headquarters at Oran.[21]

The need to release more men to staff Signal Corps installations in Italy brought about the activation of a complete WAC Signal company, which arrived in Algiers in November. The women were welcomed by the Deputy Chief Signal Officer, Allied Forces Headquarters, who promised them long hours of work under "far from ideal conditions." [22] Members were assigned as route clerks, high speed radio operators, teletypists, cryptographic code clerks, and to cutting tape in radio rooms. On the same ship with the Signal company was the first WAC contingent for duty with the Air Forces in the theater, the Twelfth Air Force Service Command at Algiers. This unit was picked up at the port of debarkation by transport planes and flown direct to its parent organization. According to its chief of staff, "We have a lot of work for them to do right now; we can't afford to have them sitting around a port for several days." [23] Enlisted members, like those in preceding shipments, were again highly qualified, ranging from a former dean of women to a translator of African dialects. Most were assigned as telephone operators, file clerks, typists and stenographers, and to aid in the air service command's work of forwarding ammunition, engines, and other supplies to the front.

In January of 1944, the theater announced the arrival of another contingent for service with the Air Forces—some 372 enlisted women and eight officers, bringing the theater total to over 1,500. These were divided into four units for assignment to four different stations in North Africa. Other shipments continued during 1944, eventually bringing the theater total to approximately 2,000. The theater thus assumed third place in the number of Wacs employed overseas, being eventually surpassed only by the European and Pacific theaters.[24]

These Wacs were processed in and out of the theater by a WAC replacement depot, operated in conjunction with the men's depot, but with its own experts in WAC clothing, equipment, records, and other processing. Here new arrivals were given orientation talks by the WAC staff director, with hints as to how best to profit from the past experiences of Wacs in the theater. New arrivals were interviewed and assigned by a WAC specialist in G–1 Division, who was aware of the probable training, capacities, and limitations of WAC personnel.[25]

Even before the full 2,000 had arrived, Wacs had begun to move out into Italy, beginning with the Fifth Army platoon in November of 1943. At the height of the Italian campaign there were fourteen WAC units with the various commands in Italy, including the Fifth Army, the Peninsular Base Section, and the Air Forces. By the end of the Italian campaign all of the theater's Wacs were in Italy.

---

[21] (1) ETO WAC Hist, p. 7. (2) Cbl W3105, 20 Jun 43. SPWA 320.2 sec 5. (3) Press Release cited n. 12. (4) Hist, 6669th WAC Hq Plat (Overhead), APO 464, Fifth Army. In possession of Lt Col Cora M. Foster. (5) Intervs cited n. 1(3).

[22] Press Release, 11 Nov 43, sub: WAC Sig Co Arrives; and Press Release 125237 AFHQ. WDBPR Env, Africa—SigC Co.

[23] Two Press Releases, subs: WAC, and WAC Air Corps Co Arrives. WDBPR Env, Africa—AAF Wacs.

[24] (1) *Stars and Stripes*, Algiers, January 20, 1944. (2) Peak strength: 1929, Jun 45 (96 offs, 1833 EW). Lowest: 206, Jan 43. Coleman file.

[25] Intervs with Col Coleman cited n. 1(3). (2) Ltr, Hq MTO to CO Hq Comd AF (Overhead), 2 Apr 45. AG 2220.3 298-P.

WACS ARRIVING AT CASERTA, ITALY, *17 November 1943.*

Fifth Army Wacs

The theater's most unusual experiment in WAC employment was that of the Fifth Army Wacs, who claimed the honor of being the first Wacs to set foot in Italy, and in fact on the continent of Europe. Although never more than sixty women were involved, the experiment was considered to be potentially more important than its size would indicate, since it might determine the degree to which women could in future emergencies make up part of tactical units.[26]

When the Fifth Army jumped off for Italy, the Wacs were not too far behind it, arriving in Caserta, via Naples, on 17 November 1943 under the command of 1st Lt. Cora M. Foster. The T/O called for 10 telephone operators, 7 clerks, 16 typists, 10 stenographers, 1 administrative clerk, and cooks and other cadre. In late January the unit split into forward and rear echelons, and the forward echelon, including all telephone operators and some stenographers, moved into the bivouac area near Presenzano. Here the women lived in pyramidal tents and worked chiefly in the Fifth Army's mobile switchboard trailer. By March the rear echelon was also in tents near Sparanise. Unit records noted

[26] All statements concerning Fifth Army Wacs are from: (1) Hist cited n. 21(4). (2) Intervs cited n. 1(3). (3) Redeployment Summary.

that the women "thrived on it; the sick call rate dropped way down."

For the rest of the Italian campaign, the units followed the Fifth Army up the peninsula, usually being located from twelve to thirty-five miles behind the front lines. From June to September the forward echelon's longest stay in any one place was five weeks, the average being two. The women lived in whatever billets were available—schools, factories, apartments, and chiefly tents. The forward section spent most of the winter of 1944–45 living in tents in the mountains above Florence. The women usually wore enlisted men's wool shirts, trousers, and combat boots, and carried only the few necessities that could be moved forward with them.

The unit proved unusually successful. It received Lt. Gen. Mark W. Clark's praise as well as being one of the few to receive both the Fifth Army Plaque with clasp, in 1944, and the Meritorious Service Unit Plaque in 1945. There were twenty-seven awards of the Bronze Star. The forward echelon included some of the most skilled telephone operators on the Continent, able in a matter of minutes to get through the complicated communications networks to the commanding officer of any unit sought by General Clark. The unit's morale and *esprit de corps* were perhaps the highest in the theater's WAC units. Its members wore the Fifth Army's green scarf. "They were," said Colonel Coleman, "Fifth Army first and Wacs second—perhaps the best-integrated unit in the theater."

The dangers of a combat area did not present any great problem in this case. During the last days in Anzio, air raids offered the nuisance of noise and falling shell fragments, but fortunately the area had just been vacated by a combat unit that had left adequate foxholes and dugouts. The Wacs were lucky in having no injuries, in spite of some close calls. During the advance up the peninsula, they were frequently within sound of long-range artillery, and almost always in an area of complete blackout, but required no guard except the usual one that patrolled the entire camp. Italian service troops ordinarily set up the WAC and Nurse Corps tents, and the Wacs took them down themselves, with the aid of two Italian laborers to load them on trucks.

In spite of the fatigue that developed from repeated moves, officers noted that the women "griped and complained less than soldiers in rear areas." Signal Corps units in rear areas were surprised, upon offering Wacs rotation to less exhausting conditions, to find that not one telephone operator would agree to quit the Fifth Army unit.

WAC commanders credited much of the sustaining sense of integration to the personal policies of General Clark, who saw to it that a stirring ceremony was made of any occasion, such as the swearing-in of Waacs to the Army, or the presentation of medals and awards. When visiting dignitaries came to the area, Wacs stood honor guard with men of the 34th Division. Women were left in no doubt that they were considered useful and valuable members of the group. Such measures inspired a loyalty among the women that long hours, fatigue, and discomfort could not shake.

It was undoubtedly significant that Wacs under such circumstances seemed exempt from the pattern of administrative needs that was invariable in sedentary units. The almost total absence of grades and ratings did not noticeably affect morale, although an expert switchboard operator of two years' service was lucky to get a pfc. rating. The longer hours, the lack of privacy, the necessity for wearing clothing designed for men, all seemed to have little effect. Telephone operators seemed to be immune from the illness and tension experienced by women on identical shift work in more permanent companies a few score miles to the rear. Rude remarks from occasional unfriendly males, which caused a morale problem among Wacs elsewhere, were more or less brushed aside by women too busy to notice them and too assured of their own usefulness to doubt it. Under the circumstances, the staff director considered it a pity that all Wacs could not be employed in such units.

WAC advisers recommended that any such groups, for best success, be carefully selected. The best-suited type of woman was believed to be one whose physical stamina was average or better, who liked outdoor life, and who was well balanced mentally and emotionally. In a small isolated group, it was found fatal to include those with irritating habits and mannerisms—overtalkative, grouchy, or erratic. The ambitious or the highly qualified woman was also not a good choice, since top supervisory jobs in a tactical headquarters did not go to women, nor did they receive the high grades that those in rear areas did, many still being privates after two years. Emotional self-control was also vital, since the women were constantly surrounded by men, especially by combat troops coming back for a rest. Besides fending off advances, said the staff director, "The Wacs had to listen to the men and let them blow off steam, and this put an additional strain on the women's nerves." A mature woman was found preferable, especially one whose only interest was not the opposite sex. The lone company officer who accompanied each section had to be especially self-reliant and self-sufficient; her conduct necessarily had to be above reproach.

It was Colonel Coleman's final judgment that WAC units similar to this Fifth Army unit might easily be used, if necessary, in army and corps headquarters, or wherever nurses could be used, provided that the women were as well qualified. The only limitation, she believed, was that too many women should not be used in any one headquarters, since the headquarters force was also the security force, and lost a combatant for every woman specialist present.

Headquarters Duties

With the exception of the small Fifth Army experiment, the remainder of the theater's Wacs were employed in ordinary headquarters duties not strikingly different from those in comparable agencies in the United States. Woman in Allied Forces headquarters and Services of Supply headquarters were employed by almost every staff section—ordnance, judge advocate, adjutant general, and others, and as secretaries to ranking officers. In addition, the headquarters of the Mediterranean Allied Air Forces and the Twelfth and Fifteenth Air Forces employed Wacs, in units stationed at various times at Caserta, Bari, Foggia, and Florence.

In organization, the most distinctive feature in the Mediterranean Theater's employment of Wacs was that all women, with the exception of Air Forces and Fifth Army Wacs, were eventually organized into a single battalion under WAC command. This unit was the 2629th WAC Battalion, stationed in Caserta, under the command of Maj. Hortense M. Boutell. The battalion included a postal company, a message center company, a WIRES company, two headquarters companies, a replacement depot detachment, and a small Signal group on detached service, first in Constantine and then in Rome. Eventual enlisted strength totaled approximately 900, about half of theater WAC strength, plus 52 WAC officers, and some 400 British ATS attached to the battalion and housed with it. With such a large concentration of female personnel under WAC command, matters of supply, administration, and discipline were considerably simplified.[27]

In all units—Air, Ground, or Service Forces—the most striking difference from employment in the zone of the interior was the higher degree of skill and ability involved. All but a negligible number of enlisted women were qualified for skilled jobs; there were only two WAC "laborers" in the theater—one basic, one repairman. The statistically average Wac in this theater was a trim, healthy young woman, unmarried, older than the average soldier, with better education than the zone of the interior Wac.[28] Thus, the theater never experienced the problems of the zone of the interior in assigning personnel of low aptitude or little training. The higher average ability was reflected in the final tally of WAC assignments in the various headquarters.[29]

Type of Duty	*Percent*
Total	100.0
Clerical	31.9
Communications	29.2
Stenographic	16.3
Miscellaneous	12.7
Headquarters Cadre	9.9

Such high percentages of skilled personnel not only made assignment easier, but minimized problems of maladjustment, mental health, discipline, and improper or too-heavy work. In most units women did not even have to perform kitchen police and fatigue duties, since local labor was available and it was considered uneconomical to waste a skilled Wac's time in such heavy tasks. The large WAC battalion alone eventually had about 250 civilian servants.

The advantages that the theater enjoyed in such high-type personnel were made clear by its reaction when, in later shipments from the United States, the quality declined somewhat, especially for officers, concerning whom an observer protested, "If we cannot give them officers who are outstanding leaders, we should at least give them officers who are ladies and whose conduct is such that the enlisted women can respect them."[30] A corresponding decline in the care of selection of enlisted women more than doubled the number who promptly had to be returned to the United States because of physical breakdowns.

The only major problem in WAC employment encountered by the theater was that of the Signal Corps Wacs.[31] By midsummer of 1944, women in this work began to suffer from fatigue and depression; their sick rate was twice as high as other Wacs', and morale was low. The theater chief signal officer diagnosed the fatigue as "resulting from working long hours over a period of several months

[27] Interv with Maj Boutell, 8 Mar 50; interv and written statement, 6 Mar 51.

[28] Redeployment Summary.

[29] *Ibid.* Statistics in Coleman file are even higher for the period Jun 43–Jun 45.

[30] (1) Memo, Col Boyce for Dir WAC, 28 Oct 44. WDWAC 320.2. (2) Cbl W1960, 5 Feb 44. SPWA 320.2 sec 9.

[31] Discussion from: (1) Memo, WAC CO 6715th WAC Com Co, APO 512, for CO 2629th WAC Bn Hq 29 May 45. Coleman file. (2) Rpt, Bn CO 2629th WAC Bn to Dep CSigO NATOUSA, 11 Sep 44. Coleman file. (3) Memo cited n. 10, Tab A, 2 Sep 44. (4) Memo, Traf Off 3141st Sig Serv Gp for CO 2629th WAC Bn, 22 Sep 44; Tab B is D/F, Theater WAC Stf Dir, 10 Sep 44; Tab C is Rpt of Surv Com, 11 Sep 44.

under pressure caused by the necessity for speed and security in handling messages, and the realization of the operational importance of every duty they perform." WAC and Medical Corps officers felt that the difficulty was due, not to the work itself, but to the lack of leaves or furloughs, the rotating shifts so short that sleeping habits never became adjusted, and the long hours of off-duty training in other Signal Corps work. Most important was the shift work; Major Boutell later surmised that this alone was the chief problem, coupled with the lack of privacy or quiet in living quarters, which made sleep difficult for night workers.

By mid-1944 about 8 percent of the women who had been in Signal Corps work for a year or more were found already unfit for further overseas service, and others were, in the opinion of their commanders, building up a nervous tension that in time would permanently end their usefulness unless transfers or leaves could be arranged. The theater chief signal officer was unable to suggest any remedy except to try to enlist large numbers of Wacs for Signal Corps work with the promise that they would be discharged after two years in the Army. He also felt that, if a 2 percent overstrength of personnel could have been provided the Signal Corps by the War Department, the Signal Corps could have permitted rotation and leaves for all personnel. Major Boutell was of the opinion that there was no real solution for the Wacs unless one could be achieved by the Signal Corps as a whole.

With this exception, there could be no doubt that most of the women were happy in their overseas assignments; they commented variously: "I love my work and enjoy being overseas"; "I wouldn't miss being here for anything"; "I like being close to the war"; "I wouldn't change places with any girl back home." One said:

> Many wonderful things happen to us. We've sailed the seas as other troops have, served on foreign soil, marched in Allied Nations parades, have seen and met some of the world's most famous men.

Another wrote:

> Mother, I have never before had the complete satisfaction of worth-while accomplishment that I get from my present work. And I greatly doubt that I shall ever be inspired to put such wholehearted energy and effort into anything which my life after the war will demand of me.[32]

There was considerable debate, as in most such theaters, whether special housing and security measures required for women outweighed their usefulness on the job. It was the opinion of the staff director that there was little truth to the allegation that women required extra guards. She added:

> There was an M.P. at each barracks in the cities, but there was also one at the officers' club, the noncoms' club, and all enlisted men's barracks, to keep out the civilians. Even where women lived in tents, they needed only the usual patrol.

The only objection, in her opinion, was that Wacs, on moving to new quarters, usually needed help in setting up bunks, latrines, and showers.

Degree of Adjustment Required

At no time was there any serious question of the ability of enlisted women to adjust to conditions of life in a theater such as this one. Possibly the greatest physical hardship was the usual absence of heat and hot water in the quarters,

[32] (1) Press Release 117715, Wacs in Ord, Env, Africa—Ord Wacs. (2) Ltrs, WDBPR Env, Africa—Ltrs.

whether tents or mansions. Except for this, the usual billet was comfortable—frequently an apartment house, school, or hotel, considerably more pretentious than barracks in the United States. Theater authorities closely co-ordinated housing assignment with WAC advisers, leaving to them the matter of determination of suitability. The theater consistently maintained the requirement that Wacs be assigned only in groups of fifty or more, in spite of attempts by numerous agencies to scatter them about in small numbers; this requirement, in the opinion of WAC advisers, greatly facilitated control of housing as well as of discipline.[33]

Wacs in North Africa or Italy were seldom long without items of clothing necessary to health, if not to comfort and good appearance. Because of the relative difficulty of maintaining a stock of women's clothing, men's clothing frequently had to be issued in emergencies. Women's supplies were hard to come by. Women went months without their full issue of clothing; large and small sizes were scarce, shoe repair slow, the stocking situation often critical. Post exchanges did not always have items needed by women. Hairdressing facilities were often inadequate, since the men's barbershops were of course of little use to women. Supplies forwarded to units near the front were often lost, particularly in the case of small units like the Fifth Army Wacs, whose usual supplies filled only a small easily misplaced box. "Some of the stuff was never found; I suppose it's still over there," commented Colonel Coleman.

It was discovered that women in all kinds of work did not have enough warm clothing. For many work-type clothing needs, neither regulation WAC coveralls nor seersucker work dresses were suitable. For lack of a warm soft work jacket, oversized wool shirts were cut off and made into battle jackets. In meeting these various emergencies, theater headquarters permitted close staff supervision by WAC specialists. The theater was always quick to requisition available new items of clothing, and protested whenever such items appeared in the United States before theater Wacs had them.[34]

Health, Discipline, and Morale

Theater Wacs were never in danger of capture, although they were frequently in air raid areas. The WAC billet in Naples was hit in one bombing, but, as members commented, "it only knocked down the ceiling and one wing." Maj. Margaret D. Craighill, inspecting for the War Department later in 1944, noted no serious tension among enlisted women as a whole. She stated, "The medical service given to the WAC is excellent," consisting of two women medical officers on full-time duty, which included a monthly trip to all units for the required monthly physical inspection. The only defect in care noted by Major Craighill was the fact that "sick in quarters" was not permitted, resulting in many subterfuges by women with temporary menstrual disorders who did not wish to be hospitalized. Major Craighill reported:

> Dysmenorrhea seems to be more frequent among both nurses and Wacs in this theater. Some who were never before subject to difficulty during menstruation now complain of severe pain. Reasons for this are not apparent; fatigue and constipation may be contributing factors.[35]

[33] NATO Ltr, 9 Jul 44. AG 324.5-O. Coleman file.

[34] (1) Cbl W-464/6920, 19 Sep 43. SPWA 320.2 sec 7. (2) Ltr, Gen White, G-1, MTO, to Dir Hobby, 16 Sep 44. WDWAC 320.2. (3) Cbl W6706, 2 Dec 43. SPWA 320.2 sec 7.

[35] Memo, Maj Craighill for SG through SG MTOUSA, 20 Sep 44. SGO Hist Div 319.1 MTOUSA, Women In.

Administrative arrangements for WAC medical care, with minor exceptions, were pronounced satisfactory, with adequate unit dispensaries and beds in general hospitals. As a result, evacuation rates were not excessive, being somewhat less for Wacs than for nurses. As for time lost from work, the theater enjoyed the distinction of being one in which the annual noneffective rate for Wacs was less than the comparable nonbattle noneffective rate for men—about 2.7 per 100 strength for Wacs, and 3.6 for all personnel exclusive of battle injuries.[36]

The rate of venereal disease was so low as to be negligible, although for men in the area rates were frequently alarming and the disease endemic. Among Wacs, cases were pronounced by Major Craighill to be "very infrequent"—less than one per month, with only fourteen cases having been reported by the end of 1944. The rate of pregnancy was less than that in comparable civilian age groups in the United States, and quite similar to that of nurses in the theater. Although no records were kept as to marital status, it was known that more than half of such women were married, with the number of pregnancies increasing toward the end of the war as the theater began to provide hotel quarters for married couples.[37]

The number of courts-martial during the entire two and a half years, through June of 1945, was extremely small, records showing no general courts, only six special courts—chiefly for censorship violations—and thirty-two summary courts for almost 2,000 women. WAC officers felt that "the worst disciplinary problems have been created by a few women who obviously should not have come overseas because of lack of adaptability or emotional instability." Some cases of misconduct were attributed to unhappiness about job malassignment.[38]

There was every indication that WAC morale was good in most cases. Company commanders considered morale to be almost entirely dependent upon full utilization of time and skills; one observed, "They wanted not only to be useful but to be inwardly sure of their usefulness." In cases where low morale did occur, such as in certain Signal Corps units, investigators were unable to find any bad effect from rough living conditions or lack of recreational facilities or the desire to return home; the only factor that had a definite effect on health, morale, and efficiency was that of job satisfaction.[39]

For the maintenance of morale, off-duty classes of interest to women were sponsored by the theater early in 1944, even before Director Hobby was able to get the U.S. Armed Forces Institute and other agencies to provide special courses for women. When rest camps and recreation centers began to be established to permit brief passes and furloughs for male troops, Wacs were included and were provided with living quarters, messing facilities, and quotas for such locations as Algiers, Capri, Rome, and later the Riviera, Venice, Switzerland, and Jerusalem. Although the average stay was only from three to five days, the rest was found to be a great relief from barracks life for women, and served in lieu of a convalescent hospital, which was not provided for women. For a restful evening off, WAC authorities, working with British women's services, were able to obtain adequate separate facilities for women's clubs, where women could enjoy an evening of

[36] (1) *Ibid.* (2) Statistical Data. Coleman file.

[37] *Ibid.*

[38] (1) Statistical Data. Coleman file. (2) Rpts, WAC COs, 6715th WAC Com Co and 6667th WAC Hq Co, 29 May 45 and 7 Jun 45, respectively. Coleman file.

[39] Documents cited n. 31.

bridge, cooking, and feminine companionship. The theater also permitted the WAC staff director to call meetings of WAC company officers and to send them monthly news bulletins, in an effort to improve leadership and administration.[40]

In the matter of decorations and citations WAC units and individuals were also included. The WAC sergeant who had supervised theater telephone communications received the Legion of Merit from General McNarney himself in a retreat ceremony. General Eisenhower awarded the Soldier's Medal to a WAC private who saved a soldier from a pool of flaming gasoline, at the expense of severe burns. Other decorations and battle stars were not unusual. Grades and ratings were likewise generally good, except among Fifth Army and Signal Corps personnel, and by the end of the war in Europe only a few hundred women who had been long overseas were still unrated.[41]

Soldier opposition had possibly a worse effect on morale than it did in the United States, since not only the Corps' existence but its presence overseas was questioned. Even before the women's arrival, letters streamed homeward advising against enlistment, and afterwards came the mass-pregnancy rumors which helped to inaugurate the slander campaign. Almost 100 percent of soldier comment was unfavorable or obscene. An Army supervisor noted:

> That sort of thing has a direct effect on the morale of the girls over here because their families write to them and want to know if they are true. . . . It bothers me because I can see the effect on the morale, spirit, and pride of the kids working for me.

Early arrivals also had to face insults in the streets. One Wac, advising her sister on enlistment, noted, "How many times I've heard some GI call us dirty filthy names, names that I would never allow anyone to call me if I were a civilian." WAC officers excited even more resentment than enlisted women, chiefly when they outstripped men in rank or promotions.

From those soldiers who worked in offices with Wacs, or were stationed in the same vicinity, the comments tended to become more favorable. The women's military bearing likewise was praised. Soldier comment noted:

> As the girls came down the boulevard, they really drew the applause. You have to hand it to them, they look good on parade. I doubt if West Pointers could appear better.

Unfortunately, only a minority of soldiers ever worked in offices with Wacs, or had any opportunity for becoming acquainted with them. Even where soldiers and Wacs were stationed in the same cities, there were not enough Wacs to permit much firsthand acquaintance between the two. Women shortly became so surfeited with requests for dates that it was necessary to provide separate recreation rooms and snack bars where they might relax and enjoy feminine companionship. The conduct of combat men, particularly young Air Forces officers, was a problem, for, said the staff director, "They would just sweep into the dayrooms and take over the place, and nothing could be done with them." There was no doubt that a Wac date, where available, meant much to many American men. One wrote:

---

[40] (1) Intervs with Col Coleman cited n. 1(3). (2) MTO Ltr, 29 Apr 45, sub: Educational Interests of WAC Pers. AG 353.8/264 TWSD-O. (3) Memo, G-1 NATOUSA for Sp Servs, 29 Dec 44; Conf Notes; Monthly Ltrs. Coleman file.

[41] (1) Statistical Data, 1943-45. Coleman file. (2) M/R, 10 Jan 44, SPWA 200.6. (3) At end of War: about 34 Pvts, 207 pfc's, according to Coleman intervs.

> Some of the boys and I have been dating some Wacs and having a grand time. They have a day room all of their own with ping-pong table, darts, and cards. They have a small library too. About eleven o'clock in the evening they get some sandwiches for us at the mess hall and we polish them off with a couple of cokes. They are a heck of a nice bunch of girls and full of pep and fun. It's nice to be able to talk to some feminine company that speaks your own language.

On the other hand, a Red Cross worker noted:

> Too many GI's get out of hand and then wonder why the Wacs won't go out with them. There are about 10 percent of the GI's that spoil things for the other 90 percent.

From the viewpoint of the War Department, the most alarming thing about soldier opinion in the North African theater was its effect on recruiting. The theater, like others not surrounded by the American public, never displayed a highly developed sense of those public relations measures that might have counteracted the soldier comment and reassured the women's families. Thus, in the early days it failed to take any positive countermeasures, although as early as February and March of 1943 some American foreign correspondents attempted to do so, notably John Lardner, Inez Robb, and Ernie Pyle.[42]

Although the first overseas Waacs were the Corps' greatest potential recruiting publicity asset, few stories or pictures came back. Recognizing that such attention could not be afforded by a combat theater for a tiny minority, Director Hobby attempted to stem the flood of bad publicity by sending a Waac photographer and a writer on temporary duty to get favorable stories. These could be given little assistance in the way of Signal supplies, laboratory facilities, or transportation, since the problem of winning the campaign was immediate and that of recruiting Waacs was remote. The two officers nevertheless managed to secure much favorable comment in various news sources, and to send back pictures of Waacs with General Eisenhower, Maj. Gen. James H. Doolittle, Lt. Gen. Carl Spaatz, and Gen. Henri H. Giraud, and with Spahis in a colorful review.[43]

Social Association

Possibly the only other theater problem in WAC employment stemmed from the lack of a firm policy on officer-enlisted relationships. General Eisenhower and General Hughes, in setting the first policy, were of the opinion that no restrictions should be placed on social engagements. Any other policy appeared difficult to enforce because of the presence at Allied Forces headquarters of large numbers of Allied officers and American naval officers, whose services permitted off-duty social association of officers and enlisted personnel. After her arrival, Major Boyce concurred, provided that officers did not date enlisted women who worked for them, thus avoiding the danger of favoritism in promotion. When General Devers took command, he reversed this policy and, over the objection of Major Boyce, directed that both Wacs and nurses observe military customs in selecting their off-duty escorts. Exceptions were made only for relatives and fiancés, who had to carry a letter to prove their status.[44]

[42] WDBPR Env, Africa—Opinion, Favorable.

[43] Rpt, Capt Ann McIlhenney and Lt Charlotte McGraw to Dir WAC, sub: Rpt on NATO Pub Relations. SPWA 320.2 sec 5.

[44] (1) Hughes Ltr to Col Hobby cited n. 14. (2) NATO Cir 23, 26 Feb 44; Med Base Sec Cir 20, 3 Mar 44. Coleman file.

GEN. HENRI H. GIRAUD REVIEWING WAACS *in a colorful ceremony in North Africa.*

Neither system worked perfectly. Mixed dating in itself caused no difficulty, so long as women did not date their supervisors, except in a few instances in which officers brought enlisted women to their clubs against the wishes of other officers who did not wish to associate with them. Enlisted men were bitterly jealous when officers monopolized the Wacs, but when the Devers ruling restricted enlisted women to dating enlisted men, the unfavorable comments continued and worse problems arose. Major Boyce asked that British and Navy officers be made subject to the ruling, but this proved impossible, and unfortunate public disturbances resulted when military police arrested enlisted Wacs escorted by British or Navy officers.

WAC commanding officers also reported difficulty with Army officers who ignored the rule and then attempted to prevent punishment of enlisted women whom they had dated. WAC commanders were loath to enforce the ruling when officer escorts were not likewise punished, but were allowed no choice under the regulation. Violations were flagrant, and commanders feared that respect for all discipline was being damaged. It was attempted to allow some latitude by issuing letters of authorization for engaged couples, but when informed that notices of engagement would be mailed to hometown papers, all but two such couples withdrew their requests for authorization.

Inasmuch as over half of the enlisted women in the theater were eligible for officer candidate school, it was impossible to hope that "natural social level" would cause them to prefer enlisted men. Only the Fifth Army experienced little difficulty, because it was possible to maintain one policy in this command.

No solution was ever found to the problem, which enlisted women frequently cited as the reason for their desire to leave the Army as soon as eligible for discharge. Upon losing an excellent secretary, General White discovered that her departure was due entirely to one incident in the theater, in which she had been arrested for dining in public with her husband, an officer.[45]

Marriage was not forbidden, and in addition to the Wacs who married officers there were a number of the theater's 4,000 nurses who had married enlisted men. Theater authorities were shortly obliged to be realistic about the situation. Finally, hotels were set aside to which officers and enlisted men might take their wives, and a wife was authorized to take the status of her husband when accompanied by him to hotels, messes, and clubs. However, WAC weddings were not frequent; only about 135 had occurred by the end of the war.[46]

Other Women's Groups

Wacs in North Africa and Italy considered themselves particularly blessed in the relative scarcity of American civilian employees, who in many other theaters occupied the best jobs. In the few cases where civilian women were employed in this area, the WAC staff director succeeded in pulling Wacs out of the offices concerned, so that conflicts were avoided, and, said Colonel Coleman, "We also successfully ran the OSS civilian women out of the WAC off-duty dress." It was not until the virtual end of hostilities that good enlisted jobs began to be taken over by that phenomenon which always annoyed Wacs irrationally: numbers of civilian women, complete with officers' privileges, fur coats, and duly authorized WAC uniforms.

In addition to the few civilian women employees of the Army, there were other women's groups in abundance—civilian groups such as the USO and the Red Cross, and military groups such as the British ATS and the French "WAC." Of these, the British ATS were the most numerous. Some months after the arrival of the first Waacs, and especially when the theater's Supreme Allied Commander was British, British women in small numbers began to arrive. In general they were housed and fed with the Wacs and worked with the British sections of the various headquarters. The British women proved most generous in sharing their recreational facilities with the Wacs, and all services mingled companionably for tea and other occasions.

In the Fifth Army two groups of ATS women lived and ate with the Wacs. The British women were pleased with the American mess and, said Captain Foster, "it was a pleasure to have them." As much

[45] Intervs and excerpts from mail cited n. 1(3) and n. 2.

[46] MTO Cir 52 sec II, 3 Apr 45. Coleman file.

could not be said for certain USO women sometimes billeted with the Fifth Army Wacs, since they habitually managed to be absent at moving time, leaving their tents for the Wacs to take down. Also living with the Fifth Army Wacs were the chief nurse, Red Cross women, and several French women officers. In general Wacs had only praise for the French women's corps, which, after a confused start on ambiguous status, achieved military standing and discipline and whose ambulance drivers were often sent farther forward than the women of any other service.

As for the American Red Cross, personal relations usually depended on the individual; as a group, it appeared to WAC commanders that "there was no place in their program for us." The older British services, on the other hand, had by this time made provision for women in their recreational services. When the American Red Cross failed to provide for the Wacs, the British YWCA set up a series of rest camps in Italy, and General Sir Harold R. L. G. Alexander backed an Allied Women's Club in Caserta. All of these served as many Wacs as ATS, and made it possible for the Wacs to have recreation and three-day passes comparable to those which American services provided for men only. In general, WAC authorities noted that "American agencies were not too well set up to do anything as far as women were concerned, and more often the Wacs had to take care of them."[47]

For a time it was feared that the Wacs would have to take over supervision of some 500 members of the German women's forces captured in the area. The uncertain status of these women, which might or might not have been that of civilians, roused some doubts in the minds of WAC authorities as to the result if the situation had been reversed. Had German captors been forced to decide WAC status, there appeared some likelihood that Wacs would have been considered civilians through confusion with the numerous civilian women in the OSS, USO, and other groups, all of whom wore the WAC uniform.

Demobilization

As the end of the war approached, WAC advisers became increasingly concerned that in wartime there was no limit to the overseas tour of duty of administrative troops. WAC company commanders noted that there were a few women in every company who became physically or emotionally unfitted for efficient overseas service but who could not be returned, for lack of any authority, until totally disabled.

Colonel Coleman felt that for certain types of work, particularly in the Signal Corps, personnel should be able to look forward to relief at a definite time, and that it would in the final analysis be cheaper to return a good worker while he or she was still capable of rehabilitation and continued service in the United States. She advised theater commanders, "Good service should not be permitted to end in hospitalization if it is possible to avoid it."

It was also Colonel Coleman's opinion that women, if left overseas past the proper time for return, were affected spiritually, not by war and death so much as by long residence in war-disrupted countries where the standards of native life were low and heartbreaking destruction was widespread. Under such circumstances, WAC commanders noted that

[47] Intervs and statements cited n. 1(3).

"women gradually hit a place where they have to live on the surface," and a certain temperamental change, of which they were often unaware, appeared to have made them less vulnerable to the sight of others' suffering. For this reason, even when women's health and efficiency were untouched, Colonel Coleman concluded that it would be better, if replacements and shipping were available, to limit an overseas tour to eighteen months.[48]

When rotation to the United States was finally authorized by the War Department, it was clear that Wacs would not be eligible in competition with men who had much longer service, and in fact most women felt that return would be improper under the circumstances.

At Director Hobby's request, the War Department late in 1944 added a microscopic quota for Wacs and nurses, enough to assure return of emergency cases. Even when Wacs became eligible under regular rules, there was no immediate relief. The theater decided that directives on rotation did not oblige it to return Wacs until a "suitable replacement" was available, and that men even in the same skill were not suitable replacements for women. The theater's new G–1, General White, commented:

> I contend emphatically that you cannot cut us off from WAC replacements on the grounds that the theater is overstrength. That is like saying you won't give us any more shoes because we have too many caps.

Eventually the theater was obliged to permit a few Wacs to be returned without replacements, since the War Department decided that these were more badly needed in more active theaters.[49]

When actual demobilization began, some 80 percent of the women were found to be eligible for discharge under the point system. In spite of some attempts to persuade them to sign up for longer service, some four fifths of eligibles indicated that they wished to be discharged. Many were retained as long as the "military necessity" clause permitted, until after the defeat of Japan, when the War Department directed return of remaining overseas Wacs. By the end of 1945, all enlisted Wacs were out of the Mediterranean theater.[50]

The Wacs' eagerness to depart indicated, by all evidence, merely a desire for discharge and no condemnation of theater employment as compared to that in the United States. When rotation first began, women who had returned to the United States wrote back comparisons highly favorable to the theater. Accustomed to being treated as essential employees, they found civilians in the best jobs in the zone of the interior, and the civilian population scarcely as respectful as that of occupied areas. Army barracks, even with heating and laundry facilities, were not as comfortable as the apartments and hotels that the Mediterranean theater had provided; neither were there civilians or prisoners to relieve the troops of kitchen police and fatigue duty. Upon receipt of such reports, an overwhelming majority of 95 percent of the remaining women in the theater stated that they preferred to stay overseas until eligible for discharge, rather than return

[48] (1) Memo cited n. 31(1). (2) Quotations from M/R, Stf Dir's Opinion, Rotation. Coleman file.

[49] (1) Memo, Col Hobby for Gen Henry, 21 Aug 44. WDWAC 320.2. Quota asked was 0.5 percent per month. (2) Memo cited n. 30(1). (3) Quotation from Ltr, Gen White to Col Hobby, 3 Jan 45. WDWAC 320.2. (4) Only 120 women were ever returned (1944–45), according to Col. Coleman. About 136 received a furlough in the United States.

[50] (1) Statistical Data, Jun 43–Jun 47; (2) MTO Ltr to all subordinate comds, 23 May 45, sub: Readjustment of WAC Pers. AG 370.01/746–A–O, Coleman file.

to face the hardships of life in the United States. When, upon the cessation of hostilities, a sizable number of individuals on temporary duty were obliged to remain for permanent duty in the United States, their protests were audible in Italy.[51]

Final comments of both Army and WAC authorities indicated that the theater's WAC experiment had been a successful one. The first theater commander, General Eisenhower, stated of the women:

> The WAC in Africa has proved that women can render definite contributions to the winning of the war, and that their capabilities in this regard extend to an actual theater of operations. . . . In some cases one Wac has been able—because of her expert training—to perform tasks that previously required the assignment of two men. The smartness, neatness, and esprit constantly exhibited has been exemplary . . . their general health and well-being have certainly been equal to that of our best enlisted units.[52]

The Air Commander in Chief of the Mediterranean Allied Air Forces, Lt. Gen. Ira C. Eaker, added:

> As a result of my observation of the WAC personnel to date, I am thoroughly convinced that it should be retained in the after-the-war peacetime Army. I believe it is an indispensable service to our Army in the present emergency. There are innumerable tasks which I have observed WAC personnel performing in the Air Forces with remarkable efficiency. . . . They have a capacity for many specialized duties essential to the Air Forces to a greater extent perhaps than any other Air Force soldier.[53]

General Clark of the Fifth Army called to tell the Director that he wanted to take his headquarters platoon into Austria with him: "They've plenty of points but I have to have them."[54] Col. Westray Battle Boyce, visiting the areas as second Director WAC, after the end of the war, noted that morale was good and the women well adjusted, with adequate provision by the theater for their welfare.[55]

---

[51] (1) Memo cited n. 30(1). (2) Coleman Interv.

[52] Statement by Gen Eisenhower, AFHQ, 9 Jun 43, Hist Background of WAAC-WAC, App. E. WDBPR file, DRB AGO.

[53] Ltr, Air CinC MAAF to Dir WAC, 27 Jan 44, in History of Mediterranean Allied Air Forces, Dec 43-1 Sep 44, by Hq MAAF, Vol. II.

[54] Memo, Exec WAC for Dir WAC, 16 Jun 45. 1945-46 WAC files.

[55] M/R, Rpt of Visit by Dir WAC, Tab B, Mediterranean. WDWAC 333.1 (1945-46).

CHAPTER XXI

The European Theater of Operations

The European Theater of Operations, the first to requisition Waacs, was the second to receive them. By the end of the war it employed 8,316 women, the largest number in any overseas theater.[1]

The first requisitions, made by the theater Services of Supply in the summer of the Corps' formation, were not pressed after Lt. Gen. Dwight D. Eisenhower's departure for North Africa. After some six months of waiting for shipping priorities, the first two units earmarked for the theater were disbanded and their members were sent either to North Africa or to stations in the United States. In the spring of 1943, the Services of Supply again sent a requisition, but again furnished no shipping priority and shortly thereafter canceled the request.[2] Impetus in the program passed to the theater's Air Forces, of which General Eaker noted later:

> Shortly after I arrived in England the problem was presented to me. . . . The commanding officer counseled me against bringing Waacs over . . . he said that I would be held responsible.[3]

Accepting the responsibility, the Eighth Air Force requested an entire WAAC battalion and set a shipping priority for the summer.[4]

In mid-April of 1943 the theater WAAC staff director, Capt. Anna W. Wilson, arrived in London to make preparations for the unit. Small advance parties from the Signal Corps and other groups were also present. Captain Wilson made plans for reception, housing, and assignment of the women, and for such feminine needs as laundry rooms, sanitary facilities, special post exchange items, and clothing main-

[1] Unless otherwise indicated, all references and quotations are from the Report of the General Board, U.S. Forces, European Theater, Study of the Women's Army Corps in the European Theater of Operations. OCMH. (See Bibliographical Note.) Hereafter cited as ETO Bd Rpt. Other major references: (1) ETO WAC Hist. (2) WAAC classified files, SPWA (later WDWAC) 320.2 secs 1, 2, 4*a*, 4*b*, 6, and 8*a*, DRB AGO. (3) History of USSTAF, 516.01, Vol. III, Ch. II, pp. 150–58. USAF Hist Div. Cited as USSTAF Hist. (4) Historical Records Rpt 315*c*, Wacs in European Division, Air Transport Command. USAF Hist Div. Cited as ETO ATC WAC Hist. (See Bibliographical Note.) (5) Interv with Col Mary A. Hallaren, Dir WAC, and written comments, 14 Mar 50, (6) Comments on draft of MS by Dr. Hugh M. Cole, author of *The Lorraine Campaign*, UNITED STATES ARMY IN WORLD WAR II (Washington, 1950).

[2] (1) Ltr, ETO WAAC Stf Dir to Dir WAAC, 16 Jun 43. AG 322.3, SPWA 320.2 sec 4*a*. (2) Rad 8786, USFOR London to WAR, 15 Apr 43. (3) Ltr, WAC Hq to CG ETO, 20 Feb 43; Rad 7329, London to WAR, 17 Feb 43. SPWA 320.2 secs 2 and 4*b*. (4) Memo, WAAC Hq for CG ASF, 17 Mar 43. SPWA 320.2 sec 4*a*. (5) D/F, OPD WDGS for CG ASF, 22 Mar 43; 1st Ind to unidentified basic Ltr, Opns Div ASF to Dir WAAC, 12 Apr 43. SPOPU 320.3 (4–6–43).

[3] Address to WAC OCS 60th Graduation Class, 17 Nov 45. WDWAC 352.

[4] Rad 9130, London to WAR, 27 Apr 43. SPWA 320.2 sec 2.

tenance supplies. The WAAC's way was smoothed by the long-accepted presence of the many British women's services, as well as of the Voluntaires Françaises, the Norwegian ATS, the Polish ATS, the Canadian WAC, and other such groups.[5]

The arrival of the American WAAC was nevertheless of considerable public interest. Captain Wilson reported:

> We were interviewed by the press and on short-wave radio, and were a little surprised to find ourselves near-celebrities. We are now settled in a very comfortable flat overlooking Hyde Park . . . two bedrooms, two baths, kitchenette, and a living room.[6]

Captain Wilson noted that a disproportionate amount of her time was consumed by the social and public relations activities required of her. As in other overseas theaters, there always existed a certain tendency for the WAC staff director to be considered dedicated to official teas, press conferences, and ceremonies instead of to her military responsibilities.[7]

First WAAC Separate Battalion

Upon receipt of the Air Forces' shipping priority, WAAC Headquarters rounded up as many members of the original units as could be found, with replacements for those who had gone to North Africa. These were rapidly organized at Fort Devens to form the 1st WAAC Separate Battalion, under the command of Capt. Mary A. Hallaren. Women were equipped with full winter uniforms, as authorized for England, plus gas masks, canteens, first-aid packets, utility packs, helmets, and pistol belts without pistols. Their training included judo, hikes with full packs, and an obstacle course which, although not required for noncombatants, every woman elected to go through. The 555 enlisted women and 19 officers were so eager for the long-awaited assignment that no case of AWOL occurred prior to sailing. On 16 July 1943, almost a year after the initial request, the battalion reached England.[8]

Upon arrival the women were divided into companies and assigned to various Air Forces stations. Housing consisted of "everything from castles to huts,"[9] generally being identical with the type used for male troops at the stations. Theater historians later noted, "It was anticipated that special housing and physical facilities would be required for the Wacs, but the problem of housing Wacs provided no particular difficulties." [10]

It was immediately apparent that WAC employment in England offered little more difficulty than employment in the United States, and had the advantages of new sights and experiences and the stimulus of closeness to combat. The regulation WAC uniform was obviously inadequate for work in unheated buildings during the approaching winter and for the projected movement to the Continent, but time permitted remedial action. The supply lanes to the United States were by this time open and were relatively short, and many items could be procured from British stocks. The dangers of the area were no more

---

[5] (1) ETO WAC Hist, pp. 7-13; also ETO Bd Rpt. (2) Ltr, Lt Gen Frank M. Andrews, Hq ETO to Dir WAC, 1 May 43. SPWA 320.2.

[6] Ltr, WAAC Stf Dir ETO to Dir WAAC, 30 Apr 43. SPWA 320.2 sec 4*a*.

[7] ETO Bd Rpt, incl App. 112.

[8] (1) Rpt of Tng and Staging, PRD, Camp Shanks, N. Y., undated. SPWA 320.2 sec 2. (2) Interv cited n. 1(5). (3) Series of documents, SPWA 320.2 secs 1 and 2. WAAC sources say 557 enlisted women.

[9] Statement by Col Hallaren, Nov 51. WAC files, OCMH.

[10] ETO WAC Hist, pp. 10-11; also ETO Bd Rpt.

DOCKING IN SCOTLAND. *Advance party of WAAC officers and enlisted women are greeted by a bagpipe band upon their arrival, 11 May 1943.*

than those confronting the civilian population, and appeared to have little effect on morale; General Eaker later noted, "One of the factors in their success was courageousness. I saw this demonstrated when German planes came over . . . they keep more calm than men in emergencies." [11]

The first arrivals were an immediate success to a degree which, in the opinion of Air Forces commanders, motivated the extensive requisitions that followed. General Eaker stated, "It was not long before they were the best we had . . . other forces followed our lead." [12] An Eighth Air Force supervisor stated:

> Their work has improved the efficiency of my office tremendously. Their attitude, discipline, and efficiency are of such value that not only enlisted men but some officers have been released to perform other duties.[13]

In a typical Air Forces flight control room, Wacs were shortly assigned as teleprinters, typists, and switchboard operators.[14] General Eaker noted his surprise in finding that "military secrets were going to have to be kept by women," and added:

> One of their most important duties was to keep secret and confidential files. They were intelligent and learned quicker. They were the best photo interpreters . . . keener, and more intelligent than men in this line of work.[15]

Build-up of Units Before D Day

Further requisitions followed from both the Air Forces and theater headquarters. Personnel for the next shipment was at first organized as the 2d WAC Separate Battalion, but at the request of the theater the battalion was inactivated and its members shipped as casuals without grade allotments. The move was made possible by the Corps' recent military status, and was pronounced by the theater to be more suitable to its needs, since the women could be integrated into existing units. These Wacs arrived in the theater in September and November of 1943, and were assigned to both Air Forces and theater headquarters, bringing the total of theater WAC enlisted strength to 1,126. In October of 1943, as a result of continuing requisitions, the War Department raised the theater WAC quota to 2,775.[16]

Before this quota could be shipped, the War Department at the end of the year informed the theater that no more Wacs could be sent for several months. Losses of the conversion period had left existing units in need of replacements, while the manpower shortage of early 1944 was already being felt. Theater protests followed, citing the approaching invasion of the Continent. General Spaatz and General Eaker appealed personally to General Arnold in Washington, asking that he prod the War Department.[17]

The theater also asked permission to make up some of the deficit by recruiting and training American civilian women living in the British Isles, believing that considerable numbers might thus be obtained. This authority was not granted for some time, and even when later put into effect proved an insignificant source of recruits, as did a similar provision for transferring to the WAC the few American citizens in

[11] Address cited n. 3.

[12] *Ibid.*

[13] ETO Press Release 1601, 14 Aug 43. ETO PR file, Adm 365 WAC, DRB AGO.

[14] ETO Press Release 1832, 6 Dec 43. Same file.

[15] Address cited n. 3.

[16] (1) See Table 7, Appendix A. (2) ETO WAC Hist, pp. 8–9; also ETO Bd Rpt. WAC sources say September and October for arrival.

[17] Rad W8189, USFOR London for USAWW–WAR, 6 Dec 43. SPWA 320.2 sec 6.

the British women's services. It became clear that the European theater must depend upon recruiters in the United States for its supply of Wacs.[18]

There followed what was, over the next months, to be a continuing skirmish over the numbers and skills of Wacs who could be sent overseas. The War Department noted that, to meet the theater's heavy demands for skilled typists and stenographers, not only would the entire recruiting intake be required, but it would be necessary for stations in the United States to resort to the embarrassing expedient of using male typists to replace Wacs "for combat" instead of vice versa. The theater in reply urged that they be so embarrassed. In its opinion, overseas areas, being obliged to support and supply troops under difficulties, required the best workers for each type of duty; the argument was that one Wac typist could replace two men while eating only half as much.[19]

At the moment part of the dispute was resolved in the theater's favor by General Arnold of the Army Air Forces, who secured permission to cut loose the Air Forces quota from the rest of the theater's, and to supply all the Wacs that General Spaatz requested, provided that only AAF stations were called upon to supply them. In February of 1944 the War Department therefore approved a new quota for the theater, 4,448 Wacs for the Air Forces and 1,727 for all other commands, a total of 6,215 for the theater.[20]

Requests from the Services of Supply in Europe were slower to come in. It was not until just before D Day that the SOS requested several large groups, so hastily that some, especially telephone operators, had to be provided with air priority.[21]

Under the new quota, WAC shipments were at once resumed and, during the first half of 1944, arrived in the theater almost faster than they could be accommodated at reception facilities. Just before D Day, theater WAC strength had reached 3,687. At this time the War Department was again forced to shut off the supply, pleading heavy shipments to the Pacific. In spite of theater protests, no further quota increase was to be granted for some months.[22]

By D Day, Wacs were assigned to Headquarters, Supreme Allied Expeditionary Force; Headquarters, ETO; to the Allied Expeditionary Force; and to other theater organizations. Within the Air Forces, they were in Headquarters, U.S. Strategic Air Forces; Headquarters, Eighth Air Force; Headquarters, Ninth Air Force; Headquarters, Allied Expeditionary Air Forces; Headquarters, VIII Fighter Command; Headquarters, IX Bomber Command; I, II, III Bomber Divisions; and the 2d, 14th, 20th, and 96th Combat Wings. There were also numerous independent organizations that used almost a thousand Wacs in Europe, including the Air Transport Command and the Office of Strategic Services.[23]

Before D Day, initial job assignment in some of these organizations suffered from the fact that only the Air Forces in Eng-

[18] (1) ETO WAC Hist, p. 9. (2) ETO Bd Rpt, Vol. I, p. 27. (3) Ltr WAC Stf Dir to Dir WAC, 7 Aug 44. WDWAC 352 British Stf College.

[19] (1) Cbl E-31714, Eisenhower to WD; (2) Cbl WAR 49888, 13 Jun 44; (3) Ltr, WAC Stf Dir ETO to Dir WAC, 24 Jun 44. WDWAC 320.2 ETO.

[20] ETO Bd Rpt; differs slightly from Ltr, TAG to CG ETO, 10 Mar 43, sub: Quota of WAC Pers for ETO. AG 320.2 WAC (2-18-44) PO-M-A, in SPWA 320.2 (1944).

[21] Sources cited n. 2.

[22] Sources cited n. 19. These say strength was almost 6,000; later TAG figures give 3,687. See Table 7, Appendix A.

[23] Later mentioned: U.S. Gp Contl Council, 12th Army Gp, First AB Army, and MIS. ETO Bd Rpt; also ETO WAC Hist, p. 19. Independent orgns: ACT, OSS, O of Mil Attaché, Governments in Exile, Br O of JAG MIRS, AACS, U.S. Strategic Bomb Surv.

land were as yet engaged in combat. Other headquarters still had plenty of enlisted men, and were still located in areas where competent civilian help was obtainable. Therefore, while Wacs would be needed in the future, and were welcomed by higher commanders who had requisitioned them for this reason, their welcome from subordinate commanders and enlisted men was less enthusiastic. In Lt. Col. Anna W. Wilson's opinion:

> Male soldiers who were firmly established in their administrative, clerical, and communications jobs and were not trained for combat naturally were fearful of the arrival of WAC personnel. Their comments on and reception of WAC personnel were derogatory and cold. The Wacs felt that they were unwelcome, which in fact they were. Months passed before this feeling of ill-will disappeared.[24]

Censorship of letters showed that there was much bitterness in the accounts written home by enlisted men and Wacs alike, which upon investigation had "no reasonable basis" but which authorities feared would damage WAC recruiting.[25]

Until D Day, as a result of the abundance of personnel, Wacs therefore at times did not replace men, or were assigned to only part-time work, or were fitted into existing organizations in duties less responsible than those of enlisted men and British civilians and less useful than the positions the women had occupied in the United States. The staff director listed incorrect assignment as first among causes of low morale. The WAC personnel officer noted, "The greatest gripe from Wacs has constantly been, 'not enough work to do.' "[26] Luckily, the serious morale problems that usually accompanied such a situation in the United States did not develop, chiefly because of the women's sense of nearness to combat and the prospect of more useful days ahead.

Office supervisors discovered no loss of efficiency during the buzz-bomb raids, and in fact, as D Day approached, there were grounds for believing that the reverse was true. One unit, which arrived in London simultaneously with the V–1, reported:

> At first the Wacs were up most of every night, spending much of their time in the improvised basement shelter. Since this area was not properly equipped, there was great loss of sleep, and the nervous tension which resulted was almost as bad as the bombs.[27]

However, the women soon became accustomed to the alarms and slept all night without going to the shelter, and they worked without interruption even when bombs fell close to their offices. Several Wacs were awarded the Purple Heart for injuries received in bombings; they were more fortunate than British servicewomen in that none died of these injuries.[28]

The Move to the Continent

Less than a year after the first group's arrival, the invasion of Normandy began and Wacs moved out into France and later into Germany. A lively dispute had raged all spring over the propriety of taking women along. The more gloomy objectors stressed the fact that the women would probably be killed or captured and, if captured, might not, under the rules of land warfare, be recognized as "protected personnel" and would not, like nurses, be entitled to officer prisoners' privileges. Some

---

[24] ETO Bd Rpt, Vol. I, pp. 10 and 12.

[25] (1) ETO Bd Rpt. (2) Morale Rpt, 1–15 May 43, to Chief Info and Censorship Sec ETO. ETO Opns Rpts, Adm 365 WAC, DRB AGO.

[26] ETO Bd Rpt, p. 55; also Vol. II, App. 17.

[27] ETO ATC WAC Hist.

[28] Six deaths of Wacs were reported as of December 1944: 2 accidental poisonings, 2 auto accidents, 1 peritonitis, 1 meningitis. ETO Bd Rpt, Vol. III, App. 114.

LANDING IN NORMANDY, *14 July 1944, above. Photographs, below, were taken in the WAC camping area on 1 August 1944.*

feared that fighting men would worry so much over the Wacs' safety that their efficiency would be impaired. Others questioned whether women could live in tents under field conditions. Officers on the other side of the argument declared that their offices required their Wacs and that conditions on the Continent would be no worse than those in London under the buzz bombs, which had not disrupted office work. The WAC staff director was of the opinion that Wacs should move with their headquarters, in view of the reported success of the Fifth Army's experiment in Italy. The Wacs themselves were, as always, inordinately eager to go.[29]

The theater's decision was that the Wacs would go along, and that 50 percent of the personnel of Forward Echelon, Communications Zone, would be WAC. Necessary plans were made for clothing and maintenance, tentage, and well-trained cadre. An initial attempt was made to select personnel carefully on the basis of ability, physical stamina, and emotional stability; the group selected to follow invasion troops onto the beaches was also given a strenuous training program, plus field clothing and equipment. Unfortunately, it was soon clear that the whole WAC population of England would shortly be on the move, and that other units coming directly from the United States would have to go to France without special screening or acclimatization.

As the time approached for movement to the Continent, certain bookkeeping difficulties had to be solved. When the WAC's own Table of Organization units had been dissolved, the War Department had specified that overseas Wacs would be asigned only to large headquarters operating under bulk allotments. Assignment to vacancies in men's T/O units was prohibited because of the notion that every T/O unit would sooner or later move out to do battle.

Actually, Air Forces and Signal Corps headquarters, although under T/O's, were able to operate from bases farther behind the lines than were some of the bulk-allotment Communications Zone headquarters. To avoid violating the T/O ban, the Air Forces for the first months employed Wacs on an impractical and confusing lend-lease basis from theater headquarters. Bookkeeping difficulties almost wrecked this program and shortly forced the War Department to relax its ban and permit Wacs to be assigned to men's T/O units, to "suitable noncombatant positions in fixed headquarters or installations." [30] With this, the stage was set for easy movement of Air Forces and Signal Corps Wacs to the Continent. As a matter of fact, the term "fixed installation" was shortly to be interpreted by the theater to mean a tent moved forward not more than once a day.

The first Wac to arrive on the Continent was a stenographer who, on 22 June, flew in and out of one of the beachheads to record a conference. The first regular WAC unit landed in Normandy on 14 July 1944, almost a year to the day after the first group had arrived in England, and thirty-eight days after D Day. These Wacs, assigned to Forward Echelon, Communications Zone, arrived on the Normandy beaches in LST's, and camped out in an apple orchard. They slept under shelter halves or in pyramidal tents on Army cots, received the usual field rations, and washed in cold water carried in hel-

[29] The entire account of the movement into France and Germany, unless otherwise stated, is from ETO Bd Rpt and ETO WAC Hist.

[30] Ltr, TAG to CG ETO, 10 Mar 44. AG 320.2 WAC (2-18-44) PO-M-A, in SPWA 320.2. Change was later written into WAC Regulations.

mets. They immediately went to work as telephone operators, typists, and clerks, working in tents, cellars, prefabricated huts, and mobile switchboard trailers. Because of the cold and mud, they wore leggings, trousers, and combat jackets; even long underwear was now popular.[31]

It was not noted that the men's anxiety over WAC safety impaired their efficiency, as had been feared; in fact, historians recorded that men were glad to have American women to talk and sing with, and "just to look at." However, since living and working conditions were equal, the normal amount of masculine griping was found to be definitely hampered.

More Wacs arrived at once, by plane and boat, some in units and some with their sections—Normandy Base Section (Supreme Headquarters, Allied Expeditionary Force), and Air Forces units, also Quartermaster, Transportation, Engineers, Ordnance, and others. Large groups of Signal Corps telephone operators arrived, many having been hastily flown in from the United States on a high-priority requisition. The most forward and mobile WAC detachment was reported to be the unit with the 12th Army Group, which advanced with its headquarters close behind the fighting lines.

On 31 August, five days after the fall of Paris to the Allied forces, a WAC advance party arrived, and on 1 September Wacs began pouring into the city from Normandy and England, by plane and by truck convoy. WAC telephone operators at once took over French switchboards just abandoned by the German army; office workers operated with captured German office supplies; cooks and mess sergeants assumed supervision of French cooks and waitresses in the various messes; and the Wacs themselves moved into and gradually filled a number of hotels in the city. By October, 3,000 Wacs were in France; by the end of 1944 over half the ETO WAC units were there.

The Transportation Corps shortly sent picked personnel from the United States to form a traffic regulating group entirely composed of Wacs; many of the map tracings for routing combat supplies were made by this group. The Signal Corps organized an all-Wac battalion, composed of switchboard teams, message center teams, and teletype teams. This could not be called a WAC battalion because its tables called for combat equipment, and Wacs were noncombatants; Signal Corps authorities therefore called it the % 3341st Signal Service Battalion, with the % mark indicating, to the initiate, Wacs. Another all-Wac battalion, composed of Negro personnel, came in a unit toward the end of 1944 and took over half of the Postal Directory for the theater. Besides these all-WAC units there were Wacs in almost every command echelon down to the Army group level.[32]

In the eight months before V-E Day, Wacs followed closely behind the fighting forces. Food, quarters, and supplies were the same as those available to the men. The only reported difficulty concerning quarters was the absence of heat and hot water; doctors estimated that 25 percent of the women in one unit had chilblains. Conditions were extremely unsettled, medical and hospital care uncertain; there was cold, rain, snow, and mud; it was seldom any longer possible to have advance inspection of housing. Nevertheless, the strenuousness of this period had no perceptible effect; authorities were in fact

[31] See also Synopsis Hist of WAC, atchd to Memo WAC Gp WDBPR for Dir WAC, 31 Jul 45, sub: Proposed ETO WAC film. WDWAC 062.

[32] After V-E Day, Wacs were also assigned to the Third, Seventh, and Fifteenth Armies, and to the 40th Bombardment Wing.

LT. COL. ANNA W. WILSON, *WAC Staff Director, European theater, lunches in the field in France, 1 August 1944.*

surprised to discover that the WAC sick rate was the lowest of the year. During the initial six weeks when Wacs were landing on Normandy beaches, only one woman had to be sent back to England because of illness. Wacs noted that after the V-bombs in England, the sound of distant artillery fire in Normandy was a relief. Morale remained high in direct proportion to the women's sense of playing an immediate and essential part in the winning of the war. With civilians temporarily left behind, and men for office work scarce, Wacs were at last the valued employees they had been in some units in North Africa. This period, in spite of its lack of physical comforts, was reportedly the one in which ETO Wacs were happiest. In December of 1944, during the dark days of the Battle of the Bulge, offices depended greatly on their women, and WAC morale and efficiency were at a peak.

At the time of the Battle of the Bulge, theater representatives went to Washington and personally sought and obtained resumption of Wac shipments. These, by V-E Day, had raised the ETO WAC strength to more than 8,000—the largest number of Wacs utilized by any foreign theater. The number was only half of the 16,000 that had been requisitioned, and much less than the 25,000 to 50,000 that the WAC staff director felt would have been requisitioned had there been any hope of receiving so many. As the Air Forces mission shrank after V-E Day, greater proportions of the total came to be employed by field and service forces.[33]

When, in the spring of 1945, the First Tactical Air Force and the 12th Army Group headquarters opened in Heidelberg and Wiesbaden, no one debated whether or not to take Wacs to Germany; they moved with their headquarters as a matter of course. After V-E Day, only a handful of Wacs were left in England; the majority came to be concentrated in Germany at Berlin, Frankfurt, Wiesbaden, Heidelberg, and other headquarters. Here and in Paris, living conditions became luxurious in comparison to wartime billets and to the average accommodations of troops in the United States.

In Paris, Wacs lived in hotels with maids, dining rooms, no work details, and good recreation and sight-seeing facilities. In Frankfurt, they were housed in a well-heated apartment building, ordinarily with only two women to a room, and ate at an excellent mess boasting small tables and tablecloths, with civilians to do all but supervisory work. There were available a complete and quick civilian laundry service and good recreational facilities. In Wiesbaden, women lived in cold but comfortable apartments and a handsome private home. In Heidelberg, a comfortable hospital building was used for housing. Visiting War Department inspectors late in 1945 reported that "with few exceptions, these Wacs were living under better conditions and with more comforts than it was possible to have during the war."[34]

WAC Job Assignments

During the entire period of employment, the most notable difference from the pattern of job assignment in the United States was that a higher type of personnel was available to the theater. As in the North African theater, the women who made up these shipments constituted, on a statistical average, the cream of the Corps. Records showed that 99 percent of ETO Wacs were skilled, and that more than 50 percent of enlisted women had the AGCT scores required of officers. Such women were not only easier to assign but, as later studies showed, were less liable to become disciplinary problems, suffer low morale, or require medical discharge. No "holding depot" to accommodate unassigned personnel was ever needed for women, since all were readily assignable.[35]

Another notable difference in the employment pattern resulted from the theater's decision to limit women's jobs to a rather narrow range. Like most overseas theaters, the European theater considered women's job superiority proven only in routine clerical, stenographic, and communications duties, and was loath to risk shipment of other skills in which a man might not be efficiently replaced. Final records of WAC assignment indicated that about 35 percent of the theater's enlisted women were employed as stenographers and typists; another 26 percent were clerks; 22 percent were in communications work. These percentages of scarce skills were considerably higher than those available to the Corps as a whole. In the United States, as in Britain, personnel shortages had admitted servicewomen to a wide variety of work and had made it necessary to accept recruits in any skill. In the European theater, only 8 percent of American servicewomen were employed

[33] ETO Bd Rpt, Vol. I, pp. 26 and 27; Vol. II, App. 10, pp. 4–5, and p. 14.

[34] Rpt, Maj Frances Clements, G-3, and Capt Velma P. Griffith, PRO. WDWAC 333.1 (1945).

[35] ETO Bd Rpt statistics.

POSTAL DIRECTORIES IN THE EUROPEAN THEATER

in the more unusual skills: medical technicians, mechanics, draftsmen, interpreters, weather observers, and photographic specialists.

The maintenance of this policy at times was carried so far as to result in the malassignment of women with "unfeminine" skills. Thus, some one hundred WAC mechanics in the Air Service Command, although pronounced "highly satisfactory" in this work, were ordered reassigned to unskilled desk jobs on the grounds that it was not appropriate for Wacs to wear coveralls. Other WAC mechanics remained unassigned at ports, and were eventually transferred to the Air Transport Command.[36]

The policy also made it increasingly difficult for the War Department to approve theater requisitions, and for shipping agencies to fill them satisfactorily. Many requisitions, while numerically reasonable in view of the total strength of the Corps, would have required skimming almost all stenographers and many typists from stations in the United States—a sacrifice which the War Department deemed unjustified in view of frequent reports that there was already widespread theater use of WAC stenographers as typists and of typists as clerks. Stations in the United States experienced difficulty in providing even the numbers approved. The theater reported receiving typists labeled stenographers and clerks called "potential typists." Also, as volunteer stenographers and typists were all shipped, it became necessary for stations to send nonvolunteers who were known to be more liable to psychiatric breakdowns. Plenty of volunteers for overseas service remained in the Corps, but not in the requested skills. Eventually the War Department was obliged to inform the theater that its requisitions could no longer be approved if they asked percentages of scarce skills higher than could reasonably be supplied. The theater's requisitions thereupon dropped a notch in skill, but continued to concentrate upon clerical fields, in spite of the staff director's attempts to interest offices in using Wacs in medical, mechanical, and other work.

An additional complication in shipment was the matter of grades. Since arriving casuals were assigned to existing units, the theater asked that they arrive with low grades in order not to outrank men and women already in those units. Expert stenographers were sought in the grade of private, and experienced company officers in the grade of second lieutenant. The Army Service Forces filled several shipments with Wacs who voluntarily relinquished their ratings for a chance at overseas service. However, when later nonvolunteer shipments arrived with ratings, a morale problem arose and the War Department ordered the ratings restored to earlier arrivals, not wishing to establish a policy of reducing to the grade of private all men and women sent overseas. Eventually the War Department was forced to apply a percentage ruling to WAC shipments, requiring the theater to absorb a percent of higher grades equal to the over-all distribution.[37]

Rather than ship women who did not meet the exacting specifications in theater requisitions, incomplete shipments were sometimes made, a practice that also met with objections from the theater, since short shipments upset assignment plans almost as seriously as did unqualified ones. At other times, shipments were delayed until after the theater's need had changed

[36] ETO Bd Rpt statistics, Vol. II, App. 12 of App. 1, and Vol. III, Apps. 95–96.

[37] (1) ETO WAC Hist, p. 26, 34. Also ETO Bd Rpt. (2) D/F, G–1 to MPD ASF, 10 Mar 45. WDGAP 320.2 WAC, in WDWAC 320.2 ETO.

or had been filled from other sources. Colonel Wilson later commented:

> The fluctuating supply of WAC personnel, engendered by the voluntary recruitment policy, prohibited firm planning . . . and resulted in an immeasurable expenditure of money, time, and effort, all of which would have been avoided by the application of selective service to women.[38]

Officer Assignments

As was generally true overseas, ETO WAC officers were frequently less fortunate than enlisted women in the variety of their assignments and the degree of responsibility delegated to them. Several WAC officers attended the British Staff College in a class with British women officers, but an assignment adequate to justify this high-level training could not be found. Wacs who had graduated from the Army's own Command and General Staff School were assigned to jobs such as photointerpreter. A generous portion of 32 percent of the WAC officer strength was made available for WAC administration. A high percentage of the others was used in routine jobs which were similar to those of enlisted personnel except that an officer was required for security reasons—code and cipher officers, watch officers, signal officers, photointerpreters, censors, and military secretaries.[39] In spite of a War Department letter prohibiting the employment of WAC officers as stenographers, aides, and chauffeurs,[40] by the end of the war, no less than forty-five WAC officers—14 percent of the total—were serving as personal assistants to ranking officers in the European theater.

Such employment of WAC officers caused other ranking officers in the theater to ask that their enlisted WAC secretaries be directly commissioned and continue on the same job. Director Hobby steadily opposed all of these proposals, maintaining that if the women were more worthy of commissions than other Wacs in the theater, they should be selected by an officer candidate board, and that to open the door to such a policy would mean yielding to the political pressure of hundreds of other prominent sponsors who still sought WAC commissions for friends and constituents. The War Department therefore never wholly sanctioned the European theater's request, but a concession was made which permitted deserving cases to be forwarded for review in Washington. Under this authority several direct commissions were eventually granted—none to secretaries, but several to technicians such as the WAC sergeant who, as an expert on bridge demolition, was already filling the job of a major in the Corps of Engineers.

Because of a peculiar loophole in the WAC legislation, there was no legal barrier to the direct commissioning of British civilian women in the WAC; however, in the European theater only one civilian, General Eisenhower's chauffeur-secretary, was commissioned in this way, over Director Hobby's protest.[41]

---

[38] ETO Bd Rpt, Vol. I, p. 27.

[39] ETO Bd Rpt statistics as of 1 Nov 44.

[40] Ltr, DCofS USA to AAF, ASF, AGF, and overseas theaters, 11 Aug 44. CofS 324.5 WAC.

[41] For details see Chs. XXII and, below. Colonel Hobby had repeatedly discouraged ETO inquiries in this matter, and General White had written General Eisenhower a long personal letter citing obstacles, but after three Allied citizens were commissioned in Australia without her consent, the Director received a note from Colonel McCarthy, Secretary to the General Staff: "After a talk with Gen Eisenhower, Gen Marshall decided that Miss Summersby should be commissioned directly. I understand that the WAC will have objections but I believe that the decision is final." Memo, Secy WDGS for Dir WAC, 31 Oct 44. WDWAC 320.2 ETO.

WAC Staff Director's Office

Some initial difficulty was reported by Colonel Wilson in establishing her own office and her relative responsibilities. In her opinion, theater authorities, while welcoming the enlisted women, were not initially receptive to an active role by WAC staff officers, particularly in the first months before General Eisenhower's return in 1944. The title of Director was objectionable to the headquarters because "by definition and practice in the civilian world [it] implies a certain amount of the command function." [42] As a result, Colonel Wilson and her subordinate staff directors in air and service forces headquarters found that

> . . . until the WAC Staff Directors became personally known in the commands, they met many difficulties in the performance of their mission, especially with respect to making staff visits and collecting information necessary for making long-range plans and policies.

Equally objectionable to theater authorities was the word *inspection* in the War Department's provision that "a principal function of staff directors will be continuous inspection . . . for the purpose of gaining information upon which to base recommendations on . . . plans and policies." In spite of the War Department directive, the theater refused to authorize staff advisers to visit WAC units, on the grounds that this usurped the inspector general's duties or those of male commanders. Theater authorities also were reluctant to allow WAC staff officers to receive statistical reports on WAC strength, losses, disciplinary action, and other problems.

Until the end of 1943 the theater did not define the position of the WAC staff director's office, which was unofficially that of a section of G–1 Division. As a result, other sections of the headquarters co-ordinated matters with the staff director only when the existence of the section was known to them.

Colonel Wilson noted that staff actions and directives which supposedly applied to both men and women frequently omitted any provision for women, with resulting confusion in the field, but that "failure to provide adequately for female personnel was never the result of intentional neglect, but of oversight resulting from the magnitude of the overall European Theater planning job."

Soon after General Eisenhower's return from North Africa at the end of 1943, a formal announcement was made of the establishment of the WAC Section, G–1 Division, and it was required that "all matters pertaining exclusively to the Women's Army Corps and necessitating the formulation or interpretation of policy will be referred directly to the WAC Section."

WAC officers also became part of the offices of the quartermaster, the provost marshal general, Special Services, Information and Education, the adjutant general, and the inspector general. This generally proved a great advantage from the Wacs' viewpoint. Not until a WAC officer was assigned to Special Services in April, 1944, was a complete recreational program designed to fit the needs and interests of women. The assignments also generally proved acceptable to the office concerned; the theater's chief quartermaster later noted that numerous early mis-

[42] This and following, including quotations, all from ETO Bd Rpt, Vol. I, pp. 16–17, 17, 19, 20, 21, 43, 80–82, and Vol. II, App. 2 of App. 1, and Apps. 7 and 10, p. 11.

takes would not have occurred except "for lack of a female advisor in OTCQM prior to the arrival of the first women troops." [43]

In the next few months, subordinate headquarters employing numbers of Wacs acquired their own WAC staff directors. A directive was published officially authorizing field "visits" by the staff director. The informal collection of a monthly statistical report through "technical channels" was also sanctioned, although only the Air Forces in the theater ever agreed to require this officially.

In spite of these great improvements, the position in G–1 Division did not prove entirely satisfactory to Colonel Wilson, who found that her office's nonconcurrences were often lost in the final G–1 decision and were not apparent to either higher echelons or other staff sections. With the establishment of Supreme Headquarters, Allied Expeditionary Force, at some distance from the rest of the headquarters, the WAC staff director was not again officially able to see the general whom she was responsible for advising; at this time there were nine intermediaries between them.

On the other hand, the staff director, who had left the United States in WAAC days, retained many operating functions that in the United States were willingly dropped as soon as military status permitted. The staff director's office had powers of assignment and reassignment of Wacs; it approved or disapproved all promotions of WAC officers, even those employed on duties unrelated to the Corps; it requested and suballotted quotas, requisitioned personnel, and took action on requests for the discharge or return to the United States of individuals in the WAC. In general, it performed for G–1 Division all personnel functions relating to women. Here, as elsewhere in WAC administration, it proved to be difficult to relinquish operating powers until some voice was obtained in the shaping of policy.

It was not to be until after the end of the war, and after Colonel Wilson was replaced as staff director by Lt. Col Mary A. Hallaren, that earlier efforts bore fruit in a directive stating:

> Although her office is an adjunct of G–1 Division, the position of the WAC Staff Director is on the same level as the assistant chiefs of staff with respect to all matters pertaining to the WAC. On such matters she is advisor to the Chief of Staff and the chiefs of the general and special staff divisions. . . . Channels of communication between . . . divisions and the WAC Staff Director are direct.

This circular also repeated War Department directives insuring "continuous inspection." [44] Colonel Hallaren later stated, "From the day this went out, we had no problem; it was a joy to be in the Headquarters." [45]

Housing, Supply, and Clothing

From the first it was clear that administrative arrangements for Wacs in a theater of this type would present no insuperable problem even in the absence of a strong staff planning section. In general, Wacs received exactly the same type of housing that was provided for male administrative troops, which was usually much superior to that provided for troops in the United States. The majority lived in hotels and houses, although a few others had barracks, Nissen huts and, temporarily, tents.

[43] Memo, Actg Theater Chief QM to Theater Gen Bd, USFET, 5 Dec 45. ETO Bd Rpt, Vol. III, App. 75.

[44] Ltr, Hq USFET to all CGs, 8 Aug 45. AG 322.01 GAP-AGO.

[45] Interv, 14 Mar 50.

While the women remained in England the theater required that all proposed WAC housing be inspected by a representative of the staff director before a unit moved in. On the Continent such inspection was not always possible.[46]

The locating of most WAC housing in civilian communities made extra protection for women's quarters unnecessary, and also eliminated the need for excessive restrictions upon the women's freedom except those required of the entire population. The theater found that Wacs resented even minor restrictions such as "march outs" to work during morning and evening dark hours, preferring to get about independently. Because of the uncertain nature of transportation facilities, enlisted women were not sent from station to station without a WAC officer escort, unless a woman's commanding officer certified that the travel concerned did not require an escort. Other than these, there were no restrictions for Wacs which did not apply to men.

Organizational supplies were also generally superior. Because of the precedent of the British women's services, the Wacs obtained bed linens and mattresses, plus a limited number of electric irons and some hairdressing equipment, although this last was never rated entirely adequate. Post exchanges were ordinarily as good as those in the United States. Mess equipment was also generally adequate. In spite of the absence of milk, fresh fruits, and other items, the WAC messes were well rated, with cooks showing ingenuity in preparing dehydrated foods and other standard Army rations, which were superior to those of the British. Wacs usually had a separate mess only if civilian employees could be obtained for all heavy labor; kitchen police and similar fatigue duty became almost as unknown as it had been in Italy.[47]

The only immediate problem presented by the area was that of suitable clothing. The standard winter uniforms that Wacs had brought from the United States were considered inadequate, especially in the Air Forces, where many women were assigned to night shifts in unheated buildings and underground operations centers. The situation was declared an emergency, and the issue of enlisted men's long-sleeved undershirts and long drawers was authorized. Since Wacs were found to be catching colds because of a certain reluctance to wear these ill-fitting garments, the theater immediately requested many extra items above the regular issue: wool underwear designed for women, wool shirts, a field jacket and wool liner, with matching trousers and trouser liner, combat boots, and long wool hose. Most of these were obtained from the United States after much delay. Because of the limited cleaning and repair facilities, all Wacs in the theater also received extra exercise dresses, wool skirts, and field shoes.

The Quartermaster-designed trousers and jacket, when finally received from the United States, were found to be bulky and unbecoming, could not be tailored to fit the figure, and shrank and faded when laundered in the field. A smart three-piece wool uniform was therefore designed by the European theater—slacks, skirt, and battle jacket—which was found to be durable, warm, lightweight, and becoming for wear either under field conditions or in the city. A request for this was for-

[46] ETO WAC Hist, pp. 10–11; also ETO Bd Rpt.

[47] (1) ETO Bd Rpt, incl Vol. III, App. 75. (2) Ltr, Asst Theater WAAC Stf Dir to Dir WAC, 5 Jun 43. SPWA 320.2 sec 4*a*.

warded to the War Department, which refused to authorize it on the grounds that all available material was being used to produce the new ETO battle jacket uniform for men. At this, the European theater found means of procuring in England enough of these uniforms to furnish each woman at least one. With the benefit of these additions, Colonel Wilson noted that ETO Wacs were better dressed than those in the United States, where cold-weather garments and other extra items were not issued except to a few outdoor workers.[48]

Maintenance supply of uniforms was also good in comparison to that of other overseas locations. The theater engaged in a continual struggle to keep up adequate stocks of women's clothes, and had chronic shortages, particularly in small sizes, but this was no more than could be said of most stations in the United States. The theater experienced an additional difficulty in that purchase of rationed clothing in civilian shops was forbidden to American troops in England. Wacs needed many items that were no longer issued, such as warm bathrobes, bedroom slippers, brassières, and handkerchiefs. As items brought from the United States began to wear out, the theater cabled to The Quartermaster General to point out that, while in the United States it might be practicable to call these items nonessential and to require a woman to buy her own, such procedure was inappropriate in an overseas area where the items required ration cards that the Wacs did not possess. Fully 60 percent of the women, the theater estimated, were in need of girdles. The Quartermaster General refused to resume issue of these items, but eventually arranged for their purchase and resale through post exchanges to women overseas.[49]

Health, Morals, and Discipline

During the entire period of employment, there was little to indicate that service in the European theater was more detrimental to women's health or behavior than service in the United States, or produced any problems that would have rendered the employment of Wacs uneconomical. For most of the war months, the WAC attrition rate remained the same as that of noncombat men—one half of one percent.[50] This was in spite of startling statistics computed in the early days, when it was reported that ETO Wacs, unlike those in the United States, appeared to be losing twice as many days from work as men. This rate seemed due chiefly to the high incidence of respiratory disorders. Upon investigation it was found that most of the Wacs at this time were newly arrived in England, and that the rate of respiratory disorders among newly arrived male troops was also twice as high as that of the theater as a whole. In addition, Wacs were chiefly office workers, and the rate of respiratory disorders among male headquarters troops was also twice that of the field forces.

An odd additional factor was the unsatisfactory model of military shoes for women, and the shortage of proper sizes, which, according to a medical report,

> . . . loomed large in the noneffective rate for Wacs during the period when they were being admitted to the hospital for adjustment of

[48] (1) Rpt, by ETO WAC Stf Dir, 12 Jul 44. WDWAC 421. (2) ETO Bd Rpt. (3) ETO WAC Hist, p. 14.

[49] (1) Cbl W8720, USFOR London to AGWAR, 16 Nov 43, SPWA 320.2 sec 6. WAAC Cl file. (2) ETO Bd Rpt, Vol. III, App. 121.

[50] This and following from ETO Bd Rpt, Vol. I, pp. 48, 58–60, 123; Vol. II, Apps. 1 and 13; Vol. III, Apps. 114, 117.

shoes. . . . The proper fitting of shoes is still an unsolved problem. Metatarsal bursitis and fallen arches are common disorders.[51]

The WAC illness rate was further increased by the fact that pregnant women, although not ill, were hospitalized and returned as patients, for their better protection. As Wacs became acclimated, and as improved administration caused less delay in processing of female patients, the final rate of time lost in hospital was found to be only 2.2 percent for Wacs as against 2.5 percent for all troops, nonbattle cases only.[52]

Women were found to develop psychological disorders somewhat less frequently than men, and records kept for the first year and a half showed that only a dozen requests for return home had been initiated. "Female complaints" caused almost one-fourth of WAC hospitalizations, and were serious enough to cause the Air Forces surgeon to recommend that "women with menstrual disorders should be eliminated prior to their dispatch to a Theater of Operations." [53]

However, this rate of time lost because of female complaints was exactly compensated for by the higher rate of venereal disease among men. The WAC venereal disease rate was variously computed at from one sixth to one ninth that of the men, and considerably less than that of civilian women in the United States. In the first year, with over 6,000 women, there were only about 10 cases, including both married and unmarried.[54] Most of the latter were from a small minority of women who, in the theater's opinion, should not have been enlisted and should certainly not have been sent overseas. After inspecting the theater Wacs for the Medical Department late in 1944, Major Craighill stated, "We have found that the medical problems as respects morals are much better controlled in WAC than in civilian units." [55] Army historians later noted, "venereal disease among the Wacs was conspicuous by its almost total absence." [56]

The WAC pregnancy rate was likewise negligible.[57] For unmarried women it was estimated as about one a month, out of 6,000 women then in the theater, while about two married women a month were returned for pregnancy, the total rate being less than one sixth of the rate among civilian women in comparable age groups in the United States. The number of Wacs returned because of pregnancy did not raise the total WAC attrition rate above that of the men. The infrequency of discharge for this cause was such that no entirely satisfactory system of processing pregnant returnees was ever worked out, with some hospitalized women awaiting return for as much as two months, and at least three babies being born in the theater.

Both the WAC pregnancy rate and the venereal disease rate compared favorably with statistics available on Wacs in the United States. Fearing that overseas conditions placed a greater strain upon accepted standards of conduct, Colonel Wilson and the ETO medical consultant

---

[51] Memo, Maj Craighill for SG, through Chief Surg, ETOUSA, 30 Nov 44. SGO Hist Div 319.1 ETO, Women In.

[52] Bull, U.S. Army Med Dept, Vol. VI, No. 3, Sep 46, p. 285.

[53] ETO Bd Rpt, Vol. II, App. 1

[54] (1) ETO WAC Hist, p. 32. (2) ETO Bd Rpt, Vol. II, App. 13, p. 9.

[55] Min, Senior Offs Mtg, Celtic Hotel, 6 Nov 44. ETO folder Adm 534, Mss—Wacs in ETO, Opns Rpts, DRB AGO.

[56] USSTAF Hist.

[57] This and the following paragraphs are based on and quote: ETO Bd Rpt, Vol. I, p. 59, Vol. II, App. 13, p. 6, Vol. III, Apps. 114, 115, 121. Also see Ch. XXXI, below, for details.

decided that Colonel Hobby's moralizing approach to the facts of life, as incorporated in hygiene manuals, was too Victorian and that it "had been a mistake" not to include very frank sex advice. Several "fairly vigorous lectures" were therefore presented to the enlisted women, but because of "reverberations" from the WAC unit, medical authorities were obliged to desist. The British women's services likewise discovered that their women preferred advice that was moral rather than chemical or mechanical in nature.

Medical care provided for women appeared to be quite similar to that available in the United States. Like the War Department, the European theater appointed a woman doctor to be medical consultant for women, suffered setbacks in its attempts to provide service for scattered WAC minorities, and did not until the end of the war get around to appointing gynecologists on the staffs of general hospitals that treated Wacs and nurses. The War Department's medical consultant for women's welfare, Major Craighill, reported:

> The medical care of Wacs in England and France is excellent . . . dispensaries, station and general hospitals having special medical facilities for women. Very few hospitals are equipped for gynecological or obstetrical cases. . . . The health of the WAC is good . . . there is no evidence of undue fatigue. Tension is apparent only among those groups having insufficient work to occupy them fully.[58]

There was no indication that ETO Wacs were appreciably less healthy than those in the United States. Some reports showed that ETO Wacs lost less time in the hospital than Wacs in the zone of the interior, but ETO authorities believed their WAC noneffective rate to be slightly higher.[59]

WAC disciplinary records were also good. Statistics compiled by the Air Provost Marshal showed that men committed various offenses from ten to a hundred and fifty times as often as women.[60] On a percentage basis, men were AWOL 89 times as often as women, drunk 85 times as often, and violated miscellaneous Articles of War 150 times as often.[61]

Commands reported that "Wacs classed among the most smartly and neatly dressed troops on the stations." Serious misconduct was lacking, and the advance plans for a group of female military police were never carried out. Court-martial cases were few; there were only two general courts reported among Wacs in the history of the theater. Historians, questioning whether this rate resulted from good behavior or from undue leniency of WAC company commanders, examined company punishment records closely and determined that there had been "no tendency to coddle or minimize punishment for misdemeanors. . . . Supervision at the detachment level was more thorough and constant than among males." [62]

Of aid in maintaining this record was the fact that the theater acted to make clear the disciplinary powers of WAC commanders by reaffirming the War Department's two safeguards: that enlisted women would be assigned only in groups, and only under the command of a WAC officer.

[58] Memo cited n. 51.

[59] Weekly Health Rpt, Vol. V, No. 9, ASF, SGO; ETO Bd Rpt, Vol. I, p. 123; Vol. III, Apps. 114, 119, 121. Hospitalization rate:
35 per 1,000 for U.S. Wacs
34 per 1,000 for Army, less evacuees
27.18 per 1,000 for Wacs in ETO

[60] ETO Bd Rpt, Vol. I, p. 40.

[61] ETO Bd Rpt, pp. 37-42.

[62] USSTAF Hist.

Civilian Competition

Final theater reports indicated only one important handicap to WAC employment in an area such as this one: the competition with civilian employees. This problem, although appearing minor at first consideration, became more prominent in the period following the slackening of hostilities and was eventually rated the European theater Wacs' number one difficulty by both WAC staff directors, as well as by returning enlisted women.

It had been originally contemplated by the Chief of Staff that no civilian women would be employed overseas, but by the time of the WAAC's formation it was already clear that such employment could not be prevented in certain areas. As early as 1942, Director Hobby had therefore proposed that difficulties between the two groups be minimized by directing that "no organization or group of women employed by the American forces . . . wear a uniform or parts of a uniform which are of a color or a pattern which would cause them to be confused with members of the WAAC."[63]

She pointed out that existing regulations provided a satisfactory uniform for American civilian employees—the blue hostess-librarian uniform—which was ordered through Special Services from commercial firms that were ready to supply them. However, the Army Service Forces refused to send this proposal to the War Department, and it was not considered. Instead, at a meeting at which the Director was not represented, it was decided to sell the WAAC officer uniform to any civilian women going overseas.[64]

The problem had small immediate importance, since until after the cessation of buzz-bomb danger very few American civilians went to the area, and British employees lived at home and seldom wore uniforms. To Colonel Hobby's inquiry, Colonel Wilson replied, "This is a situation about which we are not unduly concerned." In the early and dangerous days of the invasion, Wacs were the only female employees in most cases. Later, the theater decided to take female civilians to the Continent as soon as conditions became safe and stable, and it was necessary to provide them with quarters and identifying uniforms. Colonel Wilson concurred in a recommendation that civilians get separate messes and distinctive uniforms, but instead the theater authorized officer accommodations for them and allowed each command to prescribe its own uniform. At this, a considerable morale problem resulted among equally qualified enlisted women, mitigated somewhat in cases where British civilians were veteran employees who had shared the bombings and rationings.[65]

Early in 1945, a morale crisis was precipitated by news of plans to send overseas large numbers of American civilian women as soon as hostilities ended. At this time Director Hobby's staff learned that "a meeting was held in the Secretary of War's office, at which representatives of the Quartermaster and G–1 were present, but no representative of this office . . . those present agreed that the WAC uniform should be provided for civilians overseas." In a series of confidential memoranda, the

[63] Memo, Dir WAAC for CofS, 10 Oct 42, sub: Ltr of Instr to CG ETO. SPWA 320.2 sec 1.

[64] D/F, Dir Mil Pers ASF to Dir WAAC, 16 Oct 42. SPGA M/322.5 WAAC (10–10–42), in SPWA 320.2 sec 1.

[65] (1) ETO Bd Rpt, Vol. I, pp. 71, 21; Vol. III, App. 135; Vol. III, App. 10, p. 7. (2) Quotation from Memo, WAC Stf Dir ETOUSA, for Col Howard E. Kessinger, Exec, G–1 WDGS (on tour of ETO), 28 Aug 44. WDWAC 421.

Director protested that WAC enlisted women were as well qualified as the civilians who would wear the officer uniform; that civilians should not receive officer privileges without accompanying military responsibilties and discipline; that the social association problem would be hopelessly complicated if civilians wore the WAC uniform; and that the practice "would react to the serious detriment of the morale of the enlisted women who have served the Army well and faithfully under the rigorous conditions of overseas life without [such] pay or privileges. . . ."[66]

The Director shortly succeeded in getting a directive to the European theater that civilians would get only the enlisted uniform, and this with the addition of conspicuous colored shoulder straps, sleeve braid, and cap crown, plus a blue hostess-librarian overcoat. The theater cabled back for a reconsideration of continued issue of the unchanged WAC uniform, which would considerably simplify its supply and storage problems. This was not granted, but hundreds of unmodified uniforms were already in the hands of civilians, and War Department attempts to get the directive enforced in the theater were never wholly successful. Colonel Wilson was not asked to comment; she noted, "The problem was of such a controversial nature that it was impossible for the WAC Staff Director to initiate a paper as a branch of G–1."[67]

With the posthostilities arrival of American civilian women, the Wacs' *esprit* vanished, as well as pride in Army status and uniform. The women talked of little else; the gist of their more printable comment was that they had been hooked as sentimental suckers by a government that penalized enlisted service and handsomely rewarded those who stifled their patriotic fervor until the danger was over. The theater noted numerous "natural jealousies" that arose between the two groups. The Wacs' chief complaint was not the pay or officers' mess and quarters, but the uniform, and the fear that the Corps' hard-won reputation for neatness and good conduct was rapidly vanishing in the eyes of the occupied nations and the American soldiers alike as a result of the way in which their uniform was worn by civilian women untrained in military customs and unsupervised as to appearance or living habits.

In addition to the civilian employees in WAC uniform, there were women movie stars, newspaperwomen, Congresswomen, society women, and the entire casts of traveling shows such as *Panama Hattie.* Further indignation was provoked when these women, as well as USO workers and nurses, got priority on the scanty WAC maintenance stocks, resulting in shortages to enlisted women. Also, civilians could date officers and attend officers' social functions. They had different housing standards, less restrictions, no inspections, and no company punishment. Though holding jobs similar to enlisted women's, their top salary was $745 a month as against the WAC maximum of $138.[68]

[66] (1) Memo, Exec for Dir WAC, 24 Feb 44; (2) WD Cir 118, 22 Mar 44; (3) Memo, "KKJ" for Dir WAC, 26 Feb 45, WDWAC 319.1 Johnson (1945–46); (4) Memo, Dir WAC for Col Guenther, 1 Mar 45; (5) Memo, Dir WAC for G–1, 27 Mar 45; (6) Cbls CM–OUT 57510, WD to ETO, and CM–IN 26948, ETO to WD, 25 Mar 45. All in WDWAC 421.

[67] *Ibid.* Quotation from ETO Bd Rpt, Vol. II, App. 7: Memo, WAC Stf Dir for G–1 SHAEF, 4 May 45.

[68] Min cited n. 55. ETO Bd Rpt, Vol. III, Apps. 71, 132, 134. Monthly salaries:

	Minimum	*Maximum*
U.S. civilian in ETO	$220.00	$745.00
Continental civilian	110.00	372.00
WAC	50.00	138.00

The theater also admitted that "occasionally there was a tendency to cater to the civilians in order to keep them happy and on the job, and on the other hand to delegate to WAC personnel in the same office the more difficult and less pleasant jobs and overtime work because they were military personnel and subject to orders."[69] When questioned by WAC officers concerning the necessity for this practice, ETO's Chief of Military Personnel replied: "The matter is a command function based on the overall view which probably the individual does not understand."[70]

The civilian women themselves were equally resentful. They felt that their quarters were often worse than the WAC officers' quarters, sometimes lacking in heat when WAC quarters were heated. Civilian personnel also lacked the many military benefits and provisions for morale-building recreation and entertainment that enlisted Wacs had. No system of hearing civilian women's complaints was set up until 1947, and no action was taken on them.[71]

It was in vain that WAC officers attempted to point out to the enlisted women that the fault lay in Congress' failure to draft women as well as men, thus forcing the Army to employ civilians to make up the deficit, and to grant extra privileges to attract them. The women's hostility remained unabated through the final days of the war, and was reflected in postwar refusal of former Wacs to accept civilian employment with the Army. As Wacs became eligible for discharge, the theater, with the War Department's approval, worked out a system that would permit former Wacs to transfer to civilian status and good civilian jobs in the Army of Occupation, but of more than 8,000 Wacs, only 126 chose to accept the offer. Neither did the majority desire to serve the Army any longer on military status.[72]

Social Association

A second policy problem, which became more noticeable in 1945 as a complication of the civilian importation, was that of the social association of officers and enlisted personnel of opposite sexes.[73] Until American civilians arrived and began dating their section chiefs, Wacs had lived without too much objection under General Devers' written directive that the customs of the service would apply except to relatives and fiances carrying letters of authorization. This prohibition, during the war months, was fairly well observed in the subordinate commands and in the rural areas, but was generally ignored in cities and higher headquarters because of the anonymity which couples could easily find in large cities such as London and Paris. Enlisted men made sporadic objections when Wacs dated officers, although believing that they themselves should be allowed to date nurses, but the relative abundance of dateable women in England prevented any real hard feeling.

It was only with the move to Germany and the end of the war that feelings on the subject became more intense. For the eighteen months after his return to Europe, General Eisenhower was not informed of his theater's policy in this

[69] ETO Bd Rpt, Vol. I, p. 132.

[70] Min cited n. 55.

[71] Intervs with civilians returned from Europe.

[72] ETO Bd Rpt, Vol. III, Apps. 131, 133, 134; Vol. I, p. 133.

[73] Unless otherwise stated, discussion is from ETO Bd Rpt, Vol. I, pp. 6, 25, 135–39; Vol. III, Apps. 137–49.

regard. After V-E Day he discovered it from "gripe letters" from Army nurses to *Stars and Stripes,* complaining that it would be less evil to go out with "good respectable privates" than with officers who were always married men. General Eisenhower asked what the "alleged regulation" was that prevented a nurse from going out with an enlisted man, and added, "What is all this?" Upon learning that the theater had a written restriction against social mixing, he wrote:

> I want *good sense* to govern such things. Social contact between sexes on a basis that does *not* interfere with other officers or enlisted persons should have the rule of decency and deportment—not artificial barriers.[74]

Colonel Wilson had not brought the matter to General Eisenhower's attention because she believed that distinction was advisable and had a basis in civilian life, where, "for example, no railway company exists in which section hands or brakemen expect or want social intimacy with the president of the company." [75] In answer to General Eisenhower's query, Colonel Wilson recommended, with the concurrence of all subordinate Army commanders, that the current situation be allowed to continue. She informed General Eisenhower that Colonel Hobby favored such a policy, although actually the opposite was true.

Colonel Hallaren, upon assuming office as ETO WAC staff director, advocated a reversal of the policy. She felt that she spoke for the enlisted women, who still resented social discrimination, and that while Wacs probably would not take undue advantage of a relaxed policy, "they get claustrophobia with the door closed." [76]

In the commotion of demobilization, the theater commander failed to approve either proposal. Without rescinding the written ruling, he authorized establishment of an all-ranks restaurant in Paris where "mixed" couples could get food. Later, after civilian restaurants were opened, there was no particular problem for violators of the rule.

Policy on Marriage

Closely connected with the social association problem was that of marriage of military personnel. The theater's written orders were, "When two members of this command marry, thereafter either they will be stationed at widely separated posts, camps, or stations, or, when appropriate, one of the parties will be removed from the theater." [77] The orders applied only to nurses and Wacs, and not to American and Allied civilians and women of liberated countries. They directly contradicted the War Department circular, which stated that Wacs would not be transferred solely because of marriage to persons serving in the same station.[78]

The theater chaplain protested to General Eisenhower that, since engaged couples did not wish to be separated because of marriage, "a condition of concubinage" was resulting, and the WAC staff director put it more explicitly that "military couples lived together without the marriage ceremony for fear of being separated." General Eisenhower, upon reading these statements, personally reaffirmed his policy that "Persons in the military service will not be permitted to establish homes and families in this active theater."

[74] ETO Bd Rpt, Vol. III, Apps. 136, 146, (11-23 May 45)

[75] ETO Bd Rpt, Vol. I, pp. 136-39.

[76] ETO Bd Rpt, Vol. III, App. 147.

[77] Hq ETOUSA Cir 41, par 5*a*, 17 Apr 44.

[78] WAAC Regs, 1 Jun 43; WD Cir 462, 1944, par 12*b*.

Although medical reports indicated that "arbitrary separation of husbands and wives within the theater does not tend to reduce the incidence of pregnancy," [79] it was not until after the end of the war in Europe that the transfer requirement was relaxed. The WAC marriage rate at once quadrupled. Even so, a station was forbidden to allow married couples, military or civilian, to be domiciled together—a provision that later resulted in publication by an interested newspaper of a reprimand to an Army captain beginning, "It has come to the attention of this headquarters that you are living with your wife. This must cease at once." [80]

There were only 323 WAC marriages in the theater, of which all but eight were to Americans. Historians noted, "It has always been a rare sight in London to see a Wac with a member of any foreign service or with a civilian from any country other than her own." Women noted that "almost any man from the States seemed to think it was a crime for any Wac to cast more than a pleasant glance at a foreigner, when there were so many Americans in Europe who preferred American women. . . ." Likewise, the Wacs mixed less with the British civilian population. Reports noted:

> Though a large majority of the Wacs were entertained in British homes, few of them visited these homes more than once or twice. Wacs were in a rather different position from that of the American men in the Army. . . . [81]

Public Relations

A last management problem, which was intensified with the close of hostilities, was that of WAC public relations.[82] Wacs in the European theater, like those in other overseas areas, never had a public relations problem in the sense that Wacs at home did, where they were constantly observed by critical American civilians. ETO Wacs generally encountered only respectful Continental audiences. However, war correspondents did their best to substitute for the home folks in this respect, from the day that the 1st WAC Separate Battalion landed and was met by thirty-eight correspondents and publicized in 600 newspapers. Press relations were so time-consuming that a WAC staff officer was installed in the ETO Public Relations Office, to cope with news requests. Here her work resolved itself chiefly into a struggle to produce favorable news releases faster than enterprising reporters could produce unfavorable ones.

On the credit side, the European theater quickly reached a volume of WAC news stories equaled by few others, in spite of disagreements with Washington as to a reasonable compromise between the capacity of cable facilities and the needs of recruiters in the United States. Because of the news value of overseas items, Director Hobby sent a WAC writer and a photographer to stimulate output, and later devoted much of her own 1944 visit to England to this purpose. The theater obligingly produced many general-interest stories on the WAC, plus home-town releases on almost every Wac in the theater. The WAC Public Relations Officer, Maj. Henriette Horak, also publized events that would put the WAC in a good light:

[79] Memo cited n. 51.

[80] (1) Hq ETOUSA Cir, 9 Jun 45, par V. (2) Above passage based on ETO Bd Rpt, Vol. I, pp. 55–58, and Vol. II, Apps. 39–47. General Eisenhower's policy: App. 39. Chaplain's protest: App. 42. Newspaper quote: *Washington Post,* March 2, 1947.

[81] ETO ATC WAC Hist.

[82] All statements on public relations, unless otherwise indicated are from ETO Bd Rpt, Vol. I, pp. 115–18, and Vol. III, App. 112.

the "adoption" of British war orphans; visits of Wacs to Paris cathedrals; the story of the responsible work done by Wacs at the Postdam Conference and the Nuernberg trials. Stories were placed in *Yank, Life, Vogue,* and many others; newsreel coverage was obtained; Wacs were pictured with Generals Eisenhower, Smith, Doolittle, Eaker, Lt. Gen. John C. H. Lee, and others.[83] To promote GI-WAC friendship, much was made of the naming of a bomber the "Pallas Athene-GI Jane" and a locomotive the "WAC Blazer." By the machinations of the WAC Public Relations Officer, Wacs were slipped into Berlin ahead of other women to achieve a noteworthy "first." [84]

The size of the theater nevertheless made it necessary for the WAC Public Relations Officer to decentralize control of releases to various lower echelons not having Wacs on their public relations staffs. The unfavorable stories that slipped by male public relations officers were all of one type: derogatory comments about Wacs by the soldiers, which reporters seemed to take especial pains to evoke. The WAC Public Relations Officer admitted that one of her "chief headaches" was "the antagonism between the Army male and the Army female." As a result, there were frequent releases such as one from Eighth Air Force headquarters, which consistently refused to accept a Wac public relations officer, "Soldiers Prefer English Girls to Wacs." [85]

The worst releases of this type came toward the end of hostilities, when Field Marshal Sir Bernard L. Montgomery's headquarters released a statement that numbers of the WAC and ATS would be sent into Germany to prevent soldiers from fraternizing with German women. The women's duties, said the story, were to act as dance partners, to chat, and to drink tea.[86] This appeared in *Stars and Stripes,* in the *London Daily Mail,* and in the *New York Times.* The story was widely reprinted in the United States; some writers and one radio religious hour stated that Wacs were being transferred to camps where men visited German brothels too often, and that the brothels were now being filled by Wacs.[87]

In the United States, the National Federation of Press Women asked that the War Department take a definite stand against this type of overseas "propaganda," but the head of the War Department Bureau of Public Relations took no action because, as he informed Director Hobby, "I think the statement [from Montgomery's headquarters] was well meant, even if poorly advised . . . I smell the needling by correspondents." [88] This type of release continued, fostered somewhat by soldier letters in the European theater *Stars and Stripes,* which in August of 1945 printed what United Press reporters called "a blizzard of bitter letters" from soldiers; this was also picked up in the United States.[89]

---

[83] Ltr, PRO ETO Hq to Dir WAC, 30 May 44. WDWAC 320.2.

[84] ETO Press Releases 8296, 29 Jan 44, and 9747, 16 Nov 44. ETO Adm 365 WAC, DRB AGO.

[85] Eighth AF Hq Release, *Des Moines Register,* January 20, 1944. Also: (1) Memo, Dir WAC for CG AAF, 2 Feb 44, and reply, 21 Feb 44, from CofAS, Gen Giles, citing cbls. SPWA 001.–N. (2) *Chicago World Herald,* August 21, 1945, UP release, "Soldiers Lambaste Wacs."

[86] *New York Times,* December 5, 1944, by *Times* correspondent with British Second Army.

[87] (1) Ltr, Dr. K. Frances Scott to Dir WAC, 14 Jan 45, and atchd file. WDWAC 330.14. (2) *New York Journal-American,* December 8, 1944.

[88] Ltr, Natl Fed of Press Women to Dir Hobby, 16 Dec 44, and Inds. WDWAC 080.

[89] *Chicago World Herald,* August 21, 1945, UP release.

WAC recruiters in the United States lived in constant apprehension of these periodic outbursts, which got wide publicity because of public interest in overseas news. This particular problem never found a solution; even in the Army of Occupation damaging releases continued to be permitted and even originated by Army public relations officers.[90] From the European theater viewpoint the problem had no great effect on WAC efficiency, although it did tend to hamper recruiting in the United States.

Posthostilities Period

While the WAC's problems of public relations, civilian competition, and social discrimination were considered worthy of record by students of personnel management, they were not conspicuous in the general morale let-down that followed the end of hostilities in Europe, nor as serious as the social and moral problems of men in the occupation forces.

In this period, the efforts of the theater as a whole were turned toward the maintenance of morale and good conduct for both men and women. Women received furloughs and passes under the same rules as for men, were included in Information and Education plans, and were provided with facilities at selected leave areas. Wacs were generally less able to avail themselves of those facilities than were men, since most Wacs were employed in critical categories and could not be released by section chiefs to participate in programs designed mainly for idle troops. Wacs were permitted to take courses such as those at Shrivenham and Biarritz, but few were able to attend.

American Red Cross authorities had from the beginning endeavored to provide identical services for Wacs and enlisted men, but reported themselves handicapped by the attitude of their women field workers: "Many of the women approached [to run a WAC rest area] said they had come over to serve the men and not the women."[91] Clubs for enlisted women and WAC officers were eventually set up in London. After some delay, a WAC rest home was established near Oxford soon after the invasion, and moved to Nice after V-E Day; it proved extremely useful, since the Wacs had no convalescent hospitals.[92] The men's program of return to the United States—Rehabilitation, Recuperation, and Recovery—was never applied to Wacs, because of their shorter length of service, although Colonel Wilson felt that if a small quota had been included in the program it would have aided morale.[93]

In this period various devices were employed by WAC commanders to help in sustaining morale: unit clubs; a relaxation of regimentation; athletic tournaments; the salvage of toys, food, and clothing for homeless children; and the hospital visitation program.

Demobilization

Colonel Wilson was not included in planning conferences for the Army of Occupation, but the theater planners determined independently to employ even

[90] Memo, Actg Chief, WAC Gp WDBPR for Dep Dir WDBPR, 29 Jan 45. WDBPR WAC Gp file Organization of WAC Group.

[91] Ltr, Miss Eleanor C. Vincent, Asst National Dir, Mil and Naval Welfare Serv, Red Cross, to Miss Helen Walmsley, 7 Sep 44. WDWAC 080 Red Cross (1945).

[92] This was administered by the Army with recreation furnished by the Red Cross. ETO WAC Hist, pp. 30–32; ETO Bd Rpt, Vol. II, App. 10, p. 10.

[93] ETO Bd Rpt, pp. 52–55 and 69–70.

more Wacs than had been used in wartime. It was decided that 50,000 Wacs could be used in the Army of Occupation, but that in view of the size of the Corps, only 10,000 would be asked, these to be in addition to the thousands used by the Air Forces in Europe.

Further disagreement with the War Department followed as to the relative needs of the zone of the interior and the Army of Occupation. Theater representatives asked, in conference, "Why is it difficult to get personnel now? What are the Wacs doing over there that they can't be spared to come here as replacements?" To this, War Department spokesmen replied that the administrative load of demobilization would fall on the United States, not the theater, and that "Wacs [in Europe] gripe that there isn't enough work to do—that they would go in the Army of Occupation only for fraternization—they resent it." [94]

The War Department therefore at first refused the request for more WAC personnel, stating:

> The average limited service man in the United States has a much lesser degree of intelligence and effectiveness than the average Wac, and it would thus be most difficult to have limited service soldiers take over the jobs which Wacs are now performing.[95]

The General Staff also objected that ex-combat returnees could not be traded for Wacs in the United States because the men had too high grades, were lacking in skills, or were temperamentally unsuited for the work.

Director Hobby did not take sides in the matter, other than to recommend that all or no requests be approved, in order that Wacs not be used in the Army of Occupation except in sufficient numbers to give them a "recognized status" and to make provisions for their health and supply economically justifiable.[96] Finally, a compromise was reached: the War Department refused to exclude the AAF's quota from the requested quota of 10,000, but agreed to raise the theater quota, including Air Forces, from 8,000 to 10,000—thus in effect permitting shipment of 2,000 more women.[97]

Shipments to supply the additional women had already begun when the onset of demobilization intervened. At this time the European theater used its unfilled WAC quota of 10,000 as an excuse for refusing to release women eligible for discharge, saying, "This personnel will necessarily be considered essential if the theater quota has not been reached, and will not be eligible to return to the Zone of the Interior." [98]

As in the Mediterranean theater, male replacements in the same skill and MOS were not considered acceptable. The War Department therefore took a step that would have forced return; it cut the theater quota from 10,000 to 6,000 on the basis of "other needs" elsewhere.[99] The theater refused to comply; while returning a few women in unwanted skills, it informed the War Department that it had "frozen" the return of WAC stenographers and typists until WAC, and not male, replacements in these skills were received.

This proved a fruitless move, for after the defeat of Japan and the end of WAC

---

[94] Min cited n. 55. Also ETO Bd Rpt, Vol. II, App. 7.

[95] Memo, Secy WDGS for Dir WAC, 12 May 44. WDCSA 424.5 WAC, in CofS 324.5 WAC.

[96] (1) D/F, G-1 for CofS, 2 Feb 45. WDGAP 320.2 WAC, in CSA 324.5 WAC. (2) Memo, Exec WAC for Dir WAC, 21 Feb 45. WDWAC 201.6.

[97] ETO Bd Rpt, Vol. I; also D/F, G-1 WD to MPD ASF, 10 Mar 45. WDGAP 320.2 WAC in WDWAC 320.2 ETO. Quota was 200 offs and 10,400 EW.

[98] Memo, ETO for TAG, 6 May 45. AG 387 RCGA, in WDWAC 320.2 ETO.

[99] D/F, G-1 Div WDGS to TAG, 7 Jul 45. WDGAP 320.2 WAC (2-2-45), in WDWAC 320.2.

recruiting, the European theater, like all others, was informed that no more Wacs would go overseas. Soon after, the War Department forbade any theater to retain discharge eligibles, men or women, on the grounds of "military necessity." Thereafter, demobilization of nonvolunteer Wacs proceeded rapidly. The theater reported that the War Department's action "was a serious blow to the European theater, which considered Wacs its primary source of critically needed clerical and communications personnel, and [it] required the revision of plans for their utilization in the Army of Occupation." [100]

Appraisal

The statements of both theater commanders and the women themselves indicated that WAC service in the European theater was considered suitable and successful. General Eisenhower informed the War Department:

> During the time I have had Wacs under my command they have met every test and task assigned them. I have seen them at work in . . . England, France, and at Army installations throughout the European Theater. Their contributions in efficiency, skill, spirit, and determination are immeasurable.

General Lee added, "The work of the organization has been superior." The Chief Signal Officer, Maj. Gen. William S. Rumbough, called their work with the Signal Corps "superior." Maj. Gen. Willard S. Paul, G–1 of ETO, reported, "I have received nothing but the highest praise for the results you have achieved." [101]

The Air Forces in the theater added similar praise. General Spaatz of USSTAF stated:

> The WAC has been of inestimable value. . . . Its members have worked devotedly, often at arduous tasks requiring exceptional performance.[102]

General Eaker later added:

> Women made, in my opinion, the best soldiers in the war. I feel that Wacs should be retained as part of the postwar military plans.[103]

The Air Forces in the theater summarized their conclusions concerning WAC employment favorably:

> Wacs proved much less of a problem than had been envisaged. . . . It was found that Wacs could live under conditions substantially the same as those of male personnel. . . . Perhaps the greatest achievement of the Wacs was their triumph over the prejudices of the male military mind. The half-amused, half-scornful attitude of some officers in responsible positions was not justified by the performance of the Wacs. A balanced judgment would find that the Wacs have been deserving of any extra time and effort which might have had to be expended on them because of their sex.[104]

The verdict of the Wacs themselves was also favorable. The life of a Wac in the European theater was obviously superior with regard to housing, recreation, sightseeing, and the absence of kitchen police and other work details. ETO Wacs were more smartly and warmly dressed than Wacs elsewhere. They had experienced the morale lift of constant forward movement to new and exciting areas, as opposed to the tedium of years at one station in the United States; the element of shared dangers had added to Corps *esprit*. Also, in spite of the poor start in obtaining appropriate ratings, ETO Wacs by the end of the war had more than twice as high a percentage of the first three grades

[100] ETO Bd Rpt, Vol. I, p. 114.
[101] All from ETO Bd Rpt, Vol. I, p. 11.
[102] *Ibid.*
[103] Address cited n. 3.
[104] USSTAF Hist.

as did Wacs in the United States, even though they still had only half as many high ratings as ETO air force men.[105] WAC officers also had twice as high a percentage in the grades of major and captain as did those in the United States.[106]

Awards and decorations were plentiful. Besides three presidential citations, there were about two hundred Bronze Stars and a few other decorations, and the percentage of women receiving the Legion of Merit was considerably higher than that in the United States.[107] As a result of all these advantages, ETO Wacs were distinctly proud of their period of service, and upon discharge frequently protested turning in their battle jackets, the only means of distinguishing themselves from the 92 percent of the Women's Army Corps that had not been sent to Europe.[108]

On the other hand, only a fraction so small that it was not recorded volunteered to stay in Europe, either as Wacs or civilians, beyond the date of discharge eligibility. Only 30 percent said they would later be interested in joining an inactive WAC reserve if one should be created.[109] Theater surveys by ETO Special Services Division, attempting to find out how many women would stay, started much unrest and eventually caused Congressional investigation of charges of undue influence in forcing women to volunteer.[110] One such allegation stated, "Their officers want to keep the Wacs overseas but the girls themselves do not want to stay." [111] However, while asking discharge, some 63 percent of enlisted women and 85 percent of WAC officers preferred to stay in the theater until eligible for discharge, instead of again serving in the United States.

The approval of both ETO Wacs and the theater commanders of the conditions of WAC service, as they had known it in England and Europe, was to have an effect upon the future of the Corps far beyond that which might have been expected from their small numbers. From Colonel Wilson and theater commanders there came repeated and enthusiastic recommendations that Congress be asked to provide for "the inclusion of women as an integral part of . . . the Regular Army and Reserve Corps." These recommendations were later to be revived when the postwar War Department and the Director's Office were taken over by new staffs from the European theater. At that stage, WAC demobilization plans were to be reversed, and the Army of Occupation was to receive its Wacs. At that time also Generals Eisenhower, Devers, Spaatz, and Eaker were to spearhead the successful drive to ask Congress to continue the Women's Army Corps in the peacetime Regular Army.[112]

---

[105] The exact comparative figures are not given on the graph (ETO Bd Rpt, p. 45; also Vol. II, App. 1) from which the comparison is taken, but seem to correspond to those from Strength of the Army, STM-30, June 1945.

[106] ETO Bd Rpt, Vol. I, pp. 14, 71: ETO approximate grades as of 31 Aug 45; WAC officers, worldwide, are from Strength of the Army, June 1945.

[107] ETO Bd Rpt, p. 48 and App. 26: 4 Meritorious Unit Plaques, 3 Presidential Citations, 15 Purple Heart, 9 Legion of Merit (7 to officers), 229 Bronze Star, 2 OBE, 1 Legion of Honor, 6 Croix de Guerre, 113 Certificates of Merit, 7 MBE, 2 Luxembourg Couronne de Chene. For final TAG figures on decorations and awards, see Table 8, Appendix A.

[108] (1) Cbl WAR-65259 to CG ETO, 17 Sep 45; (2) D/F, G-1 to ASF, 25 Sep 43. WDGAP 421 WAC, in WDWAC 421.

[109] ETO Bd Rpt, Vol. I, pp. 7 and 133.

[110] (1) Rpt, Nov 44, sub: Postwar Plans of EW in WAC; (2) Rpt, Nov 44, sub: Attitude of 175 WAC Offs Toward Remaining in WAC Overseas. WDWAC 320.2.

[111] Ltr, Dir WAC to Clare Boothe Luce, 15 Jun 45, in reply to correspondence, 7 Jun 45. WDWAC 095 Forsht. Also similar letters in this file.

[112] See Ch. XXXVI, below.

CHAPTER XXII

The Southwest Pacific Area

The last of the major overseas commands to employ considerable numbers of WAC personnel was the Southwest Pacific Area (SWPA). Here, although the first shipment did not arrive until the middle of 1944, some 5,500 women eventually served—the second largest number to be employed by any overseas area.[1]

The delay in employing Wacs was readily explained: there was no shortage of female clerical workers in Australia, where some 20,000 civilians were employed by the American forces. Therefore, when Army-wide surveys were made at the time of the Auxiliary's formation, and again at the time of its contemplated expansion, the theater rejected the idea of using Waacs, stating that all available shipping was needed to bring combat personnel to its authorized level.[2]

Until late in 1943, several of the headquarters in Australia had entertained the hope that civilian employees would be permitted to move with them when the northward attack was launched through New Guinea and the Philippines. At this time, a personnel crisis was precipitated by the immediacy of the offensive combined with the final refusal of the Australian government to let its women move from the continent, in view of existing labor shortages. The theater therefore began a series of requests to Operations Division of the War Department General Staff, seeking to raise its Troop Basis to permit the shipment of general service men to replace the civilians. Limited service men were not wanted, in view of the unpromising nature of available material.[3]

These requests met with consistent refusal from the War Department. The theater was already authorized to have 3.2 percent of its military strength in overhead, as against only 1.5 percent for the

[1] No official WAC history was prepared by SWPA. (For a description of the sources used in this chapter, see Bibliographical Note.) All references, unless otherwise indicated, are to: (1) 30 folders of SWPA material, now at Kansas City Records Center, cited by folder and drawer number. (2) Historical material, WAC Assigned to SWPA, typed, undated, and unsigned, although Capt Velma P. Griffith (WAC PRO, SWPA—an eyewitness to many events and with access to theater WAC data) acknowledges authorship. Cited hereafter as Griffith Account. (3) Rpt, GHQ USAFPAC to TAG, 1 Nov 45, sub: Rpt on WAC Pers in SWPA Prepared in Accordance with AG Ltr 333.5 (9-27-45) OB-S-A, 1 Nov 45. Official copy in G-1 WD 321-333 WAC, 18 May 45, AGOF-C-333 WAC, DRB AGO. Cited as AFPAC Reply. (4) Draft of the above, prepared by WAC Staff Director, AFPAC, and approved by WAC Staff Directors, AFWESPAC and FEAF. Now in WAC Classified file. Cited as WAC Draft AFPAC Reply. (5) Official Rpt, Col Westray Boyce, Dir WAC, to CofS USA, 22 Mar 46, sub: Rpt of Visit to WAC Pers in Overseas Theaters. Also earlier draft of same. WDWAC 333.1. Cited as Boyce Rpt. (6) For list of interviews, statements, and comments, see Bibliographical Note.

[2] Rads: 1122 (CM-OUT 7466), 22 Aug 42, and reply by C-484 (CM-IN 5492), 13 Sep 42; 1631 (CM-OUT 508), 2 Mar 43, and reply by C-724 (CM-IN 2934), 6 Mar 43.

[3] Unless otherwise cited, all references to OPD action are to OPD 320.2 Australia, Sec IX, Cases 265-303, Dec-Apr 44, DRB AGO. Cited as OPD file.

European theater, and stations in the United States were by this time almost stripped of general service personnel. Instead, it was suggested that the theater render its headquarters mobile by screening the half-million troops it already had, among whom there were believed to be numbers of clerically skilled men.

This move was considered impracticable by the theater. Instead, there was some discussion within the several headquarters of the possibility of employing Wacs, under the misapprehension that they could be requisitioned to fill civilian job vacancies. The move was generally favored by Headquarters, Fifth Air Force, and by the Services of Supply, but rejected by the higher echelon, U.S. Army Forces in the Far East (USAFFE).[4]

In the early days of 1944, Operations Division of the War Department received a personal visit from Lt. Gen. George C. Kenney, Commanding General, Allied Air Forces and Fifth Air Force, who expressed a desire to get WAC personnel for his command. However, a call to ASF Military Personnel Division revealed that, of the Wacs who were expected to be available for overseas shipment during the entire year, only about 800 enlisted women and 200 officers remained unallotted. OPD at once acted to freeze this number until the Pacific's needs could be considered. Director Hobby was then overseas, but a call to her office revealed that, at the current rate of recruiting, a "reasonable number" of this 800 women could probably be obtained within a few months. At this, OPD radioed the theater asking whether General Kenney's request represented theater opinion. If so, it was promised that the theater's Troop Basis would be increased by 1,000 in order to accommodate the shipment.[5]

SWPA protested by radio:

> Proposed allotment totally inadequate for minimum theater requirements. . . . Can use 10,000 or more Wacs. . . . Theater Chief of Staff who should reach Washington today has data on theater overhead requirements and on positions which Wacs may advantageously fill.[6]

Director Hobby first learned of the new development on 11 February 1944, soon after her return from Europe, when Lt. Gen. Richard K. Sutherland, General MacArthur's chief of staff, came to her office and informed her that at least 10,000 Wacs would be immediately required by the theater. "He appeared," she stated later, "to be under the impression that Wacs were still Auxiliaries and would not count against the Troop Basis." [7]

The Director informed General Sutherland that in her opinion the shipment could not be made for two reasons: first, Wacs were military personnel and would require a military allotment just as men would; and second, the WAC had not yet recovered from its conversion losses and from heavy shipments to Europe, and could not provide women in any such numbers in the clerical skills requested.

---

[4] No supporting documents were found. The WAC Staff Director, USAFFE, recalled seeing in the files a USASOS request disapproved by Maj. Gen. Charles P. Stivers (USAFFE) without stated reason. Brig. Gen. Harry H. Baird (USAFFE) stated in 1948 that earlier staff studies questioned supply, housing, protection, etc.; these were not found, nor any evidence of why such doubts were not communicated to the War Department by Lt. Gen. Richard K. Sutherland in his 1944 request for Wacs. Ltr, Mary-Agnes Brown to OCMH, 23 Apr 51; Memo, Col Harry H. Baird for HD SSUSA, 29 Jul 48. OCMH.

[5] (1) M/R, 1 Feb 44, sub: WAC Pers in SWPA, with copy of rad atchd. OPD file. (2) For final theater strength figures, see Table 7, Appendix A.

[6] Rad C-1319 (CM-IN 4025), GHQ SWPA to WAR, 6 Feb 44.

[7] WAC Daily Journal, 11 Feb 44; quotation from tp conv, Mrs. Hobby and author, Oct 50.

General Sutherland refused to accept this rejection as final. A few days later the Director learned that, in a conference with General Somervell, a plan had been devised whereby 4,000 of the theater's civilian allotment would be given to the ASF in the United States in exchange for 4,000 of the ASF's military allotment. General Somervell also agreed to supply 4,000 Wacs from those under his control, to fill the allotment. This arrangement was approved by the War Department, and the theater was advised by radio that its Troop Basis had been increased by 4,000.[8]

This transaction naturally differed from those concerning initial shipments to Africa and England, in that plans were no longer made by the Director WAC, in accordance with the system adopted at the end of Auxiliary status. Thus, matters concerning the Troop Basis and overseas shipment were now properly planned by OPD and were carried out by the various ASF operating agencies.[9] The staff studies and directives concerning the 4,000 allotment bore only the concurrence of agencies that normally concurred in such matters for men: G–1, G–3, Military Personnel Division of the Army Service Forces, and the War Manpower Commission.

At a later date the question was to be raised as to whether the General Staff should have requested the Director WAC to make a study of the area from the standpoint of the well-being of women. However, her new small staff no longer contained inspectors, and in any case the original destination, Australia, appeared considerably more safe and stable than had North Africa at the time of initial shipments. As of this date, small groups of Wacs were also successfully serving in India, Egypt, and New Caledonia. To assist the theater in planning for any contemplated forward movement of Wacs, comparable to those into Italy and Normandy, it was determined that a WAC staff director would be sent in advance of the enlisted women, and the Director was allowed to nominate this officer. As OPD began to order shipments to the port, the Director was generally furnished with an information copy of their composition.

In three matters the Director attempted to secure retention of Auxiliary procedures different from those for shipment of men. In December of 1943, and again in the following March, the Office of the Director asked ASF's Mobilization Division to provide any Wacs going to the Pacific with the same tropical clothing and equipment already authorized for the China–Burma–India theater, since no maintenance stocks for women yet existed in Australia.[10] This was not done; Wacs being ordered to the port were authorized only winter clothing, in line with the custom for the shipment of men to Australia in its winter season. The Office of the Director also requested that WAC company commanders continue to have a voice in the approval of the stability and character of Wacs sent overseas. This request was disapproved upon the recommendation of ASF's Director of Personnel. Finally, Director Hobby requested The Surgeon General to continue giving Wacs a com-

[8] D/F, OPD for DCofS, 2 Mar 44, sub: Increase of Overseas Establishment in SWPA, with approval "OLN" for DCofS. OPD file.

[9] For a discussion of the relations between OPD and ASF operating agencies, see: (1) Ray S. Cline, *Washington Command Post: The Operations Division*, UNITED STATES ARMY IN WORLD WAR II (Washington, 1951). (2) Chester Wardlow, *The Transportation Corps: Responsibilities, Organization, and Operations*, UNITED STATES ARMY IN WORLD WAR II (Washington, 1951).

[10] Memo, Asst Exec WAC for ASF Mobilization Div, 11 Dec 43; and Memo, Actg Dir WAC for same, 21 Feb 44. SWPA 320.2 (8–18–42)(1) sec 6.

plete overseas physical examination, as had been previously done in Auxiliary days to detect gynecological disorders or pregnancy, but this was also refused as too time-consuming, since men going overseas had no such examination, only a briefer physical inspection.[11]

The assembly of shipments proceeded with unusual speed, because of the immediacy of the move from Australia. On the basis of informal lists of needed personnel furnished by General Sutherland, orders for the first shipment of 1,000 women were issued by OPD on 15 February even before the request for 4,000 was formally approved. A formal requisition for 270 WAC officers and 4,730 enlisted women was forwarded from the theater on 11 March 1944. The majority of the requests were for clerical and stenographic skills such as had already been requested by other theaters. This requisition was approved by the General Staff, plus other later requests from the Far East Air Forces in the theater, to a total of 7,500—a quota that was never to be entirely filled.[12]

Direct Commissions

The first three Wacs in the theater were appointed by local action before any shipments were made from the United States. General Kenney during his visit had requested WAC status for his Australian secretary; General Sutherland during his stay sought the same for his receptionist and for the secretary of Maj. Gen. Richard J. Marshall, Deputy Chief of Staff, SWPA. It was stated that key officers would suffer a serious lapse in efficiency if obliged to part with their experienced Australian and British personnel. Although the act of Congress limited WAC enlistments to American citizens, appointment to officer status was not mentioned; since all WAC officers were derived from the ranks, the possibility of appointing noncitizens had never been considered. Previous requests from the European theater had, up to this time, been withdrawn when the WAC policy was explained to General Eisenhower. However, General Sutherland refused to withdraw his requests, and, over Director Hobby's nonconcurrence, direct commissions were immediately given to the three Australian and British employees. Although all WAC officers to date had been initially commissioned second lieutenants, two of the women were commissioned as first lieutenants and one in the grade of captain.

This action was taken over the adverse recommendation of General White, G–1 of the War Department General Staff, who felt so strongly on the matter that, after being unable to dissuade theater representatives, he personally walked with them to the Chief of Staff's office to register his protest. General White was finally overridden when General McNarney, Deputy Chief of Staff, was informed[13] that the commissions were personally desired by General MacArthur as essential to headquarters operation.

When Director Hobby also called to register her protest, she was informed that it was the Chief of Staff's policy never to refuse any urgent personal request of a theater commander if it was possible to grant it.[14] She nevertheless placed in writing her request that the three commissions be revoked, stating:

[11] See Ch. XXIX, below.

[12] AFPAC Reply. FEAF requisition, 15 Jul 44: 84 offs, 2,916 EW.

[13] By Gen Sutherland, according to Gen White's diary, entry of 3 Jan 44.

[14] *Ibid.*, and intervs with Gen White, 5 Jan 51, and with Lt Col Mary-Agnes Brown, Jun 45 and 5 Jun 47

> I believe that the policy of appointment cited above will cause the personnel of the WAC to feel that the War Department has broken faith with them, and that the continuance of such a policy will be a great blow to the morale of the entire Corps. . . . The policy of selecting officers only from graduates of officer candidate school, and of selecting officer candidates only from the ranks, has been presented to the American public throughout the history of the Corps as a soundly democratic one.[15]

This apparently minor event received widespread and generally unfavorable publicity in Australia, which had it own women's services for qualified women, as well as British women's services. An even worse reaction came from Army men, particularly young combat officers who had not yet been promoted to equal ranks. Soldier mail, in the month before the first American Wacs arrived, showed that 90 percent of the comments about all Wacs were unfavorable, many obscene, alleging that all Wacs would be used only for "morale purposes" for officers.[16] This situation was to cause a serious morale problem among arriving Wacs.

In the United States, General White's office was at once swamped with renewed demand from Congressional and military sources for direct commissions for American civilian women with prominent sponsors.[17] The War Department held that the three commissions were irrevocable, but public reaction continued to be so adverse that G-1 Division succeeded in blocking any further appointments, except the one in Europe which was requested again in view of the exceptions made for SWPA. However, the Director secured a ruling against further such action in the future.[18]

In the theater, commanders noted continuing bad effects for months after the event, especially upon the morale of highly qualified enlisted men and women. Some commanders were obliged to forbid the women to talk or write of the subject, on threat of disciplinary action. The women's reaction was generally surprising to those not informed of the Corps' previous history and basic problems.[19]

Arrival of WAC Staff

Director Hobby released for assignment as staff director an experienced officer from her own staff, Lt. Col. Mary-Agnes Brown. Since the Southwest Pacific Area was known to be headed by ranking Regular Army officers with World War I experience, Colonel Brown was believed especially qualified by virtue of twenty-five years' acquaintance with Army channels. She had been employed in Army finance in World War I, had ten years' experience as executive secretary to ranking Army medical officers, and was thereafter an attorney for the Veterans' Administration, holding the degrees of A.B., LL.B., and S.J.D. In the Army she had been the first staff director for the Eighth Service Command and later executive officer in the Office of the Director WAC. With four officer assistants, Colonel Brown flew to the theater, arriving on 15 March 1944, to

[15] D/F, Dir WAC for G-1, 25 Feb 44. SPWA 320.2 (2-25-44)B, in SPWA 314.7 (1-7-43) sec 5.

[16] (1) See Chart, p. 000. (2) MI Rpt 107, 31 Aug 44. SFPOE, in MID files. (3) Personal ltr, Catholic priest to Col Hobby. WDWAC 330.14 (1945).

[17] Numerous ltrs, especially to and from Senator Charles O. Andrews, 10 Apr 44. WDWAC 095 and SPWA 095.

[18] D/F, G-1 for TAG, 21 Apr 44, WDGAP 210.1 WAC (4-17-44), approves attached Memo, Dir WAC for CofS, 14 Apr 44. Approved by CofS, 18 Apr 44. WDWAC 210.1.

[19] (1) Seminar comments. (2) Griffith Account, "Awards and Decorations." (3) Intervs with Col Brown, Col. A. Robert Ginsburgh, and Miss Dorothy Pat Costello, former sergeant in GHQ detachment. (See Bibliographical Note).

help prepare for the arrival of the enlisted women two months later.

Upon her arrival, the staff director was assigned to the senior American administrative headquarters, USAFFE. No WAC personnel was assigned to General Headquarters, an operational Allied headquarters, except for two of the directly commissioned officers. The staff director was not delegated the customary responsibility [20] for the WAC program. This was instead placed upon G-1 Division, USAFFE, headed by Col. Harry H. Baird. Subsequently, G-1 Division noted:

> The G-1 Section originally was charged with the handling of WAC personnel due to: (a) Their newness in the theater; (b) The fact that they were covered by a special allotment of overhead grades and ratings by the War Department; (c) The fact that it was necessary to keep in complete touch with the strength of the WAC Corps and the utilization of members during the period of arrival in the theater.[21]

Most of the duties assumed by G-1 were merely the normal staff duties now commonly performed by similar commands in the United States; however, it also held responsibility for policy matters. It believed that, so far as possible, enlisted men's policies should be applicable to the WAC. Colonel Baird recalled later that, "As far as the theater was concerned, there was no 'WAC program' except for an evident required publicity program." [22] There existed some belief that WAC advisers' experience and training were not always adequate to cope with the conditions that existed in the area. Colonel Baird later noted:

> Almost every one in top positions in GHQ, USAFFE, and SOS had served a generation in the Regular Army, as well as having seen service in World War I. For them to have formed very definite ideas on staff procedure, command, and military custom—which was the situation—was natural and logical. The Wacs lacked the background of military custom and they had little understanding of the possible far-reaching and serious effects of an action or decision.

The staff director's office was set up as the WAC Section of the Special Staff and, in the months before the Wacs' arrival, was called on chiefly to attend conferences and for public and social duties. In the matter of public relations, Colonel Brown was given a free hand and was encouraged by USAFFE to take all necessary steps to establish good will prior to WAC arrival. In the following weeks the staff director received calls from heads of the Australian Women's Auxiliary Service, YWCA, and Red Cross, and other officials. She attended theatrical performances, the Australian Arts and Crafts exhibit, a memorial mass, a WAAF ceremony. She went to receptions, gave luncheons, made formal calls at Admiralty House, Government House, and the American consulate. She broadcast radio programs, met visiting celebrities, and was photographed with koala bears. As a result, before the arrival of the first shipment in May, public attitude was believed to be definitely more favorable.[23]

[20] All WAAC and WAC Regulations without exception required every commanding general employing WAC personnel to employ also, on his staff, a WAC staff director who would be responsible for continuous inspection and for advising him on the welfare of the women. WD Cir 462 (1944) was latest wartime statement.

[21] Check sheet, G-1 USAFFE for CofS USAFFE through WAC Sec, 11 Apr 45, Tab 6 to WAC Draft AFPAC Reply.

[22] All of Col Baird's statements from Memo, Col Baird for HD SSUSA, 29 Jul 48.

[23] Monthly Historical Reports, WAC Staff Section, USAFFE, for Jun, Jul, Aug, Sep 44, Incl in Ltr, WAC Stf Dir USAFFE to Dir WAC. WDWAC 319.1 SWPA (1944); also in Folder, Monthly Historical Rpts, WAC Sec USAFFE, at Kansas City Records Center. These reports hereafter cited as WAC Sec USAFFE Hist Rpt, by month.

The only serious deficiency in planning at this date appeared to be in the matter of clothing. During a visit to the United States in March, Colonel Baird urgently recommended that two skilled WAC supply officers be sent by air to the theater. Queried for confirmation of this request, the theater authorized air shipment of one supply officer and one public relations officer. The supply officer, Capt. Lavern Bartholomew, was first ordered to OPD for indoctrination on supply needs of the area, and arrived in the theater in mid-April. Since at the same moment the first large group of enlisted women was boarding ship on the U.S. west coast, and the second was under orders, her supply planning was limited to later shipments. Planning was further delayed by the fact that Captain Bartholomew was not at once assigned to the office of the theater chief quartermaster, although Colonel Brown recommended this action on the grounds that supply planning was not a function either of the WAC Section or of G-1, which lacked the necessary data and authority.[24]

On 4 May 1944, a week before the landing of the first ship bringing Wacs, Colonel Brown and her assistant, Capt. Charlee L. Kelly, were sent to New Guinea to inspect the first advanced location, Port Moresby, to which some of the women would be sent two weeks after arrival. No unusual difficulties were expected in the area, since Port Moresby had long been regarded as a rear echelon, at which American nurses and Red Cross women had been stationed for about two years.

At the camp site, it was discovered that the quarters were entirely adequate but not prepared for troop occupancy; the staff director secured from the local quartermaster an agreement to prepare the quarters as soon as possible. Colonel Brown also discovered a fact not previously considered by supply officers—that malarial control regulations required all women in New Guinea to wear trousers both at work and off duty. The regular issue for men assigned to headquarters duty was six pairs of summer khaki. As officers, nurses had been able to purchase the trousers through commercial civilian channels in Australia, and were now required to have six mosquitoproof khaki shirts and six pairs of slacks apiece, which the chief surgeon described as "the minimum number which would take care of requirements."[25]

Returning immediately to Australia, Colonel Brown informed Brig. Gen. William F. Campbell, Chief Quartermaster, U.S. Army Services of Supply, Southwest Pacific Area (USASOS), that there was no such item as slacks in the WAC wardrobe, and that the nearest approach was the heavy herringbone twill coverall, of which arriving Wacs would have only one pair apiece. On the same day, the WAC supply officer was assigned to General Campbell's office, and a radiogram was sent to Washington requesting that the issue of two-piece coveralls to Wacs not already en route be increased to five pairs per individual, pending study of the problem.[26]

On the following day, Colonel Brown and her staff were received by General MacArthur, who, according to historical reports,

[24] WAC Sec USAFFE Hist Rpts, Apr, May 44; also in OPD file.

[25] Annual Rpt, 1944, Chief Surgeon USASOS SWPA, pp. 36-37.

[26] Rpts cited n. 24; also interv with Col William F. Campbell, 31 Jul 47.

ARRIVING IN SYDNEY, AUSTRALIA, *on 12 May 1944, above. Two days later members of the first contingent board buses for their new camp at Yeronga Park.*

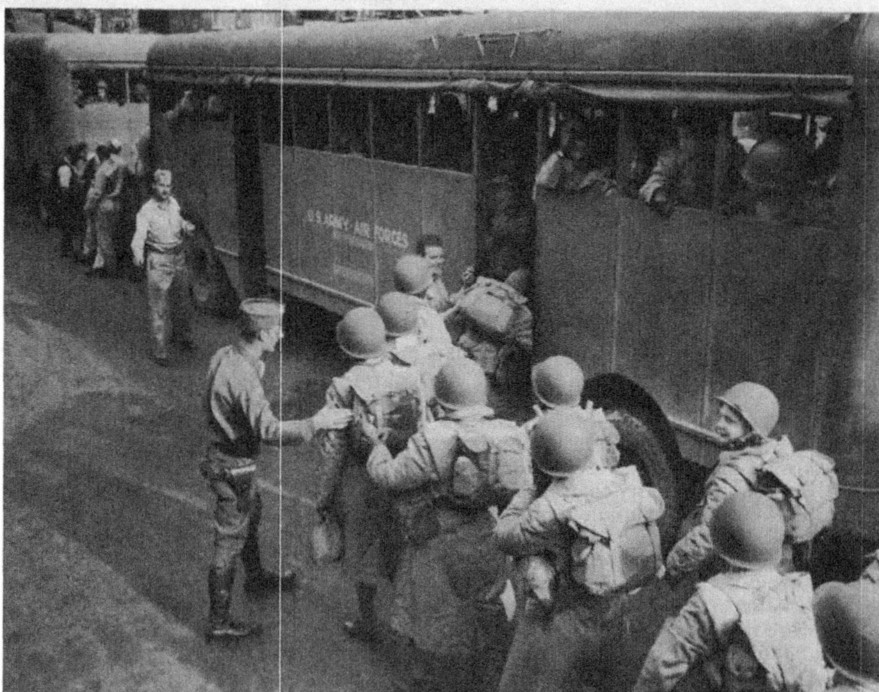

> . . . welcomed the WAC; spoke of the magnificent contribution of Australia to the war effort; praised the nurses and Red Cross women; spoke of how horrible war is; the two enemies our men have to fight—Japs and nature; and the morale factor of the presence of the WAC besides their contribution to their jobs . . . that he had asked the War Department for 10,000 Wacs and that he was getting less than half that number.[27]

This was the last official conference with General MacArthur until just before the Wacs departed at the end of the war.

First WAC Contingent in Australia

A few days later, after all preparations had been made for the reception of the first Wacs at Brisbane, it was discovered that the ship was about to put in at Sydney. Colonel Brown and other welcoming officers flew to Sydney, where it was found that USASOS personnel, long desirous of obtaining Wacs, had planned a cordial reception. On the morning of 12 May, an impressive number of general officers, heads of Australian services, and curious soldiers and sailors were on the pier to watch the first 640 Wacs disembark from the transport *West Point.*[28] The Wacs made a good military appearance in their winter uniforms, according to a public relations officer:

> With every eye trained on them, the Wacs marched off in their best military formation to pile into trucks. They were dressed in Class A uniform plus utility coat, field shoes, helmet, full field pack strapped on the back, and they carried in their hands everything from Red Cross ditty-bags to pillows, candy, radios, typewriters, and other cherished treasures. One Wac even carried a guitar. "Gee, they're real American girls," remarked one GI unbelievingly.[29]

The Wacs proceeded at once to Brisbane, which was only a two-week stopover for the first group destined for New Guinea, but where others were to stay for several months while civilian employees were replaced and the various headquarters rendered mobile. The 526 enlisted women and 114 officers were divided between USAFFE and USASOS. Most of the officers, and one company of the enlisted women, were assigned to operate the Central Postal Directory at Port Moresby, New Guinea. USAFFE's remaining 198 enlisted women were assigned to various headquarters offices, including 55 to the adjutant general, and from one to ten each to G–1, G–2, G–3, G–4, the chaplain, inspector general, headquarters commandant, judge advocate, provost marshal, special services, information and education, and finance. USASOS assignments were not recorded, but were presumably similar.[30]

Army officers responsible for the women's placement reported strong initial resistance on the part of many section chiefs, who were not enthusiastic about releasing civilian employees, but such objections were usually short-lived in view of the prospective departure from Australia.[31] An additional difficulty in replacement was presented by the fact that the first to arrive were in many cases not clerical workers, the higher clerical skills in the United States having been already well combed over by the European thea-

---

[27] WAC Sec USAFFE Hist Rpt, May 44.

[28] Paragraphs on landing and stay in Australia, unless otherwise specified, are from: (1) Griffith Account, "Australia." (2) AFPAC Reply, both draft and approved versions.

[29] Griffith Account. Numbers are from AFPAC Reply, although Colonel Brown later believed that nearer 1,000 had landed.

[30] Folder 203.1, 1st Contingent, Theater WAC Stf Dir files.

[31] Col. A. Robert Ginsburgh, "The Good Staff Officer in the Field," *Infantry Journal,* August 1946, pp. 10–11.

ter. Also, theater requisitions had included drivers, mechanics, radio operators, and other specialists who, it was now decided, could not safely be used in New Guinea. As a result, a headquarters in the urgent days of preparing for movement was faced with the necessity for retraining chauffeurs and other workers as clerks and typists—a problem which occasionally left the impression that untrained Wacs had been shipped.[32] In August the theater requested deletion from future shipments of fifty-four drivers, as well as some draftsmen, cashiers, bookkeepers, and other specialists, but the War Department reply was, "Impossible to delete . . . as personnel already en route to theater or under shipment orders." [33]

Under the circumstances, it was surprising to staff officers that the replacement was none the less accomplished speedily and successfully. General Campbell noted later:

> I was assigned 39 Wacs to replace 78 Australian girls. The Wacs didn't know one Quartermaster report from another but they quickly caught on: even those who had been drivers showed aptitude for it, and did more and better work than civilians. I never saw a bunch more willing to do a job.[34]

In one other respect, that of selection, several of the Wacs were inferior to the initial shipments to other theaters. A few days after the arrival of the first group, one member went AWOL and was finally apprehended only after conspicuous misconduct in the city of Brisbane, which came to the attention of the American consul and of some of the citizenry. Inspection of her records showed that she had been shipped overseas after a series of disciplinary offenses and courts-martial. Further investigation uncovered several more records of the same character. The women were promptly shipped back to the United States, accompanied by a report pointing out that Wacs should not be selected on the same basis as combat troops, and that as the first ambassadors and representatives of American womanhood in a foreign country, such women left something to be desired.

Using this evidence, Colonel Hobby finally secured re-establishment of the requirement—a considerable deviation from men's rules—that WAC company commanders must certify the suitability of any enlisted woman sent overseas, and that WAC staff directors must certify any officer.[35] Later selection was reported as much improved. Theater medical authorities felt that many physical and mental breakdowns could have been avoided had women with known records of instability not been sent. A subsequent report commented:

> It appears that the selection of WAC personnel sent to this theater was not made carefully enough, especially in the case of commissioned personnel. The "unloading" of less than "excellent" type personnel on an overseas requisition was a practice which flourished to the detriment of the WAC in this theater. WD Circular 362, 6 Dec 44, put a check on this practice, but a great deal of damage had already been done. Conditions of this theater put the greatest demands on women commanding other women as regards stability and judgment, and the calibre of many company officers sent here leaves much to be desired.[36]

As expected, the Australian stopover presented no important difficulties in housing or supply. Wacs were housed at

---

[32] Tech Int Rpt T/PFI-2472, Strategic Int Det, Hq 4th SvC, 7 Jul 45, to CG ASF. MID files.

[33] Folder 203.6, 6th Contingent.

[34] Interv, 31 Jul 47.

[35] (1) Ltr, Col Brown to Dir WAC, 7 Aug 44. WDWAC 320.2. (2) WD Cir 291, 11 Jul 44, Sec IV, succeeded by WD Cir 362.

[36] WAC Draft AFPAC Reply.

Yeronga Park, generally riding to work on city trams; USASOS Wacs later transferred to Victoria Park. The lack of heat in some offices and part of the quarters, customary as it was to Australians, caused a high incidence of minor pulmonary disorders, seldom disabling in character. Barracks were somewhat more primitive than those in the United States, with bucket latrines, outside showers, and limited laundry facilities, but were in general superior to those of male troops in the area. Wacs enjoyed city recreational facilities, and stated that they had plenty of Australian fruit and green vegetables and more than plenty of Australian mutton.

Arrival in New Guinea

Beginning on 28 May, only two weeks after the landing in Australia, 100 WAC censorship officers and 88 enlisted women left by air for Port Moresby.[37] As perspiring Wacs stepped from the first plane, clad in coveralls and wool-lined field coats, and carrying full field pack, they were met by an advance party with cameras and flashbulbs, intent on preserving the moment for recruiting purposes. Driving along the 14-mile road from Port Moresby to their camp site at John's Gulley, they found much of the way lined with fuzzy-haired natives, and with soldiers who shouted, waved, whistled, and called out such questions as "What State are you from?" or "How's San Francisco?"

The WAC quarters, inside a hot and dusty barbed-wired enclosure, were among the best enlisted housing in New Guinea—long wooden barracks with iron roofs, cement floors, and screened walls, with outside showers and toilets, but boasting a water-borne sewage system. The barracks were not yet fully prepared for troop occupancy, and the women, fresh from the United States, were at first baffled by the lack of furniture except Army cots, which thwarted their attempts to unpack clothing or set up the orderly room typewriter. They were also worried about prospects of doing their laundry without laundry facilities or hot water, or of eating, since the WAC mess was choked with debris and there was no drinking water.

However, a friendly men's unit nearby loaned the women a Lyster bag for drinking water, and fed them bread and jam, after which they borrowed trucks and set out to scavenge supply dumps for boxes, crates, tin cans, nails, and broken furniture. "The Quartermaster salvage dump proved a fascinating place," wrote one woman. They also borrowed a tool kit from the men, and improvised shelves and hangers for clothing and stands for the orderly room typewriter and files. By nightfall they considered themselves theater veterans. An enlisted woman wrote, "Now that we have our boxes, we are different people. We have found our sense of humor, and a sort of objectivity."

The work at Port Moresby proved heavy but routine, the operation of the Central Postal Directory in general offering nothing unfamiliar. The enlisted women sorted, checked, readdressed and forwarded soldier mail in what inspectors pronounced a highly competent manner. The WAC officers were employed as censors to replace male officers, and were also

[37] All descriptions of Port Moresby, unless otherwise specified, are from: (1) Monthly Hist Rpts, 5203d WAC Det, May–Aug 44, incl in Ltr, WAC Stf Dir USAFFE to Dir WAC. WDWAC 319.1 SWPA. (2) Griffith Account, "Port Moresby." (3) AFPAC Reply, both WAC Draft and final versions.

complimented by their supervisor, who observed:

> I don't know what there is about women that makes them so sharp-eyed in reading letters, but the ones I have here possess an uncanny knack for picking up hidden security breaches, such as tricky codes a soldier may devise to tell his wife where he is. . . . They are turning out more and better work than the male officers they released to the combat area.[38]

The only immediate difficulty in the area was that of the uniform. The skirted winter uniforms had necessarily been left behind for storage in Australia, leaving each woman, for her entire supply of outer garments, only the one pair of herringbone twill coveralls that had been issued for shipboard wear. Since there were no WAC maintenance stocks in the theater, each woman was provided with two more pairs of coveralls by the expedient of withdrawing them from other Wacs still in Australia. The heavy coveralls proved too hot for the climate and irritating to women's skin, as well as being so unsightly that nurses and other female personnel in the area were not permitted to wear them. Heat and skin diseases soon made it necessary for the headquarters to authorize the wearing of the WAC cotton shirt with the trousers of the two-piece coveralls, a practice that eventually became theater-wide, although officially forbidden by the theater surgeon, since the WAC shirt was lighter than the men's khaki, and below the weight believed safely mosquitoproof. The unit historian noted also that a few lucky women "had adopted suntan pants." Gifts of flowers and candy were scorned, and the successful applicant for a date was one who came carrying, as well as wearing, khaki trousers.

An additional problem that soon became evident was the unexpectedly restricted life of the enlisted women. WAC advisers, having been informed of the safe and civilized nature of the Port Moresby area, had believed that, as in Italy and Normandy, the ordinary camp security system would afford women sufficient protection, since there was little danger at this date from either the enemy or the natives. Instead, the headquarters directed that, in view of the large number of male troops in the area, some of whom allegedly had not seen a nurse or other white woman in eighteen months, Wacs would be locked within their barbed-wire compound at all times except when escorted by armed guards to work or to approved group recreation. No leaves or passes, or one-couple dates, were allowed at any time. The women's reaction was unfavorable; inspectors reported that they believed that higher commanders thought them "children or criminals," and therefore confined them in "a concentration camp." On the other hand, those restrictions that also applied to men were accepted by the women without comment.[39]

Within a few months the WAC area had been rendered more comfortable than most of those that were later to be found at more advanced bases. For labor beyond their strength, the women gave the natives cigarettes, candy, and even the otherwise useless WAC hats. Visitors reported considerable shock at beholding the natives dressed in WAC hats and very little else.

There were soon shelves, a graveled walk through the mud, a dayroom set up by the Red Cross, and even improvised dressing tables. The women learned from the men to wash their clothing in cold

[38] Griffith Account. Copy of commendation also in Folder, 5203d WAC Det, Drawer X-28040.

[39] AFPAC Reply, both versions. Also interviews. (See Bibliographical Note.)

water and to press it by sleeping on it or by hanging it up very wet. WAC cooks eventually cleared the debris from the WAC mess hall, overhauled the stoves, and soon had a mess which, their historian reported, "had a steady stream of daily business calls from officers who usually arrived just before dinner."

The women later obtained two washing machines, electric lights, and electric irons. In a few more months the Wacs had such luxuries as a recreation hall, a softball team, and a jukebox. They also persuaded the nurses to move their hairdressing shop so that Wacs could share it, since they could not use the men's barbershop. A Wac was put into the post exchange to order items women needed.

In spite of initial adjustments, only one woman in the detachment showed complete "inability to adjust to a tropical environment," and had to be returned to Australia. The unit grew in size, and was soon bypassed by Wacs moving north.

Headquarters Rendered Mobile

Meanwhile, with the aid of two more large shipments of WAC personnel, the various headquarters in Australia had succeeded in making themselves mobile. On 26 June 1944, the second WAC contingent landed in Brisbane—357 enlisted women and 28 officers—which was again divided, with USAFFE headquarters receiving about 99, the Port Moresby detachment 68, and the remainder going to Headquarters, USASOS. Included among the various skills were 75 clerks, 65 stenographers, and 52 typists. On 10 July a third contingent landed at Brisbane—8 officers and 371 enlisted women—again in chiefly clerical skills, and again divided among USAFFE in Australia, USAFFE in Port Moresby, and USASOS in Australia. In addition, 84 women of this last group were given to the newly formed Headquarters, Far East Air Forces (FEAF), which had not filed its own formal requisitions until July and thus had not received any on the scheduled date for movement in August.[40]

A matter which unit organization did not reveal was that not only USAFFE, USASOS, and FEAF, but also GHQ, were employing Wacs. By August of 1944, shortly before the jump-off from Australia, GHQ had borrowed 74 Wacs from USAFFE and 31 from USASOS—a total of 105, with 150 more on requisition. Colonel Brown recommended that these women be organized into their own detachment when numbers warranted, and assigned to GHQ, as were enlisted men, and Dutch, Australian, and British women. This request was refused, since it was the Supreme Allied Commander's desire that American women not be assigned to GHQ, but attached for duty from some lower administrative echelon. Of these 105 women, 56 worked in GHQ's Signal Intelligence Division, 28 as telephone operators, 10 with the Chief Regulating Officer, and lesser numbers in other offices.[41]

Scarcely any one policy, supply situation, or personnel practice was common to all of these groups—GHQ, USAFFE, USASOS, and FEAF—even when they shared one camp. Because of the poor communications and local differences in climate and facilities, it was USAFFE's practice to promulgate only broad instructions and to leave the subordinate com-

[40] (1) Folders 203.2, 2d Contingent, and 203.3, 3d Contingent. (2) List of shipments, AFPAC Reply.

[41] Correspondence folder, 5205th WAC Det, Drawer X-29853.

mands free to meet the changing situation. For this reason, Colonel Baird noted that "the theater was most reluctant to publish a fixed policy or general rule to establish uniformity in WAC control and administration. This was true for the men also." Exceptions could therefore be found to almost any general statement which could later be made concerning theater Wacs as a group. A theater housing policy for Wacs was almost the only published over-all policy, and even this was found impracticable outside of Australia.[42]

A source of some later comment was the fact that Wacs were assigned to the advanced echelons of all of these headquarters. For nurses, the theater had formerly adopted a contrary policy of "using them only in rear areas," causing them to be replaced by male corpsmen in evacuation hospitals and in field hospitals during active phases of service—a practice which, inspectors reported, had caused low morale and psychiatric casualties among the women.[43]

Theater historians later expressed some doubt as to the necessity for the opposite policy concerning Wacs, since Australia was, to the end, a great supply base at which some noncombat men remained, and certain other headquarters, such as Air Transport Command, winnowed the necessary male clerical personnel for more advanced bases from among ordinary troops.

Maj. Gen. William O. Ryan of the ATC, upon inspecting the New Guinea bases, declared the area unsuitable for women and canceled the orders of ATC Wacs about to fly in from Hawaii. Nevertheless, responsible Army officers were unanimous that the theater could not have taken similar action. Col. A. Robert Ginsburgh, who was both G–1 and G–3 of USASOS, stated later:

> USASOS could not possibly have moved from Australia without the Wacs. . . . Except for the Wacs, I had only the few clerically skilled men which higher headquarters let trickle through to USASOS, plus a few malarials and other men sent back from combat, naturally not very useful at exacting office work.[44]

Colonel Baird of USAFFE likewise indicated that the decision was a military necessity. General MacArthur himself, when later questioned as to the advisability of this course, replied:

> I moved my Wacs forward early after occupation of recaptured territory because they were needed and they were soldiers in the same manner that my men were soldiers. Furthermore, if I had not moved my Wacs when I did, I would have had mutiny . . . as they were so eager to carry on where needed.[45]

The Wacs themselves were eager to get forward, and the staff director later expressed "complete accord" with the decision.[46]

Oro Bay

Wacs began to move out of Australia early in August of 1944; by October only

[42] AFPAC Reply, both versions; also Memo cited n. 22.

[43] (1) Memo, Maj Craighill, MC, Consultant for Women's Health and Welfare, SGO, for TAG through CG USAFFE, 8 Jun 45, sub: Med and Social Conditions of Women in Mil Serv in SWPA. WDWAC 333 Pacific, and in SGO Hist Div 319.1–2 Women, SWPA. Hereafter cited as Craighill SWPA Rpt. Attached is a list of interviews with more than 200 individuals in SWPA, including chief nurses, WAC detachment commanders, hospital commanders, Army commanders, inspectors, etc.; also Table of Comparative Rates of Medical Evacuation for Wacs and Nurses, 1 Jul 44 to 1 Apr 45. (2) Data in possession of Col Brown.

[44] Seminar Comments. (See Bibliographical Note.)

[45] Statement to Col Boyce in Tokyo, 14 Oct 45, quoted by her in Boyce Rpt.

[46] Ltr, Miss Mary-Agnes Brown to OCMH, 23 Apr 51. OCMH.

COMMANDING OFFICER OF THIRD CONTINGENT OF WACS *to arrive in Australia is greeted at the pier on 10 July 1944. In the group greeting Capt. Ida M. Ross from left to right are Brig. Gen. Homer C. Brown, Lt. Col. Mary-Agnes Brown, Capt. Charlee L. Kelly and Lt. Vera Mankinen.*

a negligible number remained on the continent.[47] First to move was USASOS, which on 9 August 1944 sent an advance unit to Oro Bay for service with its Intermediate Section (Intersec). These were shortly reinforced by the arrival at Oro Bay on 6 September, direct from the United States, of the largest WAC shipment yet received—47 officers and 1,253 enlisted women. These were divided between Intersec headquarters at Cape Sudest and USASOS Base B at Oro Bay proper. On 30 October another large shipment of 21 officers and 571 enlisted women for USASOS also landed at Oro Bay, followed later by a final smaller one in January of 1945. A WAC staff director for USASOS was designated, the position being held successively in the next twelve months by Capt. Natalie Reebel, Maj. Ellen Bailey, and Maj. Annie V. Gardiner.[48]

The supply situation of the Oro Bay Wacs was not improved, for the first few months, by the fact that the entire September shipment arrived completely equipped with arctic clothing, including ski pants and ear muffs. No explanation

[47] Oro Bay discussion, unless otherwise stated, is from Griffith Account and AFPAC Reply.

[48] Sources cited n. 40.

or comment on this phenomenon was ever advanced by supply agencies in the theater or in the United States; no arctic shipment of Wacs had ever been directed by the War Department during World War II.[49]

The Oro Bay area proved generally suitable for women, with WAC camps, carved from the jungle, lying along the seashore, with beaches for front yards and palm trees for shade. Swimming was permissible under certain conditions. Barracks were made of wood and screens, with a large recreation hall and mess hall already built nearby. The staff director, Colonel Brown, accompanied by Captain Reebel, had visited and slept in these barracks before the arrival of the women, and had checked with the Base B staff on other necessary preparations.

Since Oro Bay was principally a major supply base, much of the Wacs' duties concerned the stock record reports and other paper work necessary to get matériel forwarded to combat troops. The Distribution Office was almost entirely staffed by Wacs, who kept track of the ships and supplies in New Guinea, whether mobile or at bases or consigned to the area from Australia or San Francisco. Army service command units to enter Leyte with the next wave of combat troops included men pulled out of offices at Oro Bay and replaced by Wacs.[50]

As at Port Moresby, the only unexpected difficulty in the employment of women in the area was that of restrictions not applied to men at the same base. Colonel Brown noted that before her departure the base commander, Brig. Gen. Clarence L. Sturdevant, had agreed to treat the women like other troops. General Sturdevant, one of the few members of the Pacific staff who had observed WAC administration in the United States prior to his assignment to the Pacific, believed that "overly protective measures such as barbed wire fences and armed guards were unnecessary and undesirable."[51] However, before the troops' arrival this decision was reversed. Thereafter, all USASOS bases used a system of protective custody for women. Barbed wire was put up around the Oro Bay WAC area, and the women were forbidden to leave "any area at any time" without armed guards, even being marched to approved movies in formation under guard. A list of further restrictions was published in order to prevent "regrettable incidents and unwholesome impressions of any nature."[52]

Off-duty activities were limited to approved unit parties and other mass entertainment to which women could be taken under guard, and even for these a woman's date had to be named twenty-four hours in advance, subject to disapproval by the senior WAC officer. Outside the WAC area, vehicles carrying women could not stop en route to the approved destination, or any woman leave the vehicle, or the vehicle leave the main road. Women were forbidden to board boats, ships, or craft of any sort, or to ride in aircraft. "Informal social gatherings of males

[49] Description of arctic shipment was deleted by theater from WAC draft report, but is well substantiated by interviews with participants. The only arctic shipments were those made to Alaska and Labrador by the ATC, which staged, equipped, and shipped its own Wacs.

[50] (1) Interv with Col Ginsburgh, 25 Jul 45. (2) M. Hamlin Cannon, Leyte: Return to the Philippines, a forthcoming volume in this series.

[51] (1) Marginal note on draft of this chapter. For fuller documentation see: (2) WAC Draft AFPAC Reply. (3) Memo 90, Hq Base B USASOS, 21 Nov 44. (4) Reg 10-55, Hq USASOS, 19 Dec 44. (5) Reg 10-100, Hq AFWESPAC, 2 Sep 45. (6) Ltr cited n. 46.

[52] Memo 66, Hq Base B USASOS, 22 Aug 44. This replaces an earlier Memo 15, 8 Apr 44, with C-1, 5 Jun, and C-2, 25 Jul.

and females" were forbidden unless the commanding officer designated the spot for them. All women were required to be in quarters by 2300 unless authorized to stay out until midnight, which was the absolute deadline.

Because of the growing effect of these provisions upon discipline and mental health, Colonel Brown later appealed again to USASOS headquarters to revert to the system first approved by the base commander. Although Maj. Gen. James L. Frink later agreed to authorize his base commanders to grant Wacs "the greatest possible freedom consistent with their personal safety," very few actual modifications ever resulted.[53]

Hollandia

On 31 August 1944, three weeks after the first Oro Bay shipment, a group of Far East Air Forces Wacs took off from Brisbane for the next stop, Hollandia, some eight hundred miles beyond Oro Bay. Here the environment was far less favorable and, while combat troops had landed four months before, the last Japanese raiding parties had still not been driven from the hills. The prudent FEAF Wacs, having heard that wild country lay ahead, climbed from their air transport laden with sixty baby chicks and twenty-five laying hens—which, incidentally, never laid thereafter. USAFFE and USASOS Wacs followed, until there were shortly more than a thousand in the Hollandia area.[54]

Climate and living conditions in Hollandia, except for GHQ's elevated camp site, were in general the least favorable yet encountered by the advancing headquarters. Rain was continual in some seasons, clothing was generally wet from perspiration, and heat prevented more than a few hours sleep at night. The red New Guinea clay apparently had the ability simultaneously to blow into the hair as dust and stick to the clothing as mud. During the first weeks there were no laundry facilities for men or women, and the slow-drying coveralls proved difficult to keep clean. For some time women had no hairdressing facilities, of which a medical officer noted, "They have as much need of them as men do of barbershops."[55]

Also, everyone shortly turned yellow from the required atabrine. One officer reported, "Frankly, women who had looked very well in Australia looked like hell in Hollandia." Men and women alike began to get skin diseases; some had to be sent back, including one competent WAC commander whose loss, USASOS authorities stated, "we could very ill afford." Weather conditions often prevented active recreation, while attempts at approved social entertainments bogged down under the weight of heavy coveralls and field shoes in 100-degree temperatures.

These conditions were particularly felt by USASOS personnel, both men and women, in their hot and humid camp site on low ground near the harbor. Here Wacs shared, with assorted insects, wooden-floored tents and a converted warehouse. The FEAF camp site was scarcely more healthful, with tents pitched on ground subject to flash floods of mud

[53] (1) Memo, Gen Frink, CG USASOS, for CG Intersec, 23 Oct 44, with incl Stf Dir Recommendations re WAC Pers at Intersec and Base B. GSCG 324.5. (2) Ltr, Stf Dir SWPA to all WAC Dets, 27 Oct 44, with incl Ltr. Folder, 5200th WAC Det, Drawer X-29443.

[54] Hollandia account, except as cited, from: (1) Griffith Account. (2) AFPAC Reply, both versions. (3) Interv cited n. 50(1).

[55] Major Craighill. See Table of Comparative Rates of Medical Evacuation for Wacs and Nurses, atchd to Craighill SWPA Rpt.

and water, and at the foot of a hill a strenuous hike from the offices on top. For a time the FEAF unit also housed the USAFFE Wacs, most of whom worked for GHQ, but within a few weeks this arrangement became so unsatisfactory to FEAF that, without warning, it bodily ejected the GHQ Wacs and their belongings, without a company officer.

Emergency housing was hastily worked out, and the GHQ Wacs eventually benefited by a move to tents in the GHQ area on a cool mountain top. To prevent such misunderstandings in the future, the staff director recommended in September and again in January that the unit be formally organized and assigned to GHQ as were enlisted men, but she was again informed by USAFFE that "The Commander-in-Chief did not want Wacs assigned to GHQ and under the circumstances no attempt would be made to do so." [56] Because of its ambiguous status, the GHQ unit suffered chronic supply difficulties, and eventually departed from Hollandia without ever receiving its organizational equipment.[57]

The GHQ unit, upon Colonel Brown's recommendation, attempted the experiment of housing Wacs, nurses, and Red Cross women in one camp without barbed wire, and with a minimum of restrictions. No particular difficulties were experienced, and there were only two instances of prowlers, of whom it was reported, "The Wacs took care of them." [58] The success of this experiment did not, as the staff director had hoped, persuade other commands to try it. USASOS units at all stations, employing some 60 percent of all Wacs, continued to use varieties of the unpopular "concentration camp" system. Air Forces units ordinarily relaxed the system to abolish locked compounds, requiring only armed escorts and traveling in minimum groups of four. The FEAF staff director noted that "No serious incidents were reported as a result of the wider degree of freedom exercised by Air Force units." [59]

A fourth WAC unit was eventually organized in Hollandia by the Far East Air Service Command (FEASC). Although handicapped when the departing FEAF transferred to it more than sixty WAC drivers and other unassignables needing retraining, the FEASC detachment requisitioned more Wacs and reached an eventual strength of approximately seven hundred.[60]

At this time premature optimism was general when, toward the end of 1944, the WAC medical evacuation rate for the first six months in the theater was computed and proved to be not too much higher than the men's, the loss rate being 62 per 1,000 per year for men (nonbattle causes only) and 98 per 1,000 for Wacs. This rate appeared especially good in view of the current rate of 202 per 1,000 for Army nurses, a rate that had generally prevailed since their arrival. Theater reports stated that the WAC rate compared "very favorably" with the men's, and that the difference was due to gynecological disorders, other disabilities incurred by the Wacs being generally of the same nature and degree of severity as those incurred by men.[61]

---

[56] (1) WAC Hist Rpt (shows Stf Dir check sheets on 23 Aug and 7 Sep). (2) Folder, 5205th WAC Det, Drawer X-29853. Also minutes of seminar.

[57] From 16 different items of correspondence in Folder, 5205th WAC Det, Drawer X-29853.

[58] Statements by Colonel Brown at seminar and on margin of draft of chapter.

[59] Ltr, Lt Col Mary L. Kersey, WAF, to OCMH, 21 Apr 51.

[60] Hist Rpts, Jan, Feb 45. Folder, FEASC WAC Det II.

[61] WAC Draft AFPAC Reply, cited in D/F, G-1 WD to CofS USA, 13 Sep 45. WDGAP 330.11 WAC (5-26-45).

HOLLANDIA, NETHERLANDS NEW GUINEA. *Wooden-floored tents occupied by the women.*

Other evidence tended to support even more optimistic conclusions. For one period, during which the average length of nonbattle hospitalization was computed separately, the average stay for Wacs was only 10.5 days as compared to 18 days for men. There was a slight suspicion among the Wacs themselves that they could withstand the hot weather better than the men. On the other hand, while many Wacs had gained weight in Australia, they, like many men, now began to lose; some suffered a "gradual and continuous" loss from the time of their arrival. Authorities blamed this on "loss of appetite after long periods of diet composed principally of canned and dehydrated foods."[62]

The Philippines

Wacs arrived in the Philippines on 26 November 1944, some thirty-six days behind the first wave of combat troops. The Wacs' presence was again required by USASOS' personnel plan, a repetition of its previous maneuver. To man the immediate northward push from Leyte, it was again necessary to pull men out of the organizations that had landed on Leyte and to send them forward in the island-hopping toward Manila. The mechanics of this exchange was handled by a WAC enlisted woman, later a warrant officer, Margaret Sterling, who received high commendation from USASOS authorities

[62] AFPAC Reply, both versions.

HOLLANDIA, NETHERLANDS NEW GUINEA. *Women at work in the finance office.*

for her management of the selection and training of WAC replacements in this and previous moves. GHQ and FEAF also made the decision to move their women to Leyte.[63]

Colonel Brown attempted to move the staff director's office forward from Hollandia ahead of the enlisted women, and succeeded in getting orders for a permanent change of station to Leyte. However, these orders were shortly revoked by G–1 Division, over her protest, and no representative of the WAC Section was permitted to visit the Leyte area until some weeks after the enlisted women's arrival.[64]

The first enlisted women, the GHQ unit, took off for Leyte accompanied by "Tokyo Rose's" prediction, "Of course they won't get there." As promised, trouble was encountered when the first planes were obliged to circle for some time before they could land, and were immediately strafed when they started to unload. Arriving Wacs, accustomed to being a sensation, were nevertheless surprised to see hundreds of men fleeing the airstrip and plunging into brush and foxholes. The women soon grasped the maneuver, and hit the foxholes as promptly as did the combat veterans on the same planes; from this vantage point they cheered as two

[63] (1) Interv cited n. 50(1). (2) Griffith Account, "Citations."

[64] Colonel Brown's notes on draft of chapter. WAC file, OCMH. USAFFE orders: WAC Sec USAFFE Hist Rpts.

Japanese planes were shot down in flames.[65]

Conditions during the rainy season nevertheless proved so unexpectedly bad as to raise the issue of whether or not the Wacs should be sent back, even over their objections. Proceeding through hub-deep mud to Tacloban, the GHQ Wacs were housed in a deserted mission school where Japanese soldiers had been recently garrisoned. Fortunately, the yards were already honeycombed with Japanese foxholes; the entire day and night were described as "a series of enemy air-raids and ducking for shelter."[66] USASOS men and women, arriving soon after, shared an area in a sea of mud, in which frameless tents were pitched so closely that they had to be tied together. FEAF Wacs were slightly better off. On the day after their arrival General Kenney himself inspected the area, and in the next month drainage ditching was performed by "shovel-equipped Colonels, Majors, Lieutenants, and non-coms from the A–2 Section."[67] However, the chief WAC problem on Leyte was not housing, which was the same for men and women, but rather the lack of suitable equipment. There were no overshoes for women, and only four of the GHQ Wacs had more than one pair of field shoes, which never dried out but had, the commander reported, to be "cleaned and put back on while still wet." The supply of coveralls was still so poor that, she noted, "there is no way of keeping in dry clothes."[68] A search was made for men's overshoes or field shoes in small sizes, but even after calling upon the Navy it was impossible to find enough to equip the women.

Nevertheless, the women were averse to leaving Leyte, and no move was made to return them. The History of USAFFE reported:

> They kept communications open at Leyte between alerts that sent them into foxholes, only to return to take more messages all through the night.[69]

The deputy chief ordnance officer noted:

> The Advance Section, USASOS, which went into Leyte, worked long night hours in the lantern-lighted tents when we were getting ready for an operation. We were interrupted frequently by air-raids, but they didn't panic.[70]

After a rough three months on Leyte, conditions improved as they had at previous stations; food and supply were better, laundry and hairdressing facilities were obtained, and emergency housing was improved. Troops of both sexes had more time for swimming, sight-seeing, and shopping at the local market. The chief interest was the battle for Manila; as usual, the Wacs again wished to move forward. As released prisoners, wounded, and sick began to be evacuated, some Wacs took up clothing collections among their own scarce stores for the women and children from internment camps, and volunteered off-duty help in hospitals.

Other Bases

Meanwhile, during and after the move to Leyte, Wacs had been sent to other bases.[71] The decision was made by USASOS headquarters that Wacs must go out to all of Intersec's lettered bases

[65] (1) Folder, 5205th WAC Det, Drawer X–29853. (2) Griffith Account.

[66] Descriptions of Philippines experiences not otherwise documented are from: (1) Griffith Account. (2) AFPAC Reply, both versions.

[67] Folder II, WAC Det FEAF, Drawer X–25096.

[68] Folder, 5205th WAC Det, Drawer X–29853.

[69] History of the United States Army Forces in the Far East 1943–1945, by Hist Sec, G–2 Far East Command. OCMH. Cited as USAFFE Hist.

[70] Interv with Col Merle H. Davis, 24 Jul 47.

[71] This section based on Griffith Account, AFPAC Reply, and Interv cited n. 50(1).

WAC AREA, TACLOBAN, LEYTE ISLAND, *27 December 1944. Note regulation dress shoes worn by woman.*

except those already to the rear, even where somewhat primitive conditions existed. As a result, USASOS Wacs were moved from Oro Bay not only to Hollandia but, on 24 November, to Base E at Lae and, on 6 December 1944, to Base F at Finschhafen. The USASOS staff director visited these bases in advance of troops, to help select WAC areas, and no particular difficulty was reported. Wacs at Lae lived in a hospital area, and those at Finschhafen in quarters of native construction, which were considered superior to tents. Both groups were later to be the last to leave as other bases moved forward. Jobs remained the familiar USASOS routine of cargo tonnage, troops movements, vessel routing reports, and other service paper work.

On 13 February 1945, the USAFFE mail and censorship unit likewise moved forward from Port Moresby to the coral island of Biak. Although the location presented problems of heat and glare, it eventually offered many conveniences. Except for an air raid in which no WAC casualties were suffered, the Biak Wacs' closest approach to war came with their visits to hospital ships and general hospitals.

Manila

On 7 March 1945, four days after organized resistance ended in Manila, the GHQ Wacs left Leyte and reached the area by air, across a bay still clogged with sunken ships. A WAC officer described their arrival:

> The plane landed on a black-top highway [72] that served as an air-strip. Wacs climbed out and sat cross-legged on the ground taking in the scene. Down this modern roadway passed an array of assorted vehicles . . . people and their belongings coming back to the city. . . . Heavy artillery fire could be heard in the distance routing the Japs out of the hills only a few miles away. The route into the city led past . . . a once thriving business section, now a pile of crushed buildings, twisted iron gratings, debris. . . . Dead Japs still lay along the streets awaiting burial. GHQ and USAFFE Wacs were housed in De La Salle College . . . called "atrocity college" because of the atrocities committed there by the Japanese. Bodies had been removed a day previously and floors and walls creosoted and fumigated.[73]

Fighting was still going on in the outskirts of the city, snipers were being routed out, and guards were heavy. One Japanese prowler was later caught in the WAC mess hall, and two others were shot down and died at WAC onlookers' feet, but there was, an observer reported, "no hysterics" among the women.

The disturbed conditions continued for some weeks. Water was scarce and had to be hauled. The GHQ Wacs bathed in helmets, lived by candlelight at night. Latrines were erected in the courtyard, with half a dozen shower heads. Men's mess halls lacked space to serve the women, and Wacs lived on K rations until they could get their own mess hall into operation. Dehydrated food was the chief sustenance for all troops; some women lost their appetites, preferred to sleep at mealtimes, and often ate only one meal a day. Until blood and filth could be removed, flies were so numerous that a severe epidemic of dysentery afflicted the GHQ Wacs. Their commander's greatest fear was that some of the violently ill women, groping their way in the blackout to the outdoor latrines, would fall and detonate the remaining Japanese booby traps believed to be on either side

[72] A later description refers to the highway as a city street and gives a route somewhat different from that in the present account.

[73] Griffith Account for entire Manila discussion unless otherwise cited.

of the board walks. Nevertheless, there was soon set up what was described as "an orderly efficiently operating headquarters in the midst of chaos, confusion, and destruction." In buildings full of bullet and shell holes, GHQ Wacs for the first weeks propped up broken desks and chairs, used crates for file cases, chunks of twisted metal for paperweights, and fragments of exploded shells for ash trays.

The USASOS Wacs delayed their arrival until, in the opinion of their battalion commander and later staff director, Major Gardiner, the worst of the dirt with the resulting dysentery was eliminated. They were housed in the Far Eastern University and in overflow tents on adjacent areas. Buildings were in relatively good condition, and the headquarters was quickly established.

As various New Guinea stations were closed, Wacs were moved into the Manila area until eventually most of the entire five thousand were concentrated there. Army Forces, Western Pacific (AFWESPAC), the successor to USASOS, used over a thousand women who were housed in frame-and-burlap barracks on Ascarraga Street. A small general depot group at Quezon City lived in floored tents in an area which the Wacs soon neatly landscaped. At the demolished Fort McKinley, FEAF Wacs lived in an attractive breezy camp with partitioned buildings, neat dayrooms, and their own club for dancing. At Nichols Field, FEASC Wacs had hot and dusty tin barracks. The Philippine Base Section WAC detachment drew comfortable but crowded nurses' quarters, where they lived six and eight to a two-woman compartment, and enjoyed that novelty, clothes closets. Wacs at Santa Ana Racetrack, working for the Allied Translator and Intelligence Service, lived for a time in muddy unfloored tents, later moving to thatch-roofed huts of 20-Wac capacity. Wacs with the signal intelligence section were housed at San Miguel Plantation in frame and tin barracks, and lived a very quiet country life.[74]

For some time restrictions in Manila were even tighter than elsewhere, since the city's remaining recreation facilities were generally off-limits, and a seven o'clock curfew was imposed for security. Also, the move to Manila did little to improve the supply situation. One company commander was obliged to ask that Wacs not be arrested by military police for not being dressed in the khaki uniform now required of all men and nurses, since "WAC personnel are not out of uniform when wearing coverall trousers with cotton shirts, because of the necessity for such wear as Class A uniform due to lack of supplies. Some members have only one pair of [men's] khaki slacks."[75] Just after the first Wacs' move to Manila, the WAC Section sent a memorandum to G–1 of USAFFE pointing out that while the supply difficulties for male troops in the area were great, those for Wacs had been even greater: "No supply has been available for Wacs where I have been stationed since our departure from Australia."[76]

Job Assignments

In Manila, and in the outlying bases, final records of WAC job assignment were scanty. The theater estimated that, as of the spring of 1945, 70 percent of the Wacs were engaged in administrative and office

[74] For WAC detachments at end of war see Griffith Account.

[75] Correspondence Folder, 5200th WAC Det, Drawer X–29443.

[76] Memo, Capt Juanita Stryker, WAC Sec USAFFE, for Gen Baird, 12 Mar 45. Tab 12, WAC Draft AFPAC Reply.

work. The most common WAC jobs were, in order: clerk-typist, clerk (general), stenographer, typist, message center clerk, chauffeur, cook, and file clerk.[77] The estimated percentages in each type of job were, toward the end of the war:[78]

Duties	*Percent*
Technical and Professional	2.52
Radio and Electrical	.08
Communications	2.04
Mechanical and Trade	.81
Administrative and Office	69.33
Motor Transport	6.45
Food	4.52
Supply and Stock	8.77
Miscellaneous	5.48

By comparison with world-wide employment, the Pacific had used notably fewer women in communications, and more in routine office and supply jobs.

The approximate peak distribution of WAC troops among the several commands was estimated as: [79]

Command	*Percent*
USAFFE (including GHQ)	23
USASOS	60
Air Forces: FEAF	6
FEASC	11

Concerning the work done by the 23 percent assigned to USAFFE and GHQ, the History of USAFFE noted:

> The WAC were typists, switchboard operators, administrative officers in signal centers and message centers, chief clerks, mess sergeants, translators, historians, Personnel and Transportation Officers, Top Secret Officers, Executive Officers, and Assistants to Adjutants General. They drove heavy vehicles; handled stock reports, censored the mail of the whole theater.[80]

In addition to such duties, Wacs were by 1945 also assigned to the Joint Supply and Survey Board, the Radio and Radar Unit, the Psychological Warfare Board, the USAFFE Board, the fiscal director, the Civil and Military Censorship detachments, the A-D Board, and Section 22 of GHQ.[81]

Within USASOS, its 60 percent of Wacs were assigned not only to the headquarters but to ten base depot companies including quartermaster, ordnance, signal, engineer, chemical warfare, and medical.[82] For example, the ordnance office of USASOS employed several hundred Wacs, of whom its chief noted:

> Ordnance had more than 100,000 line items on which it maintained stock control in the theater, first by cards and later by IBM. Most men do not like this type of work, and even so the women are more painstaking. Each enlisted Wac would handle one small group of items, and each WAC officer would supervise several such groups. Our tendency was to leave them on the group they had learned, but they did not show signs of weariness. Many of the women were outstanding in morale.[83]

Percentages of Wacs in the various USASOS offices were not recorded, with two exceptions: the finance office at Intersec, serving approximately 2,800 personnel, was staffed entirely by Wacs; the editorial section of USASOS headquarters was composed of one WAC officer, eight enlisted women, and two enlisted men, who reviewed all publications.[84]

The 17 percent of Air Forces Wacs did, in general, the same type of work as that in the Service Forces. No record of FEAF jobs was preserved, but as of June 1945, FEASC employed 574 Wacs.[85] Of them,

[77] "Suggested Materials for a WAC Newsmap . . ." Material in possession of Mary-Agnes Brown.

[78] AFPAC Reply.

[79] *Ibid.*

[80] USAFFE Hist.

[81] List in possession of Mary-Agnes Brown.

[82] *Ibid.*

[83] Interv cited n. 70.

[84] Newsmap material cited n. 77.

[85] Folder I, WAC Det FEAF, Drawer X-25096.

Maj. Gen. Clements McMullen of the Far East Air Service Command stated:

> The FEASC WAC Detachment rendered superior service to this command. Our Wacs typed letters, ran mimeo machines, almost entirely manned an International Business Machine Statistical Unit, filed papers, drove automobiles, worked in dispensaries, and maintained their own homes, in a wonderful manner. . . .[86]

Signal Corps duties were common to GHQ, USAFFE, USASOS, and the Air Forces. In the Intersec Signal Center, Wacs at teletype machines handled an average of 100,000 coded groups a day. During the Leyte campaign, 130,000 words a day were handled, including coding and decoding. One four-position telephone switchboard averaged 6,500 calls daily. GHQ's Signal Operations Group commended the Wacs handling the Bataan exchange, saying:

> Many favorable comments have been received by this office recently on the operation of the Bataan Switchboard. The speed, efficiency, and courtesy with which calls are now being handled have been highly commended by representatives of the Chief Signal Office, GHQ.[87]

FEASC reported that during the move of its advanced headquarters to Manila, six WAC telephone operators "without assistance, for two consecutive weeks ran the FEASC switchboard at its new headquarters on a 24-hour 7-day week." [88]

Several hundred women were employed on skilled work of a classified nature. The Allied Translator and Intelligence Service had 150 Wacs, including nisei, who did what was described as "important secret work" as interrogators and translators. This unit maintained a high morale, and was regarded as outstanding. Equally secret although more monotonous was the work of the 120 Wacs with the Signal Intelligence Service. The women were described as "highly skilled technicians," but it was admitted that "the secret nature of their work, regardless of how interesting and stimulating it was, in time became tiring due to the fact that it was confining and restricted the individual." [89]

The work of the censorship unit and the Central Postal Directory, successively at Port Moresby and Biak, continued almost unchanged. Approximately a million wrongly addressed pieces of mail were processed monthly by the Wacs, who received reports of all changes of status or location of theater personnel and worked seven days a week checking cards in huge locator files. The censorship group received the Meritorious Service Unit Plaque for having censored not only ordinary mail but "practically all the foreign-language mail written in the theater, involving a knowledge of 31 languages." [90]

Nevertheless, inspectors found that censorship work, which at first had appeared highly suitable for women, offered a definite mental hazard. One reported:

> Much of what the men wrote was so obscene that the women became demoralized from having to read it all day. . . . The WAC censors became nervous, temperamental, and complaining; first they thought they would lose their eyesight, then their minds. All came to complaint hour. By the end of the war . . . many of them had become definitely neurotic if not actually psychotic.[91]

An inspector general report in April 1945

[86] Ltr, CG FEASC to Dir WAC, 3 Jan 46. sub: Commendation. WDWAC 201.22 and 330.13. (1945–46).

[87] Correspondence folder, 5205th WAC Det, Drawer X–29853.

[88] Hist Rpt, Jul 45, Folder II, WAC Det FEASC.

[89] Boyce Rpt.

[90] GO 157, GHQ USAFPAC, sub: Meritorious Serv Unit Plaque Award (5203d WAC Det). Tab 14 to WAC Draft AFPAC Reply.

[91] Interv with Capt Jeanne K. Letallier, WAC IG SWPA, 9 Jan 46. Similar rpt in Craighill SWPA Rpt.

stated that an extremely low state of morale existed and that the health record was poor: 20 percent were on sick call daily, 50 percent suffered from eye strain, and approximately 8 percent had already been evacuated due to tropical diseases and psychoneurosis.[92] After the unit was returned to the United States, the postwar theater staff director wrote, "In any future planning, I definitely oppose the use of women in censoring male correspondence."[93]

A few other defects marred the WAC job-assignment program, of a type which in less active theaters had caused more serious difficulty than was reported in the Pacific. In some units, women, especially stenographers, were not always kept busy enough to suit them. Inspectors reported of the FEASC detachment that "the largest percentage of complaints from this group involved job utilization . . . complaints were registered in regard to the limited amount of work actually being performed."[94]

Nevertheless, women were apparently more fortunate than men in this respect, for a FEASC management control survey reported the job utilization of enlisted women to be better than that of enlisted men in the same offices.[95] On the other hand, highly skilled women technicians were more subject to malassignment than men, for the environment greatly limited the extent to which women could be scattered about, and WAC radio operators, photographic technicians, drivers and mechanics usually had to be retrained to other skills.

In almost all commands, job promotion was slow and skilled workers did not receive the ratings that the War Department, in allotting Wacs to the theater, had provided. This spread of grades and ratings would have been generally appropriate for the MOS numbers held by the women, but after a few months these were not reserved for the Wacs but made available to the men, according to the system of integration favored by WAC advisers in the United States. Any other course, according to Colonel Ginsburgh, "would have caused much ill-feeling among the men."[96] Grades received by Wacs were below the theater average for men in skilled jobs; they were also somewhat lower than those received by Wacs in Europe and Italy. On the other hand, they were higher than those of male labor battalions and unskilled service troops in the theater, and even a single stripe on a woman's sleeve provoked so much caustic comment from unrated men that Colonel Ginsburgh recommended that, in the future, women in the Army be given no grades. WAC officers' promotions were likewise slower than in other theaters, about half never being promoted in the theater.[97]

Because theater policies differed, awards and decorations were given less frequently than in other theaters. Although Pacific Wacs had moved more rapidly and lived longer under field conditions than those in Europe, they received only about one fourth as many Bronze Stars. In some cases there existed some feeling that

[92] Material atchd to Ltr, WAC Stf Dir AFPAC to Dep Dir WAC, 8 Nov 45. Hobby files.

[93] Personal ltr, Lt Col Mera Galloway, WAC Stf Dir AFPAC, to Dir WAC, 28 Dec 45. WDWAC 320.

[94] Memo, WAC Stf Dir FEAF for CG FEAF. WDWAC 319.1 Pacific.

[95] Hist Rpt, May 45. Folder II, WAC Det FEASC.

[96] WAC Draft AFPAC Reply. Confirmed by Interv cited n. 50(1).

[97] (1) AFPAC Reply, both versions. (2) STM-30, Strength of the Army, Jul 45, p. 41. (3) Morale Evaluation Rpt on Women's Servs, AFPAC, Jul 45, incl in Ltr, Actg Stf Dir AFPAC to Dir WAC, 11 Aug 45. WAC C1 files.

Bronze Stars and similar awards were inappropriate for women. One WAC sergeant noted that identical recommendations were forwarded by her section chief for herself and for the men under her supervision, but that only the men's were approved.[98] There was only one award of the Meritorious Service Unit Plaque, elsewhere quite common among WAC units.[99]

The Question of Expense

There existed from the beginning the question of whether the expense of maintaining women in such an area was not so prohibitive as to nullify any possible gain from even the most superlative efficiency. Theater authorities discounted this view almost completely, and were unable to understand how it had become so widely circulated. To some extent its spread seemed due to hundreds of statements in soldier mail, such as "As far as replacing men, that is purely a myth; it takes more poor GI's to work for them, and more to guard them, than they come anywhere near replacing"; or, from an officer, "The lady GI's haven't freed any men for the fighting fronts. They require special quarters, special handling, and so damned many MP's to guard them that it more than eats up the manpower relieved." Some thought that it took "three or four GI's" for every Wac, or even, "For every Wac that is sent overseas it required approximately eight soldiers to take care of her."[100]

Actual employers of Wacs could find little basis for such charges. "Wacs needed some extra care and help in setting up their housing," the USASOS G-1 stated, "but a headquarters has to have administrative troops, and where will you find male stenographers?"[101] A section chief noted, "When we moved, we had to provide areas for women, but we would have had to set them up for the nurses anyway; it was only a few days' work, and didn't have to be done again for months."[102] The general view was that it was not an uneconomical procedure to employ service troops to set up facilities for key office workers, regardless of sex, although the service battalions concerned could not be expected to concur in the opinion.

Some guards had also undoubtedly been required, much against the women's wishes, but in daylight hours only one or two were employed to march the whole unit to work, which took a very few minutes. At night, for attendance at unit parties, the duty was a purely voluntary one. As for gate guards, the comment of Colonel Boyce, the second WAC Director, after a visit late in 1945, was:

> It was noted that guards were considered necessary and were therefore assigned to all the other groups in the forward area, i.e., nursing personnel, medical groups, and general headquarters areas. It was established that no more guards were required for WAC personnel than for other types of noncombat personnel assigned to duty in these forward areas.[103]

Apart from the labor involved, there remained the question of whether WAC housing, messing, and other accommodations had not been prohibitively expensive, or cost much more than the men's. There was little evidence that they had, although

[98] Interv with Miss Dorothy Pat Costello, 4 Aug 48.

[99] AFPAC Reply, both versions: 1 Meritorious Service Unit Plaque; 4 Legion of Merit; 3 Soldier's Medal; 50 Bronze Stars.

[100] All except second from GHQ AFPAC Summary, Comments on Women's Servs, 16 Jun 45, in possession of former WAC Stf Dir; second from WDBPR file.

[101] Interv with Col Ginsburgh, 24 Jul 44.

[102] Interv with Col Merle H. Davis, Chief OO WESPAC, 24 Jul 47.

[103] Boyce Rpt.

Wacs were often allotted the best of the enlisted accommodations already available to a headquarters.[104]

The choice of camp site went according to the priority rating of each organization. Thus, USASOS usually rated housing, on a priority basis, after GHQ, USAFFE, FEAF, and the Sixth and Eighth Armies. The housing of USASOS Wacs was therefore ordinarily less desirable than that of the men of those groups, although possibly superior to that of USASOS men. The History of USAFFE noted briefly, "The WAC were not accorded special quarters and special transportation or special supplies. In New Guinea, their quarters, mess, and recreation were comparable to those of male troops." [105]

The general impression that Wacs had more expensive quarters seemed due largely to the fact that visitors and historians arriving several months after the advanced echelons found some comforts present. Reliable sources ascribed these only to the persistent scavenging of the women themselves, abetted by the ever-willing labor of the endless supply of GI suitors. In a typical camp on Leyte, Wacs were assigned the usual dirt-floored tents, but the women's dates brought them floors for the tents, a piece at a time. Wacs also traded with the natives for heavy labor and useful or decorative items.

A far more useful source than either Army men or natives was the Navy, particularly the Seabees, who were no less willing to assist and far better supplied with the wherewithal. A settled WAC unit with reasonable access to land, sea, and air forces was therefore likely to have a landscaped camp site, graveled paths, floors in the tents, dressing tables, and, overhead, forming a heat-resisting space below the iron or canvas roof, the billowing nylon of a discarded parachute. Recreation halls were likewise decorated with native thatch, palms, trophies, and bright-colored supply parachutes.[106]

WAC messes, equipment, and rations were identical with those of male units. The Southwest Pacific Area was the only overseas theater that could not ordinarily relieve skilled office workers of kitchen police duty, since civilian labor in New Guinea was limited, although some was available in Manila. Wacs staffed their own messes or, if eating at the consolidated mess, generally furnished their proportionate share of kitchen police, as they did in the United States. Two men were usually assigned to large WAC messes to service the field ranges and do heavy lifting.[107] Again, it appeared that any appearance of superiority was due only to improvised decorations or well-prepared food, which often ranked a WAC mess with higher headquarters messes. The Biak historical report stated, "The WAC Mess Hall soon became the envy of the post . . . the rumor flew that Wacs had better rations, but the Wacs refuted this." [108]

As for the matter of clothing, there would have been no question of unusual expense even had the Wacs received any. WAC staff members estimated that, even if the Army's usual policy of issuing all needed items had been followed, the cost of a Wac's clothing in malarious areas would have been only $124, in nonmalarious areas $143, as against $233 for comparable items for Wacs in the United States.[109]

[104] D/F cited n. 61.

[105] USAFFE Hist.

[106] (1) Boyce Rpt. (2) Griffith Account.

[107] AFPAC Reply, both versions.

[108] Folder, 5225th WAC Det, Drawer X-28368.

[109] *Ibid.* Also (1) Memo, Capt Stryker for Gen Baird, 12 Mar 45. (2) Check Sheet, Capt Stryker to CofS, USAFFE, 2 Apr 44. Tab 12, WAC Draft AFPAC Reply. (3) Personal ltr, Maj Gen R. J. Mar-

Medical Evacuation Rate

The greatest question concerning the efficiency of WAC employment in the theater did not concern expense of maintenance. It began to come to light only after the move to Manila: this was the increasing rate of loss for medical reasons. The earlier favorable rate had continued through the first months of 1945, even during the emergency period in Leyte. The change in health was at first not startling; in April the GHQ detachment noted a high percentage on sick call, but no serious diseases, and concluded, "Apparently many are in a somewhat run-down condition." In May the trend continued, again without serious disorders but with "the usual complaint [of] boils and other skin ailments, colds, or other manifestations of fatigue."[110]

When the evacuation rate was computed theater-wide, it was obvious that in these few months it had suddenly jumped from 98 per thousand per year to 267 per thousand. The chief surgeon's office informed Colonel Brown in February of 1945 that "At this rate, almost 20 percent of the command will be lost per year by evacuation."[111] By April, it was clear that the figure was more like 25 or 30 percent, and as much as 60 percent per year in some units, such as FEASC. The significant fact in this connection was that the 6 percent rate for men had not increased to any comparable degree. The rate was even more startling when compared to that of the Wacs in Europe and to those in India and China, whose losses were the same as the men's in those areas. The Pacific WAC rate was now virtually the same as that which Army nurses had experienced since their arrival, and four or five times that of the men.[112]

The onset of this loss rate had come too late to damage the efficiency record of the WAC's first year, but at this time commanders believed that this efficiency would have to be sustained for still another year of war in the Pacific. It therefore appeared urgent to discover, and if possible remove, the cause of the increased loss rate before it began to affect WAC strength more seriously. Brig. Gen. Guy B. Denit, Chief Surgeon, USASOS, stated emphatically that the increase had no medical or geographic cause: "Under the sanitary system we had by that time, a white woman could live in the area for any length of time as easily as a white man."[113]

Neither did the chief surgeon agree with some medical officers who believed that the Pacific theater, unlike the European, had no facilities for treating disorders peculiar to women, which thus required evacuation, whereas a man could be treated and returned to duty. One doctor alleged that he and other medical officers were

> . . . unable to do anything for Wacs reporting to the hospital with such diseases, as there are no instruments or medications furnished to Army hospitals overseas for their treatment. It seemed a waste of space and manpower to send Wacs overseas and then not have medical instruments and medications available to treat them.[114]

shall, CofS AFPAC, to Gen Stivers, OPD, 14 May 45, (4) Ltr, Asst AG USAFPAC to CG USAFWESPAC, 31 Jul 45. AG 420 (7-31-45) WA. Tab 12, WAC Draft AFPAC Reply.

[110] Hist Rpts, Apr, May 45. Folder, 5200th WAC Det, Drawer X-29443.

[111] Memo, Asst Chief Surg for WAC Dir USAFFE, 16 Feb 45. REMD 704.11. In possession of Mary-Agnes Brown.

[112] All rates and other health statistics cited, unless otherwise specified, are from Craighill SWPA Rpt or AFPAC Reply.

[113] Interv, 24 Jul 47.

[114] Tech Int Rpt T/PFI-2614, SID, Hq 4th SvC, to CG ASF, 23 Jul 45. G-2 MID files.

No women medical officers were employed in the Pacific as they had been by the European and Mediterranean theaters.

This explanation of the loss rate was contradicted by General Denit, who stated that he had available every supply and facility, could get any needed gynecological instrument by air, and had deep X-ray machines, radium, and other equipment. General Denit's view was for the most part supported by the statistics, which showed that less than one fifth of the total loss was due to disorders peculiar to women. The War Department's medical consultant, Major Craighill, after a visit, noted, "Gynecological conditions are usually mild in character and transitory in occurrence." [115]

Neither could the high rate be attributed to undue medical lenience in the return of women to the United States. Major Craighill indicated that "an even more lenient policy in regard to return . . . would seem indicated for the good of the group and to salvage individuals for continued service in the States."

The only notable medical deficiency which was cited by theater authorities was that of poor initial medical screening in the United States. Major Craighill reported:

> Medical officers in all overseas theaters commented on the inadequacy of screening of women for overseas duty. It was thought to be the primary cause of medical evacuation rates for women, especially in the Southwest Pacific, where they were far higher than for other personnel.[116]

A survey in the FEASC WAC detachment indicated that as many as 50 percent of the medical returns were attributed to poor medical screening in the United States, with a few evacuees having been in the theater less than a month.[117] On the other hand, the screening of women for service in the Pacific had been as good as that of men, with identical selection procedures. Also, four fifths of the European theater's Wacs, and most of those for India and other areas, had been selected, like the Pacific's, after the end of the Auxiliary requirements for physical examinations.

With no medical deficiency thus entirely sufficing to explain more than a part of the loss, General Denit stated instead that in his opinion the primary cause of the higher loss rate was to be sought among theater policies other than medical. Independently, WAC inspectors also arrived at the conclusion that the cause of the higher WAC rate of loss did not lie in the environment but in theater policies toward women that did not apply to men. Colonel Boyce, the second WAC Director, stated after an inspection visit, "It is believed that the experiences of Wacs in the Pacific which adversely affected their effective utilization were not inherent in the geographical location." [118] The same view was expressed by Major Craighill after an inspection trip in 1945. Major Craighill felt that there was nothing necessarily dangerous about the area for women, and made recommendations for improvement that were all in the fields of supply and administration.

Deficiencies in Uniform and Supply

Among nonmedical deficiencies which had possibly contributed to the medical loss rate, the most conspicuous was that of

[115] Unless otherwise cited, all of Major Craighill's comments are from Craighill SWPA Rpt.

[116] Hist of Women's Med Unit, draft by Lt Col Margaret D. Craighill, 23 Jul 46. SGO Hist Div.

[117] WAC Sec USAFFE Hist Rpt, Mar, Apr 45. Folder II, WAC Det FEASC.

[118] Boyce Rpt.

the clothing supply.[119] Women had in fact faced a difficult and dangerous environment less well equipped than the men in the same headquarters. It was noted particularly that dermatitis, sometimes popularly called "jungle rot," led all other causes for evacuation of women, even though medical officers had originally believed that women would prove less susceptible because of higher standards of personal cleanliness and more frequent laundering of garments. There appeared to be some obvious connection with the fact that the WAC service uniform had been the heavy herringbone twill, well known to cause skin irritation to women even in the United States. In fact, in the India–Burma theater the chief surgeon had refused to let Wacs wear this garment at any time, pronouncing it a major health hazard in a tropical climate. On the other hand, some medical officers noted that even had the Wacs possessed summer khaki like the men's, women might possibly still have had a higher dermatitis rate because of their more sensitive skins.

In addition, it appeared possible that the rate of loss for malaria had increased wherever Wacs substituted their cotton shirts for the coverall top, since the shirts were lighter in weight than the men's and not believed mosquitoproof by The Quartermaster General. On the other hand, as both General Denit and Colonel Brown pointed out, experimentation had never conclusively indicated what weight of shirt was really mosquitoproof, and Navy nurses in the area wore skirts instead of slacks, so that the Wacs' uniforms might possibly have offered as good protection as any other type of garment.

Other minor possibilities were also noted. It appeared that the rate of respiratory disorders might have been increased by lack of protective footwear. The shortage of all garments had apparently increased illness during rainy seasons when not enough dry coveralls were available to permit a change. There was also some evidence that the regular WAC wool socks, heavy rayon underwear, low oxfords, and other issue items were not particularly suitable to the area.

Up to this time, all attempts at remedial action had merely demonstrated that, if women were launched into a campaign improperly equipped, no "business as usual" would produce the requisite supplies. Theater action to secure mosquitoproof slacks for women, and all other needed items, was begun in 1944 in the month of the Wacs' arrival in Australia.[120] The months of May and June were consumed in Quartermaster staff studies as to the proper amendments that should be recommended in theater Tables of Equipment. The matter was not as simple as it appeared to nonexperts in supply matters, since there were many items of WAC uniform, undergarments, and footgear to be considered, as well as the opinions of medical and technical specialists on the proper weight and design of slacks and shirts. On 6 July 1944, a large-scale conference on WAC supply was held by representatives of USAFFE and USASOS with visitors from Operations Division of the War Department General Staff. This produced no immediate results, and later in July the assistant staff director appealed by personal letter to Colonel Hobby to expedite matters by securing immediate War Department authorization for ordinary khaki slacks. In reply she was informed that The

[119] Interv with Capt Letallier, 9 Jan 46. Entire following discussion is based on interviews and statistics already cited.

[120] Except as otherwise documented, the supply discussion is from the monthly WAC Sec USAFFE Hist Rpts.

Quartermaster General was already developing specifications for a slacks suit, but could not expend funds to procure or ship any item without requisitions and shipping priorities from the theater. It was not until after the movement of most of the Wacs from Australia had begun in August that the theater requisition for slacks went forward.[121]

Among women moving out to Hollandia and other bases, only the 6 percent working for FEAF were reported as equipped as well as men, since General Kenney had refused to let them board planes until they had been issued men's khaki trousers from the ample stocks of men's clothing in Australia.[122] Issue of men's clothing was considered by some other commanders, but generally rejected because the garments were ill-fitting and required "unsightly" alterations; it was decided to await arrival of women's slacks from the United States.

However, following the theater's requisition in August, there ensued another two months' delay for queries and replies concerning specifications. On 14 October 1944, the War Department gave its final approval to specifications for women's slacks, and informed the theater that the slacks would be available in only ninety days—one fourth the usual procurement time for men's clothing. This schedule was very nearly met, and in only four more months the first shipment of slacks and proper-weight shirts was on the pier in California. Not only was this still a considerable distance from New Guinea, but it was by this time February of 1945, and the Wacs were moving on to Manila.

It was confidently expected that the slacks would reach the women in Manila; some 75,000 pairs were promptly shipped from the United States, 5,000 of them by air. However, upon reaching the theater, they were off-loaded at way stations and never reached the Wacs. When queried by the War Department, the theater explained that it lacked transportation to lift the newly received slacks from various depots, so that assembly in the Manila area was never accomplished.[123]

A similar difficulty accounted for the lack of normal maintenance stocks of standard items such as field shoes, coveralls, and undergarments. Maintenance stocks had been routinely authorized for shipment even before the first Wacs left the United States for Australia, and numerous shipments had been made, few of which reached the Wacs.[124] The difficulty was the same as that already discovered by the Fifth Army in Italy—that the supplies for a tiny group became lost or misplaced when handled in the routine supply channels for men's clothing. In this case the difficulty was a hundred times intensified by the distances and shipping problems in the New Guinea area, which had no highways or railroads and few good docks.

The problem was illustrated by six WAC clothing shipments that were eventually traced.[125] Two of these shipments reached the areas before the Wacs, were put on sale in officers' stores, and were "badly depleted" by unidentified custom-

[121] (1) Personal ltr, Capt Juanita Stryker, Asst WAC Stf Dir SWPA, to Dir Hobby, 12 Jul 44, and reply from Capt Patricia Lee Chance, Exec WAC, 16 Aug 44. WDWAC 320.2 SWPA. (2) WAC Draft AFPAC Reply, Tab 12 (omitted from theater-approved version).

[122] According to his memoirs, *General Kenney Reports* (New York, 1949).

[123] AFPAC Reply.

[124] Ltr, TAG to COMINCH SWPA, CG SFPOE, and CofT, 21 Apr 44. AG 400 (4-18-44) OB-S-E-SPMOT; Tab 8 of WAC Draft AFPAC Reply.

[125] Discussion from theater-approved version, AFPAC Reply. Six examples are in WAC Draft, deleted by theater.

ers before the Wacs arrived. Two others went to a base where Wacs were never stationed, and remained there. General Campbell noted, "You don't run a Liberty ship up and down the coast to pick up a few WAC shoes." [126] One went to a station where about one half the theater's Wacs were then located, and was well utilized. The remaining one also reached a WAC detachment, although months after its arrival. General Campbell added, "Supply for a million men spread over 4,000 miles was difficult, but for 5,000 Wacs in small scattered groups it was desperate." [127]

Even if supplies were received by the women, it was difficult to keep them, in view of the usual system of movement of Wacs by air. This explained part of the situation of the GHQ detachment on Leyte. As early as September of 1944, Colonel Baird had unsuccessfully recommended to USAFFE's deputy chief of staff that, when a WAC unit moved by air, its equipment accompany it, since "Experience has shown that otherwise it is a long time before such items become available." [128] This was not considered practicable, but upon departure from Hollandia the WAC supply officer had personally watched the crating and loading on a ship of whatever clothing and equipment could not be carried by air. When this did not arrive in Leyte, investigation disclosed that the vessel was at another base, towing two barges, and could not be turned around for a few crates of WAC clothing.

Similarly, when the censorship unit moved by air from Port Moresby to Biak, some forty-eight tons of equipment and personal baggage had to be left for shipment by water. Two months later, when this still had not arrived, the detachment commander protested that it was "critically needed . . . especially clothing of personnel," but the regulating officer informed her that there was no prospect of any ship calling at Port Moresby to pick it up.[129]

As for nonissue items such as girdles, brassières, and pajamas, purchase of these was not possible in Australia because of rationing restrictions. Wacs got 25 clothing coupons a year, but the purchase of a girdle required 10 coupons, pajamas 12, and a brassière 5. In New Guinea and the Philippines these were unprocurable. Colonel Brown recommended to USASOS that brassières be added to allowances:

> These items are absolutely necessary and are peculiar to the needs of women. Enlisted men are furnished all necessary items; therefore enlisted women should not be expected to buy items which are necessary for wear. Women are continually requesting families and friends to mail these articles, but with the uncertainty of arrival by mail, they are often without these necessary articles.

The theater therefore four times submitted a request for the issue of such items, but was four times refused; it never achieved the solution, found successful in other overseas theaters, of procuring the items for sale in post exchanges. In fact, post exchange supplies of any sort for women were extremely scarce. Several large groups had never to this date been able to purchase any sanitary napkins—these also being considered "nonessential" items which were not issued to women, but in other theaters usually purchased at post exchanges.[130]

In the final analysis, most observers concluded that the factor of supply had been a major one in the WAC loss rate,

[126] Interv, 31 Jul 47.

[127] Unless otherwise documented, entire discussion above is from: (1) Monthly WAC Sec USAFFE Hist Rpts. (2) Tp Conv, 2 Feb 51.

[128] Check Sheet, 30 Sep 44. Folder, 5205th WAC Det, Drawer X-29853.

[129] Folder, 5203d WAC Det, Drawer X-28040.

[130] AFPAC Check Sheet, WAC Sec to G-3, 19 Apr 45. In possession of Mary-Agnes Brown.

particularly in the rate of loss for dermatitis, malaria, and respiratory disorders. Had it been possible to retain the women in Australia until they were fully equipped for field conditions, as had been possible in England, the educated guess was that the rate in excess of men's losses might have been lowered.

Morale

There was every indication that the women's morale had held up excellently at most stations. All observers agreed that, as in Italy, the one factor which had done most to sustain morale was the nearness to actual combat. Units in the most forward areas were consistently found to have the highest morale. The race to get forward had become a matter of competition, and WAC officers reported that the panacea for supply shortages and primitive living conditions consisted in "excitement and pride in being the first Wacs."[131] A typical group upon disbandment wrote sentimentally of the Leyte experiences—tents, candles, no supply, mud, foxholes, air alerts, rain—and added, with no apparent sense of incongruity, "We'll never have a detachment so ideal and so perfect again."[132]

Even when Wacs were not actually near the combat area, the constant presence of combat-wounded and released prisoners gave a sense of immediacy to their efforts. Wherever possible—as for example at the 51st General Hospital—women were allowed to visit hospital wards. They saved their beer and cigarette rations, cookies from home, and flowers grown around their tents to share with the patients, with resulting good effect not only on the wounded, but on the women themselves.[133]

Units everywhere had adopted monkeys, dogs, chickens; smuggled a duck on shipboard; cherished unidentifiable native animals. Songs were invented for all subjects ("If you meet a looie with big blue eyes, you can look but you can't fraternize".)[134] The women took pride in improvising: a Christmas tree was made of native thatch tied to a bamboo trunk, decorated with scrap metal from enemy planes. When most Christmas packages failed to get through, inspectors noted that "they opened the few that came and shared them like sisters," while one supply sergeant filled their shoes with "gifts" of pistol belts and towels. Desiring a green-and-gold WAC guidon for the funeral of a member of the FEASC WAC detachment, the women improvised one from parachute lining, dyed with atabrine and green ink.[135]

By comparison with problems of morale among noncombat men, the Wacs' good morale appeared even more striking, for women seldom complained about, or were depressed by, factors that were high on the men's list—mail service, rotation, officers' superior privileges, and the lack of good movies and other entertainment. The only common complaint factor was that of food. Particularly noteworthy was the mildness of WAC complaints about their company officers' deficiencies as compared to the complaints by noncombat men about their officers.[136]

---

[131] AFPAC Reply, both versions. See also Hist Rpts. Folder, 5225th WAC Det, Drawer X-28368.

[132] Correspondence folder, 5205th WAC Det, Drawer X-29853.

[133] Hist Rpts, Jan, Feb 45. Folder II, WAC Det FEASC.

[134] Interv with Lt Mary B. Warner, 3 Oct 47.

[135] (1) Folder I, WAC Det FEAF, Drawer X-25096. (2) Hist Rpts, Jan, Feb 45. Folder II, WAC Det FEASC. (3) Quotation in Folder, 5225th WAC Det, Drawer X-28368.

[136] For typical noncombat male complaints, see History of South Pacific Base Command by Hist Sec, G-2, Hq SOPACBACOM. OCMH.

However, there appeared to be a possibility that the generally high morale had accounted for the moderate loss rate of the first year. Company commanders voiced the opinion that there was a general "end of the line" feeling in Manila, at which time women who had actually been ailing for months suddenly gave up to their illness. Medical officers noted that, with women, illness was more likely to mark the breaking point in tension, rather than the spree of some sort in which men might seek escape from prolonged strain.

In addition, there was some evidence that the factor of social pressure, which had helped to sustain morale, had simultaneously damaged health. At all stations not under actual attack, there continued to be more unit dances and approved group parties than the women had energy to attend. The enlisted Wacs were the chief source of American dates for the enlisted men, who vied for invitations to the WAC recreation halls and in providing unit entertainments approved for WAC attendance, with bands, decorations, refreshments, and other inducements. WAC bulletin boards in Manila were filled with bright-colored posters carrying mass invitations. Some listed the menu, promising ice cream or some other rare delicacy, for the Wacs soon became blasé about masculine company and refused to stir without the guarantee of food.

On one occasion fraternal bloodshed was narrowly averted when one unit's trucks appeared, flying false colors, and carried off the Wacs destined for another unit's dance. No Wac was so old or unattractive as to be neglected, although commanders feared that this would some day cause a painful readjustment to the facts of life in the United States. While such attention was of great assistance to morale, the constant social pressure increased the ever-present problem of fatigue. Group or mass entertainment had always proved tiring to WAC units, but in most Pacific bases there was often no alternative recreation. Colonel Brown early recommended that, at well-established camps, service clubs for joint use be provided, but this was seldom considered practicable.[137]

Length of Working Day

In accordance with tropical custom, working hours generally began at seven or eight in the morning and ended at nine or ten at night, in order to permit time off in the heat of the day for leisurely meals. Thus, at Intersec headquarters the working day covered fifteen hours, from seven in the morning to ten at night, with three-hour breaks for meals.

While the relative merits of the system on a six-day week were not primarily of WAC concern, since all personnel was affected, inspectors noted adverse results in certain organizations staffed almost completely by Wacs and which, because of the pressure of work, were obliged to work these hours for seven days a week. The all-Wac Central Mail Directory was reported as working such hours for seven days a week, with no day off for five months; the same was true of the all-Wac Casualty Section and the Machine Records Unit.[138] In some cases it appeared that women did not avail themselves of the full time off for meals when work was heavy or an operation being launched. An FEASC management control survey noted, "In general the enlisted women were working longer

[137] Memo and Ltr cited n. 53.

[138] (1) Interv cited n. 119. (2) AFPAC Reply, both versions, except that official version omits names of specific offices having no time off.

hours than enlisted men performing similar skilled duties." [139] Also, women's time-consuming chores such as laundry were usually not done during the mealtime breaks but after hours, by their own preference.[140]

Inspectors were unable to determine whether adverse effects were due to lesser endurance than men in the same offices, or to the women's tendency to put in overtime. In any case, even without overtime, the hours of work were longer than those used successfully with civilian women in either the United States or Britain.

Among reported instances, the greatest damage to health occurred among the Hollandia Wacs in General McMullen's Far East Air Service Command. These stood reveille every morning at 0530, walked back and forth some distance to work three times a day in all weather, worked an official minimum of ten hours a day, until ten o'clock at night, seven days a week, with only Sunday afternoons off, these being reportedly spent standing in line to get to laundry facilities. After less than a year of this schedule, the medical loss rate in the FEASC WAC detachment jumped alarmingly, from 3 per month in January of 1945 to 30 per month in June.

The Air Surgeon reported that "an increasing number of cases are on record for nervousness and exhaustion," and recommended that personnel be given one full day off per week to relieve "nervous tension." The June rate, if continued, would total more than 62 percent medical evacuation per year, an unprecedented loss even for combat troops. The staff director for FEAF supported the Air Surgeon's recommendation, noting that offices were not really busy; it was believed that the day off could profitably be applied to all personnel to avoid discrimination in favor of women. However, General McMullen rejected the request for Sunday mornings off, and informed the surgeon, "Medical records do not indicate any situation . . . to cause me serious concern." [141]

Discipline and Morals

In the minds of some observers a question arose as to whether the medical loss rate could have been affected by the type of moral disintegration popularly supposed to afflict men and women in the tropics. However, if such disintegration took place, it was more subtle than could be measured statistically, and did not consist of the gross sexual immorality about which most rumors centered. On the contrary, the pregnancy rate for the Wacs' chief period of service in the Pacific was, like that of other overseas theaters, less than one half the world-wide WAC rate, which was itself low by civilian standards. During the period after the arrival in Manila, with the relaxation of previous restrictions on marriage, the WAC pregnancy rate rose considerably, but was still less than the world-wide rate, and about the same as that for nurses in the Pacific. Of the theater's 5,500 Wacs, only 111 were reported returned home pregnant, with an estimated half of these being married women. However, in spite of the frequent restriction to compounds, the Pacific Wacs' rate was considerably higher than that of the European theater, which placed no special restrictions on women.[142]

[139] Hist Rpt, May 45. Folder II, WAC Det FEASC.

[140] Interv with Maj Charlee L. Kelly, 28 Jul 47.

[141] Memo, WAC Stf Dir FEAF for CG FEAF through CG FEASC, 19 Jun 45, with Rpt of Surg and Ind, CG FEASC to CG FEAF, 22 Jul 45. AG 333.1. Correspondence folder, WAC Det FEAF, Drawer X-25096. See also Ltr cited n. 86.

[142] For world-wide rates, see Ch. XXXI, below. For SWPA rates, see Craighill SWPA Rpt.

Venereal disease rates were likewise so low as to be negligible. One commander of a unit of over 700 women stated that she did not, in the entire time, have any new cases.[143]

Further evidence that no gross moral breakdown occurred was offered by the fact that the Wacs were gradually able to convince the enlisted men that their mission in the area was a military one. Statistics on the trend of enlisted men's comments showed that a remarkable decline in the number of derogatory comments took place as soon as the men actually met the Wacs. From a high of 90 percent unfavorable comments in the month before the Wacs arrived, the number fell slowly until, in the month in which demobilization began, only 28 percent of comments were still unfavorable, many of these being from units that had never seen Wacs.[144]

WAC detachment commanders frequently expressed an uneasy belief that the discipline or perhaps merely the *esprit* of their detachments was not as high as that which they had known outside the theater. Some felt that the women's habitual evasion of petty restrictions had undermined respect for more important rules and for those in authority. Others noted a lessened sense of conventionality, decorum, or discreet behavior; the WAC inspector general added, "Dress them like tramps and they may act like tramps." [145] However, detachment disciplinary records seldom revealed other than exceedingly minor offenses attributed chiefly to tension, exhaustion, or loss of respect for authority. One detachment's worst case was a woman who was AWOL from midnight until three in the morning. Another detachment's worst was a woman who, after fifteen months on a particularly exhausting job, was "on the verge of a complete physical and mental breakdown," went AWOL, and "spent her entire absence in a native hut sleeping." [146]

A number of minor offenses concerned violation of the ban on officer-enlisted dating, although the theater experienced fewer problems in this respect than the European, since most Wacs were already restricted to approved group entertainment. Restrictions on officer-enlisted association were frequently severe; at some USASOS bases in New Guinea, even husbands and blood relatives were for a time not permitted to see Wacs of a different status. WAC staff officers stated, "It is believed that this has resulted in resentment of persons concerned and in flagrant violations." [147]

The prohibition was originally adopted only over the nonconcurrence of the staff director, in line with a previous ruling that nurses would not be allowed to date corpsmen. WAC company officers noted that they were obliged to punish their enlisted women while officer dates were not punished; officers sometimes removed their insignia to attend unit dances, and at other times flagrantly violated the rule. Combat units in the area merely tore up such notices; while in one case the officer concerned was the head of the local military police who were charged with enforcement. Air Forces and USAFFE headquarters ordinarily enforced nonassociation rules less strictly than USASOS, thus fur-

[143] Craighill SWPA Rpt.

[144] (1) Tab 13, WAC Draft AFPAC Reply. Also Surv, Comments on Women's Servs, 16 Jun 45, GHQ USAFPAC, copy in possession of former WAC Stf Dir. (2) Morale Evaluation Rpt, Attitude Towards Women's Servs, Jul 45, Hq AFPAC, incl in Ltr, Actg Stf Dir AFPAC, to Dir WAC, 11 Aug 45. WAC Cl file SWPA.

[145] Interv cited n. 119.

[146] Folders, 5200th, 5203d, 5205th, and 5215th WAC Dets. Quotation from 5215th WAC Det, 21 Aug-20 Sep 45.

[147] WAC Draft AFPAC Reply.

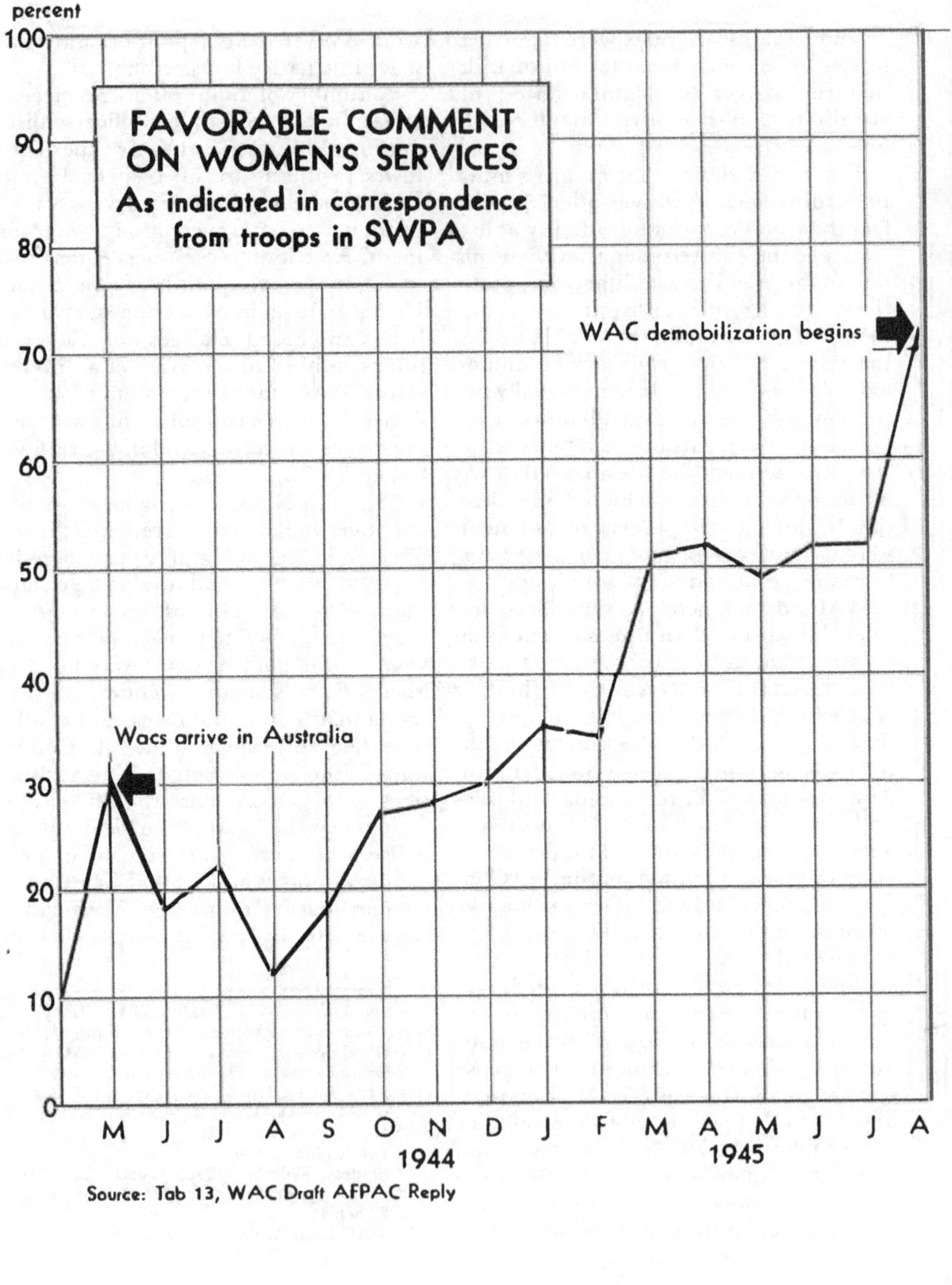

Source: Tab 13, WAC Draft AFPAC Reply

ther complicating the problem of unequal punishment.

After Wacs reached Manila, the problem began to assume added importance, and the WAC staff director again requested permission for the social association of "relatives and friends of long standing" carrying passes from their WAC commanding officers. The theater again refused to set an over-all policy. WAC company officers, when asked to make suggestions for the improvement of discipline, included "a definite policy for officer-enlisted relationships"; "fewer rules and more rigid enforcement of same." [148]

An additional restriction, more severe than any in the European or Mediterranean theaters, was FEASC's customary prohibition against marriage of its members; as applied to women, it allowed them to marry only if they became pregnant. In this command, women engaged to men about to leave for combat often quite frankly consulted the WAC staff director and the chaplain to ask why they should not take that course. To prevent this, the staff director repeatedly attempted to secure more lenient terms for both men and women of American citizenship; this was eventually granted by General McMullen only after the headquarters reached Manila.[149] Only 121 of the theater's 5,500 Wacs married in the theater during the war.[150]

Restrictions on Daily Life

In the opinion of many observers, one of the most serious handicaps to the health of women in the area had been the restriction to locked compounds in the majority of stations in New Guinea, a factor which did not affect male troops. It was noted that, among medical causes of evacuation for women, psychoses and neuropsychiatric disorders were second only to dermatitis. It was always to remain a matter for debate whether the unusual restrictions placed upon women were necessary and inherent in the dangers of the Pacific area, or were an unnecessary creation of arbitrary theater policies. Colonel Brown continued throughout the stay in New Guinea to recommend the abolition of the system, but few modifications were ever reported with the exception of the GHQ unit in Hollandia, the smaller Air Forces group, and minor adjustments elsewhere.[151]

Three resulting unfavorable factors were noted. The most obvious was the noise, crowding, and lack of privacy which, in the opinion of medical officers, fostered mental disorders. Even if the area was not noisy, women experienced a "sense of confinement" which made them unable to relax after strenuous office hours or tiring mass entertainment. Major Craighill reported that women appeared to value privacy more than men, and that even among nurses, who as officers were less crowded in tents than enlisted women, the rate of loss for psychiatric reasons was increased by lack of privacy.

More important, in medical officers' opinions, were the physical disorders fostered by simple lack of physical exercise, recreation, and rest. Major Craighill wrote, "Facilities for exercise are lacking in most places. . . . In the tropics, more

[148] (1) Marginal note by Colonel Brown on draft of this chapter. OCMH. (2) Fraternization in FEAF was forbidden by Stf Memo 15, 30 Jul 44. Folder, WAC Det FEAF I. (3) Interv with former CO of a USASOS WAC Det, 22 Jul 47. (4) Check Sheet, Hq USAFFE, WAC Stf Dir, to DCofS, 23 Jul 45, Tab 15 to WAC Draft AFPAC Reply. (5) Quotation from Correspondence folder, 5200th WAC Det, Drawer X-29443.

[149] Memo cited n. 141.

[150] Griffith Account, "Statistics May 44–Sep 45."

[151] Sources cited n. 51.

than elsewhere, proper exercise is important." According to one WAC adviser,

> The greatest deficiency in recreational facilities since the Wacs left Australia has been that at no time has there been a place to which they could go for a three-day pass and get away from the mental strain of group living.[152]

The Pacific area obviously offered few advantages in this respect. Leave areas for men existed in Australia, operated by the Red Cross, but in spite of Colonel Brown's repeated recommendations, the handful of Wacs remaining in Australia were refused permission to use them until near the end of the war, when the Red Cross replaced one of its workers who had blocked admission of women.[153] The Air Transport Command also operated a regular "leave shuttle" to Australia, carrying men from New Guinea on 3-day passes. Although this shuttle was organized in Hollandia by the date of the Wacs' arrival, and although ATC historians noted that it often operated half empty, it appeared to be seldom employed by most headquarters, and in any case Wacs were not given a quota.[154]

Army nurses were originally allowed leaves from forward areas every six months, and were transported by special planes to leave areas,[155] but as the headquarters advanced, this privilege was also considerably restricted, and it never applied to Wacs. During the entire stay in New Guinea, no leaves or passes were reported as given to enlisted women at most installations, because of the absence of any approved place to spend them. Major Craighill noted that, for both Wacs and nurses:

> Recent restrictions against nurses' leave to Australia completely deprive personnel of all rest areas. None are provided in the islands for women for even short periods. Such facilities are especially needed because of the unusual restrictions on women's activities and the lack of outside places for recreation.

In those areas where men also had no rest areas or were not allowed to use the leave shuttle, the situation appeared unavoidable, although some WAC advisers questioned whether a local commander's action in restricting women to compounds did not obligate him to find some means of equalizing the resulting mental tension and physical sluggishness.

Also increased, in company officers' opinion, was the women's resentment, disobedience, and immature conduct. As the women were well aware, "restriction to quarters" was ordinarily employed by the Army only for punishment. When restriction to quarters was imposed by court-martial, Army Regulations contained the caution that it "will not be imposed in excess of three months." [156] Even though the Wacs' custody was protective and not punitive, it was difficult to find a previous occasion in Army history in which a soldier had been kept in protective custody for a year or eighteen months. If the system was indeed required, as theater authorities insisted, to prevent rape of Wacs by Negro troops in New Guinea, the War Department was not in a position to protest command policies. However, to WAC leaders in the United States, viewing the problem from a distance, it appeared that, if an area was so dangerous as to require the long-term confinement of soldiers of one sex, such soldiers should not be considered for employment in that area.

[152] AFPAC Reply, both versions.

[153] Ltr cited n. 46.

[154] (1) Interv with Dr. Samuel Milner, author of Victory in Papua, a forthcoming volume in this series. (2) The ATC in the Pacific, 7 Dec 41–1 Aug 44, p. 148. USAF Hist Div.

[155] Annual Rpt, 1944, Chief Surg, USASOS, p. 35. SGO R-3232 (1-7-47).

[156] A Manual for Courts-Martial, U.S. Army, pp. 94, 104.

In her final report to the War Department, Major Craighill stated that in her opinion the rate of WAC loss was definitely increased by the lack of healthful recreation, the undue restriction in compounds, and the failure to treat women as responsible adults. This opinion coincided closely with that reached by The Surgeon General's Office in Washington concerning the rate of loss for men in the area:

> The health of troops is directly affected by command policies and procedures governing rotation, discipline, recreation, and the provision of incentive and motivation. It seems probable that these are influential factors.[157]

Inferior WAC Commanders

Inspectors' reports indicated that in certain detachments the sick rate had possibly been aggravated by inferior WAC command practices. It was known that some company officers with low efficiency ratings had been unloaded on early shipments by commands in the United States. However, these were not unduly numerous, and a number of excellent company officers were also in the theater. Unfortunately, by a combination of circumstances, quite often the inferior officers were placed initially in company work from which staff directors were later unable to get them removed, while trained company officers were assigned to office work.

Thus, the first shipment of Wacs had contained a number of trained company officers, intended to staff future companies. Instead, since USAFFE headquarters desired to set up its censorship detachment immediately, these officers were included, with the promise of reassignment to company work when more enlisted women arrived. However, in spite of the efforts of the staff director, their later release was never concurred in by the central directory.[158]

In several cases, officers were assigned to company work who had already failed at it in the United States. Possibly the worst example cited in inspection reports was that of the FEAF WAC detachment. Soon after its establishment, the USAFFE inspector general discovered improper assignments, low morale, irregularities in detachment records, undue delay of mail, partiality and lack of interest in enlisted women's needs, and profane language on the part of the WAC commander, as well as threats to enlisted women in detachment meetings.[159] Unsatisfactory conditions were attributed by the inspector general chiefly to the WAC unit commander, who had been sent overseas by the Fourth Service Command after being relieved of two different WAC commands in the United States, one with an efficiency rating of Unsatisfactory. The situation also reflected the initial lack of a staff director. Because of the unit's unexpectedly sudden beginnings with borrowed personnel, the FEAF WAC staff director, Maj. Mary L. Kersey, did not arrive from the United States until the unit was already in Hollandia, and was then unable to correct the already-established situation.

At this time Brig. Gen. Harry H. Baird informed the USAFFE chief of staff:

> A recent inspection of the FEAF WAC Detachment by the Inspector General of this headquarters makes it clearly evident that adequate supervision of the WAC Company Commander and administration of the WAC Detachment has not been exercised. . . . It

[157] Memo, Dir Contl Div SGO for Gen Kirk, 12 Jul 45, with Tab A from Chief Consultant in Medicine, Tab B from Chief Consultant in Surgery, and Tab C from Dir Neuropsychiatry Consultants Div. SGO Classif MEDC-701.

[158] Boyce Rpt.

[159] WAC Draft AFPAC Reply only; certified copy of USAFFE Inspection Rpt atchd.

is apparent that full use had not been made either of Major Kersey's abilities or her prescribed duties as WAC Staff Director. It is the opinion of G–1 that Headquarters FEAF should either be required to conform to War Department and Theater policy or that the Women's Army Corps should be withdrawn from that command.[160]

As a result, in January of 1945 FEAF headquarters belatedly required its various sections to co-ordinate with the staff director when setting WAC policy. Although the inspector general also recommended removal of the WAC detachment commander, General Kenney refused to effect a replacement, since he had personally selected the officer. Instead, she remained with the detachment until its inactivation, was promoted to the rank of major, and received the Bronze Star.[161]

In another case, the WAC commander of a USASOS detachment had been removed from a "large" command of 500 in the United States in the belief that she would be "all right in a small command overseas." Instead, she was assigned to command almost 1,000 women at one of the most difficult stations. As in the previous case, difficulties developed and the inspector general recommended her removal, but she was retained and promoted.[162] In one instance, in which a commander was sent direct from a WAC training center, she was found to know little of field life, and inspectors discovered her women to be near collapse from keeping the floors polished, barracks shining, shoes and clothing in basic-training spotlessness, while working long hours in hot weather.[163] In still another case, an officer without previous company experience was assigned as commander of a large detachment, over Colonel Brown's protests. Inspectors noted, "Many problems, which would have been routine for an experienced company commander, arose and had an adverse effect upon the detachment." [164]

Not all WAC units were so afflicted; in many cases WAC commanders were outstanding. Colonel Boyce, on a visit to the area, particularly cited the commanders of the FEASC detachment, which, after the end of the war, she found still suffering from illness and a long working day. Colonel Boyce found the unit living in tin barracks in an area without shade or trees, very muddy or very dusty by turns, with the women drenched by torrential rains as they attempted to get to mess or to the outside latrines. Dead Japanese were buried so close to the surface that women had to scatter lime around the mess hall daily to control the stench. Nevertheless, since the unit's establishment it had had WAC commanders who were sincerely interested in the women's welfare, and Colonel Boyce reported the unit's state of morale and discipline at that time far better than that of more healthfully situated units.[165]

However, even where good company officers were assigned, they were frequently found under a strain caused by overwork and shortage of assistants. The FEASC WAC detachment had 714 en-

[160] Check Sheet, G–1 USAFFE to CofS USAFFE through WAC Dir, 29 Dec 44. Folder, WAC Det FEAF.

[161] (1) FEAF Stf Memo 4, 15 Jan 45. (2) WAC Draft AFPAC Reply only. Records of FEAF Detachment are incomplete; files noted destroyed by authority of company officers included: WAC Disciplinary Problems, Questionnaire Rpt to Maj Kersey, Leyte to Manila Move, Check Sheet re AAF Policy, Conditions of Latrine, Unauthorized Departure, Policy of Beach and Facilities, etc. Remaining are only Commendations, PRO file, vehicular reports, etc. (Envs 1 and 2, Folder, Correspondence, General, Drawer X–25096. (3) Interv with Lt Col Mary L. Kersey, 23 Jul 47.

[162] Insp Rpt, atchd to WAC Draft AFPAC Reply; Boyce Rpt.

[163] Interv cited n. 119.

[164] Boyce Rpt; Rpt cited n. 162.

[165] Boyce Rpt.

listed women and only three officers, the number commonly assigned to a unit one fourth that size in the United States. Inspectors recommended more, since work was heavy, but more were not assigned, although ten other WAC officers at the same base were assigned operational duties. The GHQ detachment for a time had only one officer for 386 women; eventually one more was added. The huge AFWESPAC battalion of some 3,000 women never had more than seven officers, even when it was scattered over three widely separated stations with the commanding officer alone in Manila attempting to prepare for reception of the women. In all, only 15 percent of the WAC officers in the Pacific were assigned to WAC administration, as against 32 percent in the European theater. There was only one troop or staff officer for approximately 100 women, as against the one per 20 believed suitable in the United States.[166]

Command of Enlisted Women by Male Officers

In several detachments, although WAC officers of proven ability were in charge, they were considered to have no powers other than those of providing food and housing, command and discipline being the prerogative of male commanders to whose units or offices the women were assigned. The situation was parallel to that of enlisted men assigned to one command and merely attached for quarters and rations to another. Even the ordinary Army unit, formally activated and with its own grades, encountered some friction between commander and section chief over such matters as promotions. Most WAC units in the area, with a few exceptions such as those in the Air Forces, were even more vulnerable, being gradeless and without "assigned" personnel other than cadre.

Noting that the command of enlisted women by male officers was in her opinion contrary to War Department regulations, Colonel Brown soon after her arrival recommended publication of a clarification such as that used in the European theater and in the Army Air Forces in the United States, to the effect that the WAC commander had full disciplinary powers even though her unit was gradeless or not activated. When this was refused, she attempted to send a letter through channels to the War Department, asking clarification of its ruling under the circumstances, and stating that she considered the current system "dangerous" in view of certain factors involved in the command of enlisted women by male officers. This letter was stopped by USAFFE on the grounds "that it did not pertain to the health, well-being, or morale of women," and Colonel Brown was informed by Maj. Gen. Charles P. Stivers, "Matters of assignment of Wacs, determination of this headquarters."[167]

Discipline in the GHQ USAFFE WAC detachment eventually reached a crisis. The WAC commanding officer, although responsible for almost 400 enlisted women, had no "assigned" personnel except one assistant, and had in the detachment not only Wacs working for GHQ but personnel assigned to nine other separate detachments as well as to Air Evaluation Board and Philippine Base Section.[168] After women guilty of disciplinary offenses in

[166] (1) Memo, WAC Stf Dir FEAF for CG FEAF, sub: Inspection of WAC Det FEASC, 4–6 Feb 45. WDWAC 319.1 Pacific. (2) Folder, 5200th WAC Bn, cx-447. (3) AFPAC Reply. Griffith Account gives 5 percent of officers on WAC administration, 95 percent elsewhere.

[167] (1) WAC Sec USAFFE Hist Rpts, particularly Dec 44 and Jan 45. (2) Check Sheets in same folder, particularly Check Note to CofS USAFFE, 2 Feb 45, re Stivers decision. (3) Colonel Brown's marginal note on draft of chapter. OCMH.

[168] Folder, 5200th WAC Bn, cx-447.

the detachment were promoted by their male commanders, one only a few days after conviction by court-martial, General Baird informed General Stivers:

> It has come to our attention that there have been a number of instances where enlisted women have taken up matters concerning WAC detachment internal administration with section chiefs. In several cases, staff officers have suggested that enlisted women follow a course of action directly in opposition to the orders of their company commander. It is obvious that a complete breakdown of internal company command will result if the above situation is allowed to continue within this headquarters.[169]

USAFFE headquarters therefore published a memorandum, for its own detachment only, stating:

> Enlisted women seeking advice or decision as to internal detachment administration will be referred to the Commanding Officer, WAC Detachment, by the section chiefs concerned.[170]

This directive, published without co-ordination with the staff director, still was not effective, and squabbles ensued, such as one in which the USAFFE board informed a WAC commander:

> You, as commander of the detachment serving this, as well as other units, in an administrative capacity, have no discretion in the matter [of WAC discipline in an off-duty offense.] [171]

In any case, the ruling was applicable to the USAFFE unit only, and did not solve the problem elsewhere. WAC company commanders, asked to comment on the causes of delinquency, noted, "Most of our troubles arose from a divided command which created and fostered a disregard for authority," and "Many Army Section Chiefs tend to encourage infraction of rules and try to protect the enlisted women working for them." [172]

Major Kersey, the FEAF staff director, noted later:

> The Air Force WAC Detachments, FEAF, and FEASC, had all personnel assigned to the WAC Squadrons, and consequently the WAC commanders did not encounter the same problem of disciplinary control.[173]

Some WAC commanders of proven ability eventually asked to be removed from company work because of their ambiguous status. One capable officer quit after eight months because, "I could not do anything about the things that were being done and I could not stay there and see what was happening to [the women]." Another WAC commander developed a mental condition that required a disability discharge. Still another, of previously irreproachable character, had to be relieved of command after she formed the habit of drinking too much and spending the evenings on the beaches in a state of blissful unconsciousness.[174]

It eventually came to be Colonel Brown's opinion that no mere publication concerning disciplinary authority would correct the situation, and that the enlisted women should be formally assigned to WAC units, the device adopted in the Mediterranean theater and in the Army Ground Forces. In spite of the cumbersome bookkeeping involved, such a command system was considered more nearly foolproof.

---

[169] D/F, G-1 USAFFE to DCofS, 26 Mar 45, atchd to WAC Draft AFPAC Reply.

[170] Stf Memo 31, Hq USAFFE, sec III. Atchd to WAC Draft AFPAC Reply.

[171] Memo, Det Comdr, USAFFE Bd, to CO 5200th WAC Det, 1 May 45, sub: Restriction of WAC Pers, with 4 inds. Atchd to WAC Draft AFPAC Reply.

[172] Folders, 5200th WAC Det, Drawer X-29443, and 5205th WAC Det, Drawer X-29853.

[173] Statement on draft of chapter, incl to Ltr, Col Kersey, WAF Stf Dir, Hq Strat Air Comd, to Chief, Mil Hist, 21 Apr 51. OCMH.

[174] Interviews listed in Bibliographical Note.

Transfer of Responsibility to WAC Section

In attempting to solve the various WAC problems, Colonel Brown's office came to the opinion that "in all headquarters, full advantage was not taken of the experience and knowledge of WAC staff officers." [175] Her own office was, in her opinion, handicapped not only by lack of formal responsibility but also by its inability to secure transfer to the advanced echelon where were located higher commanders and many of the enlisted women. Shortly before the move to Manila Colonel Brown informed a critic:

> Until it is possible for me to talk to General Richard Marshall [Deputy Chief of Staff] personally, we shall have to do the best we can. I have every confidence that the General will make it possible for me to carry out the functions of a Staff Director as listed in the War Department Circular 462.[176]

In the spring of 1945, the WAC Section finally overtook the advanced echelon in Manila, and Colonel Brown was able to confer with General Richard Marshall. She noted later, "General Marshall was more than willing to seek and act upon the Staff Director's recommendations." [177] However, no written directive from General Marshall proved necessary, for shortly thereafter G-1 Division of USAFFE recommended of its own accord that "all policy and supervision of the WAC corps in this theater be immediately taken over by the WAC Section." It was stated that G-1 Division had felt it necessary to retain control during the experimental stages of the WAC program, which now, being safely established, could be turned over to WAC officers.[178]

This transfer was only a part of the general reorganization going forward at the time, during which GHQ and USAFFE were consolidated to form the new Headquarters, Army Forces in the Pacific (AFPAC). At this time the USAFFE WAC Detachment working for GHQ at last found itself working for the headquarters of assignment, AFPAC.

The powers transferred to the staff director in no case included command. G-1 Division and the adjutant general's office retained all normal operating functions such as promotion and assignment. Even in policy matters it developed that not too many of the WAC Section's recommendations were accepted.[179] Nevertheless, it was Colonel Brown's opinion that her office was now in a position to take action against the several probable causes of the medical evacuation rate.[180]

One immediate action was the publication of a directive concerning the disciplinary power of company commanders:

> Where enlisted women are assigned to a unit or organization other than a WAC detachment, the commanding officer of the WAC detachment to which they are attached for rations, quarters, and administration will exercise disciplinary control.[181]

This directive applied only to the USAFFE GHQ detachment, its promulgation to other units being prevented by the command policy of noninterference. Colonel Brown was never entirely satisfied

[175] WAC Draft AFPAC Reply only.

[176] Personal ltr, CO 5205th WAC Det to Col Brown, 11 Mar 45, and reply, 14 Mar 45. Folder, 5205th WAC Det, Drawer X-29853.

[177] Written comments on draft of chapter; also comments in seminar.

[178] Check Sheet, Hq USAFFE, G-1 for CofS through WAC Sec, 11 Apr 45, signed H.H.B. Approved by Comment 2, CofS to G-1, 12 Apr 45. Tab 6 to WAC Draft AFPAC Reply.

[179] USAFFE Check Sheet, WAC Sec to G-1 Div, 27 May 45, sub: Critical Score, Deployment Plan, and comment 2 from G-1. FEWAC.

[180] List of "Subjects for Discussion with the Dir WAC." Copy in possession of Mary-Agnes Brown.

[181] Atchd to WAC Draft AFPAC Reply: (1) D/F, WAC Stf Dir USAFFE for G-1 Div USAFFE, 8 May 45, sub: Restriction of WAC Pers. (2) Stf Memo 52, Hq USAFFE, 18 May 45.

with the resulting situation, and in her final recommendations proposed that, regardless of accounting difficulties, all women be clearly assigned to WAC companies. Otherwise, War Department regulations were ignored and "the placing of enlisted women under the immediate command of male officers . . . had an adverse effect upon their well-being." [182]

A second issue immediately forced by the WAC Section was that of supply. A memorandum to General Richard Marshall asked that the theater take decisive action to get proper clothing, and that a representative be sent to the United States to trace supplies. Much activity resulted in the spring and summer of 1945, when General Richard Marshall personally wrote friends in the War Department to try to expedite shipment. The matter eventually came to be considered important enough to merit air priority, which had to date proved the only means by which minority supplies had any chance of reaching their goal. In May of 1945, Colonel Brown was ordered to return to the United States and personally check the status of each requisition. By these means, Manila Wacs obtained some supplies before the end of the war with Japan. With the end of hostilities, it was believed that a reasonably satisfactory supply system would have been perfected had any Wacs remained in the theater.[183]

War Department Investigation

A few days after transfer of responsibility to the WAC Section, the theater was the object of a War Department "Investigation Concerning the Morale, Health, and Living Conditions of WAC Personnel in the Southwest Pacific Area." [184] Disquieting reports on the women's health and well-being had been reaching the War Department since the beginning of the move to New Guinea. While many were obviously exaggerations—particularly those from parents alleging that all Wacs were being eaten up by "jungle rot"—yet the area in any case was believed to merit a visit from Director Hobby such as the two that she had already made to Europe and Africa.

From May of 1944, when the first Wacs went to New Guinea, Director Hobby therefore planned to visit the area, but was prevented by repeated hospitalization. In October of 1944, she placed Deputy Director Rice on orders to go in her place. Again at the last moment the Director's illness intervened to cancel the trip, since during these months the Director had only a deputy and one assistant for her entire staff.

At this time, the Director secured permission for an inspection trip by Major Craighill of The Surgeon General's Office. This visit proved of little immediate assistance to the women, as it became part of a tour of all theaters.

It was not until after the end of the war that the second WAC Director, Colonel Boyce, reached the area. Colonel Hobby's resignation at this time was in fact hastened by the need for the Director to visit the Pacific and her own inability to do so.

Meanwhile, the War Department's concern was spurred by the condition of Wacs who began to trickle back from the Pacific, of whom it was reported, "Not only do these women arrive without adequate clothing and equipment, but their state of

[182] AFPAC Check Sheet, WAC Sec to CofS through G-1, 28 May 45. FEWAC.

[183] Sources cited n. 109.

[184] (1) Ltr, TAG to CINC SWPA, 9 Jun 45. sub: Request for Investigation Concerning Morale, Health, and Living Conditions of Members of WAC in Pacific Areas. AGOB-C 333 WAC (5-18-45). (2) See also n. 1(3).

mind is very low and their health below par." [185] Immediate action was precipitated when similar reports reached Congress and resulted in Congressional demands for an explanation. At this, General George Marshall requested of General MacArthur a full confidential report of the condition of Wacs in his command.[186]

The reply to this inquiry proved unsatisfactory to the War Department; it consisted chiefly of a copy of a press release made by Colonel Brown, who was then in the United States checking supplies, plus the 1944 statistics on evacuation, which indicated that Wacs were in good health. The War Department immediately demanded a second report, stating sharply, "A release for publicity purposes made here in the War Department is not considered adequate as a confidential report to the War Department. . . ," and adding that other statements were "very general in content and the code indicates that it was prepared by G-1 and not by the WAC Staff Director's office or the Inspector General." [187]

A more detailed reply was then prepared in the WAC Section of AFPAC, but was considerably censored by G-1 Division, so that the version which reached the War Department was rather short. The staff director noted:

> G-1 here decided to scrap my report. . . . They had some Major who has not long been in the theater do the principal amount of work on the one that was sent. So far, I have not been able to get a copy of their finished product.[188]

The theater-approved version gave only two adverse items: (1) the failure of Christmas presents to arrive on time, and (2) the death of eight of the Wacs in a plane crash.[189]

As a result, before illness finally forced Colonel Hobby's resignation in the summer of 1945, several decisions were reached by the Army's highest authorities, with her concurrence. One was a general decision to return all Wacs from all overseas theaters as soon after the end of hostilities as they became eligible for discharge, and not to send replacements. This measure would have been required in any event to permit demobilization of the Corps as required by the current legislation. To cover the period before demobilization, a second plan was prepared, which would have permitted voluntary rotation of Wacs after 12 months in the Southwest Pacific, and mandatory rotation after 18 months. These decisions effectively prevented WAC employment in the Army of Occupation in Japan. It was the theater's verdict that it would not be worthwhile to prepare housing and move the units to Japan in view of the few remaining weeks. A computation of point scores for discharge eligibility showed that by the end of 1945 all but 500 of the Pacific's Wacs would be eligible for return under the Army's Adjusted Service Rating System, and that even these 500 would be eligible in a few more months.[190]

The War Department also refused to allow the Pacific theater to keep dis-

[185] Memo, Lt Col Helen H. Woods, Deputy Dir WAC, for G-1, 10 Oct 45. WDWAC 330.11. She refers to returning Wacs of previous months.

[186] Ltr cited n. 184(1). Inclosure 1 is Congressional request.

[187] D/F cited n. 61. Incls show: Congressional Inquiry, 18 May 45; first request to Pacific, 26 May 45; interim Ltr to Congress from CofS, 8 Aug 45; first SWPA Rpt, 21 Aug 45, rejected for reasons given above.

[188] Ltr, WAC Stf Dir AFPAC to Col Boyce, 28 Dec 45. WDWAC 320.

[189] Rpt cited n. 1(3). All supporting documents were omitted in theater version.

[190] (1) See Ch. XXXVI, below. (2) Memo, KKJ (Maj Katherine Johnson, Exec WAC) for Dir WAC, 11 Jun 45, sub: Rpt of Action 6-11 Jun 45. WDWAC 319.1. (3) Boyce Rpt, draft version.

charge-eligibles who volunteered to remain. These were not many. Surveys before the defeat of Japan showed that 57 percent of WAC eligibles—80 percent of the AFWESPAC (formerly USASOS) group, 82 percent of FEAF—wanted to go home; by comparison, only 5 percent in Italy and 38 percent in England had expressed a desire to go home before the fighting ended.[191]

Extensions and exceptions were later to be given to the European and China theaters, so that their volunteer Wacs were not forced to return; no such extension was allowed the Pacific theater. Colonel Brown, herself a victim of amoebic dysentery that had already required hospitalization, believed an extension should have been granted, but was not called on to comment in the matter and did not return to the Pacific after demobilization began. She was awarded the Legion of Merit and a high efficiency rating personally signed by General MacArthur.

Lt. Col. Mera Galloway, formerly WAC Staff Director, Transportation Corps, was sent to the theater with the assigned mission "to close WAC History in the Southwest Pacific in an efficient and successful manner." [192] Colonel Boyce also arrived in the area as the closing-out began; until her report was received there remained the possibility that an extension of the return date of WAC volunteers would be granted.

Colonel Boyce was accompanied on her visits by the new staff director, by a War Department representative of G–3 Division, and by Dr. Marion Kenworthy, a prominent psychologist and member of the National Civilian Advisory Committee. The team talked individually with hundreds of enlisted women, and visited remaining WAC units on Hollandia, Biak, and Luzon.[193]

Their impression was not favorable. From Manila, Colonel Boyce cabled to the War Department: "I recommend all enlisted women in this theater be enroute USA for discharge or reassignment not later than 1 January 1946." [194] The medical loss rate had declined only slightly and Colonel Boyce, although a veteran of the North African crisis, was disturbed by the apparent poor health of members still on duty. Most were underweight, and all were highly fatigued. When the headquarters attempted to hold a parade formation in her honor, and delay resulted in an hour's wait, eight women fainted. Many were still working seven days a week with additional evening duty. Food was still poor, some quarters still unhealthful with beds standing in an inch of water, while few women yet had adequate clothing and footgear for protection against the daily rainstorms.

Colonel Boyce advised theater commanders that the women's condition required immediate action: less overtime office work, granting of quotas for rest areas, and more suitable clothing, such as that provided enlisted men for protection from the rain. Her advice was not accepted, although a service club and a rest camp were eventually opened to women a few weeks before they left. Colonel Boyce, after her encounter with theater authorities, wrote to Colonel Brown: "I want you to know that my observations in the theater simply confirmed the faith I have always had in your ability to perform

[191] "Readjustment Rating Scale Figures, Manila Area." In possession of Mary-Agnes Brown.

[192] Boyce Rpt, draft version.

[193] Ltr, Lt Col Mera Galloway Ferrin to OCMH, 22 May 51.

[194] Cbl, CinC AFPAC Comd, Manila, Boyce for WD, 27 Oct 45. C–19184, WDWAC 320.01.

difficult assignments in a superior manner." [195]

Many section chiefs, as well as Colonel Brown, felt that the refusal to permit WAC volunteers to be retained was unwarranted and deprived them of needed office help. The Commanding General, AFWESPAC, Maj. Gen. Wilhelm D. Styer, requested that the War Department at least allow a short extension; he proposed to make all Wacs warrant officers if this would allow them to remain. Colonel Boyce replied from Washington that the War Department was extremely sorry not to grant the request but that the January deadline could not be extended. Women were extended the same privilege offered in the European theater, that of remaining as civilian employees if they desired, but only about 200 elected to remain.[196]

Demobilization

A final problem was presented: the shipment of the women to the United States. Colonel Boyce while still in the theater found the first large group already boarding the ship *Evangeline* for return. The ship appeared unsuitable—vermin-ridden, ill ventilated, with men obliged to walk to mess through the WAC sleeping quarters. The women's physical condition left much to be desired. They had no woolen clothing, such as was issued to men for a comparable voyage, the original clothing in which they arrived having long since deteriorated in distant warehouses.

Colonel Boyce, however, was unwilling to stop the shipment in view of the crowded state of port housing, the scarcity of ships of any sort, and the women's eagerness to get home. She radioed her office that she was "extremely disturbed" at the condition of the women, and feared that they "will make a very poor impression upon their communities unless their general condition can be improved upon their arrival." Accordingly, the War Department directed that the AAF WAC Officer and the ASF WAC Officer, Colonels Bandel and Goodwin, hurry to a west coast port to meet the shipment. This was accomplished, but obviously little could be done in the few days before the women were discharged, and the precautions did not prevent unfavorable publicity and headlines, such as VERMIN-RIDDEN TROOPSHIP, in both the Associated and the United Presses. WAC authorities feared the Corps' whole hard-won public relations victory might be jeopardized if the entire 5,500 women returned to their forty-eight states in a physical or mental condition alarming to the public.[197]

The second WAC group was being loaded on another ship, previously used to transport Chinese laborers, when the commanding general of the Hollandia Base saw it and refused to allow his Wacs to join those already on it. Colonel Boyce hurried to the scene and, without authority, ordered the remaining Wacs off, and port commanders thereupon turned the ship back to its owners as unfit for military use.

Before she left the theater in November, Colonel Boyce conferred with the chief of the transportation section and secured the assignment of a WAC officer to the office of the port director, to co-ordinate future shipping arrangements for Wacs and nurses. The WAC staff director or her assistant thereafter personally inspected shipping space assigned to women before

[195] Ltr, Dir WAC to Mary-Agnes Brown, 12 Dec 45. Personnel Abstracts, WAC file, OCMH.

[196] (1) Ltr, Dir WAC to WAC Stf Dir, AFPAC, 11 Dec 45. WDWAC 320. (2) Ltr cited n. 193.

[197] (1) Memo cited n. 185. (2) UP Release, 23 Oct 45, with headlines such as that in the Denver, Colo., *Rocky Mountain News*.

the date of loading. No further difficulties were noted.[198] On 10 January 1946, the last major shipment of returning enlisted women was loaded onto the *West Point,* the ship that had brought the first group to Australia, and from which the last now waved farewell to Manila.

Conclusion

The Director WAC, when later asked whether she would recommend that women ever again be stationed in such areas, was exceedingly doubtful as to the correct reply. Some of her advisers felt that the greater sacrifices of combat troops in the area made it impertinent even to question whether women should be sent to any area in which men were required to fight. Others commented:

> Men had to be there; nurses had to be there so long as the Army had no male nurses; but Wacs didn't have to be there; and it seems important to discover whether a decision to use enlisted women in such an area is wise and economical or not.[199]

Theater authorities with one exception agreed that the WAC experiment was an unqualified success. General MacArthur stated to WAC leaders that he was satisfied with the results of WAC employment. When Colonel Boyce consulted him in Tokyo, he praised Wacs highly, calling them "my best soldiers," and alleging that they worked harder than men, complained less, and were better disciplined. He informed Colonel Boyce that he would take any number of the Wacs the War Department would give him in any future command he might ever have.[200]

General Styer, commanding general of the Army Forces in the western Pacific, called the Wacs' service "a matter of everlasting credit to the Women's Army Corps," and said that the Wacs were "courageous soldiers," who from Australia to Manila "more than carried their share of the burden."[201]

General Kenney, commander of the Far East Air Forces, stated that his Wacs' work "has been of the highest caliber. Each has better than replaced a soldier."[202] General McMullen of FEASC said, "The detachment in every way upheld the fine traditions of American womanhood and their departure is a distinct loss to the command."[203]

Colonel Ginsburgh of USASOS added, "The Wacs did a job. We never had the sense of frustration which a discussion of their problems gives you; we only had a sense of achievement."[204]

On the other hand, General Baird, G-1 of USAFFE, later wrote:

> I cannot truthfully say that their contribution, great as it was, outweighed the difficulties. On the other hand, nothing I could say would quite do justice to the loyal and efficient effort they put forth under all conditions and circumstances; to their courage and fortitude. . . . I do not believe that Wacs should be sent outside the limits of the United States until every general service man has been replaced by a Wac, and until every other available limited service man has been replaced. . . . The hardships, isolation and privation of jungle theaters are jobs for men. Women should be employed there only as a last resort.[205]

[198] Ltr, WAC Stf Dir AFPAC to WAC Stf Dir SFPOE, 8 Nov 45.

[199] Interv with Lt Col Mera Galloway, second AFPAC Stf Dir. (See Bibliographical Note.)

[200] (1) In interv with Col Boyce, 14 Oct 45, quoted in Rpt to Gen Marshall. Also see his statement in booklet "The 5205th WAC Det, GHQ," published in theater. WDWAC 000.7. Incl in Ltr, CO, 5205th WAC Det to Dir WAC. (2) Statement by Colonel Boyce to author and others, 19 November 1946, upon her return from SWPA.

[201] Griffith Account, "Commendations."

[202] *Ibid.*

[203] Ltr, CG FEASC to Dir WAC, 3 Jan 46, sub: Commendation. WDWAC 201.22 and 330.13.

[204] Interv, 13 Aug 47.

[205] Memo, Col Harry H. Baird for Hist Div SSUSA, 29 Jul 48. OCMH.

An opposite view was offered by the first WAC staff director, Colonel Brown, who wrote later:

> The fact that the mission was successfully performed is much more important than a hind-sight consideration of whether it was wise to use Wacs in SWPA. . . . In my opinion the service of WAC members was essential to the success of Army operations in New Guinea and the Philippines. I deplore the failure of those immediately over the Staff Director . . . to accept my recommendations on procedures to assure the well-being of WAC personnel, but the WAC mission was accomplished in spite of these and other obstacles and at not too great a sacrifice.

Colonel Brown joined with the staff director of the European theater in advocating a permanent WAC in the Regular Army, and pointed out:

> Colonel Hobby's recommendation to General Marshall that Wacs be withdrawn from SWPA was made without consulting the Staff Director. . . . When General Marshall decided to send Colonel Boyce to the theater to "put the lid on," the recommendations or opinions of the Staff Director . . . were not sought.

Colonel Brown also was of the opinion that the heavy loss rate would have slackened eventually with improved supply and other provisions.[206]

From the War Department's Army-wide viewpoint, the verdict was less positive. Even in the few brief months between the incidence of the loss rate and the recall of the women, heavy casualties had been suffered, while the future usefulness of many of the other women was questionable.[207] The War Department inspection team noted, of the entire group:

> It was evident that in their forced adaption they had no realization of their physical and mental sacrifices. It is possible that not until they return to the Zone of the Interior will they acknowledge the changes which have taken place within them.[208]

These were, to the War Department, expensive losses. A skilled Wac was valuable Army property, representing a considerable investment in recruiting and training, and was needed for assignment in four or five jobs for every one that could be filled. Had the same 5,500 women been assigned to any of the other numerous overseas or domestic commands that had unfilled requisitions, few of these losses would have occurred under prevailing loss rates.

It therefore appeared to have been poor economy to employ women in circumstances where the loss among an equal number of men would have been only one fourth as great. If men in the proper skills were unavailable, the loss was obviously required in winning the war. The question of whether the entire Army of 7,700,000 men was not at this stage able to afford 5,500 men in clerical skills was one to be answered by students of the world-wide replacement system, and was a matter beyond the scope of either WAC or theater knowledge. Operations Division's decision to send Wacs had been made in the belief that, in the moment of manpower shortage, there existed no such 5,500 men. Thus, the original decision to send Wacs needed no justification; any future decision concerning the use of women in such an area would require consideration of the existing manpower shortage versus the probable loss rate.

It came to be the final opinion of most WAC leaders that women not only should be sent to such an area under equally desperate circumstances, but that they could

[206] Memo, Lt Col Brown for OCMH, 23 Apr 51, and comments at seminar, 1948.

[207] All from AFPAC Reply, both versions, except nonmedical evacuations, which are from WAC Stf Dir's file.

[208] Boyce Rpt.

probably be employed without any comparable loss rate by application of the principles of WAC employment used elsewhere. These included, particularly, advance planning, a definition of the place of WAC advisers, command of women by female company officers, specialist attention to tracing the supplies of the relatively small group of women, and consideration of women's psychological need for treatment as responsible adults. In her final report to General George Marshall, Colonel Boyce summarized the WAC impression:

> Women in the Army should be required and permitted to serve wherever the Army serves when their services are needed. However, in . . . New Guinea, Biak, and the Philippines, it is apparent that essential safeguards to the effective utilization of WAC personnel were not applied.[209]

A clue to an explanation of this lack of safeguards could be found in world-wide WAC experience, which indicated that the fault lay in the time factor as much as in any human element. In both other major theaters, a considerable shakedown period was required after WAC arrival in order to set up a satisfactory administrative system and specialist attention to supply and other needs. The Fifth Army experiment had not been attempted until eleven months after the first Wacs' arrival; the Normandy move, until twelve months after arrival. In the Pacific, the world-wide norm of a WAC administrative system had also been worked out at the end of a year. It was the theater's misfortune that, because of the lateness of its requisitions for Wacs, the achievement of this system had coincided with the end, and not with the beginning, of a risky experiment.

[209] *Ibid.*

CHAPTER XXIII

Other Overseas Theaters

In addition to being employed in the three major overseas theaters, Wacs were stationed in every other active theater of war. In each case the employment was relatively minor, never exceeding a few hundred women. Nevertheless, remaining records of the various experiments often provided clear-cut indications of the practicability of employment of womanpower in differing climates and administrative situations.

Southeast Asia Command

Pioneer in the employment of American Wacs in India was Vice-Adm. Lord Louis Mountbatten's Southeast Asia Command. The headquarters already employed nurses, British servicewomen, members of the Indian WAC, and local civilians, so that no great venture was undertaken when a small unit of sixty-two enlisted Wacs and four officers was added to the staff in October of 1943. After six months in New Delhi, the unit was transferred in April of 1944 to Southeast Asia Command headquarters in Ceylon, where it remained for the duration.[1]

Fourteen of the enlisted women were assigned to Signal Corps duties, and the remainder to routine clerical work in offices of the headquarters, both British and American. Officers, in addition to company duty, were assigned operational jobs in the same offices.

No great difficulties were encountered by the unit in the performance of its jobs. The two stations of assignment, while by no means pleasure resorts, were not the most trying in India. New Delhi, the capital, was one of India's cleaner and more healthful cities in spite of temperatures ranging up to 110° F. The Wacs nevertheless soon discovered the ailment theater veterans designated as "Delhi belly." The climate of Ceylon was tropical, without extremes of heat but with excessive humidity during the rainy seasons. Enlisted women were housed in barracks in the WRNS area, which inspectors pronounced excellent in spite of the customary bucket latrines and absence of hot water. Both men and women ate at a joint mess, which lacked cold-storage facilities and thus had only small amounts of meat and few fresh vegetables. Drinking water was boiled and stored in Lyster bags. The working day generally followed the nonstrenuous British schedule, with an 8-hour day and one day off per week.

Major Craighill, who visited the unit after Wacs had been in India for some

[1] No records of this group were found, with the exception of Memo, Maj Craighill for The SG through SEAC and IBT, 6 Apr 45, sub: Med and Social Conditions of Women in Mil Serv in Ceylon. SGO Hist Div 319 Special, Ceylon, Women In. In addition, comments on drafts of the present chapter were given by Charles F. Romanus and Riley Sunderland, authors of *Stilwell's Mission to China*, UNITED STATES ARMY IN WORLD WAR II (Washington, 1953).

eighteen months, stated that "the general health of the detachment is excellent." Except for dengue fever, mild gastrointestinal conditions, and skin infections, there had been no notable illnesses, and especially no amoebic or bacillary dysentery, or malaria. One serious exception was the Signal Corps women, who were on around-the-clock shifts, rotating every four days. Major Craighill noted:

> A high incidence of sickness among the group working on irregular hours with the Signal Corps is significant. Of the 58 admissions for all causes, 20 were from this group of 14 women. This includes the hospitalization of ten of the fourteen: the other four have all been on sick call. These conditions are attributable to the irregularity of meals and the loss of sleep due to constantly shifting hours of work.

In spite of Major Craighill's recommendations, the headquarters did not modify the shift changes, since they were "in accordance with the present British system." [2]

No final recommendation was on record from the headquarters concerning the success of employment of Wacs. There appeared to be no very valid reason why women could not successfully fill many of the jobs in any such high administrative headquarters. On the other hand, there was no proof that local employees could not have done the work adequately without the importation of American women.

The China–Burma–India Theater

The intrinsic suitability of the China–Burma–India (CBI) theater for WAC employment was by no means identical with that of the Southeast Asia Command.[3] Principal supply installations were in Calcutta, a city noted for oppressive heat, disease, and filth, and generally considered even more unhealthful for unacclimated white women than the more sparsely populated New Guinea islands. However, there were certain advantages not possessed by the Southwest Pacific Area, notably the fact that in Calcutta, a city approximately the size of Chicago, there were available civilian shops and tailoring facilities, as well as some approved recreational facilities. Nevertheless, the experience of nurses and Red Cross women already in the area was not encouraging; nurses had previously suffered from clothing shortages, emotional strain, tension, and a medical evacuation rate four times that of the men.[4] "Wacs in India were definitely on trial," a later report stated, "as many Army officials believed that women would not be able to stand for long the climate and diseases found in Asiatic countries." [5]

It was therefore not until July of 1944, two years after the Corps' establishment, that the Air Forces headquarters in Calcutta received its first Wacs. Earlier attempts to bring in Wacs had been blocked by the theater commander, Lt. Gen. Joseph W. Stilwell, who eventually yielded to the Air Forces' requests only after receiving the express pledge of the Air

[2] 1st Ind to Memo cited n. 1.

[3] (1) CBI WAC files, by folder number and specific document therein. (See Bibliographical Note.) Other important sources are: (2) Rpt, AAF, IBT to TAG, 5 Jul 45, sub: Surv of WAC Health and Living Conditions in India–Burma Theater. AG 321 WAC, Hobby files, also in CBI WAC files. Hereafter cited as IBT WAC Survey. (3) Rpt, Maj Craighill to The SG through CG IBT, 30 Mar 45, sub: Med and Social Conditions of Women in Mil Serv in IBT. SGO Hist Div 319.1-2, Women, India–Burma; hereafter cited as Craighill IBT Rpt. (4) Comments on draft of present chapter by Maj. Beryl Simpson, wartime WAC Staff Director, China Theater, and Maj. Betty Clague, wartime WAC Staff Director, CBI Theater.

[4] Incl 3 to Craighill IBT Rpt.

[5] Rpt, WAC Stf Dir AAF CT, 25 Oct 45, sub: Narrative Account of WAC Activities in Chungking, China, 15 Jul–13 Oct 45. WDWAC 319.1; also in Folder, Chungking, CBI WAC Sec files.

Forces commander, Maj. Gen. George E. Stratemeyer, that any Wacs brought in by the AAF would never under any circumstances be assigned to other than Air Forces headquarters.[6]

Scrupulously careful advance preparation was the keynote of the entire Air Forces project. Wacs for shipment to India were selected by the Army Air Forces from among women proved capable and emotionally stable by service at airfields in the United States. The result was so good that General Stratemeyer, upon later checking their job performance, congratulated Director Hobby, who congratulated General Arnold, who congratulated air commands in the United States, who passed the commendation on to selecting officers at airfields—the whole project ending in a glow of mutual admiration.[7]

This success was the more remarkable in that the theater, with typical overseas optimism, asked percentages of scarce skills never remotely approached in the United States—one fourth stenographers, one fourth typists, and the rest all highly skilled in other fields.[8] It was possible to supply these satisfactorily only because the total original request was not large, less than 300. Selection was marred only by seven cases in which the routine physical inspections failed to detect serious physical disqualifications that caused the women to be returned to the United States. Much to the theater's chagrin, the first report of its WAC experiment was given to the press by one of these women, who had never set foot in India except to enter the comfortable port hospital but who was described by reporters as the heroine of a bout with jungle, heat, and disease-bearing insects, as she personally battled her way through Burma side by side with General Stilwell.[9]

The actual reception of the Wacs was considerably less alarming. "Never has any WAC contingent received a more cordial welcome than this group," wrote the WAC staff director, Maj. Betty Clague.

> Everyone from the Commanding General on down has apparently had a personal interest in our welfare and happiness. . . . Everyone is impressed with the fact that they must not let down these men who have expressed such a satisfaction in having us as an integral part of their headquarters.[10]

Accommodations were unique, consisting of a section of a huge eight-acre jute mill, which also sheltered headquarters' offices and many of the officers' and enlisted men's quarters. Space allotment per individual was somewhat greater than that of the zone of the interior, and Wacs had not only single-decked beds with sheets and pillows, but dressers, chests of drawers, overhead fans, and mosquito nets. Latrines had toilets, showers, tubs, and mirrored washbasins. For the less than 400 enlisted women[11] who eventually lived in these quarters, there were 23 Indian sweepers and about 50 ayahs, who

[6] Ltr, CG AAF IBT to Dir WAC, 15 Nov 44, sub: WAC Utilization, (Incl 1 to IBT WAC Survey).

[7] *Ibid.* Also Ltrs, Dir WAC to CG AAF and CG AAF IBT, 18 Dec 44. WDWAC 320.2.

[8] IBT WAC Survey, p. 6. The 257 enlisted women on duty at that date: 67 stenographers, 61 clerk-typists, 69 clerks, and the rest qualified in such jobs as key punch operator, draftsman, radar specialist, intelligence specialist, photolaboratory technicians, classification specialist, telephone operator, medical technician, cadre, and the like.

[9] *Ibid.*, p. 9. See also (1) Memo, CO WAC Det for WAC Stf Dir, 14 Feb 45; (2) Ltr, WAC Stf Dir to Air WAC Off, 22 Feb 45. Folder IBT, CBI WAC files.

[10] Ltr, WAC Stf Dir AAF CBI to Air WAC Off, Hq AAF, 21 Aug 44. WDWAC 320.2 CBI.

[11] IBT WAC Survey shows 257 EW as of that date. In her comment on a draft of this chapter, Major Simpson shows 397 as the highest number ever on Morning Report. For final theater strength figures, see Table 7, Appendix A.

TAJ MAHAL *is seen by four women of the Southeast Asia Command on a sightseeing tour.*

made beds, shined shoes, washed underwear, and policed the area. There was also provided a hairdressing shop, a dispensary in the WAC area, and a dayroom with piano and radio. The only obvious objection to the arrangement was, according to medical inspectors, the noise in the big building, which made it difficult for night workers to get any sleep. In the future, living accommodations for not exceeding 75 in a group were recommended. With this exception, quarters were possibly more comfortable than the average barracks in the United States, although members expressed a mild initial surprise at observing large rats frisking through the machinery overhead, and birds roosting on the beds.[12]

Other advance preparations had been carefully made. The post exchange stocked needed women's items—cosmetics, stockings, brassières, girdles, and garter belts. Wacs were admitted to the Red Cross club and to the enlisted men's club, where tea was served daily. There was a free movie theater, and free hourly bus service to Calcutta, with its Army-approved swimming pools and dining and dancing places. Provision was made for Wacs to share in the theater authorization for yearly 15-day rest leaves in mountain resorts, with breakfast in bed and sports such as golf, riding, and tennis.[13]

Wacs ate at the men's large consolidated mess, where supervision was taken over by an experienced WAC mess officer, Capt. Roslyn Katz, a WAC mess sergeant, and other Wacs who supervised Indian cooks and kitchen laborers. Some 8,000 meals a day were served at this mess, which was later rated as "well above the average in this theater." Food was generally considered to be as attractive and appetizing as was possible in the absence of salads, milk, fresh butter, and other uncooked items. Besides regular meals, a midnight snack of hot coffee and sandwiches was available.[14]

Air Forces authorities believed that they had been equally careful in making advance provision for suitable uniforms, but immediately discovered that in fact they had shown scarcely better foresight than had the Southwest Pacific Area. At the written request of the theater, each woman had been issued six off-duty dresses for office wear, since this, although not the ideal dress that could have been designed for the area, seemed the coolest garment already stocked by quartermaster depots. It was soon discovered that the dresses were "washable" only by methods seldom known to local ayahs, and "couldn't stand beating on a rock for two or three changes a day"; even when later reserved for dress, they did not prove durable. Furthermore, although malaria control restrictions required wearing of slacks after 1800, the need for them had not been foreseen and again the only available items were herringbone twill coveralls. These, however, were pronounced totally impossible by theater authorities, since the weather was too hot to make them endurable and they "would have increased alarmingly the incidence of skin infections."[15]

The theater was saved from consequences similar to those in the Southwest Pacific Area by the fact that local shops could provide the necessary items if the Wacs had the money to purchase them. The theater could find no Army funds for the issue of men's slacks, but medical offi-

[12] (1) Craighill IBT Rpt. (2) IBT WAC Survey, pp. 1–2. (3) Intervs with CBI Wacs.

[13] IBT WAC Survey, pp. 2–5, 10 and Incl 3.

[14] *Ibid.*

[15] Comments of theater staff director on draft copy of chapter, and in interview.

cers refused to let the women leave the barracks for their evening meal until so equipped. Therefore, most enlisted women bought slacks tailored of an Indian mesh cotton. Mosquito boots were also purchased, as well as cotton anklets, sandals, and sometimes bush jackets. Also, Air Forces headquarters proved prompt in remedying deficiencies, and some months later succeeded in getting shipment for each woman of five cotton skirts and seven short-sleeved shirts for office wear, as well as three pairs of women's khaki slacks for wear after 1800. For the two or three colder months, wool clothing was provided. Eventually, CBI Wacs, by their own and theater efforts, had about double the amount provided for Wacs in semitropical climates in the United States, although no more than medical authorities deemed necessary in view of the frequent changes of perspiration-soaked garments that were required to prevent illness.[16]

All of the provisions of Air Forces headquarters for its men and women, although frequently exceeding the average for military facilities elsewhere, paid off in respect to the health rate. At the time of Major Craighill's inspection in March of 1945, no Wacs had been evacuated to the United States for medical reasons, except the original seven, who had disqualifying defects upon arrival. If these were counted, the loss rate was still only about 2 percent per year as compared with 2.1 percent for the theater.[17]

Major Craighill reported, "The medical care and condition of women is in general, good. They seem to stand the strain of overseas service as well as do men."[18] Both WAC and theater authorities attributed the low loss rate to the fact that Wacs were not only encouraged but ordered to report to the sick bay in their barracks at the slightest sign of any disorder, or even for advice. As a result, as many as 10 percent of the women in winter months, and 20 percent in summer, reported each month to the sick bay for cases not requiring hospitalization, as against 5 and 9 percent, respectively, for men. Theater authorities urged women to patronize medical facilities even more frequently if necessary, and provided not only a nurse for the well-equipped dispensary, but a gynecologist in the local hospital.

There was every indication that, without such constant care, the health hazards of the area would have been comparable to those of the Southwest Pacific Area, for illnesses included both amoebic and bacillary dysentery, impetigo, hepatitis, various fevers, and respiratory disorders. Major Craighill recommended that two years' service in the area be the limit, since "Among nurses who have been overseas two years or more, there is an obvious increase in tension, manifested by irritability, depression, insomnia, and loss of effective energy."[19]

An additional health problem was what Major Craighill called the "great emotional strain" from the social demands of "a group of isolated and lonesome men." While Wacs were by no means the only white women in the theater, they were nevertheless overwhelmed upon arrival with such numerous offers of social engagements that not all could be accepted. In the interests of health and working efficiency, WAC commanders from time to time prescribed dateless nights, quiet

[16] *Ibid.* Also (1) IBT WAC Survey, pp. 9–10. (2) Ltr, WAC Stf Dir to Air WAC Off, 7 Oct 44. Folder CBI, CBI WAC files.

[17] From Stat Div SGO: total over-all loss by medical evacuation in 1944 was 2.1 percent; in 1945, 2.7 percent.

[18] Craighill IBT Rpt.

[19] (1) *Ibid.* (2) IBT WAC Survey, pp. 10–13: Comparative statistics for men and women.

hours, and other expedients. After the first months, most Wacs were fatigued by the constant masculine attention, and found organized parties and mass invitations especially tiring.[20]

Coupled with the theater's precautions concerning health were, in the early days, numerous protective restrictions which, although not comparable to those in the Southwest Pacific Area, the Wacs found annoying. Wacs could not participate in airplane flights, had bed check even though men did not, could not accept overnight invitations from local families, could not spend the night away from their quarters even in approved service centers, could not date officers, and could not marry.[21] These restrictions caused a certain number of Wacs to inform the chaplain that they were being "treated as children."[22]

Most of these restrictions were therefore quickly abolished as initial misgivings concerning the employment of Wacs subsided. Although bed check was retained as a health safeguard, hours were usually lenient, and Wacs were eventually allowed to stay overnight at the women's service center. Wacs were also soon allowed in aircraft, and at least one technician was placed on regular flying duty for which she received the Legion of Merit and the Air Medal.[23] The restriction on dating officers was not removed until Wacs reached Shanghai, and some difficulty was caused by the fact that Air Forces Wacs were liable to punishment while officers of theater headquarters, which had no such rule, were immune.[24]

The restriction on marriage, a theater ruling, was the most objectionable to the women. The rule had been designed to prevent the marriage of soldiers to noncitizens, and had been modified just before the Wacs' arrival to permit marriage only if a soldier was leaving the theater within thirty days. Cases involving pregnancy were considered individually on their merits. At the time of Major Craighill's visit, Wacs still had a pregnancy rate of zero, but WAC commanders considered the system, for obvious reasons, incredibly shortsighted. In April of 1945, after receipt of Major Craighill's recommendations, and after the first Wac achieved marriage by the only theater-approved method, Lt. Gen. Daniel I. Sultan liberalized the policy to permit marriages between two United States citizens after a three-month waiting period. With this one case, the WAC pregnancy rate was only one fourth of one percent, or considerably less than the average. No cases of venereal disease were discovered among any WAC personnel.[25]

As always, factors of physical surroundings proved less important to the women than that of job satisfaction. Initial job assignment was far from perfect, a condition that General Stratemeyer attributed chiefly to "a failure on the part of officers

[20] (1) Rpt cited n. 5. (2) Memo, WAC Stf Dir for all offs, 26 Mar 45. Folder IBT, CBI WAC files.

[21] Memo, WAC Stf Dir for CO WAC Det, 16 Sep 44, sub: Policies Applicable to WAC Pers. Folder IBT, CBI WAC files.

[22] Memo, Stf Chaplain for CG AAF IBT, 10 Jul 45, sub: Morale of Wacs; also Memo, WAC Stf Dir for CG, 15 Jul 45. Folder IBT, CBI WAC files.

[23] WAC citations, 7 Dec 45, Folder Shanghai II, CBI WAC files.

[24] Entire lengthy dispute: (1) Ltr, Hq SEAC, 5322 D/U.S. Army Sec, 14 Jul 44. WDWAC 421. (2) Draft of Statement, 23 May 45. Folder IBT; (3) Routing Slip, Comment I, WAC Stf Dir AAF CT to AC, 28 Aug 45, Folder Chungking I; (4) Memo, WAC Stf Dir Hq USFCT for CofS, 25 Jan 46, Folder Theater WAC III; (5) Cir 7, Hq USAF China, 9 May 46, sec 1, Folder Shanghai V. All in CBI WAC files. (6) Interv with Maj Simpson, 23 Jun 47. (7) Boyce Rpt, Tabs: India-Burma, and China. WDWAC 333.1 (1945-46).

[25] (1) Memo, Capt Chance, Exec WAC, for Col Hobby, 3 May 45, sub: Marriage Policy. WDWAC 291.1. (2) MS History of India-Burma Theater, 25 Oct 44-23 Jun 45. OCMH.

to appreciate capabilities, and overstaffing during the period of orientation prior to the transfer of enlisted men replaced by Wacs."[26] General Stratemeyer took immediate and vigorous action to correct the situation.[27] As a result, the expected 30-day on-the-job training was reduced to 15 days.

A few months later General Stratemeyer directed his WAC staff director to make a survey of job utilization. It was found that, "with a very few exceptions, the qualifications of Wacs were being fully employed in jobs commensurate with their training and experience," and the exceptions were at once corrected. Some problems still remained. The offices were not exceptionally busy, at least not enough to satisfy the Wacs. Also, civilian competition was present from female citizens of Calcutta. The staff director remarked:

> I believe that the biggest hindrance toward the proper utilization of WAC personnel in this Headquarters is the usage of inefficient civilian help. Because of their employment in the majority of stenographic positions, the Wac is compelled to watch someone else do a job which would ordinarily take her one-half of the time to complete.

Because of the constant attention to the matter of correct assignment, however, lack of work never became a morale problem. An Army observer confirmed this assertion about Wacs' high morale and its cause, stating that:

> The women look well, work efficiently, are universally cheerful and seem almost without exception to feel that they are doing an important war job. . . . The WAC personnel are universally admired and respected by male officers and enlisted men here. They have made a definite contribution to the morale of the base and the efficiency of the headquarters here, and I believe that their excellent attitude reflects their pride in this fact.[28]

General Stratemeyer, at the time of the separation of the China theater from the India-Burma, reported to Generals Stilwell and Arnold:

> Experience of the past several months has proved any doubts concerning the propriety of the experiment to be false. It is with enthusiasm that this Headquarters reports the substantial overall improvement in efficiency as a result of the placement of Wacs. . . . Officers and enlisted women alike have quickly adjusted themselves to the climate, food, and the somewhat rugged living conditions. . . . Deportment is superior. Not one untoward incident has occurred as a result of their presence in this Headquarters. . . . This Headquarters recommends without hesitation to other Headquarters overseas the excellent benefits to be obtained through the placement of Wacs in theaters of operations.[29]

The newly organized Services of Supply headquarters in Kunming, China, immediately petitioned the commander of the U.S. forces in the China theater, Lt. Gen. Albert C. Wedemeyer, for permission to bring in a WAC company almost identical with the Air Forces' company, saying:

> It is virtually impossible to obtain competent civilian clerical employees in this theater. . . . Male enlisted personnel trained in typing and stenographic duties are not available in sufficient numbers. . . . Where the civilian employee is not a citizen of the United States, as is almost invariably the case in this theater, effective security is largely unknown.[30]

---

[26] Ltr cited n. 6.

[27] D/F, Gen Stratemeyer to Col David Barret, 21 Sep 44. Folder IBT, CBI WAC files.

[28] (1) D/F, WAC Stf Dir to Orgn and Plng, 8 Nov 44, Comment I. Folder IBT, CBI WAC files. (2) Statement by Maj Philip Boyer, Jr., Actg PRO. Incl 2 to IBT WAC Survey.

[29] Ltr cited n. 6.

[30] Ltr, Hq SOS, USFCT, APO 627, to CG USFCT, APO 879, 26 Nov 44, sub: Establishment of WAC Co at SOS Hq, 13525-322. Folder 5, Hq SOS USFCT file, AG 322.

However, theater headquarters refused this: "The decision of the Theater Commander is that no Wacs will be brought into the China Theater for the present." Six months later, the SOS asked again and was again refused on the grounds that "we cannot quarter these people . . . including mess, toilet, and living space." [31]

There was also some objection to the Air Forces' proposal to take Wacs to China when its headquarters moved to Chungking in July of 1945, but because of General Stratemeyer's insistence, the objection was overcome. It was known that living conditions would be far less favorable than in India, but two hundred of the Wacs volunteered to go and about a hundred were chosen, this being all that the reduced headquarters could employ. Again, preparation was careful. Maj. Beryl Simpson, who had replaced Major Clague as staff director, visited the new site to check on all accommodations before the order was given to send the enlisted women. The first WAC contingent flew the Hump on 3 August 1945, closely followed by others.[32]

The China Theater

As expected, living conditions in Chungking were less settled than in India.[33] With much work and close sanitary supervision of native help, an attractive mess hall was maintained, and within three days the Wacs had set up their own laundry and trained local help to give good personal service. The hundred women eventually employed some sixty Chinese servants for mess, laundry, and yard work. Housed in dormitories similar to the civilian women's, Wacs eventually obtained a few fans, an improved water supply, and better electric lights. In spite of predictions, the headquarters reported that a "low rate of sickness" prevailed among Wacs, and that "the sweltering heat failed to impair efficiency." As in India, Wacs were urged to seek medical advice frequently. Records showed that in six weeks over half of the women had been on sick call, although much of the rate resulted from the fact that almost all members were getting daily treatments for prickly heat, an annoying but nonfatal malady, which caused little loss of work time.[34]

In accordance with General Wedemeyer's policies, the headquarters was careful to allow no appearance of favoritism to women, who were trucked to work as were men, issued supplies in the same manner, and generally treated as other troops. In this respect, the move to China accomplished several goals that the WAC staff director had never been able to attain in India. The extreme friendliness of the Air Forces headquarters in India had always been such that WAC staff directors and unit officers were occasionally hard put to maintain strict discipline and military appearance. Before the Wacs arrived, nurses had been allowed to wear civilian evening dress in spite of Army Regulations and rules on malarial control, and it did not prove easy to convince the headquarters that the WAC was best treated as a fully military organization.[35]

[31] *Ibid.*, 1st Ind, 7 Dec 44; 2d Ind, 2 Apr 45; 3d Ind, 25 Apr 45.

[32] (1) Memo, WAC Stf Dir for Col J. S. Clark, 9 May 45, sub: China Survey—WAC. CBI WAC files. (2) Rpt cited n. 5.

[33] Account of China tour of duty is, unless otherwise specified, from: (1) Rpt cited n. 5. (2) All CBI references. (3) Memo, Maj Craighill for The SG, through CG CT, 11 Mar 45, sub: Conditions Affecting Women Pers in China. SGO Hist Div 319.1 China. No Wacs were in the area at the time of this report.

[34] *Ibid;* also interv with Maj Simpson, 23 Jun 47.

[35] Craighill IBT Rpt.

Late in 1944 the staff director discovered a move by junior WAC officers to override the regulations on conduct and uniform by securing from senior Army officers contradictory statements, which were "usually made in a spirit of conversation at mixed parties." [36] These attempted violations the staff director was able to quash, but there remained the problem that Army officers were often "accessible and sympathetic to personal requests." [37]

Some months later the staff director reported that "a good many enlisted women are relaxing their efforts with regard to their personal appearance and individual conduct." She also called the women's attention to the fact that some Wacs were not "promptly and properly rendering the salute." Detachment regulations thereupon required that hair must be neater, T shirts invisible, shirt sleeves up or down but not halfway; no flowers or "other articles of non-military nature" could be worn in the hair; short or tight skirts must be let down or out. "It is essential," the staff director informed the women, "that the reputation thus far enjoyed by all members be maintained, that no relaxation be permitted that might adversely affect the generally accepted idea that American women soldiers are well-dressed, personable, and conduct themselves as ladies." [38]

A few weeks after the stricter detachment regulations had been published, they were partially canceled by General Stratemeyer, who, on a visit to the Southwest Pacific Area, had noted that General MacArthur's Wacs and General Kenney's Wacs "were wearing decorations such as flowers and ribbons in their hair during working hours." Being ignorant of the fact that General MacArthur's Wacs by this date had nothing resembling a military uniform, General Stratemeyer was favorably impressed. Over the staff director's protests, new official uniform regulations contained the provision—probably unique in the history of the Army—that "Flowers or ribbons are authorized and encouraged for wear in the hair." [39] The staff director conceded defeat, with the notation:

> There is absolutely nothing General Stratemeyer would not do for his Wacs to make them happier . . . we will therefore accede to his wishes, though they are contrary to the military indoctrination we received. . . . [40]

The move to China, combined with the location of the Air Forces headquarters in the same compound with General Wedemeyer's headquarters, caused considerable change in policies in this regard. Corrective action was first undertaken informally by a memorandum:

> General Stratemeyer this morning expressed a personal hope that all his Wacs would be so prompt and snappy in their saluting that General Wedemeyer would voluntarily say to him, "Boy! Are those Wacs good!" He pointed out that sometimes we get so interested in the conversation that we fail to notice an officer, and often that very officer is just the type to report the failure to General Wedemeyer. [41]

This document having had all the effect that might have been expected, a month

[36] D/F, Chief of Air Stf to WAC Stf Dir, 29 Dec 44, sub: Comd Channels. Folder IBT, CBI WAC files.

[37] Ltr, WAC Stf Dir AAC CBI to Air WAC Off, 6 Aug 45. CBI WAC files.

[38] Memo, WAC Stf Dir for CO WAC Det, 17 Mar 45, Folder IBT, CBI WAC files.

[39] (1) Memo, Exec for CO WAC Det, 23 Jun 45. (2) Mimeographed Memo, Chief of Air Stf to all Wacs, 8 Aug 43. Folder IBT, CBI WAC files. (3) Ltr cited n. 37.

[40] Ltr cited n. 37.

[41] Memo, WAC Stf Dir AAF CT for all EW of Det, 10 Aug 45. Folder Chungking I, CBI WAC files.

later General Stratemeyer noted, "Saluting by enlisted women was still extremely lax." He therefore issued more unmistakable orders that "the situation be corrected at once." At this the staff director, expressing considerable pleasure, showed great alacrity in informing the women that they would be court-martialed in a most military manner if that was the only way to get them to salute and to wear hats on the street.[42]

Happily for all concerned, the women reverted to their military training without noticeable loss of efficiency or morale, and impressed General Wedemeyer sufficiently to cause him to request General Stratemeyer to contribute Wacs to theater headquarters. At the close of the stay in Chungking, General Stratemeyer informed the women:

> You have . . . established an enviable record of efficiency which has immeasurably facilitated the accomplishment of the mission of my command. . . . Your performance stands as a worthy goal for any military unit in the Armed Forces of the United States.[43]

With the defeat of Japan, the headquarters moved to Shanghai, taking the Wacs with it. With the approaching inactivation of the Air Forces headquarters, General Wedemeyer's headquarters took over all of the women, including the staff director. He later requisitioned and obtained double their number direct from the United States.

Even in the postwar wave of demobilization fever, less than half of the Wacs eligible for discharge desired it, and the headquarters strongly urged them to stay.[44] In Shanghai, the Wacs were visited by the new WAC Director, Colonel Boyce, who had intended to direct their return to the United States, as had been necessary in the Pacific. Upon seeing the women, Colonel Boyce changed her mind, pronouncing them a carefully picked, seasoned group, all volunteers, who felt that their services were needed by the headquarters, and whose morale was correspondingly high. Most had by this time worked their way into key positions, and many were section chiefs, so that the headquarters was reluctant to part with them. The women themselves stated that they were willing to remain either as Wacs or as civilian workers. As a result, Colonel Boyce agreed to leave them indefinitely and to support General Wedemeyer's requisition for more.[45]

The China theater Wacs later moved from Shanghai to Nanking and Peiping; the group in Peiping staffed General Marshall's mission to China until its return. In both cities, the staff director noted that "their living arrangements and job assignments were as carefully prepared in advance as was their original entry into China." As the American forces quit China, some women moved to Japan and others returned to the United States.[46]

The employment of Wacs in India and China, although on a small scale, had done much to counteract the impression that women could not be successfully em-

[42] Memo, WAC Stf Dir AAF CT for CO WAC Det, 17 Sep 45, sub: Mil Customs and Courtesies of EW. Folder Chungking I, CBI WAC files.

[43] Ltr, CG AAF CT to WAC Stf Dir, 14 Sep 45, sub: Commendation. AAF 201.22, in WDWAC 319.1.

[44] Of 77 enlisted women, 50 were eligible for discharge, but only 23 desired it. Routing slip, Exec for Adm Serv AAF CT, Comment 1, 31 Aug 45. Folder Chungking I, CBI WAC files.

[45] Boyce Rpt, Tab China. WDWAC 333.1 (1945–46).

[46] (1) Interv with Maj Simpson, 19 Jun 43. (2) Ltr, WAC Stf Dir CT to Dir WAC, 18 Mar 46. WDGAP 320. (3) Memo, Exec, O Dir WAC for Maj Gen William E. Bergin, G–1, 25 Apr 46. WDWAC 320. (4) Quotation from comment by Maj Simpson on draft of chapter.

ployed in a tropical or disease-ridden area. The low medical evacuation rate, and the willingness of the women to remain for indefinite periods, both indicated that they were adaptable to such employment. However, all WAC advisers noted, in conclusion, that such success was likely to occur only in headquarters that made careful advance preparation and maintained a constant friendly alertness to detect developing problems.

The Middle East Theater

An example of the use of Wacs in an inactive theater was furnished, on a small scale, in the Middle East theater.[47] Here were assigned two WAC companies, totaling slightly less than three hundred members—one located in the city of Cairo, the other several miles outside it at Camp Russell B. Huckstep.

The first request for Wacs in Cairo was received in the fall of 1943 from Maj. Gen. Ralph Royce. General Royce at this time asked only for twelve WAC secretaries "to prevent the constant and sometimes serious leaks in information." [48]

The request was approved, and in December of 1943 ten enlisted women and two WAC officers were flown to the area. Capt. Josephine Dyer headed the group and later served as staff director. Other requests followed: one for 240 women for theater headquarters, which was approved; one for 600 women for Tehran, which was disapproved on the grounds of nonavailability of WAC personnel; and two others, of 150 more Wacs for theater headquarters and 104 for the 19th Weather Squadron, which were disapproved by Operations Division of the War Department on the grounds that no military vacancies existed in the command. Before making these requisitions, the theater had entertained some hopes of on-the-spot enlistment of American women living in Cairo, but it was found that 49 out of 50 eligible civilian women refused to enlist, and the idea of local recruiting was abandoned as it had been in Europe.[49]

The shipment of 240 women for the Cairo area was accordingly organized in the United States in February of 1944. It reached the theater only after an exhausting four-month struggle to get shipping space, during which the unit spent a month at Fort Oglethorpe, a month in one area at Camp Shanks, three weeks in another area, a week at Camp Patrick Henry, three weeks on shipboard, and ten days in Naples. The delay proved trying for the women, who were so eager to get overseas that 118 of them, 13 in the first three grades, voluntarily relinquished their ratings in order to comply with the theater's requisition for unrated personnel. The women were described as "dejected and disillusioned" during the months of waiting, but their morale rose rapidly as soon as they were in transit.[50]

Arriving in Cairo on 16 June 1944, the women were sent to Camp Huckstep,

[47] (1) Historical Data, WAC Det, Middle East SvC, Camp Huckstep, 11 Sep 44, by Lt. Jane A. Riebesell, WAC Hist Off, with appendixes for Jul, Aug, Sep, and Oct 44. Hereafter cited as Unit Hist Camp Huckstep, by month. (2) Hist Review, WAC Det Cairo Mil Dist, 28 Aug 45; last 13 pages pertain to Huckstep unit, Sep 44–Apr 45. Hereafter cited as Unit Hist Cairo Det. (3) Summary: WAC–AMET, prepared by Sgt Birdie Weisbrod, Jun 45, based on file of G–2 Hist Sec AMET. Hereafter cited as WAC AMET Summary.

[48] Memo, MPD ASF for TAG, 10 Nov 45, sub: WAC Pers for Gen Royce per Msg JM 1814, 27 Oct (CM–IN 16628). SPGAR 220.3 WAC, Dir Pers ASF, DRB AGO.

[49] (1) Cbl CM–IN 3072, 4 Feb 44; Cbl 2442, 9 Feb 44. (2) Memo, Stf for Dir WAC, 31 Jan 44. WDGAP 320.2 sec 8*a*. (3) WAC AMET Summary.

[50] Unit Hist Cairo Det; WAC AMET Summary; and esp Unit Hist Camp Huckstep, p. 4.

where living conditions were pronounced by detachment officers to be "far superior to any experienced by the unit since its inception." The brick-and-wood barracks were clean, modern, and airy, and latrines were in the same building. There was plenty of hot water, and a laundry with ironing boards and washing machines. In this the women's barracks were superior to those of some of the men's units, which did not all at this time have hot water. Barracks had single-decked beds, pillows, bedding, and shelves and hangers for clothing; there was also a well-furnished dayroom with recreation equipment. On the whole, the accommodations were described as being "in striking and overwhelming contrast to the haphazard facilities encountered in the United States." The only difficulty was a temporary one when, on the first morning, it was announced that Maj. Gen. Benjamin L. Giles was to review the Wacs immediately, and 240 travel-wrinkled women descended simultaneously on ten electric irons.[51]

Within a week, 126 of the women moved to Cairo for duty with Headquarters, Africa–Middle East Theater; the remaining 114 at Camp Huckstep were assigned to the Middle East Service Command. The unit in Cairo was billeted in the New Hotel, two to five women to a room. At the end of the year, after winter rains had flooded parts of the building, and after the arrival of more Wacs left it overcrowded, the Wacs moved to new quarters in a modern building with a bedroom for every three women and a complete post exchange and hairdressing shop.[52]

As the women were interviewed and assigned to duty, a fundamental problem emerged: the theater was already at full authorized strength and not in desperate need of personnel. If Wacs replaced enlisted men, the men were idle, and if Wacs were merely added as supernumeraries, there was only part-time work for all concerned. The majority of the women were stenographers, clerks, and typists; others were messengers, bookkeepers, telephone operators, and teletypists. Many were assigned as secretaries to ranking officers, and others were placed in almost every staff section of the two headquarters, with the largest numbers in the adjutant general's office, the Signal Corps, and the Censorship Section.[53]

While most women were well satisfied with the type of work assigned them, the fact that it was not full-time caused a continuing morale problem among both men and women. At Camp Huckstep, the WAC historian reported:

> Morale was affected by the existent morale at Camp Huckstep, which was poor and contagious. The majority of the men, and consequently of the women, felt that they were not actually needed and that what they were doing was not important to the war effort. Wacs were resented by some of the men.[54]

Promotions could not be given to any of the Wacs, even to a few who had replaced master sergeants, so long as the command remained overstrength. The Cairo report added:

> Comfortable living quarters and opportunities for enjoyment seem to be secondary in building morale . . . the important factor is the feeling that they are needed and that individual assignments keep them busy.[55]

[51] *Ibid.* All quotations from Unit Hist Camp Huckstep.

[52] Unit Hist Cairo Det, p. 3.

[53] Asgmts at Hq not listed; July asgmts at Camp Huckstep given in Unit Hist.

[54] Unit Hist Camp Huckstep, p. 20.

[55] Unit Hist Cairo Det, Aug–Sep 44.

During the first summer, only 56 of the 240 Wacs replaced enlisted men. Toward the end of this time, the theater appealed to the War Department for authority to return the surplus men to the United States. By October, enough men had been removed from the theater so that 157 promotions were made among the Wacs, chiefly to the grades of private first class and corporal. These were followed by others at intervals. By the following spring the Cairo detachment reported:

> The morale for the past period has improved . . . due partly to the fact that most of the women are now fairly busy, and also to the fact that they have a feeling of being generally accepted as a part of Headquarters.[56]

The theater consistently directed its efforts toward devices to maintain troop morale, and in these the Wacs were allowed to share fully. At Camp Huckstep, women were admitted to the excellent recreational facilities—a theater, a service club with cafeteria, game room, and library, and sports equipment. In Cairo, women were allowed to join a club with swimming, tennis, and golf facilities. At both stations, Wacs shared athletic programs in league baseball, volleyball, hockey, basketball, and even touch football, although such programs appealed only to the younger women. Women were also admitted to off-duty classes offered by the Armed Forces Institute in such subjects as French, Arabic, shorthand, and photography. Social activities were also encouraged, with organized parties and holiday celebrations, but the Wacs reported that such activities soon became burdensome and most women declined unit invitations.[57]

Wacs likewise were given the same amount of leaves and passes as the men, and were provided with equal transportation and accommodations to visit various points of interest for sightseeing or recreation. There were frequent conducted tours of the environs of Cairo, and visits by some of the women to Bengasi, Alexandria, Cyprus, Palestine, and other areas.

Restrictions were seldom sufficient to cause complaint. Marriages were permitted, suitable married quarters were provided, and women were allowed and even encouraged to spend furloughs with husbands stationed elsewhere. Nevertheless, WAC marriages in the theater were not frequent.[58]

Considering the climate and the sanitary hazards of the surrounding cities, the women's health remained good. A sanitary mess was maintained, and the nature of the area caused only minor discomforts. During the winter a water shortage was suffered, and cold weather made the unheated buildings chilly, but kerosene stoves were soon obtained and men's pile jackets were issued to the women. Heavy winter rains found a small Nile coursing through the WAC barracks; the boiler broke down; a high wind blew down the reed blind fence around the WAC area and the improved view was enjoyed by both women and men until the fence was put back up. None of these minor environmental difficulties had any appreciable effect on morale or health.

Major Craighill, who visited the area late in December of 1944, found the women's health good and illnesses "relatively less than in other units in the area."

[56] (1) WAC AMET Summary. (2) Ltr, ME to TAG, 15 Aug 44. AG 320.2, Hobby files. (3) Quotation from Unit Hist Cairo Det, p. 7.

[57] Unit Hist Camp Huckstep, Jul and Oct 44 appendixes.

[58] WAC AMET Summary gives 10; Unit Hist Cairo Det gives 17.

In the Cairo unit, women suffered the usual respiratory and gastrointestinal conditions, but in less degree than the men, and there had been no cases of venereal disease, pregnancy, malaria, or the hepatitis that was prevalent in the area. Emotional maladjustments had been minor, only one case requiring return to the United States.

Camp Huckstep Wacs had likewise had no unusual medical conditions except an incidence of appendectomies to a total of almost 10 percent of the group. There had been no cases of venereal disease, and only one pregnancy, which had existed before arrival in the theater. However, Camp Huckstep Wacs had experienced a number of minor maladjustment problems and two serious psychiatric cases, one of which had to be sent back to the United States. These cases were not more numerous than similar cases among men and, according to Major Craighill, "were reflections of the enlisted men's attitude of frustration in an area removed from the combat zone." Improvement in this respect was noted when surplus personnel was shipped out. Gynecological conditions were minor and caused no loss of time from work or any evacuation of personnel from the area.[59]

The only policy problem noted by historians was a minor version of the Pacific and European theaters' major problem: the question of direct commissions for enlisted women who were secretaries to ranking general officers. As soon as the news of such commissions elsewhere reached the theater, General Royce repeatedly applied to the War Department for permission to commission his secretary. The request, with others like it, was refused upon the insistence of Director Hobby, who had secured a ruling against any further such action as a result of the bad publicity accompanying earlier cases. The Director suggested instead that the woman, if qualified, be selected by regular board procedure and returned to officer candidate school with promise of return to the area upon graduation. General Royce instead preferred to appoint the secretary a warrant officer, which he did. No further comment resulted except sharp newspaper criticism of the commanding general's wisdom in using public funds to make his personal plane available to the secretary and her husband for an extended honeymoon tour of the Mediterranean.[60]

As the end of the war neared, restlessness increased, and both men and women were heard commenting that they wished they were in Italy, France, America, or almost any other place. Luckily the waiting period was not long, and soon after V-J Day shipment began, with the last Wacs returning home in October 1945. Colonel Boyce, who stopped in Cairo shortly afterward on her way home from India, interviewed the chief of staff and stated that "The Wacs were highly commended for their work and they were released only when there was no longer a job to be done." [61]

Although recommending a two-year limit on the tour of duty, Major Craighill stated that the theater's experience with the employment of Wacs and nurses had

[59] Passage on health, including quotations and statistics: (1) Memo Maj Margaret Craighill for The SG USA through Chief Surg USAFIME, 8 Jan 45, sub: Health Conditions of Nurses and Wacs in USAFIME. SGO Hist Div 319.1 Spec Rpt—Middle East. (2) Incl 3 to Craighill IBT Rpt.

[60] Memo, Dir WAC for CofS USA, 14 Apr 44. WDWAC 314.7. Approved by CofS, 18 Apr 44; forwarded to TAG, 21 Apr 44. WDWAC 210.1.

[61] Boyce Rpt of visit to WAC Pers in Overseas Theaters, 21 Sep-26 Nov 45, Tab F: AMET. Hobby files, also OCMH.

proved that "women can adjust satisfactorily to difficult environmental problems under proper administrative control." Other authorities concurred, with the additional comment that employment in an inactive theater would possibly always present problems of full-time employment and therefore of morale.

Other Overseas Areas

At various times requisitions and inquiries were received from other overseas areas, all of which were disapproved or withdrawn. In general, refusals were due to a desire to reserve the WAC's services for more active theaters, or to the fact that civilian help was plentiful in the areas concerned.

Hawaiian Department

A tentative requisition from the Hawaiian Department was received in late 1942 for one or more WAAC companies, and was favorably considered by WAAC Headquarters.[62] However, further investigation revealed that Waacs were chiefly wanted to replace civilian women in the Aircraft Warning Service. With the decision to count Waacs against the Troop Basis, such requisitions were automatically eliminated. No Wacs served in Hawaii except those with the Air Transport Command, which employed them in military duties.

Canada and Alaska

Requisitions were received in 1943 for about 700 women for stations in Canada and Alaska, including Edmonton, Whitehorse, and Dawson Creek. These were refused, not by the WAC, but by OPD and by ASF's Military Personnel Division. At a later date, and after arctic clothing for women was standardized, Air Transport Command Wacs served in both areas without reference to the War Department for approval.[63]

New Caledonia

Inquiries were received from New Caledonia in 1943 concerning the possibilities of getting some 40 WAC telephone operators and an unspecified number of other workers. Both were disapproved by OPD, for the stated reason of shortage of WAC personnel. A few WAC officers were later stationed in the area for varying periods.[64]

Puerto Rico

When the WAAC was organized, the Antilles Department rejected the idea of employing its members, as a large Puerto Rican labor supply was already available, and housing would present problems. Therefore, for the rest of World War II, Wacs were not requisitioned or employed. In April of 1944, the Antilles Department recommended that a WAC recruiting party be sent to Puerto Rico, inasmuch as the WAVES had already been there. The War Department accordingly sent one WAC officer and three enlisted women, authorized to recruit not more than 200 women for service in the United States. There was an enthusiastic response—some 1,500 immediate applications—but many had to be rejected for failure to pass the aptitude test, or because of parental ob-

[62] 1st Ind, TAG to CG Hawaiian Dept, 23 Dec 42. AG 320.2 WAAC (11-19-42) PR-W, Folder, WAAC, Contl Div ASF Sp Coll DRB AGO.

[63] Memo, Actg Dir WAC for CG ASF, 3 Nov 43. SPWA 320.2 sec 6.

[64] *Ibid.*

jections. Also, later employment was handicapped by language difficulties.[65]

Independent Commands Overseas

In addition to overseas personnel under the command of the various theaters, there were several thousand overseas Wacs who were not assigned to any theater but to the numerous independent commands authorized to send part of their personnel overseas. Neither the allotment nor the command of such personnel was a prerogative of the theater, but remained with the parent organization. Such commands included various War Department boards and groups, the Office of Strategic Services, the AAF Weather Wing and Army Airways Communications System, and especially the Air Transport Command. Most such commands seldom employed more than five to ten women at any one station, and ordinarily attached them to the nearest WAC unit for housing.

Just as in the United States, unit problems and schisms frequently resulted from the presence of these privileged outsiders with different schedules, rules, ratings, and status in general. No particular solution was reported by the theaters, other than to suggest that full disciplinary control was essential to the unit commander. As a result, theaters did not encourage most independent commands to more widespread employment of women.[66]

Of the independent commands, the ATC was by far the largest, employing Wacs in such numbers as to make possible establishment of its own detachments to a total of some 1,300 women overseas. By special War Department authorization, ATC was empowered to determine suitable overseas stations, and select, stage, and ship WAC personnel without reference to Air Forces headquarters or to the War Department. By the end of the war it had some 241 Wacs in Alaska, 236 at two stations in Africa, 460 at five stations in the European theater, 156 in Labrador, and 227 in the Pacific.[67]

On the whole, the results of assigning Wacs overseas in such independent and often unsupervised situations compared unfavorably with the experiences of those under theater jurisdiction. While efficiency seldom was affected, morale, conduct, and the full utilization of skills suffered. Theater experiences had, with one exception, indicated that Wacs overseas were uniformly successful and presented few problems, but the example of the independent commands, as well as of the North African theater before the arrival of its staff director, strongly suggested that such success was by no means as inevitable as its universality suggested, but instead depended to some extent on close and careful selection and supervision.[68]

---

[65] Antilles Dept Hist Studies, Sec VI: Puerto Rican Induction Program and Use of Puerto Rican Troops, Ch. IX, p. 70. OCMH.

[66] ETO Bd Rpt, Vol. I, p. 21.

[67] Directory of the AUS Outside Continental U.S., AG 461 (6-1-45) OB-I-M, gives ATC WAC strength.

[68] (1) ATC Hist Rec Rpt, West Coast Wing, 1 Jan 43-30 Jun 44, pp. 50-53. (2) ATC Hist Rec Rpt 308, Alaskan Div, Nov 44-Sep 45. (3) Hist of 1306th AAF Base Unit, ICD ATC, Jul-Aug 45, pp. 18-21. USAF Hist Div. (4) Memo cited n. 59(1). (5) ETO ATC WAC Hist.

The Office of the Director WAC

At the top of the world-wide WAC organization was the Office of the Director WAC—a small advisory group of from three to five officers located in G–1 Division of the General Staff. For the eighteen months between its establishment and the end of the war, the office, although too small for field inspection or supervision, was the Corps' nerve center for policy and information. Director Hobby later noted, "The need for placing the Director's office on the General Staff level cannot be emphasized too highly. It is absolutely necessary."[1]

It was from the first obvious that the duties of the Office of the Director could not be crowded under the G–1 roof, in spite of the fact that G–1 proved a co-operative host. The part of the Director's responsibilities concerning "personnel as individuals" fitted neatly into G–1 Division—induction, transfer, discharge, morale, uniform design, and insignia. Left unhoused were all other responsibilities, which had to be co-ordinated not with G–1 but with G–2, G–3, G–4, and the Special Staff divisions. These proved a continuing puzzle not only to G–1 Division, but to other General and Special Staff offices, which were unaccustomed to co-ordinating their business with a branch of G–1.

A first attempt to solve the problem was made by splitting the Director's Office into two unequal parts, the larger one to handle G–1 matters only, the other for the rest of her responsibilities. The Director was given two titles and two sets of procedures: as Chief, WAC Branch, G–1, she sent papers to G–1 for approval and signature; as Director WAC she corresponded directly with all other staff divisions and the Air, Ground, and Service Forces on matters that did not concern G–1 Division.[2]

This peculiar organization lasted only three months, and never in reality functioned as two halves. Many of the Director's projects were an indissoluble mixture of G–1 and non-G–1 elements; her staff debated, consulted the G–1 staff, and was still unable to decide who should sign which paper. Also, the volume of work upon the Director WAC part of the office was so much greater than that upon the larger G–1 part that members of one had no desire to sit idly upon their organizational charts while members of the other worked overtime.

---

[1] Interv, 23 Aug 48.

[2] G–1 Office Memo, printed, 26 Feb 44. CofS 324.5 WAC.

In June of 1944 the G–1 part of the office was merged with other G–1 personnel branches, and its members moved to other offices, in a general G–1 consolidation.[3] Remaining in the Director's office were, for most of the following year, only Director Hobby, Deputy Director Rice, and Capt. Patricia Lee Chance, Executive Officer.

This—the number assigned to run an average WAC company—was the smallest number of officers with which the office was ever to attempt to operate. With the separation of the WAC Branch, G–1, the peculiar situation of the Office of the Director was that of having no real connection with G–1 Division, in which it was located. In the ensuing months, some confusion was inevitable. As G–3 Division had earlier pointed out, the Office of the Director was in nature, if not in size, actually a Special Staff division, and, while the position in G–1 proved more suitable than that in the Army Service Forces, it was never as appropriate to the military organization as that of a Special Staff division. The location nevertheless lent itself to relatively smooth operation during the remaining months in which the position of chief of G–1 Division was held by General White, an officer with wide knowledge and experience in WAC administration dating from the Corps' earliest days. After his departure, there was noted an increasing tendency toward activity by G–1 Division in projects more properly the concern of other staff divisions. No final solution to this dilemma was achieved at the War Department level, although some air commands began to place their WAC advisers on the special staff level, while the nine service commands moved them from G–1 to the immediate office of the commanding general.

MAJ. GEN. MILLER G. WHITE

Removal of Operating Duties

The assumption that the Office of the Director could function with so small a staff was based upon the belief that all operating duties could be removed and delegated to the new specialist groups in other divisions. To some extent this hope was justified when the WAC specialist groups in other offices, established at General Marshall's direction, came into full operation. For the last months of the war, a complete and harmoniously operating network of WAC specialists covered most of the headquarters offices, so that inquiries, investigations, calls, and necessary studies could be referred rapidly, and successfully, to recruiting experts, public relations experts, The Inspector General, and other

[3] G–1 Office Memo to Div Chiefs, 25 May 44, Incl 1. WDWAC 314.7.

staff divisions, or to the major command or overseas theater having responsibility in a given case.

The accord and unity of action existing among such offices were, in the opinion of WAC staff officers, among the most important achievements of the new organization. The invariable practice of the Air WAC Officer, the AGF WAC Officer, and the ASF WAC Officer was to act in closest co-ordination with one another and with the Director and her deputy. Not only were personal relations friendly, but advances and improvements in one command were communicated to others before being put into effect, rather than afterward. The Director never at any time during Colonel Hobby's tenure of that office approached individual cases or policy concerning Wacs in the United States except through the WAC advisers to the three major commands, thus maintaining not only military channels but the prestige of these officers within their own headquarters.[4]

Unfortunately for the high hopes of complete delegation, the next months were to demonstrate that—as was also discovered by the British and the WAVES—the efficient operation of such a network depended upon an efficient nerve center, and that there were certain operating duties that could never be removed from a director's office by any amount of effort. These duties seemed to be inherently inseparable from the head of a woman's service in the twentieth century. In General White's opinion the office's situation was comparable to that of his own G–1 Division, which also found itself performing many onerous operating duties required of it by the Deputy Chief of Staff that were not, until after the end of the war, reflected in its personnel allotments.

Few if any of these duties were military functions of any other office. Deputy Director Rice ruthlessly swept out every duty that could be delegated, and that any other agency could or would accept. Out also went routine indorsements, records, reports from the field, and other paper work not essential to planning, including much useful material that no one had time to read.

For example, reports from WAC training centers were discontinued, since the Director no longer had any jurisdiction over their activities in any case. Bulletins from field units, even if sent, could no longer be read. No more historical material could be collected, and the only agency that could be persuaded to continue the historical file was the Bureau of Public Relations, which added a few press releases from time to time. When a WAC officer wrote to ask whether material from the field was being collected for a WAC history, Colonel Rice replied:

> The Director has now only myself and Captain Lee assigned to her immediate office. Naturally we cannot do anything about collecting such data: we can only hope it is finding its way into the proper files and is being kept by the commands.

The Director constantly received communications from Wacs in the field which indicated that they believed her to be responsible for supervising a much wider range of activities. One asked why she did not send out a touring group of artists from her office to sketch WAC life in the field; others thought she should visit every one of the approximately 500 WAC units every year; one requested a chart of her

[4] According to statement of three officers concerned.

"headquarters," only to be told, "There is no organizational chart . . . Colonel Hobby and her staff of two officers are located on the War Department General Staff, G-1, Personnel Division." [5]

However, in spite of strenuous efforts to simplify the office's duties, certain operating duties proved impossible either to discard or to delegate. One of these was the *WAC Newsletter,* a monthly 8-page pamphlet that went to all commands and units in the field, much like that later adopted by G-1 Division to keep field personnel officers abreast of War Department policy. The value of the *Newsletter* was attested by hundreds of letters from field units, which pronounced it invaluable in promoting discipline, morale, appearance, and understanding of the Corps' mission and work in the Army. The Director therefore asked the new WAC Group, Bureau of Public Relations, which had twelve officers, to take over this project, but the group refused on grounds of insufficient personnel. Captain Chance of the Director's Office was therefore obliged to continue writing the *Newsletter* in addition to her other duties.[6]

Another operating duty resulted from the fact that the Director's Office was required from time to time to compile, check, or certify special Army-wide summaries of information or policy, as requested by particular offices.[7] The second WAC Director later noted, "some uniformity in policies has been necessary; consequently, the Director WAC's office has been used as the co-ordinating office by the major commands and all staff divisions." [8]

However, the most troublesome of the "operating" duties of which the Director's Office was never able to rid itself did not concern such responsibilities within the Army, but rather liaison with outside agencies and the public as a whole. This type of duty, although never recognized upon any organizational chart or in personnel allotments, in actual fact consumed more than half of the office's time.

Liaison With Other Women's Services

One frequent liaison duty outside the War Department was that of co-ordination with other women's services of the United States and Allied nations. Liaison with the WAVES, SPARS, and Women Marines was more time-consuming than the general public realized. The directors of the four services formed a special subcommittee of the Joint Army-Navy Personnel Board, and met regularly. A major disciplinary or personnel policy was seldom adopted for one service until the other three had opportunity to consider

[5] (1) Ltr, ASF to all SvCs, 12 Aug 44. SPXPO-A 210.31 WAC, in WDWAC 300.4. (2) Memo, Dir WAC for WDBPR, 5 May 44. WDWAC 000.7. (3) Ltr, Lt Harriet Barr to Lt Col Jess Rice, 8 Sep 44. WDWAC 007. (4) Chart quotation in Ltr, Lt Freddie Boyle to Dir WAC, 5 Sep 44. WDWAC 000.7. (5) Query on WAC hist from Ltr, Lt Virginia Smithson to Lt Col Jess Rice, 30 May 44, and reply. WDWAC 323.3.

[6] (1) Note, 1st O Louise Goeden to 1st O Gretchen Thorpe, re action of ASF Publications Review Bd, 22 Apr 43. (2) Memo, TAG for Dir WAAC, 26 Apr 43. SPWA 062.1. (3) Memo, Dir WAC for Dir, Bur of Budget, 20 Mar 44. WDWAC 000.76 *Newsletter.* (4) Memo cited n. 5(2). (5) Memo, Dir WAC for G-1, 12 May 44, with Tab B, Gen Surles' refusal. WDWAC 000.7. (6) Memo, Dir WAC for AGO Publications Div, 28 Jun 44. WDWAC 000.7. (7) Correspondence in WDWAC 000.76 *Newsletter,* 1945-46.

[7] (1) Memo, Dir WAC for I&E Div ASF, 2 May 45. WDWAC 201.22. (2) Memo, Exec WAC for CGs AAF, AGF, ASF, 14 Jun 44. WDWAC 000.7. (3) Other requests in SPWA 300.3 (1943).

[8] Draft (not used), 14 Nov 45, sub: WAC Hist, with Note, Dep Dir WAC, re intended use in speech. WDWAC 314.7.

it. Often identical policies were arrived at and announced jointly. The three other services, having been established later, were at times several months behind the WAC in administrative detail, and often copied almost verbatim the directives and circulars that had been used successfully by the WAC. Literature was regularly interchanged. The four directors were on the friendliest of terms and met often for lunch or informal conferences. The WAVES' assistant director later wrote to Colonel Hobby, "We have leaned heavily on the experience of the Wacs and we are extremely grateful for the assistance your staff members have given us." [9]

Liaison duties with Allied women's services were somewhat less frequent, although British and Canadian services had sizable units in Washington. Information of common interest was regularly exchanged and the Director WAC or her deputy was the expected representative at many ceremonial functions of these services, and was hostess to them in return. Arrangements were made for British and American women officers to be exchanged for attendance at their respective armies' highest staff courses. The Consulate General of the Republic of Poland arranged for a tour of Polish Wacs (called, unfortunately, *Pestkas*) to WAC installations. French authorities requested full information on WAC administration for use by their own women's corps, as well as a supply of WAC uniforms. The Netherlands Embassy requested and obtained permission to train Netherlands women at WAC training centers to form the nucleus of a Netherlands women's corps for service in the Pacific. All such negotiations with high-ranking Allied officials frequently required the personal presence of the highest-ranking WAC officer the Army was able to offer. [10]

Liaison With the National Civilian Advisory Committee

Equally official in nature, and required of the Director by War Department directive, was liaison with the WAC's National Civilian Advisory Committee. This organization had been established, upon General Somervell's recommendation, at the time of the Director's transfer to G–1 Division, in order to bring to bear the influence of prominent civilian women "to advise on WAC matters, assist in disseminating news about the Corps, and aid in recruiting." [11] This committee consisted of the chairman of each service command committee, plus twelve others selected by the Director, representative of various occupations, faiths, and races. These women maintained contact with the Office of the Chief of Staff through that of the Director WAC. It was necessary for the Director's staff to collect information and write the committee a monthly bulletin on events in the WAC, to receive and act on their recommendations, to call twice-yearly conventions, and in general to render appropriate recognition to the interest and helpfulness of these prominent women. [12]

Such a group, with its energies properly directed, proved invaluable in improving the community standing of the WAC and thus the number and type of recruits. On the other hand, to keep its members prop-

[9] (1) See Part Three, below, for instances of joint action. (2) Ltr, SPARS to WAC, 3 Nov 44, thanking WAC for info on Negro dets. WDWAC 080. (3) Ltr, Dep Dir Rice to Capt Mildred McAfee, 7 Jul 44. WDWAC 080. (4) Ltr, Asst Dir WAVES to Dir WAC, 14 May 45. WDWAC 370.01.

[10] ATS and Polish correspondents, WDWAC 080 (1944).

[11] ASF Summary, ASF Cir 101, 15 Apr 44.

[12] Ltr, Dir WAC to CG ASF, 9 Oct 44, sub: National Civilian Advisory Committee. WDWAC 334, in SPAP 319.1 WAC.

erly informed, to give proper individual attention to their suggestions, and to keep up the close relationship directed by the Chief of Staff, all could well have absorbed the entire time of one staff officer, had liaison with this organization been reflected in personnel allotments.

Liaison With the American Red Cross

The Director's Office held a similar responsibility in regard to liaison with the American Red Cross. This organization was committed to the principle of employing its funds and staff for the assistance of military personnel regardless of the race or sex of the persons assisted. Actually, in most overseas theaters, the organization had not proved sufficiently flexible to adapt to the needs of feminine newcomers to the armed forces, although not for lack of effort on the part of the national headquarters, which received high War Department praise for its efforts and cooperativeness. Frequent conferences were held by the Director's Office with national representatives, during which a maternity care plan was worked out and field recreational facilities for women discussed, as well as other needs of women on Army stations, which the local Red Cross field director might help to meet. The Red Cross likewise fulfilled for women certain functions in connection with dependency investigations which, for men, were handled by Selective Service boards. A Certificate of Appreciation was recommended by the Director WAC for Miss Helen Walmsley, assistant director of the Military and Naval Welfare Service, for her effective liaison with the WAC. Again, while certain discussions could be delegated to the Air, Ground, or Service Forces, on an Army-wide level there remained no liaison agency able to coordinate final plans with the Red Cross except the Director's Office.[13]

Liaison With Other National and Local Organizations

A similarly unrecognized but frequent duty of the Director's Office was that of liaison with the many other national organizations—civic, professional, social, and patriotic—which desired to consult someone in the War Department in connection with WAC affairs.

The United Service Organizations' national headquarters, in an effort to overcome the somewhat chilly initial reception of Wacs in USO facilities, asked for lists of all WAC locations, and in repeated conferences made plans to assist each unit. The American Women's Voluntary Services offered assistance to Wacs, and asked for messages for the AWVS Bulletin. The American Association of University Women was always a strong supporter of the women's services, and AAUW representatives frequently consulted or wrote the Director's staff. The same was true of the National Federation of Business and Professional Women's Clubs, a consistent advocate of universal service laws for women. The Association of Junior Leagues of America arranged for all local Junior League presidents to assist in recruiting, saying, "The number of Junior League members who are able to enlist is not very large," but that members would be urged to join or to help recruit nonmembers. The WAC Mothers Association desired to contribute parties and dayroom furniture, and wrote for detailed information. The

[13] (1) See overseas and policy chapters for details of Red Cross activity. (2) Memo, Exec WAC for AGO, 3 Jan 44. SPWA 001. (3) Memo, Dir WAC for Civ Awards Bd, 16 Apr 46. WDWAC 201.

Theater Wing of the Stage Door Canteen requested WAC representatives and a WAC band for a bond rally. The National Press Club desired to arrange parties for enlisted women.[14]

An extremely brief sampling of the list of other organizations corresponding with the Director's Office during this period included the National Education Association, the National Editorial Association, the National Council of Jewish Women, the National Jewish Welfare Board, the National Association for the Advancement of Colored People, the American Legion and other veterans' groups, the American Association for Health, Physical Education, and Recreation, Zonta International, the Women's Overseas Service League, the American Social Hygiene Association, and the National Federation of Press Women.[15] The national representatives of organizations of this stature, when desiring an interview with the Director or Deputy Director, obviously could not be shunted by the Army to any other source of information even if one had existed, nor could their offers of assistance be rebuffed without at least a hearing or a reply to their correspondence.

Local civic and patriotic groups were in a different category from national representatives; nevertheless, the Director's Office was obliged to reply to their inquiries, even when their requests had to be refused. For example, the Saturday Afternoon Club of Vanderbilt, Pennsylvania, asked a statement on "What America Means to Me"; the W.I.V.E.S., a group of married women dedicated to urging people to donate money and blood, asked permission to name a chapter for Colonel Hobby; the Las Vegas Chamber of Commerce asked that a WAC hospital be located in its city.

Still more difficult to refuse gracefully were hundreds of requests for personal appearances—many of them in connection with recruiting, but the majority for non-WAC patriotic activities such as bond drives, plant security awards, ceremonies to recognize civilian women's volunteer work, and other similiar festivities. These were almost all refused "because of the press of military activities" or "because of previous engagements," but nevertheless required a prompt acknowledgment.[16]

Liaison With Congress

Liaison with Congress was another delicate and time-consuming duty. Supposedly all such work was done by the War Department offices that handled Congressional liaison for the Army, but it seemed difficult to educate Congress to this belief. Although the WAC was a part of the strength of the Army, the War Department required the Director personally to prepare material for yearly budget hearings, and to attend them at the call of Congressional committees to justify and explain WAC activities separately. It was also necessary for her to answer the daily Congressional letters and telephone calls that asked for a full explanation of WAC policies. These, even if properly addressed to the Legislative and Liaison Division or the Chief of Staff, were ordinarily routed to her for consideration as to possible

[14] (1) Ltr, Secy Joint Army-Navy Committee on Welfare and Recreation, to Lt Col Rice, 21 Apr 44; (2) Ltrs from AWVS; (3) Ltrs from AAUW, 18 Jul 44, and others; (4) Ltr, Exec Secy, Assn of Jr Leagues of America, to Dir WAC, 16 Feb 44. WDWAC 080.

[15] WDWAC 080 (1944). The 080 files (1942–46) are full of letters from such groups.

[16] 1944 files, WDWAC 080, 000.7, and 350.001. Also Ltr, Mrs. William Bacon, President, Veterans Guild of Association of Army and Navy Wives, to Col Hobby, 3 Aug 44. WDWAC 080.

policy changes or explanations of existing policy. The bulk of Congressional mail asked for special favors for constituents, which could not be routinely refused without a polite explanation of the policy concerned. Hundreds of these asked waivers of entrance requirements. Dozens of congressmen asked the Director to waive the act of Congress that barred 18-year-olds and aliens from the WAC. Hundreds of other letters concerned the desire of women constituents to receive direct commissions or to attend officer candidate school—a favorite request. Others asked promotion, transfer, investigation, award of contract, and so on.[17]

Liaison With Private Citizens

All such calls, correspondence, and liaison with Congress, civic groups, and women's services were insignificant when compared with those involving the general public. No Dorothy Dix was ever taken more thoroughly into the troubled confidence of the average American than was Director Hobby. Such attention was not unknown to the Army's highest-ranking leaders, who ordinarily had large staffs equipped to handle it; but it was almost unique in War Department history that an officer of the grade of colonel, and in a subordinate staff position, should be deluged with public correspondence which, for scope and variety of subject, seemed comparable only to that received by the Chief of Staff and others with Army-wide responsibility.

Parents, husbands, and other relatives of Wacs formed the largest group of these writers. Distraught parents accused the Director of every station commander's failure to protect, promote, or otherwise care for their daughters; others feared that harm had come to daughters overseas who had not written recently. A heartbroken mother could not understand her daughter's reduction in grade, and asked that Colonel Hobby restore the rank; the father of a prospective recruit wrote to discuss with Colonel Hobby whether his daughter should join.

Husbands and would-be husbands were particular nuisances. A soldier overseas wrote that he would probably soon be sent into combat, and if so he would probably be wounded, and if so he would probably be transferred home to recuperate, and if so he would want his wife with him; therefore, would Colonel Hobby discharge her from the WAC immediately so that she could be ready to rush to his bedside as soon as all this occurred? An Army chaplain asked that a soldier's wife, who was about to get a divorce, be transferred within persuading distance of her husband. A husband wrote that his wife had joined the WAC after a fight with him, but that they had made up a few days later, and would the Army please discharge her at once before she had to report for training? Another husband said that he realized that his wife was in the Army when he married her, but that her job was "breaking up our marriage"; he asked that she be allowed to come home and care for him properly.[18]

Hundreds of letters also came from women who had been rejected by recruiters. One woman protested, in nine pages, that she was "thwarted by a psychiatrist"

[17] (1) Yearly Folders, Material for Budget Hearings, WAC C1 file. (2) Ltrs in 1942, 43, 44, and 45–46 files, under 080, 032.2, 095, 330.14, and others. (3) Ltr, Rep. Frances P. Bolton to Oveta Culp Hobby, 7 Jun 44. WDWAC 032.2.

[18] These and following personal letters unless otherwise indicated, are in 1944 files, SPWA and WDWAC 095 and 330.14. There are dozens like each case cited.

in getting into the Corps. Another, "a honorable discharge Waac," asked, "Why do a discharge Wac has to may a mental test over again?" Others wrote, "Dear President Rosevelt, I am a girl of 17 year old and would love to joind the Wacc," or "I am 62 years old. Please don't let my age interfere." Many begged pitifully that physical defects, "nervousness," and past records not be held against them. One distracted eligible, in doubt as to whether to enlist, bared her entire financial, emotional, and physical condition in a long letter and asked Colonel Hobby to make up her mind for her.

Many of the letter writers sought personal advantages—promises of commissions, promotion, or special assignments. Most of these went on to say that, while they preferred to enlist as ordinary privates, their special qualifications were such that it would not be fair to the nation for them to come in unless they were made officers.

Some of the personal requests were more or less matters of business. *The Army-Navy Journal,* the *Britannica Book of the Year,* and others asked for personal statements or articles. Emily Post wrote to inquire how a wedding invitation should be worded if the bride's mother was a lieutenant and her father only a mister. Clothing manufacturers desired to know where to write for models and materials. A New York carnival owner asked for a WAC band, WAC exhibits, and WAC personnel, and, without waiting for a reply, advertised that they would be a carnival attraction. A gentleman asked that he be employed to go to Latin America and recruit "senioritas" for the WAC. Approximately a hundred persons sent WAC songs, such as the man who said, "A good popular WAC song to make the Wacs more popular would do the trick for you and step up recruiting"; then they asked her to get music publishers to accept their manuscripts.[19]

There were numbers of friendly citizens who either asked or contributed small gifts. A kindly housewife donated a good recipe for scrambled eggs for WAC messes. Others sent poems, scrapbooks, and pictures of their baby girls in miniature WAC uniforms. A grandmother sent a square of cloth to be autographed and returned to her for part of a quilt. The numbers who wanted autographs and autographed pictures were in the thousands, ranging from private citizens to a Boy Scout museum. Another began, "I am a Senior in High School and all Seniors must write 7,500 words on some subject we prefer," and asked for 7,500 words. Several museums asked for one of Colonel Hobby's uniforms for their collections.[20]

Many letters asked no favors but were merely generally abusive. Vague charges of immorality still were received. Many berated the uniform; others objected to the Army's class distinctions or other policies. Local and long-distance calls were received and had to be listened to; registered letters were received and signed for. Mentally deranged individuals were especially threatening. Vindictive women demanded harsh punishments for erring Wacs, especially after each case of newspaper publicity such as the one concerning a Wac who became engaged to a German prisoner. When one of the service commands was forced into the unpopular role of strikebreaker, it was its command-

[19] (1) Tp Conv, Chief, Plng Br for WAC Rctg and Exec WAC, 22 Jun 44. WDWAC 341. (2) 1944 files, WDWAC 000.7, 007, and 095. Song writer quoted, 4 Sep 44, WDWAC 000.7. The Director declined to express an opinion of any song.

[20] 1944 files, WDWAC 000.7, 007, 062, and 095.

ing general who put a Wac at the telephone switchboard of a strike-bound plant, but it was the Director WAC who received the complaints.

Somewhat lost among the requests, complaints, and abuse were a few lonely letters of thanks or praise, and descriptions from pleased parents of how their daughters had benefited by Army training.

Almost all letters betrayed a misunderstanding common to both civilian and military personnel: that Director Hobby was in immediate and supreme command of all Wacs. Correspondents regularly assumed that the Director was the final authority upon all of these matters, and could grant any favor she pleased. They assumed also that she had in her office large files of information on every Wac, and was personally responsible for all the events that occurred to Wacs world-wide, or was at least equipped to discover and explain what had befallen a daughter or wife.

Answering this public correspondence proved to be the greatest of the semimilitary operating duties that the Director was never able to get out of her office. Many solutions were tried and found wanting. A plausible one, for example, was to make no acknowledgment to correspondence before forwarding it to the responsible Army command. This proved highly unsuccessful in view of the time lapse involved in the command's investigation and reply, which required several weeks for cases on domestic stations and months for overseas theaters—by which time the average distressed parent had several times more written and telegraphed the Director's Office, his congressman, the President, and any other available dignitary, thus involving the office in more negotiation than would have been required for a simple acknowledgment. Another and equally ineffective approach was to make a brief acknowledgment by form letter; this, particularly when a refusal was involved, also brought down upon the office the wrath of correspondents.

Furthermore, there were some letters that could not be forwarded to any other command. There appeared to be no other office that was responsible for placing the Director's signature upon a quilt square, resolving Emily Post's dilemmas, or writing repeatedly, "Upon behalf of the Corps, may we express our appreciation of your interest." Even had there been, the mere task of sorting and forwarding communications was a formidable one, since it appeared impossible to learn the nature of a letter or telegram without opening it, or of a telephone call without accepting it.

The final solution, perforce adopted by the Office of the Director, was highly unsuccessful from the standpoint of work load, but highly successful from the public viewpoint. Upon receipt of the daily accumulation of correspondence, it was sorted and certain routine items were sent directly to The Adjutant General or other responsible agencies, without acknowledgment from the Director. Items that could be handled in this fashion included thousands of requests from lawyers and friends for the addresses of Wacs or women believed to be Wacs, since no litigation could be taken against a missing woman who might be in the Armed Forces.

Requests for recruiting information went to the Planning Branch of The Adjutant General's Office, and requests for stories or articles were sent to the WAC Group of the Bureau of Public Relations. Gifts and other expressions of interest were acknowledged by Captain Chance with a note of thanks, and routed to the proper agency for storage or file. Inquiries from distressed relatives were referred to the Army command having jurisdiction over the Wac concerned, but with a letter of acknowledgment stating that this agency, and not the Director, had command of the Wac concerned, would investigate and make the required reply, and should receive any further inquiries.

There still remained, after such actions, a formidable daily pile of correspondence which required the drafting by Deputy Director Rice of a carefully worded individual reply, suitable for the signature of the Director, the Chief of Staff, the Secretary of War, or whatever office had received the request. These replies of necessity constituted formal Army policy and required a phrasing that might subsequently appear in print without causing embarrassment. Typical explanations required by Congress and private citizens concerned the reasons why more WAC officers could not be commissioned, why some enrollment requirements could not be waived, why the Army could not transfer or discharge all Wacs who requested it, what spiritual or medical facilities were available to protect ailing members, why British or Australian civilian women had been commissioned, and, particularly, what the current moral, venereal, and maternity situation was in the Corps.

The inevitable result of such voluminous correspondence was a load of work that was staggeringly heavy for the three staff members. In view of the apparent relation of the experiences of the year 1944 to the Director's hospitalization and later retirement, and to the Deputy Director's death a few years later,[21] it was essential to understand the motives that caused the War Department neither to increase the staff nor to decrease the duties until shortly before the end of the war. In effect, the delay was chiefly caused by a hope that the load of operating duties was merely a temporary hangover from Auxiliary status, and would diminish as soon as the public and Congress realized that the Director was not the women's commander.

Thus, General Somervell wrote to General Marshall, at the time of the office's transfer, that the WAVES had no such "operating" problem, having never gone through an auxiliary status or been allowed much voice in their own management, and that "All that is necessary to do is to place Colonel Hobby in approximately the same position as Captain McAfee." [22] As a matter of fact, postwar studies revealed that the WAVES had an identical problem. The Bureau of Naval Personnel, in a final report, stated:

> Public opinion would tend to watch jealously the health, safety, and morale of women serving with the Navy, and it would naturally tend to address its inquiries to the "Director." . . . This public position of responsibility had to be backed by some substance of authority in the Bureau, else the sham would sooner or later appear, and not just the Women's Reserve but the Navy as a whole would suffer from the resulting lack of confidence. Even if another war should bring Selective Service for women, with consequent elimination of a recruiting campaign, policy would probably again require that a woman represent women in the Navy to the public.[23]

An identical conclusion was reached by British investigators, who noted, "It may be that parents of girls . . . like to think that a woman is finally responsible for

[21] See Ch. XXXV, below, on resignation of Director Hobby. Deputy Director Rice died 11 September 1949 of cancer which was first detected and treated during the months under discussion.

[22] Memo, CG ASF to CofS USA. SPAP 324.5 (3–29–44).

[23] WAVES Hist.

their well-being and not a government department."[24]

The fact that certain duties could not be delegated to any other existing agency was also well known to experienced WAC staff officers. After months of effort to rid the office of the load, the second Director WAC likewise noted that it was impossible to do so. Not only was the Director's Office, a policy agency, unable to refer certain letters to operating agencies, but they continued to refer to her for reply many inquiries which had been addressed to them.[25] Colonel Boyce summarized the experience of years as:

> All efforts to route these inquiries to other divisions or agencies for original handling result in return to the Office of the Director, WAC for a statement of policy, for information seemingly not available elsewhere, or in some cases in second letters condemning the Army for "buck passing." No amount of effort ever succeeded in eliminating this load of work from the Office of the Director, WAC.[26]

The actually intolerable nature of the situation was masked for a time by the concentrated activity of the Director's staff, which until well into 1945 worked 12-, 14-, and 16-hour days of 7-day weeks. At this time, upon Colonel Rice's departure to other duties, three officers were eventually necessary to replace her. In the interim, only the weight of collected carbon copies in the files made credible the fact that any one individual could dictate and sign in one day as many pieces of correspondence as did the Deputy Director. Her only apparent rebellion was a brief scribble on one particularly insulting letter of complaint, "I guess we shan't live long enough for this to stop."[27]

At the same time, the Director's major official duty—her policy responsibilities to the War Department—had not ceased to be fulfilled, even though staff studies were generally written by Colonel Rice in the late evening hours after visitors and conferees had been cleared from her office. These policy solutions, which came freely after the office's removal from the ASF, constituted in some respects the real history of the wartime Women's Army Corps, and its legacy to the future. Although any one of the major policy fields would require a separate chapter, if not a volume, for proper discussion, the problems arising within them did not occur separately, but were handled simultaneously with each other and with outside liaison duties during the last eighteen months of the war.

---

[24] *Conditions in the Three Women's Services.*

[25] For example: (1) D/F, TAG to Dir WAC, 8 Feb 44, sub: Request for Info re WAC. SPWA 095. (2) Ltr, Mrs. M. V. Moore to TAG, 1 Feb 44, asking for a copy of WAC regulations; TAG sent this to Dir WAC.

[26] Draft, WAC Hist, 14 Nov 45. WDWAC 314.7.

[27] Undated handwritten note to Executive WAC in Colonel Rice's easily identifiable handwriting. WDWAC 421.

Made in the USA
Monee, IL
13 September 2026

62976616R00285